W0043982

String Gravity and Physics
at the Planck Energy Scale

NATO ASI Series

Advanced Science Institutes Series

A Series presenting the results of activities sponsored by the NATO Science Committee, which aims at the dissemination of advanced scientific and technological knowledge, with a view to strengthening links between scientific communities.

The Series is published by an international board of publishers in conjunction with the NATO Scientific Affairs Division

A Life Sciences	Plenum Publishing Corporation
B Physics	London and New York
C Mathematical and Physical Sciences	Kluwer Academic Publishers
D Behavioural and Social Sciences	Dordrecht, Boston and London
E Applied Sciences	
F Computer and Systems Sciences	Springer-Verlag
G Ecological Sciences	Berlin, Heidelberg, New York, London,
H Cell Biology	Paris and Tokyo
I Global Environmental Change	

PARTNERSHIP SUB-SERIES

1. Disarmament Technologies	Kluwer Academic Publishers
2. Environment	Springer-Verlag / Kluwer Academic Publishers
3. High Technology	Kluwer Academic Publishers
4. Science and Technology Policy	Kluwer Academic Publishers
5. Computer Networking	Kluwer Academic Publishers

The Partnership Sub-Series incorporates activities undertaken in collaboration with NATO's Cooperation Partners, the countries of the CIS and Central and Eastern Europe, in Priority Areas of concern to those countries.

NATO-PCO-DATA BASE

The electronic index to the NATO ASI Series provides full bibliographical references (with keywords and/or abstracts) to more than 50000 contributions from international scientists published in all sections of the NATO ASI Series.
Access to the NATO-PCO-DATA BASE is possible in two ways:

– via online FILE 128 (NATO-PCO-DATA BASE) hosted by ESRIN,
Via Galileo Galilei, I-00044 Frascati, Italy.

– via CD-ROM "NATO-PCO-DATA BASE" with user-friendly retrieval software in English, French and German (© WTV GmbH and DATAWARE Technologies Inc. 1989).

The CD-ROM can be ordered through any member of the Board of Publishers or through NATO-PCO, Overijse, Belgium.

Series C: Mathematical and Physical Sciences – Vol. 476

String Gravity and Physics at the Planck Energy Scale

edited by

N. Sánchez

Observatoire de Paris,
Demirm,
Paris, France

and

A. Zichichi

CERN,
Geneva, Switzerland

Kluwer Academic Publishers

Dordrecht / Boston / London

Published in cooperation with NATO Scientific Affairs Division

Proceedings of the NATO Advanced Study Institute on
String Gravity and Physics at the Planck Energy Scale
Erice, Italy
8–19 September 1995

A C.I.P. Catalogue record for this book is available from the Library of Congress

ISBN-13: 978-94-010-6589-4   e-ISBN-13:978-94-009-0237-4
DOI: 10.1007/ 978-94-009-0237-4

Published by Kluwer Academic Publishers,
P.O. Box 17, 3300 AA Dordrecht, The Netherlands.

Kluwer Academic Publishers incorporates the publishing programmes of
D. Reidel, Martinus Nijhoff, Dr W. Junk and MTP Press.

Sold and distributed in the U.S.A. and Canada
by Kluwer Academic Publishers,
101 Philip Drive, Norwell, MA 02061, U.S.A.

In all other countries, sold and distributed
by Kluwer Academic Publishers Group,
P.O. Box 322, 3300 AH Dordrecht, The Netherlands.

TABLE OF CONTENTS

PREFACE   vii

The String Equation and Solitons
S. P. NOVIKOV...1

Lectures on String Theory in Curved Spacetimes
H. J. DE VEGA and N. SÁNCHEZ..11

Strings and Multi-Strings in Black Hole and
Cosmological Spacetimes
A. L. LARSEN and N. SÁNCHEZ..65

Renormalization Constants from String Theory
P. DI VECCHIA et al...105

Closed Superstrings in Magnetic Field: Instabilities
and Supersymmetry Breaking
A. A. TSEYTLIN..121

Solution of the SL(2,R) String in Curved Spacetime
I. BARS..151

Internal Physics of Black Holes: Recent Developments
W. ISRAEL...171

Black Hole Entropy and Physics at Planckian Scales
V. FROLOV...187

Black Hole Condensation and Duality in String Theory
A. STROMINGER...209

Correlation Dynamics of Quantum Fields
and Black Hole Information Paradox
B. L. HU...219

Topology and Time Reversal
A. CHAMBLIN and G.W. GIBBONS...233

Polymer Geometry at Planck Scale and Quantum Einstein Equations
A. ASHTEKAR..255

Modular Cosmology
T. BANKS...277

Status of String Cosmology: Basic Concepts and Main Consequences
G. VENEZIANO..285

Status of String Cosmology: Phenomenological Aspects
M. GASPERINI..305

Predictions from Quantum Cosmology
A. VILENKIN...345

Statistics of the Microwave Background Anisotropies caused
by Cosmological Perturbations of Quantum-Mechanical Origin
L. P. GRISHCHUK...369

Cosmological Inflation and the Nature of Time
D. S. SALOPEK..409

Aspects of Quantum Cosmology
D. N. PAGE..431

New Aspects of Reheating
D. BOYANOVSKY et al...451

The Electroweak Phase Transition: A Status Report
L. G. YAFFE...493

Aspects of Cosmic-Ray Astrophysics in the Galaxy and Beyond
M. M. SHAPIRO..507

Quantum Field Theory for Dynamical Systems with Curved Phase Space
E. S. FRADKIN...521

PHOTOGRAPHS OF THE INSTITUTE...523

AUTHOR INDEX...543

PREFACE

The contemporary trends in the quantum unification of all interactions including gravity motivate this Course. The main goal and impact of modern string theory is to provide a consistent quantum theory of gravity. This Course is intended to provide an updated understanding of the last developments and current problems of string theory in connection with gravity and the physics at the Planck energy scale. It is also the aim of this Course to discuss fundamental problems of quantum gravity in the present-day context irrespective of strings or any other models. Emphasis is given to the mutual impact of string theory, gravity and cosmology, within a deep a well defined programme, which provides, in addition, a careful interdisciplinarity.

Since the most relevant new physics provided by strings concerns the quantization of gravity, we must, at least, understand string quantization in curved space-times to start. Curved space-times, besides their evident relevance in classical gravitation, are also important at energies of the order of the Planck scale. At the Planck energy, gravitational interactions are at least as important as the rest and can not be neglected anymore.

Special care is taken here to provide the grounds of the different lines of research in competition (not just only one approach); this provides an excellent opportunity to learn about the real state of the discipline, and to learn it in a critical way.

All Lectures took place in the "P.A.M. Dirac" Lecture Hall at the San Domenico Institute. There were two lectures in the morning and two lectures in the afternoon. Each Lecture has been followed by a 30 minutes Discussion. The Discussion Sessions were as important as the Lectures themselves. An special visit to the "P.A.M. Dirac" Museum and to the "Daniel Chalonge" Museum took place at the Clossing Session of this Course.

We wish to express our grateful thanks to all the lecturers who did so much to make this Course successful, and to both participants and lecturers for contributing so much to the outstanding discussions and to create such a stimulating atmosphere at the Course.

We wish to express our gratitude to the Scientific Affairs Division of NATO (North Atlantic Treaty Organization) and to Dr. Luis Veiga da Cunha, for their generous and efficient support.

We thank the staff members of the Ettore Majorana Centre, for their help in the organization of this Course and for their kind hospitality in the beautiful setting of Erice.

We thank the reception secretaries, N. Letourneur and N. Grabar, and the scientific secretaries: M. Campanelli, A. Campos, M. Dorca, I. Egusquiza, M. C. Falvella, A. Larsen and C. O. Lousto, for their efficient assistance all along the Course.

We extend our appreciation to Kluwer Academic Publishers, Science and Technology Division, for their cooperation and efficiency in publishing these proceedings.

N. Sánchez
Director of the Course

GALILEO GALILEI FOUNDATION
WORLD FEDERATION OF SCIENTISTS
ETTORE MAJORANA CENTRE FOR SCIENTIFIC CULTURE
INTERNATIONAL CENTRE FOR THEORETICAL PHYSICS
Four Hundred Years Since the Birth of MODERN SCIENCE

GALILEO GALILEI CELEBRATIONS

INTERNATIONAL SCHOOL OF ASTROPHYSICS «D.CHALONGE»

4th Course: *STRING GRAVITY AND PHYSICS AT THE PLANCK SCALE*

A Nato Advanced Study Institute

ERICE-SICILY: 8-19 SEPTEMBER 1995

*In collaboration with the World Laboratory
and under the auspices of the Presidency of the Council of Ministers of Italy*

Sponsored by the: • European Physical Society • French Ministry of Foreign Affairs • French Ministry of Higher Education and Research • Italian Ministry of Education • Italian Ministry of University and Scientific Research • Sicilian Regional Government

---

PROGRAMME AND LECTURERS

A New Geometry for Quantum Gravity
• A. ASHTEKAR, Pennsylvania St. Univ., University Park, PA, USA

Modular Cosmology
• T. BANKS, Rutgers University, Piscataway, NJ, USA

String Theory in Curved Space-time
• H.J. DE VEGA, University of Paris VI, France

Field Theory Quantities from String Theory
• P. DI VECCHIA, Nordita, Copenhagen, Denmark

Topology and Time Reversal
• G.W. GIBBONS, DAMTP, University of Cambridge, UK

Non-perturbative Aspects of String Theory
• M.B. GREEN, DAMTP, University of Cambridge, UK

Recent Advances in our Understanding of Black Holes
• W. ISRAEL, University of Alberta, Edmonton, Canada

Black Hole Entropy
• V.P. FROLOV, University of Alberta, Edmonton, Canada

Wormholes and the Time Machine
• I. NOVIKOV, Center of Theoretical Physics, Copenhagen, Denmark

String Equation and Solitons
• S.P. NOVIKOV, Landau Inst. for Theor. Physics, Moscow, Russia

Aspects of Quantum Gravity
• D. PAGE, University of Alberta, Edmonton, Canada

String Quantum Gravity
• N. SANCHEZ, Observatoire de Paris, France

Black Holes and Quantum Gravity
• A. STROMINGER, University of California, Santa Barbara, CA, USA

Strings in Magnetic Fields
• A. TSEYTLIN, Imperial College, London, UK

Status of String Cosmology
• G. VENEZIANO, CERN, Geneva, Switzerland

Predictions from Quantum Cosmology
• A. VILENKIN, Tufts University, Medford, MA, USA

PURPOSE OF THE COURSE

The contemporary trends in the quantum unification of all interactions including gravity motivate this Course. The main goal and impact of modern string theory is to provide a consistent quantum theory of gravity. This Course is intended to provide an updated understanding of the last developments and current problems of string theory in connection with gravity and the physics at the Planck energy scale. It is also the aim of this Course to discuss fundamental problems of quantum gravity in the present-day context irrespective of strings or any other models. Emphasis is given to the mutual impact of string theory, gravity and cosmology, within a deep and well defined programme, which provides, in addition, a careful interdisciplinary

Since the most relevant new physics provided by strings concerns the quantization of gravity, we must, at least, understand string quantization in curved space-times to start. Curved space-times, besides their evident relevance in classical gravitation, are also important at energies of the order of the Planck scale. At the Planck energy, gravitational interactions are at least as important as the rest and cannot be neglected anymore.

POETIC TOUCH

According to legend, Erice, son of Venus and Neptune, founded a small town on top of a mountain (750 metres above sea level) more than three thousand years ago. The great historian Thucydides (~500 B.C.) said that the Elymi founders of Erice — were survivors of the destruction of Troy. Ancient historians agreed that Erice was the oldest city in Europe: Homer (~1000 B.C.), Theocritus (~300 B.C.), Polybius (~200 B.C.), Virgil (50 B.C.), Horace (~20 B.C.), and others, have celebrated this magnificent spot in Sicily in their poems. In Erice you can admire the Castle of Venus, the Cyclopean Walls (~800 B.C.) and the Gothic Cathedral (~1300 A.D.). Erice is at present a mixture of ancient and medieval architecture.

Other masterpieces of ancient civilization are to be found in the neighbourhood at Motya (Phoenician); Segesta (Elymian); and Selinunte (Greek). On the Aegadian Islands - theatre of the decisive naval battle of the first Punic War (264-241 B.C.) - suggestive neolithic and paleolithic vestiges are still visible: the grottoes of Favignana, the carvings and murals of Levanzo.

Splendid beaches are at San Vito Lo Capo, Scopello, and Cornino, and a wild and rocky coast around Monte Cofano all at less than one hour's drive from Erice.

N. SANCHEZ
DIRECTOR OF THE COURSE

GENERAL INFORMATION

Persons wishing to attend the Course should apply in writing to:

• Prof. N. SANCHEZ
Ministère de l'Enseignement Supérieur
et de la Recherche
Observatoire de Paris Demirm
61, Avenue de l'Observatoire
F-75014 PARIS, France
Tel ++33.1. 40 51 22 21
Fax: ++33.1. 40 51 20 02
Telex: 270 776 OBSPARIS

They should specify:

i) date and place of birth, together with present nationality;
ii) degree and other academic qualifications;
iii) list of publications;
iv) present position and place of work.

Young persons with only a few years of experience should enclose a letter of recommendation from their research group leader or from another senior scientist active in the field.

The total fee, which includes full board and lodging (arranged by the School), is US $1000.

Early application is strongly encouraged. Closing date for application : 15 June 1995 No special application form is required

A letter will be sent to successful applicants by 30 June 1995. Participants experiencing difficulties with travel documentation and who need to know, before 30 June whether or not their applications have been accepted may get an earlier special decision upon request.

Admission to the Course will be decided on the basis of scientific excellency, in consultation with the Advisory Committee of the Course consisting of Professors W. Israel, N. Sanchez, G. Veneziano and A. Zichichi.

Arrival day is 8 September 1995; departure day is 19 September 1995. Participants must arrive in Erice on 8 September 1995 (no later than 5 p.m.). More detailed information will be sent to successful applicants together with the letter of acceptance.

A. ZICHICHI
DIRECTOR OF THE CENTRE

INTERNATIONAL SCHOOL OF ASTROPHYSICS " D. CHALONGE"

4th Course :" STRING GRAVITY AND PHYSICS AT THE PLANCK ENERGY SCALE"

ERICE SICILY : 8-19 SEPTEMBER 1995

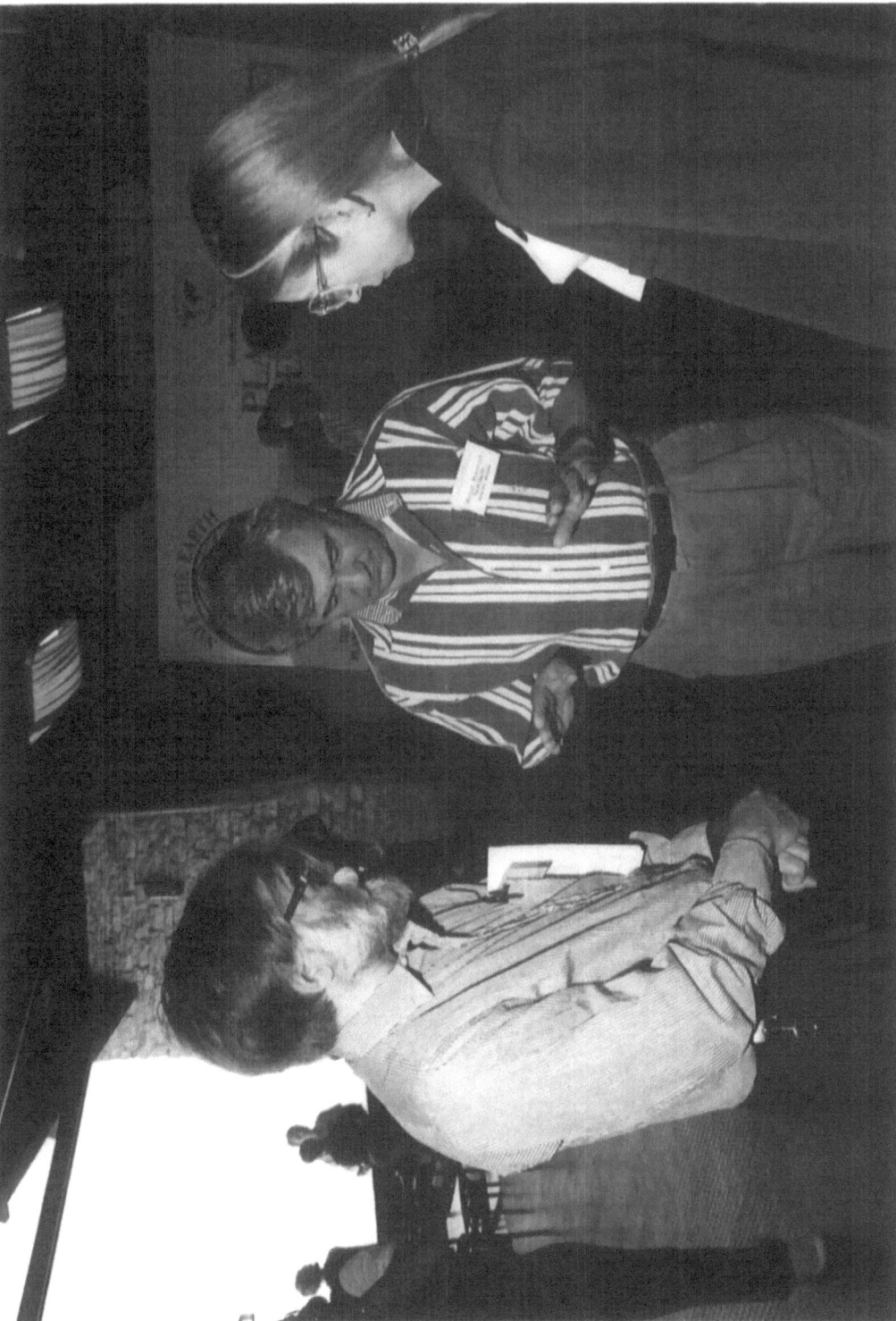

1995

INTERNATIONAL SCHOOL OF ASTROPHYSICS "D. CHALONGE"

Observatoire de Paris, DEMIRM, 61, Avenue de l'Observatoire, 75014 Paris,
France. e-mail: chalonge@meslob.obspm.fr

Ettore Majorana Centre, Via Guarnotta 26, 91016 Erice, Italy.
e-mail: HQ@CCSEM.CCSEM.INFN.IT   CCSEM::HQ

PROGRAMME

• 3rd COLLOQUE COSMOLOGIE, Paris, Observatoire de Paris, 7-9 June 1995

• 4th Course "STRING GRAVITY AND PHYSICS AT THE PLANCK SCALE", Erice, Ettore Majorana Centre, 8-19 September 1995

• 5th Course "CURRENT TOPICS IN ASTROFUNDAMENTAL PHYSICS", Erice, Ettore Majorana Centre, 4-15 September 1995

COLLOQUE COSMOLOGIE

The purpose of this meeting is to bring selected topics of high current interest in the interplay between cosmology and fundamental physics. This is an informal meeting bringing together physicists, astrophysicists and astronomers. The Colloque is intended to allow easy and fruitful mutual contacts and communications. Sessions last for three days, leaving enough time for private discussions and to enjoy the beautiful particular charms of Erice, Discussions build on the results from Colloque and are planned to prepare the next ones (every 18/2). This Colloque series, among several others, is one of those devoted to high energy physics and cosmology. These Topics covered this year 1995 include: Phase Transitions in Cosmology and Evolution out of the Equilibrium of Quantum Fields, Dark Matter and Large Scale Structure, Fundamental Strings and Cosmic Strings in Cosmology, Black Holes and Quantum Gravity.

STRING GRAVITY AND PHYSICS AT THE PLANCK ENERGY SCALE

The contemporary trends in the quantum unification of all interactions including gravity motivate this Course. The main goal and impact of modern string theory is to provide a consistent quantum theory of gravity. This Course is intended to provide an updated understanding of the last developments and current problems of string theory in connection with gravity and the physics at the Planck energy scale. It is also the aim of this Course to discuss fundamental problems of quantum gravity in the present-day context irrespective of strings or any other models. Emphasis is given to the mutual impact of string theory, gravity and cosmology, within a deep a well defined programme, which provides, in addition, a careful interdisciplinarity. Since the most relevant new physics provided by strings concerns the quantization of gravity, we must, at least, understand string quantization in curved space-times to start. Curved space-times, besides their evident relevance in classical gravitation, are also important at energies of the order of the Planck scale. At the Planck energy, gravitational interactions are at least as important as the rest and can not be neglected anymore.

ASTROFUNDAMENTAL PHYSICS

In the last years, the cross-section between fundamental physics, astrophysics and cosmology has been increasing, both at theoretical and experimental levels: particle physics experiments, astronomical observations, space satellite data. Such interplay fruitfully influenced research activity setting up Astrofundamental physics. By the nature of the domain itself, there are different aspects, approaches and points of view (sometimes complementary to each other, sometimes in contradiction) for the same topic or subject. Special care is taken to provide to the students the grounds of the different lines of research in competition (not only one approach). In this way, participants have an excellent opportunity to learn about the real state of the discipline, and to learn it in a critical way. Topics are covered with both theory and observations and with the perspective of unified knowledge: the early universe, cosmic microwave background radiation, large scale structure, the dark matter problem, and the interplay between them. Special Sessions are devoted to neutrinos in astrophysics, high energy and gamma ray astrophysics. These Courses bring together physicists, astrophysicists and astronomers; they are addressed to young scientists at post-doctoral level; they are useful to senior scientists and to advanced graduate students as well.

SCIENTIFIC CULTURE

The Daniel CHALONGE Museum and The Paul DIRAC Museum, 'ERICE, the City of Science', inaugurated during the Courses of this School, exhibit permanent collections of pictures, documents and instruments, on astrophysics and on particle physics. These Museums, together with the ERICE SEISMOLOGICAL STATION of the Ettore Majorana Centre, will participate in the 5th WEEK OF THE SCIENTIFIC CULTURE which will take place in Italy from 3 to 8 April 1995, organized by the Italian Ministry of University and Scientific Research.

JANUARY						
S	M	T	W	T	F	S
1	2	3	4	5	6	7
8	9	10	11	12	13	14
15	16	17	18	19	20	21
22	23	24	25	26	27	28
29	30	31				

FEBRUARY						
S	M	T	W	T	F	S
			1	2	3	4
5	6	7	8	9	10	11
12	13	14	15	16	17	18
19	20	21	22	23	24	25
26	27	28				

MARCH						
S	M	T	W	T	F	S
			1	2	3	4
5	6	7	8	9	10	11
12	13	14	15	16	17	18
19	20	21	22	23	24	25
26	27	28	29	30	31	

APRIL						
S	M	T	W	T	F	S
						1
2	3	4	5	6	7	8
9	10	11	12	13	14	15
16	17	18	19	20	21	22
23	24	25	26	27	28	29
30						

WEEK OF THE SCIENTIFIC CULTURE, Erice

MAY						
S	M	T	W	T	F	S
	1	2	3	4	5	6
7	8	9	10	11	12	13
14	15	16	17	18	19	20
21	22	23	24	25	26	27
28	29	30	31			

JUNE						
S	M	T	W	T	F	S
				1	2	3
4	5	6	7	8	9	10
11	12	13	14	15	16	17
18	19	20	21	22	23	24
25	26	27	28	29	30	

COLLOQUE COSMOLOGIE, Paris

JULY						
S	M	T	W	T	F	S
						1
2	3	4	5	6	7	8
9	10	11	12	13	14	15
16	17	18	19	20	21	22
23	24	25	26	27	28	29
30	31					

AUGUST						
S	M	T	W	T	F	S
		1	2	3	4	5
6	7	8	9	10	11	12
13	14	15	16	17	18	19
20	21	22	23	24	25	26
27	28	29	30	31		

SEPTEMBER						
S	M	T	W	T	F	S
					1	2
3	4	5	6	7	8	9
10	11	12	13	14	15	16
17	18	19	20	21	22	23
24	25	26	27	28	29	30

STRING GRAVITY AND PHYSICS AT THE PLANCK SCALE, Erice

OCTOBER						
S	M	T	W	T	F	S
1	2	3	4	5	6	7
8	9	10	11	12	13	14
15	16	17	18	19	20	21
22	23	24	25	26	27	28
29	30	31				

NOVEMBER						
S	M	T	W	T	F	S
			1	2	3	4
5	6	7	8	9	10	11
12	13	14	15	16	17	18
19	20	21	22	23	24	25
26	27	28	29	30		

DECEMBER						
S	M	T	W	T	F	S
					1	2
3	4	5	6	7	8	9
10	11	12	13	14	15	16
17	18	19	20	21	22	23
24	25	26	27	28	29	30
31						

The String Equation and Solitons

Landau Institute for Theoretical Physics, Moscow, RUSSIA

Lecture 1: Periodic 1- and 2-dimensional Schrodinger Operators Riemann surfaces, Nonlinear Equations

Introduction. We are going to present here some brief survey of the results of Theory of Solitons (see [1–3]) from the viewpoint of periodic theory including some new results in the theory of 2-dimensional periodic Schrodinger Operators.

A remarkable connection of some very special but highly nontrivial nonlinear (especially one-dimensional) systems with spectral properties of one-dimensional linear Schrodinger Operators was discovered in 1965–68 in the series of works [4–6] for the famous KdV equation

$$u_t = 6uu_x - u_{xxx}\,.$$

This connection is based on the identification of KdV with Heisenberg type equation for the linear operators ("Lax representation"):

$$L_t = [A, L], L = -\partial_x^2 + u(x,t), A = -4\partial_x^3 + 3(2u\partial_x + u_x)$$

This equation generates an effective GGKM integration procedure ("Inverse Scattering Transform") for KdV in the class of rapidly decreasing functions $u(x) \to 0, |x| \to \infty$, using the solution of inverse scattering problem for the one-dimensional Schrodinger Operator. A lot of important results were extracted from this method in the 70-s, using traditional methods in the modern Theoretical and Mathematical Physics like exact multisoliton solutions, asymptotics for $t \to +\infty$, action-angle type variables for the rapidly decreasing KdV, new classes of systems integrable by the same trick (see [2,1]). Many groups participated in this development, including Zakharov, Shabat, Lamb, Faddeev, Ablowitz, Kaup, Newell, Segur, Henon, Flashka, Manakov, Moser, Calogero and others. They found a lot of new important ODE-s and PDE-s (0+1,1+1 and even 2+1)- systems with Lax-type representation, including some special cases of Einstein equations and Self-Duality equation for the Yang-Mills fields (the last system is 4-dimensional).

Periodic Solitons and Algebraic Geometry.

We discuss here a different part of development of this theory, based on the "Triangle" with following vertices:

1.Algebraic Geometry, associated with Riemann surfaces and their Θ-functions (which never has been used in Applied Mathematics and Physics before);

2.Spectral theory of Schrodinger operator on the real line x with periodic (quasiperiodic) potential;

3.Periodic problems for KdV and higher analogues.

This development started from the work [7] in 1974 and was realized completely (in the case of KdV) by the present author, Dubrovin, Its, Matveev, Lax, Mckean, van Moerbeke in 1974-75 (see full survey of this theory in [8,1,3]).

1

In the periodic case (I remind that the potential u is periodic here, not the eigenfunction) the corresponding inverse spectral problem for Schrodinger operators never has been solved before the KdV theory. It was solved only in 1974-5 as a part of KdV theory, using the deep connection of KdV and Schrodinger operator. This solution was based on the new ideology, considering KdV and its higher analogs as a symmetry theory for the Schrodinger operator (for its spectral theory): eigenvalues of Schrodinger operators are the integrals of motion for all higher KdV systems. We have an infinite-dimensional commutative symmetry group for any functional $Fu(x)$, if it depends on the spectrum of the operator $L = -\partial + u$ only. Such functionals played an important role in the definition and studying of finite-gap potentials (below). Completely different important example was found later: a very well known Peierls Free Energy functional in the mean field approximation for some electron-phonon systems, describing the so-called "charge density waves" in some quasi-one-dimensional media. Its exact integrability was discovered by the group of physicists in collaboration with experts in the soliton theory: Belokolos, Dzyaloshinski, Gordyunin, Brozovski, Krichever (see, for example, in the survey [3]).

An important generalization of finite-gap solutions for some special 2+1 systems (like KP) was done by Krichever (see [10]), who extended very far an algebraic part of periodic theory. Many people worked in this area later. The references may be found in the books [1,3].

This approach is based on the special finite-dimensional families of exact solutions, whose x-dependence is specified by the Commutativity Relation of 2 different linear OD operators $[C, B] = 0$. For 1+1 (or x, t)-systems the corresponding operator C is necessarily equal to the operator L in the Lax pair for our system. The operator $B = \sum_i c_i A_i$ is some linear combination of corresponding A-operators for the so-called "Higher KdV" systems, associated with the same 1-dimensional Schrodinger Operator in the case of KdV. Commutativity Relation is equivalent to the family of Completely Integrable Hamiltonian OD systems in the variable x, admitting some very useful "Lax-type representations"

$$\Lambda(\lambda)_x = [\Lambda, Q]$$

for 2×2-matrices, depending on the additional parameter λ. Riemann surface Γ appears as a polynomial equation

$$\det[\Lambda(\lambda) - \mu] = 0$$

whose coefficients are the integrals in x.

Example: In the simplest case of stationary waves for KdV we have $C = L, B = A$, a Riemann surface Γ with genus 1, extracted from the matrices:

$$\Lambda = \begin{pmatrix} -u_x & 2u + 4\lambda \\ -4\lambda^2 + 2\lambda u - u_{xx} + 2u^2 & u_x \end{pmatrix}, Q = \begin{pmatrix} 0 & 1 \\ u - \lambda & 0 \end{pmatrix} \tag{0.1}$$

Generic finite-gap solutions $u(x)$ are periodic (or quasiperiodic) in x. They can be written by the formula

$$u = -[2 \log \Theta(Ux + Wt + U_0)]_{xx} + Const$$

with Riemann Θ-function, vectors U, W and constant C determined by the Riemann surface Γ; these potentials have a remarkable Spectral Property: Corresponding Schrodinger Operator $-\partial_x^2 + u(x) = L$ has only a finite number of gaps in the Spectrum on the line, whose endpoints are exactly the branching points of Riemann surface above.

The Spectral Problem of Bloch

$$L\Psi = \lambda\Psi, \Psi(x + T) = e^{ipT}\Psi(x)$$

is completely solvable in Θ-functions.

This class of functions generates the "Finite-Gap Solutions" of KdV [8]. This Family is dense in the class of all smooth periodic functions –see [9].

For the 2+1-dimensional KP–system the corresponding Lax-Zakharov-Shabat operators are

$$L = \sigma\partial_y - \partial_x^2 + u(x, y, t), \sigma^2 = \pm 1, A = -4\partial_x^3 + 3(2u\partial_x + u_x) + w$$

There is no relation here between the form of linear OD operators $[C, B] = 0$, describing special solutions, associated with Riemann surfaces, and Lax operators L, A. All family of "Krichever Solutions" for KP is much more broad than finite–gap families in the case of 1+1 systems, They have more or less the same analytical form as above, but the class of parameters (compact Riemann surfaces with marked point) is unrestricted. This class of solutions was used later several times for the different goals: for the solution of the classical problems of the theory of Θ–functions (like new approach to the Riemann–Schottki Problem, started in [11] and finished by Shiota), for some applications in the Conformal 2-d Field Theory and in the theory of bosonic strings on the base of new beautiful algebraic [12] and functional [13] interpretations. An extension of this class, associated with such modern aspects of Algebraic Geometry as holomorphic vector bundles over algebraic curves and their deformations, was constructed in [14].

Solitons and strings: In particular, in the works [13] we realized the following program. As everybody know, in the late 60-ies and early 70-ies a large group of physicists (Veneziano, Virasoro, Alessandrini, Mandelstam and many others) developed the very beautiful theory of the bosonic quantum strings. They used standard operator quantization, decomposing fields in the Fourier series and replacing c-numbers by operators with standard canonical commutators. This program was effectively realized for the "zero-loop" or "tree-like diagrams" only (i.e. for the processes, described by the Riemann surfaces of the zero genus). The so-called Virasoro algebra and its representations played an important role in these constructions. This program stopped because nobody was able to quantize fields in such a way for the multiloop case (i.e. for the Riemann surfaces of nonzero genus). In early 80-ies Polyakov solved the problem of quantization of bosonic strings using a functional ("path") integral. No objects like Virasoro algebra appear in this approach. In the works [13] we constructed a right analog of the Fourier-Laurent series on the Riemann surfaces, using the analytical constructions of the Soliton Theory. After that, an operator quantization of strings was done very easily. Some beautiful analogs of the Virasoro algebra appeared here. This area was not active in the last 3 years, so we shall not discuss it here.

In the second lecture we shall discuss a completely different deep connection between solitons and strings.

Let me start now the main subject of this lecture.

Topologically Trivial Periodic 2-dimensional Schrodinger Operators and Riemann surfaces..

I have no intention to discuss here all the subjects above. My goal is to explain some less popular ideas, associated with 2-dimensional Schrodinger Operator, in connection with new work of the present author and A.Veselov (in preparation).

It is more or less obvious, that there is no nontrivial Lax equations associated with 2-dimensional Schrodinger operator $-2L = (\partial + A_c)(\bar{\partial} + B_c) + 2V$. Here A_c, B_c are the components of vector-potential, V is a scalar potential, $\partial = \partial_x - i\partial_y, z = x + iy$

However, nontrivial integrable nonlinear systems can be obtained from the different equation ($"L - A - B$-**triple**"), which appeared and was investigated since 1976 (see [15,16]). The inverse spectral problem for double–periodic Schrodinger operators L is associated with one energy level only ($L\Psi = 0$) in the approach. $L - A - B$ triple equation has a form:

$$L_t = [A, L] + BL = LA + (A + B)L$$

which implies something like Lax representation, corresponding to one energy level:

$$(L_t - [A, B])\Psi = 0, L\Psi = 0$$

This representation leads to some beautiful 2-dimensional analogs of KdV, containing KP as some degenerate limit:

$$u_t = (\partial^3 + \partial\Gamma)u(x, y, t) + C.C., \bar{\partial}u = 3\partial\Gamma, u = \bar{u}$$

and corresponding analogs of "Higher KdV" systems. Nontrivial exact solutions of this nonlinear systems and periodic Schrodinger operators with zero magnetic field $-2L = \partial\bar{\partial} + 2V$ and solvable Bloch problem $L\Psi = \epsilon_0\Psi$ for one energy level were found by the present author in collaboration with Veselov in 1984 (see [17]. Our Riemann surface Γ in this case is exactly a Complex Fermi Curve. Our Bloch wave function Ψ can be expressed through the so called "Prym" Θ–functions, which are more complicated than the standard "Jacobian" Θ-functions in the case of KP above. Generic complex Fermi curve has infinite genus. This theory was developed by Krichever in 1989–90, who proved that our exactly integrable class (with Complex Fermi Curve of finite genus) is dense in the class of all double periodic potentials. Rapidly decreasing class also was investigated by Grinevich, Manakov, R.Novikov and the present author in 1987–89. It is interesting that for the two-dimensional Schrodinger operator periodic inverse problem was solved earlier than rapidly decreasing inverse scattering problem (based on the data, associated with one energy level). There exist simple rational potentials (found by Grinevich in 1988), for which the scattering amplitude is identically equal to zero for one energy level.

Topologically Nontrivial Schrodinger Operators.

All this class of integrable Schrodinger operators, associated with nonlinear systems, Riemann surfaces and Θ-functions, contains only Schrodinger operators with "topologically trivial" magnetic field: it has a "Chern class" (i.e. magnetic flux through the elementary cell in the double periodic case) equal to zero.

Completely different class of Schrodinger operators with exactly integrable ground level (which is highly degenerate) was found 15 years ago in rapidly decreasing ([18]) and periodic ([19]) cases. It corresponds to the nonrelativistic Pauli operator for spin 1/2 in the magnetic field, orthogonal to the plane, and zero electric potential. The ground energy level is equal to zero in this case. In particular, for the periodic case [19], this level is isomorphic to the first Landau level in the constant magnetic field with the same magnetic flux through the elementary cell. If this flux is an integer, the so called "Magnetic Bloch functions" were found analytically through the elliptic functions for all this class in [19]:

$$\Psi(x,y) = e^{\phi}\sigma(z - a_1)\ldots\sigma(z - a_n)e^{az}$$

$$\Delta\phi = -H$$

Here H is a a magnetic field, a is expressed through the constants $a_1,\ldots,a_n$ and H.

Cyclic and semicyclic chains of the Laplace transformations. New results of the present author and A.Veselov.

For the unification of these two theories the present author in collaboration with Veselov used an idea of "Cyclic Chains" of Laplace transformations. Let me point out that the theory of cyclic chains of Backlund transformations for 1-dimensional Schrodinger operator was developed in the beautiful work of Shabat and Veselov [20] (some first observations were found in [21]). In the early XIX century, Laplace constructed the transformations of second order linear PDE for some goals in geometry. I would like to point out that the "Laplace transformation" acts on the solutions of the equation $L_0\Psi_0 = 0$ for the two-dimensional Schrodinger operator L_0

$$L = L_0 = -1/2(\bar{\partial} + B_0)(\partial + A_0) + V_0$$

by the formula

$$\Psi_1 = (\partial + A_0)\Psi_0, A_1 = A_0 - (\log V_0)_z, B_1 = B_0, V_1 = V_0 + H_1$$

Here the magnetic field H_0 is equal to $2H_0 = B_{0z} - A_{0\bar{z}}$ and H_1, V_1 are the magnetic field and scalar potential for the operator L_1 respectively, such that

$$L_1\Psi_1 = 0, H_1 = H_0 + 1/2(\log V_0)_{z\bar{z}}$$

The requirement that the chain of Laplace transformations is periodic leads to the beautiful elliptic partial differential equations for the magnetic fields and scalar potentials of all operators in the chain. This problem was posed and studied for the first time in the XIX century by Darboux (the operator L and the corresponding nonlinear systems are always hyperbolic in geometry, but formal calculations are the same). Some useful formal calculations were done by Tsiseika in 1920-s. We applied this stuff to the theory of (elliptic) 2-dimensional Schrodinger operator. Globally nonsingular double periodic solutions of this system in our "elliptic" case give two-dimensional Schrodinger operators with Complex Fermi Curve of finite genus (Algebro-Geometric operators, as above, in the case of topologically trivial magnetic field.)

Especially beautiful well-known integrable systems appear in the case $H_0 = H_n = 0$ for the periods $n = 3, 4$. Let $V_0 = \exp f$. We have:

$$\Delta f = -2(e^f - e^{-2f}), n = 3, \Delta f = -4shf, n = 4$$

In the topologically nontrivial double periodic case, when the magnetic flux $[H]$ is nonzero, no cyclic chain is possible. Let be $[H] > 0$. Instead of cyclic chains we consider the **Semicyclic Chains** and **Quasicyclic chains**, satisfying one of the two following conditions:

1.Semicyclic chains

$$H_0 = H_n, V_n = V_0 + n[H_0]$$

It leads to some operators with special algebraic properties:

Eigenfunctions of two different energy levels 0 and $n[H_0]$ are connected by the operator

$$\Psi_n = (\partial + A_{n-1}\ldots)(\partial + A_0)\Psi_0$$

For $n = 2$ this condition leads to the equation

$$\Delta f_0 = a - bsh f_0, b > 0 ,$$

which has a lot of double periodic nonsingular solutions. However, these levels, equal to zero and to $n[H_0]$ can be out of the spectrum, so this connection is formal.

2.Quasicyclic chains

Another, more interesting analog of cyclic chains, we obtain from the condition, that for $n = 0$ and for some other value of n the operators L_0, L_n belong to the class [18,19] up to the shift of energy

$$V_0 = H_0, V_n = H_n + n[H_0]$$

For $n = 1$ there exists only a constant solution of this equation, but for $n = 2$ it leads to the elliptic PDE:

$$\Delta g = 4([H_0] - c^g), c^g = V_0$$

This equation has a lot of real nonsingular double periodic solutions on the plane. The corresponding Schrodinger operators L_n have two highly degenerate energy levels:

1.Ground level, equal to zero, with magnetic Bloch eigenfunctions written above (by the results of [19])

2. Second integrable level, equal to $n[H_0]$, with magnetic Bloch eigenfunctions of the form

$$\Psi_n = (\partial + A_{n-1})\ldots(\partial + A_0)\Psi_0$$

Here Ψ_0 is a magnetic Bloch eigenfunction with zero energy level for the operator L_0, written by the same formula, but with different magnetic field:

$$H_0 = c^{f_0}, H_2 = 2[H_0] - c^{f_0}, n = 2, [H_0] > 0$$

These levels are isomorphic to the "Landau levels" with numbers 0 and n of the operator L_0 with constant homogeneous magnetic field.

In both levels the magnetic Bloch functions can be calculated through the elliptic functions, written above. H. de Vega pointed out that our nonlinear equation for $n = 2$ appeared already in the 70-s as an "Instanton Reduction" for the Landau-Ginzburg equation for the critical value of parameter [separating between superconductors of the first and second kind (see [22])] .

Lecture 2: Theory of Solitons and String Equation

In the modern terminology, "String Equation" means exactly the equation

$$[L, A] = 1$$

for some linear OD operators (people call it also a "Heisenberg Relation").

This strange terminology appeared in 1989-90 years after the well-known works of Gross, Migdal, Brezin, Kazakov, Douglas, Shanker, David and many others (see [23–26]).

A partition function and Free energy of $N \times N$–Matrix Models in Statistical Mechanics in some very special "Double-Scaling Limit", when the size of matrices is going to infinity $N \to \infty$, have probably a beautiful interpretation in the String theory, which was conjectured by the above-mentioned physicists. These Matrix Models and their "string limit" have a deep connection with the Theory of Solitons, which was the most interesting mathematical discovery of these authors. An analysis of this limiting process was done by Its and others (see [27]). Famous KdV type Systems of the Soliton Theory play here a role of "Renormalization Group", like in Quantum Field Theory. However, completely different classes of special solutions are needed here. It is good to point out, that the Integrability in the sense of Lax–type representation leads to the effective results only for 2 classes: periodic (quasiperiodic) and rapidly decreasing. Sometimes in the theory of Solitons people needed in the self-similar solutions for asymptotic methods and so on. A beautiful idea to study them was invented by Flashka and Newell about 1979 (it was developed by the Japanese school and later used by the Leningrad and Ufa schools for asymptotical studyings)–see [28–32]. However, this approach is complicated; it gives a very few number of the effective results. We have here exactly this case. In particularly, the computation of Free energy can be reduced to the Painleve'-1 equation in the simplest nontrivial case

$$u_{xx} - 3u^2 = x$$

In fact, it is equal in this limit to the special real "Physical solution" on the positive halfline $x \leq 0$ with asymptotics

$$u(x) \sim +\sqrt{(-x)/3}$$

This nonlinear equation is equivalent to the algebraic "String Equation" or "Heisenberg Relation" above $[L, A] = 1$ for the same OD operators which give a Lax pair for the ordinary KdV equation (see Lecture 1). This observation leads to some analog of the Lax representation for this Painleve'-1 equation. Several new approaches were developed for the investigation of this equation on the base of technics of the Theory of Solitons (see [33–37]). I presented in this lecture the ideas of the last joint work of myself with Grinevich ([37]), where a special isomonodromic method for the studying of the physical solution was developed.

REFERENCES

[1] Novikov S.P., Manakov S.V., Pitaevski L.P., Zakharov V.E. *Theory of Solitons.Plenum Press, 1984*

[2] Ablowitz M., Segur H. *SIAM, Philadelphia, 1981*

[3] Dubrovin B.A., Krichever I.M., Novikov S.P. *Integrable Systems. Encyclopedia Math Sciences, Dynamical Systems, vol 4 (Edited by V.Arnold and S.Novikov), Springer.*

[4] Kruskal M., Zabusky N. *Phys Rev Lett., 1965, vol 15 pp 240-243*

[5] Gardner C., Green J., Kruskal M., Miura R. *Phys Rev Lett., 1967, vol 19 pp1095-1097*

[6] Lax P. *Comm Pure Appl Math, 1968, vol 21 iss 5 pp 141-188*

[7] Novikov S.P. *Functional Analysis Appl., 1974, vol 8 iss 3 pp 54-66*

[8] Dubrovin B.A., Matveev V.B., Novikov S.P. *Russian Math Surveys, 1976, vol 31 iss 2 pp 55-136*

[9] Marchenko V.A. *Kiev, Naukova Dumka, 1977*

[10] Krichever I.M. *Soviet Math Doklady, 1976, vol 227 iss 2 pp 291-294; Russian Math Surveys, 1977, vol 32 iss 6 pp 180-208*

[11] Dubrovin B.A. *Russian Math Surveys, 1981, vol 36 iss 2 pp 11-80*

[12] Sato M., Miwa T., Jimbo M. *Publ. RIMS: 1978,vol 14 p 223; 1979, vol 15 pp 201, 577, 871; 1980, vol 16 p 531*

[13] Krichever I.M., Novikov S.P. *Functional Analysis Appl., 1987 vol 21 iss 2 pp 46-63 ; 1987, vol 21 iss 4 pp 47-61; 1989, vol 23 iss 1 pp 24-40; in the collection of papers "Physics and Mathematics of Strings", dedicated to the memory of V.Kniznik (edited by Friedan, Brink and Polyakov), Singapour*

[14] Krichever I.M., Novikov S.P. *Russian Math Surveys, 1980, vol 35 iss 6 pp 47-68*

[15] Manakov S.V. *Russian Math Surveys-Notes of the Moscow Math Society, 1976, vol 31 iss 5 pp 245-246*

[16] Dubrovin B.A., Krichever I.M., Novikov S.P. *Soviet Math Doklady, 1976, vol 229 iss 1 pp 15-18*

[17] Novikov S.P., Veselov A.P. *Soviet Math Doklady, 1984 vol 279 iss 1 pp 20-24 ; Physica D, 1986, vol 18 pp 267-273, dedicated to the 60-th birthday of Martin Kruskal*

[18] Aharonov Y., Casher A. *Phys Rev. A(3), 1979 vol 19 p 2461*

[19] Dubrovin B.A., Novikov S.P. *Soviet Phys. JETP, 1980, vol 52.; Soviet Math Dokl. , 1980, vol 253 p 1293*

[20] Shabat A.B., Veselov A.P. *Functional Analysis Appl., 1995 vol 29 iss 1*

[21] Weiss J. *Journ, Math Phys.,1987 vol 28(9) pp 2025-2039*

[22] de Vega H. J. , Schaposnik F.A. *Phys Rev D14, 1100, 1976*

[23] Gross D., Migdal A.A. *Princeton preprint PUPT-1159, 1989*

[24] Brezin E., Kazakov V. *Preprint ENS, Paris, 1989*

[25] Douglas M., Shenker S. *Rutgers preprint RU-89-34*

[26] David F. *MPLA 5(13), 1019-1029, 1990*

[27] Its A.R. ,Novokshenov V.Yu. *Lecture Notes Math, vol 1191, Springer*

[28] Flashka H., Newell A. *Comm Math Phys, 1980, vol 76 p 67*

[29] Jimbo M., Miwa T., Ueno K. *Physica D vol 2 ,1981, p 306*

[30] Its A.R., Kitaev A.A. *MPLA, 1990, vol 5 iss 13 pp 1019-1029*

[31] Kapaev A.A. *Differential Equations, 1988, vol 24 iss 10 p 1684*

[32] Kitaev A.V. *Zap LOMI , 1991, vol 187 iss 12 p 53*

[33] Novikov S.P. *Functional Analysis Appl.,1990, vol 24 iss 4 pp 43–53*

[34] Moore G. *Comm Math Phys, 1990, vol 133 pp 261–304*

[35] Krichever I.M. *Preprint IHES, 1990*

[36] Its A.R.,Fokas A.S., Kitaev A.V. *Comm Math Phys, 1992, vol 147 pp 395–430*

[37] Grinevich P.G., Novikov S.P. *Algebra and Analysis, 1994, in the volume, dedicated to the 60-th birthday of L.D.Faddeev*

LECTURES ON STRING THEORY IN
CURVED SPACETIMES

H.J. de Vega[a] and N. Sánchez[b]

(a) Laboratoire de Physique Théorique et Hautes Energies, Université Pierre et Marie Curie (Paris VI) et Université Denis Diderot (Paris VII), Tour 16, 1er. étage, 4, Place Jussieu 75252 Paris, cedex 05, France. Laboratoire Associé au CNRS URA 280.

(b) Observatoire de Paris, DEMIRM, 61, Avenue de l'Observatoire, 75014 Paris, France. Laboratoire Associé au CNRS URA 336, Observatoire de Paris et École Normale Supérieure.

Abstract

Recent progress on string theory in curved spacetimes is reviewed. The string dynamics in cosmological and black hole spacetimes is investigated. The different methods available to solve the string equations of motion and constraints in curved spacetimes are described. That is, the string perturbation approach, the null string approach, the τ-expansion, and the construction of global solutions (for instance by inverse scattering methods).

The classical behaviour of strings in FRW and inflationary spacetimes is now understood in a large extent from the various types of explicit string solutions. Three different types of behaviour appear: **unstable, dual** to unstable and **stable**. For the unstable strings, the energy and size grow for large scale factors $R \to \infty$, proportional to R. For the dual to unstable strings, the energy and size blow up for $R \to 0$ as $1/R$. For stable strings, the energy and proper size are bounded. (In Minkowski spacetime, all string solutions are of the stable type).

Recent progress on self-consistent solutions to the Einstein equations for string dominated universes is reviewed. The energy-momentum tensor for a gas of strings is then considered as source of the spacetime geometry and from the above string behaviours the string equation of state is **derived**. The self-consistent string solution exhibits the realistic matter dominated behaviour $R \simeq T^{2/3}$ for large times and the radiation dominated behaviour $R \simeq T^{-1/2}$ for early times (T being the cosmic time).

We report on the **exact integrability** of the string equations plus the constraints in de Sitter spacetime that allows to systematically find **exact** string solutions by soliton methods and the multistring solutions. **Multistring solutions** are a new feature in curved spacetimes. That is, a single world-sheet simultaneously describes many different and independent strings. This phenomenon has no analogue in flat spacetime and follows to the coupling of the strings with the geometry.

Finally, the string dynamics next and inside a Schwarzschild black hole is analyzed and their physical properties discussed.

N. Sánchez and A. Zichichi (eds.), String Gravity and Physics at the Planck Energy Scale, 11–63.
© 1996 Kluwer Academic Publishers.

12

CONTENTS

- I. Introduction

- II. Strings in Curved and Minkowski Spacetimes.

  A A brief review on strings in Minkowski spacetime.
  B The string energy-momentum tensor and the string invariant size.
  C Simple String Solutions in Minkowski Spacetime

- III. How to solve the string equations of motion in curved spacetimes?

  A The τ-expansion.
  B Global Solutions.

- IV. String propagation in cosmological spacetimes.

  A Strings in cosmological universes: the τ-expansion at work.
  B The perfect gas of strings.

- V. Self-consistent string cosmology.

  A String Dominated Universes in General Relativity (no dilaton field).
  B Thermodynamics of strings in cosmological spacetimes.

- VI. Effective String Equations with the String Sources Included.

  A Effective String Equations in Cosmological Universes
  B String driven inflation?

- VII. Multi-Strings and Soliton Methods in de Sitter Universe.

- VIII. Strings next to and inside black holes.

  A String Equations of motion in a Schwarzschild Black Hole.
  B Strings Near the Singularity $r = 0$
  C String energy-momentum and invariant size near the singularity.

I. INTRODUCTION

The construction of a sensible quantum theory of gravitation is probably the greatest challenge in theoretical physics for the end of this century and most probably for the next century too.

Another problem (the most often discussed in connection with gravity quantization) is the one of the renormalizability of the Einstein theory (or its various generalizations) when quantized as a local quantum field theory. Actually, even deeper conceptual problems arise when one tries to combine quantum concepts with General Relativity. For example, statistical phenomena like Hawking's radiation arise when free fields are quantized in black-hole backgrounds. This points out a lack of quantum coherence even keeping the gravitational field classical.

It may be very well that a quantum theory of gravitation needs new concepts and ideas. Of course, this future theory must have the today's General Relativity and Quantum Mechanics (and QFT) as limiting cases. In some sense, what everybody is doing in this domain (including string theories approach) may be something analogous to the developpment of the old quantum theory in the 10's of this century. Namely, people at that time imposed quantization conditions (the Bohr-Sommerfeld conditions) to hamiltonian mechanics but keeping the concepts of classical mechanics.

The main drawback to develop a quantum theory of gravitation is clearly the **total lack of experimental guides** for the theoretical developpment. Just from dimensional reasons, physical effects combining gravitation and quantum mechanics are relevant only at energies of the order of $M_{Planck} = \sqrt{\hbar c/G} = 1.22 \, 10^{16}$Tev. Such energies were available in the Universe at times $t < t_{Planck} = 5.4 \, 10^{-44}$sec. Anyway, as a question of principle, the construction of a quantum theory of gravitation is a problem of fundamental relevance for theoretical physics . In addition, one cannot rule out completely the possibility of some "low energy" ($E \ll M_{Planck}$) physical effect that could be experimentally tested. One may speculate about effects analogous to the presence of magnetic monopoles in some grand unified theories. [Monopoles can be detected by low energy experiments in spite of their large mass].

Let us now see what are the consequences of Heisenberg's principle in quantum mechanics combined with the notion of gravitational (Schwarzschild) radius in General Relativity. Assume we make two measurements at a very small distance Δx . Then,

$$\Delta p \sim \Delta E \sim 1/\Delta x$$

where we set $\hbar = c = 1$. For sufficiently large ΔE, particles with masses $m \sim 1/\Delta x$ will be produced. The gravitational radius of such particles are of the order

$$R_G \sim Gm \sim \frac{(l_{Planck})^2}{\Delta x}$$

where $l_{Planck} \sim 10^{-33}$ cm. Now, General Relativity allows measures at a distance Δx , provided

$$\Delta x > R_G \quad \rightarrow \quad \Delta x > \frac{(l_{Planck})^2}{\Delta x}$$

That is,

$$\Delta x > l_{Planck} \quad \text{and} \quad m < M_{Planck} \tag{1.1}$$

This means that no measurements can be made at distances smaller than the Planck length and that no particle can be heavier than M_{Planck} . This is a simple consequence of relativistic quantum mechanics combined with General Relativity. In addition, the notion of locality and hence of spacetime becomes meaningless at the Planck scale. Notice that the equality in eq.(1.1) means that the Compton length equals the Schwarzschild radius of a particle. Since M_{Planck} is the heaviest possible particle scale, a theory valid there (necessarily involving quantum gravitation) will also be valid at any lower energy scale. One may ignore higher energy phenomena in a low energy theory, but not the opposite. In other words, a theory of quantum gravity will be a 'theory of everything'. We think that this is the **key point** on the quantization of gravity. A theory that holds till the Planck scale must describe **all** what happens at lower energies including all known particle physics as well as what we do not know yet (that is, beyond the standard model) [1]. Notice that this conclusion is totally independent of the use or not of string models. A direct important consequence of this conclusion, is that it may not make physical sense to quantize **pure gravity**. A physically sensible quantum theory cannot contain only gravitons. To give an example, a theoretical prediction for graviton-graviton scattering at energies of the order of M_{Planck} must include all particles produced in a real experiment. That is, in practice, **all** existing particles in nature, since gravity couples to all matter.

In conclusion : a consistent quantum theory of gravitation must be a finite theory [1] and must include all other interactions. That is, it must be a theory of everything (TOE). This is a very ambitious project. In particular it needs the understanding of the present desert between 1 and 10^{16} TeV. There is an additional dimensional argument about the inference Quantum Theory of Gravitation $\rightarrow$ TOE. There are only three dimensional physical magnitudes in nature: length, energy and time and correspondingly only three dimensional constants in nature: $c, \hbar$ and G. All other physical constants like $\alpha = 1/137, 04..., M_{proton}/m_{electron}, \theta_{WS}, \ldots$ etc. are pure numbers and they must be calculable in a TOE. This is a formidable, but extremely appealing problem. From the theoretical side, the **only serious candidate** for a TOE is at present string theory. This is why we think that strings deserve a special attention in order to quantize gravity.

String theory is therefore an appropriate arena to work out the quantization of gravity consistently. It provides an unified theory of all interactions overcoming at the same time the nonrenormalizable character of quantum fields theories of gravity.

As a first step in the understanding of quantum gravitational phenomena in a string framework, we started in 1987 a programme of string quantization on curved spacetimes [2,9]. The investigation of strings in curved spacetimes is currently the best framework to study the physics of gravitation in the context of string theory, in spite of its limitations. First, the use of a continuous Riemanian manifold to describe the spacetime cannot be valid at scales of the order of l_{Planck}. More important, gravitational backgrounds effectively provide classical or semiclasical descriptions even if the matter backreaction to the geometry is included through semiclassical Einstein equations (or stringy corrected Einstein equations) by inserting the expectation value of the string energy-momentum tensor in the r.h.s. One would want a full quantum treatment for matter and geometry. However, to find a formulation of string theory going beyond the use of classical backgrounds is a very difficult (but fundamental) problem. One would like to derive the spacetime geometry as a classical and low energy ($\ll M_{Planck}$) limit from the solution of (quantum) string theory.

After a short introduction on strings in Minkowski and curved spacetimes, we focus on strings in cosmological spacetimes.

Substantial results were achieved in this field since 1992. The classical behaviour of strings in FRW and inflationary spacetimes is now understood in a large extent [3]. This understanding followed the finding of various types of exact, asymptotic and numerical string solutions in FRW and inflationary spacetimes [5] - [12]. For inflationary spacetimes, the exact integrability of the string propagation equations plus the string constraints in de Sitter spacetime [4] is indeed an important help. This allowed to systematically find **exact** string solutions by soliton methods using the linear system associated to the problem (the so-called dressing method in soliton theory) and the multistring solutions [5] - [8].

In summary, three different types of behaviour are exhibited by the string solutions in cosmological spacetimes: **unstable, dual** to unstable and **stable**. For the unstable strings, the energy and size grow for large scale factors $R \to \infty$, proportional to R. For the dual to unstable strings, the energy and size blow up for $R \to 0$ as $1/R$. For stable strings, the energy and proper size are bounded. (In Minkowski spacetime, all string solutions are of the stable type). The equation of state for these string behaviours take the form

- (i) **unstable** for $R \to \infty$ $p_u = -E_u/(D-1) < 0$

- (ii) **dual to unstable** for $R \to 0$, $p_d = E_d/(D-1) > 0$.

- (iii) **stable** for $R \to \infty$, $p_s = 0$.

Here E_u and E_d stand for the corresponding string energies and $D-1$ for the number of spatial dimensions where the string solutions lives. For example, $d-1 = 1$ for a straight string, $d-1 = 2$ for a ring string, etc.

As we see above, the dual to unstable string behavior leads to the same equation of state than radiation (massless particles or hot matter). The stable string behavior leads to the equation of state of massive particles (dust or cold matter). The unstable string behavior is a purely 'stringy' phenomenon. The fact that in entails a negative pressure is however physically acceptable. For a gas of strings, the unstable string behaviour dominates in inflationary universes when $R \to \infty$ and the dual to unstable string behavior dominates for $R \to 0$.

The unstable strings correspond to the critical case of the so called *coasting universe* [19,32]. That is, classical strings provide a *concrete* realization of such cosmological models. The 'unstable' behaviour is called 'string stretching' in the cosmic string literature [20,21].

It must be stressed that while time evolves, a **given** string solution may exhibit two and even three of the above regimes one after the other (see sec. III). Intermediate behaviours are also observed in ring solutions [7,12]. That is,

$$P = (\gamma - 1) E \quad \text{with} \quad -\frac{1}{D-1} < \gamma - 1 < +\frac{1}{D-1} .$$

We also report here on the exact integrability of the string equations plus the constraints in de Sitter spacetime which allows to systematically find **exact** string solutions by soliton methods and the multistring solutions. **Multistring solutions** are a new feature in curved spacetimes. That is, a single world-sheet simultaneously describes many different and independent strings. This phenomenon has no analogue in flat spacetime and appears as a consequence of the coupling of the strings with the spacetime geometry.

The world-sheet time τ turns out to be an multi-valued function of the target string time X^0 (which can be the cosmic time T, the conformal time η or for de Sitter universes it can be the hyperboloid time q^0). Each branch of τ as a function of X^0 corresponds to a different string. In flat spacetime, multiple string solutions are necessarily described by multiple world-sheets. Here, a single world-sheet describes one string, several strings or even an infinite number of different and independent strings as a consequence of the coupling with the spacetime geometry. These strings do not interact among themselves; all the interaction is with the curved spacetime. One can decide to study separately each of them (they are all different) or consider all the infinite strings together.

Of course, from our multistring solution, one *could* just choose only one interval in τ (or a subset of intervals in τ) and describe just one string (or several). This will be just a **truncation** of the solution.

The really remarkably fact is that all these infinitely many strings come **naturally together** when solving the string equations in de Sitter spacetime as we did in refs. [5] - [7].

Here, interaction among the strings (like splitting and merging) is neglected, the only interaction is with the curved background.

The multistring property appears associated to the presence of a cosmological constant (whatever be its sign) [14]. Multistring solutions have not been found in black-hole backgrounds (without cosmological constant). More recently, new classes of dynamical and stationary multistring solutions in curved spacetimes have been found and classified and their physical properties analyzed [14]. Multistrings has been found for all inflationary spacetimes [15] but not in FRW universes.

The study of string propagation in curved spacetimes provide essential clues about the physics in this context but is clearly not the end of the story. The next step beyond the investigation of **test** strings, consist in finding **self-consistently** the geometry from the strings as matter sources for the Einstein equations or better the string effective equations (beta functions). This goal is achieved in ref. [3] for cosmological spacetimes at the classical level. Namely, we used the energy-momentum tensor for a gas of strings as source for the Einstein equations and we solved them self-consistently.

To write the string equation of state we used the behaviour of the string solutions in cosmological spacetimes. Strings continuously evolve from one type of behaviour to another, as is explicitly shown by our solutions [4] - [7]. For intermediate values of R, the equation of state for gas of free strings is clearly complicated but a formula of the type:

$$\rho = \left(u_R\, R + \frac{d}{R} + s \right) \frac{1}{R^{D-1}} \tag{1.2}$$

where

$$\lim_{R\to\infty} u_R = \begin{cases} 0 & \text{FRW} \\ u_\infty \neq 0 & \text{Inflationary} \end{cases} \tag{1.3}$$

This equation of state is qualitatively correct for all R and becomes exact for $R \to 0$ and $R \to \infty$. The parameters u_R, d and s are positive constants and the u_R varies smoothly with R.

The pressure associated to the energy density (1.2) takes then the form

$$p = \frac{1}{D-1} \left(\frac{d}{R} - u_R R \right) \frac{1}{R^{D-1}} \tag{1.4}$$

Inserting this source into the Einstein-Friedman equations leads to a self-consistent solution for string dominated universes (see sec. VI) [3]. This solution exhibits the realistic matter dominated behaviour $R \simeq T^{2/(D-1)}$ for large times and the radiation dominated behaviour $R \simeq T^{2/D}$ for early times.

For the sake of completeness we analyze in sec. IV the effective string equations [3]. These equations have been extensively treated in the literature [28] and they are not our central aim.

It must be noticed that there is no satisfactory derivation of inflation in the context of the effective string equations. This does not mean that string theory is not compatible with inflation, but that the effective string action approach *is not enough* to describe inflation. The effective string equations are a low energy field theory approximation to string theory containing only the *massless* string modes. The vacuum energy scales to start inflation are typically of the order of the Planck mass where the effective string action approximation breaks down. One must also consider the *massive* string modes (which are absent from the effective string action) in order to properly get the cosmological condensate yielding inflation. De Sitter inflation does not emerge as a solution of the the the effective string equations.

In conclusion, the effective string action (whatever be the dilaton, its potential and the central charge term) is not the appropriate framework in which to address the question of string driven inflation.

Early cosmology (at the Planck time) is probably the best place to test string theory. In one hand the quantum treatment of gravity is unavoidable at such scales and in the other hand, observable cosmological consequences are derivable from the inflationary stage. The natural gravitational background is an inflationary universe as, for instance, de Sitter spacetime. Such geometries are not string vacua. This means that conformal and Weyl symmetries are broken at the quantum level. In order to quantize consistently strings in such case, one must enlarge the physical phase space including, in particular, the factor $\exp \phi(\sigma, \tau)$ in the world-sheet metric [see eq.(2.6)]. This is a very interesting and open problem. Physically, the origin of such difficulties in quantum string cosmology comes from the fact that one is not dealing with an **empty** universe since a cosmological spacetime necessarily contains matter. In the other hand, conformal field theory techniques are till now only adapted to backgrounds for which the beta functions are identically zero, i. e. sourceless geometries. A (quantum) string theory treatment of early cosmology necessarily implies **excited** states, not just string vacua. This problem is completely open today.

The outline of these lectures is as follows. Section II presents an introduction to strings in curved spacetimes including basic notions on classical and quantum strings in Minkowski spacetime and introducing the main physical string magnitudes: energy-momentum and invariant string size.

Section III deals with the several methods of resolution of the string propagation in curved spacetimes. (In sections III.A and III.B we treat the perturbative approaches, the τ-expansion and the global solutions.

Section V deals with strings in cosmological spacetimes, the τ-expansion at work there and we present the perfect gas of strings as a model for string matter.

In section V we treat self-consistent string cosmology including the string equations of state. (Section V.A deals with general relativity, V.B with the string thermodynamics).

Section VI discuss the effective (beta functions) string equations in the cosmological perspective and the search of inflationary solutions.

In sec. VII, we briefly review the systematic construction of string solutions in de Sitter universe *via* soliton methods and the new feature of multistring solutions.

Section VIII contains the string dynamics next and inside Schwarzschild black-holes, the string behaviour near the $r = 0$ singularity and their physical properties.

II. STRINGS IN CURVED AND MINKOWSKI SPACETIMES.

Let us consider bosonic strings (open or closed) propagating in a curved D-dimensional spacetime defined by a metric $G_{AB}(X), 0 \leq A, B \leq D - 1$. The action can be written as

$$S = \frac{1}{2\pi\alpha'} \int d\sigma d\tau \sqrt{g} \, g_{\alpha\beta}(\sigma, \tau) \, G_{AB}(X) \, \partial^\alpha X^A(\sigma, \tau) \, \partial^\beta X^B(\sigma, \tau) \tag{2.1}$$

Here $g_{\alpha\beta}(\sigma, \tau)$ ($0 \leq \alpha, \beta \leq 1$) is the metric in the worldsheet, α' stands for the string tension. As in flat spacetime, $\alpha' \simeq (M_{Planck})^{-2} \simeq (l_{Planck})^2$ fixes the scale in the theory. There are no other free parameters like coupling constants in string theory.

We will start considering given gravitational backgrounds $G_{AB}(X)$. That is, we start to investigate *test* strings propagating on a given spacetime. In section IV, the back reaction problem will be studied. That is, how the strings may act as source of the geometry.

String propagation in massless backgrounds other than gravitational (dilaton, antisymmetric tensor) can be investigated analogously.

The string action (2.1) classically enjoys Weyl invariance on the world sheet

$$g_{\alpha\beta}(\sigma, \tau) \to \lambda(\sigma, \tau) \, g_{\alpha\beta}(\sigma, \tau) \tag{2.2}$$

plus the reparametrization invariance

$$\sigma \to \sigma' = f(\sigma, \tau) \quad , \quad \tau \to \tau' = g(\sigma, \tau) \tag{2.3}$$

Here $\lambda(\sigma, \tau), f(\sigma, \tau)$ and $g(\sigma, \tau)$ are arbitrary functions.

The dynamical variables being here the string coordinates $X_A(\sigma, \tau)$, $(0 \leq A \leq D - 1)$ and the world-sheet metric $g_{\alpha\beta}(\sigma, \tau)$.

Extremizing S with respect to them yields the classical equations of motion:

$$\partial^\alpha [\sqrt{g} \, G_{AB}(X) \, \partial_\alpha X^B(\sigma, \tau)] = \frac{1}{2} \sqrt{g} \, \partial_A G_{CD}(X) \, \partial_\alpha X^C(\sigma, \tau) \, \partial^\alpha X^D(\sigma, \tau) \tag{2.4}$$
$$0 \leq A \leq D - 1$$

$$\begin{aligned} T_{\alpha\beta} \equiv \ & G_{AB}(X)[\, \partial_\alpha X^A(\sigma, \tau) \, \partial_\beta X^B(\sigma, \tau) \\ & - \frac{1}{2} g_{\alpha\beta}(\sigma, \tau) \, \partial_\gamma X^A(\sigma, \tau) \, \partial^\gamma X^B(\sigma, \tau)] = 0 \ , \quad 0 \leq \alpha, \beta \leq 1. \end{aligned} \tag{2.5}$$

Eqs. (2.5) contain only first derivatives and are therefore a set of constraints. Classically, we can always use the reparametrization freedom (2.3) to recast the world-sheet metric on diagonal form

$$g_{\alpha\beta}(\sigma,\tau) = \exp[\phi(\sigma,\tau)] \; \text{diag}(-1,+1) \tag{2.6}$$

In this conformal gauge, eqs. (2.4) - (2.5) take the simpler form:

$$\partial_{-+}X^A(\sigma,\tau) + \Gamma^A_{BC}(X)\,\partial_+X^B(\sigma,\tau)\,\partial_-X^C(\sigma,\tau) = 0 \;, \quad 0 \le A \le D-1, \tag{2.7}$$

$$T_{\pm\pm} \equiv G_{AB}(X)\,\partial_\pm X^A(\sigma,\tau)\,\partial_\pm X^B(\sigma,\tau) \equiv 0 \;, \quad T_{+-} \equiv T_{-+} \equiv 0 \tag{2.8}$$

where we introduce light-cone variables $x_\pm \equiv \sigma \pm \tau$ on the world-sheet and where $\Gamma^A_{BC}(X)$ stand for the connections (Christoffel symbols) associated to the metric $G_{AB}(X)$.

Notice that these equations in the conformal gauge are still invariant under the conformal reparametrizations:

$$\sigma + \tau \to \sigma' + \tau' = f(\sigma + \tau) \quad , \quad \sigma - \tau \to \sigma' - \tau' = g(\sigma - \tau) \tag{2.9}$$

Here $f(x)$ and $g(x)$ are arbitrary functions.

The string boundary conditions in curved spacetimes are identical to those in Minkowski spacetime. That is,

$$X^A(\sigma + 2\pi, \tau) = X^A(\sigma,\tau) \quad \text{closed strings}$$

$$\partial_\sigma X^A(0,\tau) = \partial_\sigma X^A(\pi,\tau) = 0 \quad \text{open strings.} \tag{2.10}$$

A. A brief review on strings in Minkowski spacetime

In flat spacetime eqs.(2.7) become linear

$$\partial_{-+}X^A(\sigma,\tau) = 0 \;, \quad 0 \le A \le D-1, \tag{2.11}$$

and one can solve them explicitly as well as the quadratic constraint (2.8) [see below]:

$$\left[\partial_\pm X^0(\sigma,\tau)\right]^2 - \sum_{j=1}^{D-1}\left[\partial_\pm X^j(\sigma,\tau)\right]^2 = 0 \tag{2.12}$$

The solution of eqs.(2.11) is usually written for closed strings as

$$X^A(\sigma,\tau) = q^A + 2p^A\alpha'\tau + i\sqrt{\alpha'}\sum_{n\neq 0}\frac{1}{n}\{\alpha_n^A \exp[in(\sigma-\tau)] + \tilde{\alpha}_n^A \exp[-in(\sigma+\tau)]\} \tag{2.13}$$

where q^A and p^A stand for the string center of mass position and momentum and α_n^A and $\tilde{\alpha}_n^A$ describe the right and left oscillator modes of the string, respectively. Since the string coordinates are real,

$$\bar{\alpha}_n^A = \alpha_{-n}^A \quad , \quad \bar{\tilde{\alpha}}_n^A = \tilde{\alpha}_{-n}^A$$

This resolution is no more possible in general for curved spacetime where the equations of motion (2.7) are non-linear. In that case, right and left movers interact with themselves and with each other.

In Minkowski spacetime we can also write the solution of the string equations of motion (2.11) in the form

$$X^A(\sigma, \tau) = l^A(\sigma + \tau) + r^A(\sigma - \tau) \tag{2.14}$$

where $l^A(x)$ and $r^A(x)$ are arbitrary functions. Now, making an appropriate conformal transformation (2.9) we can turn any of the string coordinates $X^A(\sigma, \tau)$ (but only one of them) into a constant times τ. The most convenient choice is the light-cone gauge where

$$U \equiv X^0 - X^1 = 2\,p^U \alpha' \tau. \tag{2.15}$$

That is, there are no string oscillations along the U direction in the light-cone gauge. We have still to impose the constraints (2.12). In this gauge they take the form

$$\pm 2\alpha' p^U \, \partial_\pm V(\sigma, \tau) = \sum_{j=2}^{D-1} \left[\partial_\pm X^j(\sigma, \tau)\right]^2 \tag{2.16}$$

where $V \equiv X^0 + X^1$. This shows that V is not an independent dynamical variable since it expresses in terms of the transverse coordinates $X^2, \ldots, X^{D-1}$. Only q^V is an independent quantity.

The physical picture of a string propagating in Minkowski spacetime clearly emerges in the light-cone gauge. The gauge condition (2.15) tells us that the string 'time' τ is just proportional to the physical null time U. Eqs.(2.13) shows that the string moves as a whole with constant speed while it oscillates around its center of mass. The oscillation frequencies are all integers multiples of the basic one. The string thus possess an infinite number of normal modes; $\alpha_n^A, \tilde{\alpha}_n^A$ classically describe their oscillation amplitudes. Only the modes in the direction of the transverse coordinates $X^2, \ldots, X^{D-1}$ are physical. This is intuitively right, since a longitudinal or a temporal oscillation of a string is meaningless. In summary, the string in Minkowski spacetime behaves as an extended and composite relativistic object formed by a $2(D-2)$-infinite set of harmonic oscillators.

Integrating eq.(2.16) on σ from 0 to 2π and inserting eq. (2.13) yields the classical string mass formula:

$$m^2 \equiv p^U p^V - \sum_{j=2}^{D-1} (p^j)^2 = \frac{1}{\alpha'} \sum_{j=2}^{D-1} \sum_{n=1}^{\infty} \left[\alpha_n^j \alpha_{-n}^j + \tilde{\alpha}_n^j \tilde{\alpha}_{-n}^j\right] \tag{2.17}$$

We explicitly see how the mass of a string depends on its excitation state. The classical string spectrum is continuous as we read from eq.(2.17). It starts at $m^2 = 0$ for an unexcited string: $\alpha_n^j = \tilde{\alpha}_n^j = 0$ for all n and j.

The independent string variables are:
the transverse amplitudes $\{\alpha_n^j, \tilde{\alpha}_n^j, n\varepsilon Z, n \neq 0, 2 \leq j \leq D-1\}$,
the transverse center of mass variables $\{q^j, p^j, 2 \leq j \leq D-1\}$,
q^V and p^U.

Up to now we have considered a classical string.

At the quantum level one imposes the canonical commutation relations (CCR)

$$[\alpha_n^i, \alpha_m^j] = n\, \delta_{n,-m}\, \delta^{i,j} \quad , \quad [\tilde{\alpha}_n^i, \alpha_m^j] = n\, \delta_{n,-m}\, \delta^{i,j} \ ,$$

$$[\tilde{\alpha}_n^i, \alpha_m^j] = 0 \ ,$$

$$[q^i, p^j] = i\, \delta^{i,j} \quad , \quad [q^V, p^U] = i \tag{2.18}$$

All other commutators being zero. An order prescription is needed to unambiguously express the different physical operators in terms of those obeying the CCR. The symmetric ordering is the simplest and more convenient.

The space of string physical states is the the tensor product of the Hilbert space of the $D-1$ center of mass variables $q_V, p_U, \{q^j, p^j, 2 \leq j \leq D-1\}$, times the Fock space of the harmonic transverse modes. The string wave function is then the product of a center of mass part times a harmonic oscillator part. The center of mass can be taken, for example, as a plane wave. The harmonic oscillator part can be written as the creation operators $\alpha_n^{j\,\dagger}, \tilde{\alpha}_n^{j\,\dagger}$, $n \geq 1$, $2 \leq j \leq D-1$ acting on the oscillator ground state $|0>$. This state is defined as usual by

$$\alpha_n^j\, |0> = \tilde{\alpha}_n^j\, |0> = 0, \quad \text{for all } n \geq 1, \ 2 \leq j \leq D-1$$

Notice that a string describes **one particle**. The kind of particle described depends on the oscillator wave function. The mass and spin can take an infinite number of different values. That is, there is an infinite number of different possibilities for the particle described by a string.

Let us consider the quantum mass spectrum. Upon symmetric ordering the mass operator becomes,

$$m^2 = \frac{1}{2\alpha'} \sum_{j=2}^{D-1} \sum_{n=1}^{\infty} \left[\alpha_n^j\, \alpha_{-n}^j + \alpha_{-n}^j\, \alpha_n^j + \tilde{\alpha}_n^j\, \tilde{\alpha}_{-n}^j + \tilde{\alpha}_{-n}^j\, \tilde{\alpha}_n^j \right] \ . \tag{2.19}$$

Using the commutation rules (2.18) yields

$$m^2 = \frac{D-2}{\alpha'} \sum_{n=1}^{\infty} n + \frac{1}{2\alpha'} \sum_{j=2}^{D-1} \sum_{n=1}^{\infty} \left[\alpha_n^{j\,\dagger} \alpha_n^j + \tilde{\alpha}_n^{j\,\dagger} \tilde{\alpha}_n^j \right] \tag{2.20}$$

The divergent sum in the first term can be defined through analytic continuation of the zeta function

$$\zeta(z) \equiv \sum_{n=1}^{\infty} \frac{1}{n^z} \tag{2.21}$$

One finds $\zeta(-1) = -1/12$ [30]. Thus,

$$m^2 = -\frac{D-2}{12\alpha'} + \frac{1}{2\alpha'} \sum_{j=2}^{D-1} \sum_{n=1}^{\infty} \left[\alpha_n^{j\,\dagger} \alpha_n^j + \tilde{\alpha}_n^{j\,\dagger} \tilde{\alpha}_n^j \right] \tag{2.22}$$

Hence, the string ground state $|0>$ has a negative mass squared

$$m_0^2 = -\frac{D-2}{12\alpha'} \tag{2.23}$$

Such particles are called tachyons and exhibit unphysical behaviours. When fermionic degrees of freedom are associated to the string the ground state becomes massless (superstrings) [29].

Notice that the appearance of a negative mass square yields a dispersion relation $E^2 = p^2 - |m_0^2|$ similar to classical waves when gravity (even newtonian) is taken into account (Jeans unstabilities) [31].

Let us consider now excited states.

The constraints (2.12) integrated on σ from 0 to 2π impose

$$\sum_{j=2}^{D-1} \sum_{n=1}^{\infty} (\alpha_n^j)^\dagger \alpha_n^j = \sum_{j=2}^{D-1} \sum_{n=1}^{\infty} (\tilde{\alpha}_n^j)^\dagger \tilde{\alpha}_n^j \tag{2.24}$$

on the physical states. This means that the number of left and right modes coincide in all physical states.

The first excited state is then described by

$$|i,j> = (\tilde{\alpha}_1^i)^\dagger (\alpha_1^j)^\dagger |0> \tag{2.25}$$

times the center of mass wave function. We see that this wavefunction is a symmetric tensor in the space indices i, j. It describes therefore a spin two particle plus a spin zero particle (the trace part).

From eqs.(2.22-2.25) follows that

$$m^2 |i,j> = -\frac{D-26}{12\alpha'} |i,j> \tag{2.26}$$

This state is then a massless particle only for $D = 26$. In such critical dimension we have then a graviton (massless spin 2 particle) and a dilaton (massless spin 0 particle) as string modes of excitation. For superstrings the critical dimension turns to be $D = 10$ [29].

We shall consider, as usual, that only four space-time dimensions are uncompactified. That is, we shall consider the strings as living on the tensor product of a curved four dimensional space-time with lorentzian signature and a compact space which is there to cancel the anomalies. From now on strings will propagate in the curved (physical) four dimensional space-time. However, we will find instructive to study the case where this curved space-time has dimensionality D, where D may be 2, 3, 4 or arbitrary.

B. The string energy-momentum tensor and the string invariant size

The spacetime string energy-momentum tensor follows (as usual) by taking the functional derivative of the action (2.1) with respect to the metric G_{AB} at the spacetime point X. This yields,

$$\sqrt{-G}\, T^{AB}(X) = \frac{1}{2\pi\alpha'} \int d\sigma d\tau \left(\dot{X}^A \dot{X}^B - X'^A X'^B \right) \delta^{(D)}(X - X(\sigma,\tau)) \tag{2.27}$$

where dot and prime stands for $\partial/\partial\tau$ and $\partial/\partial\sigma$, respectively.

Notice that X in eq.(2.27) is just a spacetime point whereas $X(\sigma,\tau)$ stands for the string dynamical variables. One sees from the Dirac delta in eq.(2.27) that $T^{AB}(X)$ vanishes unless

X is exactly on the string world-sheet. We shall not be interested in the detailed structure of the classical strings. It is the more useful to integrate the energy-momentum tensor (2.27) on a volume that completely encloses the string. It takes then the form [17]

$$\Theta^{AB}(X^0) = \frac{1}{2\pi\alpha'} \int d\sigma d\tau \left(\dot{X}^A \dot{X}^B - X'^A X'^B \right) \delta(X^0 - X^0(\tau,\sigma)). \tag{2.28}$$

When X^0 depends only on τ, we can easily integrate over τ with the result,

$$\Theta^{AB}(X^0) = \frac{1}{2\pi\alpha'|\dot{X}^0(\tau)|} \int_0^{2\pi} d\sigma \left[\dot{X}^A \dot{X}^B - X'^A X'^B \right]_{\tau=\tau(X^0)} \tag{2.29}$$

Another relevant physical magnitude for strings is the invariant size. We define the invariant string size ds^2 using the metric induced on the string world-sheet:

$$ds^2 = G_{AB}(X)\, dX^A dX^B \tag{2.30}$$

Inserting $dX^A = \partial_+ X^A\, dx^+ + \partial_- X^A\, dx^-$, into eq.(2.30) and taking into account the constraints (2.8) yields

$$ds^2 = 2\, G_{AB}(X)\, \partial_+ X^A \partial_- X^B \left(d\tau^2 - d\sigma^2 \right) = G_{AB}(X)\, \dot{X}^A \dot{X}^B \left(d\tau^2 - d\sigma^2 \right). \tag{2.31}$$

Thus, we define the string size as the integral of

$$S(\sigma,\tau) \equiv \sqrt{G_{AB}(X)\, \dot{X}^A \dot{X}^B} \tag{2.32}$$

over σ at fixed τ.

Notice that the trace of the energy-momentum tensor eq.(2.27) is just the integral of $S(\sigma,\tau)^2$,

$$\sqrt{-G}\, T_A^A(X) = \frac{1}{\pi\alpha'} \int d\sigma d\tau\, G_{AB}(X) \dot{X}^A \dot{X}^B\, \delta^{(D)}(X - X(\sigma,\tau))$$

$$= \frac{1}{\pi\alpha'} \int d\sigma d\tau\, S(\sigma,\tau)^2\, \delta^{(D)}(X - X(\sigma,\tau)) \tag{2.33}$$

C. Simple String Solutions in Minkowski Spacetime

Let us now consider a circular string as a simple example of a string solution in Minkowski spacetime.

$$X^0(\sigma,\tau) = \alpha' E\tau \quad , \quad X^3(\sigma,\tau) = \alpha' p\tau$$

$$X^1(\sigma,\tau) = \alpha' m \cos\tau \cos\sigma = \frac{\alpha' m}{2} \left[\cos(\tau+\sigma) + \cos(\tau-\sigma) \right] \tag{2.34}$$

$$X^2(\sigma,\tau) = \alpha' m \cos\tau \sin\sigma = \frac{\alpha' m}{2} \left[\sin(\tau+\sigma) - \sin(\tau-\sigma) \right]$$

This is obviously a solution of eqs.(2.11) where only the modes $n = \pm 1, j = 1, 2$ are excited. The constraints (2.12) yields

$$E^2 = p^2 + m^2$$

Eqs.(2.34) describe a circular string in the X^1, X^2 plane, centered in the origin and with an oscillating radius $\rho(\tau) = \alpha' m \cos \tau$. In addition the string moves uniformly in the z-direction with speed p/E. (That is, p is its momentum in the z-direction). The oscillation amplitude m can be identified with the string mass and E with the string energy. Notice that the string time τ is here proportional to the physical time X^0 [this solution is not in the light-cone gauge (2.15)].

It is instructive to compute the integrated energy-momentum tensor (2.29) for this string solution. We find in the rest frame ($p = 0$) that it takes the fluid form

$$\Theta_A^B = \begin{pmatrix} \rho & 0 & 0 & 0 \\ 0 & -p & 0 & 0 \\ 0 & 0 & -p & 0 \\ 0 & 0 & 0 & 0 \end{pmatrix} \tag{2.35}$$

where

$$\rho = E = m \quad , \quad p = -\frac{m}{2} \cos(2\tau) \tag{2.36}$$

We see that the total energy coincides with m as one could expect and that the (space averaged) pressure oscillates around zero. That is, the string pressure goes through positive and negative values. The time average of p on a period vanishes. The string behaves then as cold matter (massive particles).

The upper value of p equals $E/2$. This is precisely the relation between E and p for radiation (massless particles). (Notice that this circular strings lives on a two-dimensional plane). The lower value of p correspond to the limiting value allowed by the strong energy condition in General Relativity [16]. We shall see below that these two extreme values of p appear for strings in general cosmological spacetimes.

The invariant size of the string solution (2.34) follows by inserting eq.(2.34) into eq.(2.31). We find

$$ds^2 = (\alpha' m)^2 \left(d\tau^2 - d\sigma^2 \right) \tag{2.37}$$

Therefore, this string solution has a constant size $2\pi\alpha' m$.

Another simple but instructive solution in Minkowski spacetime is a rotating straight string (a rotating rod) given by

$$X^0(\sigma, \tau) = \alpha' m \tau \quad ,$$

$$X^1(\sigma, \tau) = \alpha' m \cos \tau \cos \sigma \ , \tag{2.38}$$

$$X^2(\sigma, \tau) = \alpha' m \sin \tau \cos \sigma \ ,$$

or in polar coordinates

$$\rho = \alpha' m \, |\cos \sigma| \quad , \quad \phi = \tau \ .$$

That is, a straight string on the $X^1 - X^2$ plane rotating around the origin with an angular speed $\frac{1}{\alpha' m}$. Eqs. (2.38) identically fulfil the string equations and constraints (2.11-2.12).

The energy momentum tensor for this rotating string takes the form:

$$\Theta_A^B = \begin{pmatrix} m & 0 & 0 & 0 \\ 0 & \frac{m}{2}\cos(2\tau) & \frac{m}{2}\sin(2\tau) & 0 \\ 0 & \frac{m}{2}\sin(2\tau) & -\frac{m}{2}\cos(2\tau) & 0 \\ 0 & 0 & 0 & 0 \end{pmatrix}. \tag{2.39}$$

This means an energy density $\rho = m$. The spatial part of Θ_A^B becomes diagonal in a rotating frame and has $\pm\frac{m}{2}$ as eigenvalues. That is, we again find $\pm\frac{\rho}{2}$ as extreme values for the pressure in two space dimensional string solutions. The time average of the stress tensor Θ_i^j here vanishes indicating zero pressure (cold matter behaviour).

One obtains the invariant size for a generic solution in Minkowski spacetime inserting the general solution (2.13) into $\partial_+ X^A \partial_- X_A$. This gives constant plus oscillatory terms. In any case the invariant string size is always **bounded** in Minkowski spacetimes. We shall see how differently behave strings in curved spacetimes.

III. HOW TO SOLVE THE STRING EQUATIONS OF MOTION IN CURVED SPACETIMES?

There is no general method to solve the string equations of motion and constraints for arbitrary curved spacetime.

The so-called τ-expansion method provides **exact local** solutions for any background. The basic idea goes as follows. Suppose one is interested on the string behaviour near a given point of the curved spacetime. Then, one chooses a conformal gauge such that $\tau = 0$ corresponds to such a point. For example, to study strings in cosmological spacetimes near the initial singularity (cosmic time $T = 0$), one chooses $T(\sigma, \tau)$ such that $T(\sigma, 0) = 0$. It is shown below that this is indeed possible for generic string solutions. This expansion was developped first in ref. [10,11] for inflationary universes.

Similarly, in order to study strings near the black hole singularity $r = 0$, one chooses $r(\sigma, \tau)$ such that $r(\sigma, 0) = 0$.

Once this gauge choice is done, the string equations of motion and constraints can be solved in powers of τ. These powers may not be integer powers. For example, one finds powers of $\tau^{-\frac{2}{\alpha+1}}$ in FRW universes with scale factor $R(T) = c\,T^\alpha$. For Schwarzschild black holes powers of $\tau^{\frac{2}{5}}$ appear.

An approximate but general method is the expansion around center of mass solutions [2,9]. In this method one starts from an exact solution of the geodesic equations

$$\ddot{q}^A(\tau) + \Gamma_{BC}^A(q)\,\dot{q}^B(\tau)\,\dot{q}^C(\tau) = 0 \tag{3.1}$$

The world-sheet time variable is here identified with the proper time of the center of mass trajectory. [Notice that eqs.(3.1) just follow from the string equations (2.7) by dropping the σ-dependence].

Then one develops in perturbations around it. That is, one sets

$$X^A(\sigma,\tau) = q^A(\tau) + \eta^A(\sigma,\tau) + \xi^A(\sigma,\tau) + \ldots \qquad (3.2)$$

Here $\eta^A(\sigma,\tau)$ obeys the linearized perturbation around $q^A(\sigma,\tau)$ and $\xi^A(\sigma,\tau)$ the second order perturbation equations [2]. These fluctuations obey coupled ordinary differential equations that can be written systematically inserting eq.(3.2) into eqs.(2.7-2.8). [See ref. [9] for more details].

Another general approximation method is the null string approach [22]. In such approach the string equations of motion and constraints are systematically expanded in powers of c (the speed of light in the world-sheet). This corresponds to a small string tension expansion. At zeroth order, the string is effectively equivalent to a continuous beam of massless particles labelled by the parameter σ. The points on the string do not interact between them but they interact with the gravitational background.

For several spacetimes one can construct explicit string solutions using specific properties of the background. This is the case of singular plane waves, shock-waves, conical spacetimes and the de Sitter universe. The string equations of motion and constraints in the de Sitter spacetime are integrable in the inverse scattering sense as shown in ref. [4]. [The de Sitter universe is a symmetric space and hence the string equations there correspond to a two dimensional integrable sigma model].

A. The τ-expansion

Let us consider the intersection of the world-sheet with a singular or non-singular point (or surface) of the spacetime like $T(\sigma,\tau) = 0$ or $T(\sigma,\tau) = T_o$ in a cosmological spacetime or $r(\sigma,\tau) = 0$ or $r(\sigma,\tau) = r_o$ in a Schwarzschild black-hole.

We can write the curve describing such intersection with the world-sheet as

$$x_+ = \chi(x_-), \qquad (3.3)$$

whenever this intersection is nondegenerate. (Here $x_\pm \equiv \tau \pm \sigma$).

Upon a conformal transformation,

$$x_+ \to x'_+ = f(x_+) \quad , \quad x_- \to x'_- = g(x_-), \qquad (3.4)$$

we can map the curve (3.3) into $\tau' = 0$ by an appropriate choice of f and g [18]. For example, we can choose

$$f(x_+) = x_+ \quad , \quad g(x_-) = -\chi(x_-).$$

This defines our choice of gauge. From now on, we rename τ' and σ' by τ and σ, respectively. Notice that this choice does not completely fix the gauge. We can still perform transformations that leave the line $\tau = 0$ unchanged. This is the case for the following class of conformal mappings

$$x_+ \to x'_+ = \varphi(x_+) \quad , \quad x_- \to x'_- = -\varphi(-x_-), \qquad (3.5)$$

where $\varphi(x)$ is an arbitrary function. Eq. (3.5) can be written as,

$$\tau' = \frac{1}{2}\left[\varphi(\tau+\sigma) - \varphi(\sigma-\tau)\right] = \tau\,\varphi'(\sigma) + \frac{1}{6}\tau^3\,\varphi'''(\sigma) + O(\tau^4)$$

$$\sigma' = \frac{1}{2} \left[\varphi(\tau + \sigma) + \varphi(\sigma - \tau) \right] = \varphi(\sigma) + \frac{1}{2} \tau^2 \, \varphi''(\sigma) + O(\tau^4) \tag{3.6}$$

The transformations (3.5) represent a diagonal subgroup of the set of left-right conformal transformations (3.4).

In summary, any (non-degenerate) intersection of the world-sheet with a spacetime submanifold $T(\sigma, \tau) = T_o$ can be mapped into $\tau = 0$. This mapping is not unique, it is invariant under the diagonal conformal transformations.

Once this gauge has been imposed one studies the string equations of motion and constraints (2.7-2.8) in powers of τ. The equations themselves determine the precise values of the powers [10,11,18].

B. Global Solutions

There is no general method to find solutions valid in the whole world-sheet. However, many global solutions have been found in physically relevant spacetimes.

First, there are spacetimes where the **general** solution of the string equations and constraints has been found. That is, shock-waves [44], singular plane waves [46] and conical spacetimes [47].

Second, by making specific ansatz according to the symmetry of the background, the string equations of motion can be reduced to ordinary differential equations. Then, these ordinary differential equations can be solved globally by numerical methods. In this way, solutions valid in the whole worldsheet has been found in cosmological spacetimes and black holes [7,12] - [14].

Third, the de Sitter spacetime can be treated by inverse scattering methods. In this way exact string solutions has been constructed systematically (see sec. VII and refs. [5] - [8]).

In all cases where global solutions can be found, the τ-expansion results are confirmed.

IV. STRING PROPAGATION IN COSMOLOGICAL SPACETIMES

We obtain in this section physical string properties from the string solutions in cosmological spacetimes.

We consider strings in spatially homogeneous and isotropic universes with metric

$$ds^2 = (dT)^2 - R(T)^2 \sum_{i=1}^{D-1} (dX^i)^2 \, , \tag{4.1}$$

where T is the cosmic time and the function $R(T)$ is called the scale factor. In terms of the conformal time

$$\eta = \int^T \frac{dT}{R(T)} \, , \tag{4.2}$$

the metric (4.1) takes the form

$$ds^2 = R(\eta)^2 \left[(d\eta)^2 - \sum_{i=1}^{D-1} (dX^i)^2 \right] \, . \tag{4.3}$$

The classical string equations of motion can be written here as

$$\partial^2 T + R(T) \frac{dR}{dT} \sum_{i=1}^{D-1} (\partial_\mu X^i)^2 = 0 , \tag{4.4}$$

$$\partial_\mu \left[R^2 \partial^\mu X^i \right] = 0 , \qquad 1 \le i \le D - 1 ,$$

and the constraints are

$$T_{\pm\pm} = (\partial_\pm T)^2 - R(T)^2 \, (\partial_\pm X^i)^2 = 0 . \tag{4.5}$$

The most relevants universes correspond to power type scale factors. That is,

$$R(T) = a \, T^\alpha = A \, \eta^{k/2} , \tag{4.6}$$

where $\alpha = \frac{k}{k+2}$ and $k = \frac{2\alpha}{1-\alpha}$.

For different values of the exponents we have either FRW or inflationary universes.

$$\text{FRW} : 0 < k \le \infty, \ 0 < \alpha \le 1 = \begin{cases} \alpha = 1/2, \ k = 2, & \text{radiation dominated,} \\ \alpha = 2/3, \ k = 4, & \text{matter dominated} \ ; \\ \alpha = 1, \ k = \infty, & \text{`stringy'.} \end{cases}$$

$$\text{Inflationary} : \ -\infty < k < 0, \ \alpha < 0 \text{ and } \alpha > 1 = \begin{cases} \alpha = \infty, \ k = -2, & R(T) = e^{HT}, \text{ de Sitter,} \\ \alpha > 1, \ k < -2, & \text{power inflation }, \\ \alpha < 0, \ -2 < k < 0, & \text{superinflationary .} \end{cases}$$
$$\tag{4.7}$$

The denomination 'stringy' comes from the fact that such backgrounds follow as solution of the string effective equations [24]. Inflationary universes are those with accelerated expansion. That is,

$$\frac{d^2 R(T)}{dT^2} > 0 .$$

The string equations of motion and constraints take then the following form using the conformal time η:

$$\ddot{\eta} - \eta'' + \frac{k}{2\eta} \left\{ \dot{\eta}^2 - \eta'^2 + \sum_{i=1}^{D-1} \left[(\dot{X}^i)^2 - (X'^i)^2 \right] \right\} = 0 ,$$

$$\ddot{X}^i - X''^i + \frac{k}{\eta} \left[\dot{\eta} \dot{X}^i - \eta' X'^i \right] = 0 , \qquad 1 \le i \le D - 1 , \tag{4.8}$$

$$(\dot{\eta} \pm \eta')^2 - \sum_{i=1}^{D-1} (\dot{X}^i \pm X'^i)^2 = 0 ,$$

where prime and dot stand for ∂_σ and ∂_τ, respectively.

As we will see below, once appropriately averaged $\Theta^{AB}(X)$ takes the fluid form for string solutions in cosmological spacetimes, allowing us to define the string pressure p and energy density ρ :

$$< \Theta^B_A >= \begin{pmatrix} \rho & 0 & \cdots & 0 \\ 0 & -p & \cdots & 0 \\ \cdots & \cdots & \cdots & 0 \\ 0 & 0 & \cdots & -p \end{pmatrix} . \tag{4.9}$$

Notice that the continuity equation

$$D^A \, \Theta^B_A = 0 \,,$$

takes here the form

$$\frac{d\rho}{dT} + (D-1) \, H \, (p + \rho) = 0 \tag{4.10}$$

where $H \equiv \frac{1}{R}\frac{dR}{dT}$.

For an equation of state of the type of a perfect fluid, that is

$$p = (\gamma - 1) \, \rho \quad , \quad \gamma = \text{constant}, \tag{4.11}$$

eqs.(4.10) and (4.11) can be easily integrated with the result

$$\rho = \rho_0 \, R^{\gamma(1-D)} \,. \tag{4.12}$$

For $\gamma = 1$ this corresponds to cold matter ($p = 0$) and for $\gamma = \frac{D}{D-1}$ this describes radiation with $p = \frac{\rho}{D-1}$.

A. Strings in cosmological universes: the τ-expansion at work

Let us consider strings in inflationary universes with scale factor (4.6) and $k < 0$. In order to apply the τ-expansion we fix the gauge such that

$$\eta(\tau = 0, \sigma) = 0 \,. \tag{4.13}$$

As explained above, this is always possible for generic string solutions and it leaves still the freedom of the transformations (3.5). Notice that $\eta \to 0$ corresponds in the inflationary case to large scale factors $R \to \infty$.

The behaviour of $\eta(\tau, \sigma)$ and $X^i(\tau, \sigma)$ for $\tau \to 0$ follows from eq.(4.8) where we use eq.(4.13) and assume that $X^i(\tau, \sigma)$ is regular at $\tau = 0$. One finds [11]

$$\eta(\tau, \sigma) \stackrel{\tau \to 0}{=} \eta_o(\sigma) \, \tau \, \left[1 + O(\tau^2)\right] + \lambda_o(\sigma) \, \tau^{2-k} \, \left[1 + O(\tau^2)\right]$$
$$+ \, \zeta_o(\sigma) \, \tau^{1-2k} \, \left[1 + O(\tau^2)\right] + O(\tau^{1-3k}) \,,$$

$$X^i(\tau, \sigma) \stackrel{\tau \to 0}{=} A^i(\sigma) \, \left[1 + O(\tau^2)\right] + \tau^{1-k} \, B^i(\sigma) \, \left[1 + O(\tau^2, \tau^{1-k})\right] \,, \tag{4.14}$$
$$1 \leq i \leq D - 1 \,.$$

The solutions appear as a series in powers of τ^2 and τ^{1-k}. In the special case where k is rational, say $k = -\frac{l}{n}$, l, n =integers, logarithmic terms in τ appear in addition. This happens, for example in de Sitter spacetime ($k = -2$) and in Minkowski spacetime ($k = 0$).

The coefficients in eq.(4.16) result related as follows

$$\sum_{i=1}^{D-1} B^i(\sigma)\, A^{\prime i}(\sigma) = 0 \ , \ \eta_o(\sigma) = \sqrt{\sum_{i=1}^{D-1} [A^{\prime i}(\sigma)]^2}$$

$$\lambda_o(\sigma) = \frac{2k}{(2-k)(1-k)} \ \frac{\sum_{i=1}^{D-1} B^{\prime i}(\sigma)\, A^{\prime i}(\sigma)}{\eta_o(\sigma)} \ . \tag{4.15}$$

Moreover, one can use the residual conformal invariance (3.5) to set $\eta_o(\sigma) \equiv 1$. One finally obtains for inflationary universes [11],

$$\eta(\tau,\sigma) \stackrel{\tau \to 0}{=} \tau \left[1 + O(\tau^2)\right] + \lambda_o(\sigma)\, \tau^{2-k} \left[1 + O(\tau^2)\right]$$
$$+ \ \zeta_o(\sigma)\, \tau^{1-2k} \left[1 + O(\tau^2)\right] + O(\tau^{1-3k}) \ ,$$

$$\tag{4.16}$$

$$X^i(\tau,\sigma) \stackrel{\tau \to 0}{=} A^i(\sigma) \left[1 + O(\tau^2)\right] + \tau^{1-k}\, B^i(\sigma) \left[1 + O(\tau^2, \tau^{1-k})\right] \ ,$$
$$1 \le i \le D-1 \ .$$

where

$$\sum_{i=1}^{D-1} B^i(\sigma)\, A^{\prime i}(\sigma) = 0 \quad , \quad \sum_{i=1}^{D-1} \left[A^{\prime i}(\sigma)\right]^2 = 1$$

$$\lambda_o(\sigma) = \frac{2k}{(2-k)(1-k)} \sum_{i=1}^{D-1} B^{\prime i}(\sigma)\, A^{\prime i}(\sigma) \quad , \quad \zeta_o(\sigma) = \frac{(1-k)^2}{2(1-2k)} \sum_{i=1}^{D-1} \left[B^{\prime i}(\sigma)\right]^2 \ . \tag{4.17}$$

Here $A^i(\sigma)$, $B^i(\sigma)$, $1 \le i \le D-1$ are the initial $(\tau = 0)$ string coordinates and momenta.

We see that the solution depends on $2(D-2)$ independent functions among the $A^i(\sigma)$ and $B^i(\sigma)$, $1 \le i \le D-1$. All coefficients (including the higher orders not written in eq.(4.16)) express in terms of the $A^i(\sigma)$ and $B^i(\sigma)$. Therefore, the counting of degrees of freedom turns to be the same as in Minkowski spacetime: only the $2(D-2)$ **transverse** coordinates are physical.

It must be noticed that

$$\dot{X}^i(\tau,\sigma) \stackrel{\tau \to 0}{=} (1-k)\, \tau^{-k}\, B^i(\sigma) \left[1 + O(\tau^2, \tau^{1-k})\right] \to 0$$

$$X^{\prime i}(\tau,\sigma) \stackrel{\tau \to 0}{=} A^{\prime i}(\sigma) \ne 0$$

That is, $\partial_\sigma X^i$ is **larger** than $\partial_\tau X^i$ for $R \to \infty$. This is the opposite to a point particle behaviour. For a point particle, $\partial_\sigma X^i \equiv 0$ and $\partial_\tau X^i = p^i \ne 0$.

Let us now apply the τ-expansion to strings in FRW universes. That is $k > 0$ in the scale factor (4.6).

We fix again the gauge according to eq.(4.13). It must be noticed that now $\eta \to 0$ corresponds to $R \to 0$ since $k > 0$. That is the τ-expansion applies near the singularity (big bang) of the spacetime.

After calculations analogous to the inflationary case, one finds from eqs.(4.8) for FRW universes $(k > 0)$ [11]

$$\eta(\tau,\sigma) \stackrel{\tau \to 0}{=} \tau^{\frac{1}{k+1}} \left[1 + O(\tau^2)\right] + \eta_1(\sigma)\, \tau^{2-\frac{1}{k+1}} \left[1 + O(\tau^2)\right]$$

$$(4.18)$$

$$X^i(\tau,\sigma) \stackrel{\tau \to 0}{=} A^i(\sigma)\left[1+O(\tau^2)\right] + \tau^{\frac{1}{k+1}} B^i(\sigma)\left[1+O(\tau^2)\right] + \tau^{2-\frac{1}{k+1}} C^i(\sigma)\left[1+O(\tau^2)\right],$$
$$1 \le i \le D-1.$$

where the residual conformal invariance (3.5) has been used.

The string equations of motion and constraints impose the following relations on the coefficients:

$$\sum_{i=1}^{D-1}\left[B^i(\sigma)\right]^2 = 1 \quad , \quad \sum_{i=1}^{D-1} B^i(\sigma)\, A'^i(\sigma) = 0,$$

$$C^i(\sigma) = -\eta_1(\sigma)\, B^i(\sigma) \quad , \quad \eta_1(\sigma) = \frac{(k+1)^2}{4(2k+1)}\sum_{i=1}^{D-1}\left[A'^i(\sigma)\right]^2. \tag{4.19}$$

$A^i(\sigma)$, $B^i(\sigma)$, $1 \le i \le D-1$ turn to be the initial ($\tau = 0$) string coordinates and momenta.

We again find that the string solution is determined by $2(D-2)$ independent functions indicating that only the transverse coordinates are physical. As we see, the τ-expansion produces for FRW universes the string solution as a series in powers of τ^2 and $\tau^{\frac{2k}{1+k}}$.

For large R the spacetime curvature tends to zero as $T^{-2} \simeq R^{-2/\alpha}$ in FRW spacetimes. That is, for $T \to \infty$. In order to analyze the string behaviour in such regime it is convenient to choose the gauge

$$\eta = \eta(\tau) \to \infty \quad , \quad \text{for } \tau \to \infty \tag{4.20}$$

[This is a slight generalization of the previous gauge choices at finite τ].

We then find from eq.(4.8)

$$\eta(\tau,\sigma) \stackrel{\tau \to \infty}{=} \tau^{\frac{2}{k+2}} \to \infty,$$

$$\tag{4.21}$$

$$X^i(\tau,\sigma) \stackrel{\tau \to \infty}{=} \frac{1}{k+2}\, \tau^{-\frac{k}{k+2}}\left[f_i^+(\sigma+\tau) + f_i^-(\sigma-\tau)\right],$$
$$1 \le i \le D-1.$$

where the $f_i^\pm(x)$ are arbitrary periodic functions of x

$$f_i^\pm(x+2\pi) = f_i^\pm(x),$$

obeying the pair of constraints:

$$\sum_{i=1}^{D-1}\left[f_i'^\pm(x)\right]^2 = 1. \tag{4.22}$$

We obtain from eq.(4.21)

$$T(\tau) \stackrel{\tau \to \infty}{=} \frac{2\sqrt{A}}{k+2}\,\tau \to \infty \quad \text{and} \quad R \stackrel{\tau \to \infty}{=} \tau^{\frac{k}{k+2}} \to \infty. \tag{4.23}$$

In short, the string solutions in this regime are asymptotically Minkowski solutions (2.13) *scaled* by a factor R^{-1}. This is not unexpected since the spacetime curvature vanishes for this regime. The counting of degrees of freedom is again as in Minkowski spacetime.

We have determined the string behaviour for $R \to \infty$ in inflationary and FRW universes and for $R \to 0$ in FRW universes. Let us now compute for such regimes the string physical properties, energy-momentum and size.

The calculation of $\Theta^{AB}(T)$ in the different limiting regimes follows from eq.(2.29) since $\eta = \eta(\tau)$ asymptotically (both for $\tau \to 0$ and $\tau \to \infty$).

Let us start by considering the inflationary universes for $R \to \infty$. We find for the (integrated) energy-momentum tensor from eqs.(2.29) and (4.16)

$$\rho(T) = \Theta^{00}(T) \stackrel{\tau \to 0}{=} \frac{R}{\alpha'} \to +\infty ,$$

(4.24)

$$\Theta^i_j(T) \stackrel{\tau \to 0}{=} \frac{R}{\alpha'} \int_0^{2\pi} \frac{d\sigma}{2\pi} A'^i(\sigma) A'^j(\sigma) \to \infty ,$$

(4.25)

$$\Theta^0_i(T) \stackrel{\tau \to 0}{=} -\frac{1}{\alpha'} \int_0^{2\pi} \frac{d\sigma}{2\pi} B^i(\sigma)$$

where we also used the spacetime metric $G_{00} = 1$, $G_{ij} = -R^2 \delta_{ij}$.

Recall that $\sum_{i=1}^{D-1} [A'^i(\sigma)]^2 = 1$.

We see that the energy density $\rho(T)$ diverges for $R \to \infty$. The stress tensor $[\Theta^i_j(T)]$ is not diagonal but it is given by a **positive definite** matrix. Such matrix has then positive eigenvalues and therefore tells us that the pressure is **negative** [compare with eq.(4.9)]. This is the **unstable** string behaviour. That is, the energy tends to $+\infty$ and the pressure to $-\infty$. At the same time the energy flux density $\Theta^0_i(T)$ stands bounded.

The string size S in cosmological spacetimes (4.1) takes the form

$$S^2 = G_{AB}(X) \dot{X}^A \dot{X}^B = \dot{T}^2 - R^2 \sum_{i=1}^{D-1} (\dot{X}^i)^2$$

(4.26)

For $\tau \to 0$, $R \to \infty$ in inflationary spacetimes, we find using eq.(4.16) that the first term dominates in eq.(4.26)

$$S \stackrel{\tau \to 0}{=} c \tau^{k/2} \simeq R \to \infty ,$$

(4.27)

where c is a constant. We see that the string grows infinitely big when the universe inflates. The string size being proportional to the scale factor and also to the string energy [ρ in eq.(4.24)]. These explosive growings characterize the string unstable behaviour.

Let us now consider FRW universes ($k > 0$) for $\tau \to 0$, $R \to 0$. The (integrated) energy-momentum tensor in such regime takes the form

$$\rho(T) = \Theta^{00}(T) \stackrel{\tau \to 0}{=} \frac{1}{\alpha'(k+1) R} \to +\infty ,$$

(4.28)

$$\Theta^i_j(T) \stackrel{\tau \to 0}{=} -\frac{1}{\alpha'(k+1) R} \int_0^{2\pi} \frac{d\sigma}{2\pi} B^i(\sigma) B^j(\sigma) \to \infty ,$$

(4.29)

$$\Theta_i^0(T) \stackrel{\tau \to 0}{=} \frac{1}{\alpha'(k+1) R} \int_0^{2\pi} \frac{d\sigma}{2\pi} B^i(\sigma)$$

where we used eqs.(2.29) and (4.18). [Recall that $\sum_{i=1}^{D-1} \left[B^i(\sigma)\right]^2 = 1$].

We see in eq.(4.28) that the stress tensor $(\Theta_j^i(T))$ is not diagonal but it is given by a **negative definite** matrix. Such matrix has then negative eigenvalues and therefore tells us that the pressure is **positive**. This string behaviour is dual to the previous unstable behaviour.

We find from eq.(4.26) for the string size S in FRW universes for $R \to 0$,

$$S \stackrel{\tau \to 0}{=} \sqrt{\sum_{i=1}^{D-1} [A'^i(\sigma)]^2 \; \tau^{\frac{k}{2(k+1)}}} \simeq \sqrt{\sum_{i=1}^{D-1} [A'^i(\sigma)]^2} \; R \to 0 \tag{4.30}$$

In this dual to unstable behaviour, the string size **vanishes**. That is, the string starts at the big bang with zero size.

In summary, the energy tends to $+\infty$, the pressure also to $+\infty$ and the size tends to zero in the **dual to unstable** behaviour.

In this regime strings behave as radiation (massless particles). Recall that the string size is proportional to the trace of the energy momentum tensor [see eq.(2.33)].

Let us finally consider FRW universes $(k > 0)$ for $\tau \to \infty$, $R \to \infty$. There, the (integrated) energy-momentum tensor takes the form

$$\rho(T) = \Theta^{00}(T) \stackrel{\tau \to \infty}{=} \frac{2\sqrt{A}}{\alpha'(k+2)} \; ,$$

(4.31)

$$\Theta_j^i(T) \stackrel{\tau \to \infty}{=} \frac{R}{\alpha'} \int_0^{2\pi} \frac{d\sigma}{2\pi} \left[f_i'^+ f_j'^- + f_i'^- f_j'^+ \right] \; ,$$

(4.32)

$$\Theta_i^0(T) \stackrel{\tau \to \infty}{=} 0 \; .$$

where we used eqs.(2.29) and (4.21). The string energy here tends to a bounded constant. Since the $f_i'^\pm(\sigma \pm \tau)$ are periodic functions, their average on a period of time vanishes:

$$\int_0^{2\pi} d\tau d\sigma \; f_i'^+(\sigma + \tau) \; f_i'^-(\sigma - \tau) = \frac{1}{2} \int_{-2\pi}^{+2\pi} dx_- \int_{|x_-|}^{4\pi - |x_-|} dx_+ \; f_i'^+(x_+) \; f_i'^-(x_-) = 0 \; .$$

Hence, the pressure vanishes when averaged over a string oscillation. This is the **stable** string behaviour. Here strings behave as dust (cold matter) with $p = 0$ as equation of state.

The string size S follows eq.(4.26) and eq.(4.21),

$$S^2 \stackrel{\tau \to \infty}{=} \frac{2}{(k+2)^2} \left[1 + \sum_{i=1}^{D-1} f_i'^+ f_i'^- \right] \tag{4.33}$$

The string size is thus bounded for $R \to \infty$. Moreover, averaging over a period of time, we find

$$\bar{S} = \frac{\sqrt{2}}{k+2} \; .$$

B. The perfect gas of strings

Our aim is to provide a string description of matter appropriate to the early universe.

Let us consider classical strings interacting with the cosmological spacetime background and neglect their mutual interactions. That is, we consider a perfect gas of strings under the cosmic gravitational field. The energy-momentum of such gas is just the sum over individual string solutions. For each string the results of section IV.A apply.

We assume arbitrary initial data for the strings. Therefore, summing over solutions is equivalent to *average* over the initial data $A^i(\sigma)$, $B^i(\sigma)$, $1 \leq i \leq D-1$.

For inflationary spacetimes the relevant quantity to average is the integral

$$\int_0^{2\pi} \frac{d\sigma}{2\pi} A'^i(\sigma) A'^j(\sigma)$$

that appears in the energy-momentum tensor eq.(4.24).

The average will vanish for $i \neq j$ and will be i-independent for $i = j$ because we treat independently the different $A^i(\sigma)$. Taking into account the constraint (4.17)

$$\sum_{i=1}^{D-1} \left[A'^i(\sigma) \right]^2 = 1 ,$$

finally yields,

$$< \int_0^{2\pi} \frac{d\sigma}{2\pi} A'^i(\sigma) A'^j(\sigma) > = \frac{\delta_{ij}}{D-1} . \tag{4.34}$$

Therefore the stress tensor (4.24) takes for unstable strings the fluid form for $R \to \infty$,

$$< \Theta^i_j(T) > \stackrel{R \to \infty}{=} -\frac{p}{D-1} \delta_{ij} ,$$

with

$$p \equiv -\frac{R}{\alpha'(D-1)} = -\frac{\rho}{D-1} \to -\infty , \tag{4.35}$$

where we used the expression for ρ in eq.(4.24). Recall that the string size also grows as R for $R \to \infty$ [eq.(4.27)].

This equation of state exactly saturates the strong energy condition in general relativity.

The unstable string behaviour corresponds to the critical case of the so-called coasting universe [19,32]. In other words, the perfect gas of strings provide a *concrete* matter realization of such cosmological model. Till now, no form of matter was known to describe coasting universes [19].

The quantity to average in FRW spacetimes for small R is

$$\int_0^{2\pi} \frac{d\sigma}{2\pi} B^i(\sigma) B^j(\sigma)$$

which appears in the energy-momentum tensor eq.(4.28).

Following the same argument as above for the average over $A^i(\sigma)$ and recalling that (4.19)

$$\sum_{i=1}^{D-1} \left[B^i(\sigma) \right]^2 = 1 \,,$$

we find

$$< \int_0^{2\pi} \frac{d\sigma}{2\pi} \, B^i(\sigma) \, B^j(\sigma) > = \frac{\delta_{ij}}{D-1} \tag{4.36}$$

Hence the string energy-momentum tensor (4.28) for dual to unstable strings takes the fluid form for $R \to 0$,

$$< \Theta^i_j(T) > \stackrel{R \to 0}{=} -\frac{p}{D-1} \, \delta_{ij}$$

with

$$p \equiv \frac{1}{\alpha'(D-1)(k+1) \, R} \stackrel{R \to 0}{=} +\frac{\rho}{D-1} \to +\infty$$

where we used the expression for ρ in eq.(4.28). Recall that the string size vanishes as R, as R vanishes [eq.(4.30)].

Therefore in the dual to unstable case, strings behave as radiation (massless particles).

For large R in FRW spacetimes we must average independently over the functions f_i^+ and f_j^-, $1 \le i, j \le D-1$. We find then from eqs.(4.31)

$$< \Theta^i_j(T) > \stackrel{R \to \infty}{=} 0 \,.$$

That is, the equation of state

$$p = 0 \,.$$

The string energy and size are bounded in this regime. The strings behave for the stable regime as dust (cold matter). That is, they behave as massive particles.

In conclusion, an ideal gas of classical strings in cosmological universes exhibit three different thermodynamical behaviours, all of perfect fluid type:

- (1) For inflationary universes and $R \to \infty$ unstable strings: negative pressure gas with $p = -\frac{\rho}{D-1}$.

- (2) Dual behaviour in FRW universes and $R \to 0$: positive pressure gas similar to radiation, $p = +\frac{\rho}{D-1}$.

- (3) Stable strings in FRW universes and $R \to \infty$: positive pressure gas similar to cold matter, $p = 0$.

Tables 1 and 2 summarize the main string properties for any scale factor $R(T)$.

TABLE 1. String energy and pressure as obtained from exact
string solutions for various expansion factors $R(T)$.

STRING PROPERTIES FOR ARBITRARY $R(T)$

D-Dimensional spacetimes: three asymptotic behaviours (u, d, s)	Energy	Pressure	Equation of State:
(i) unstable for $R \to \infty$	$E_u \overset{R \to \infty}{=} u\,R \to \infty$	$P_u = -\frac{E_u}{D-1} \to -\infty$	'stringy'
(ii) dual to (i) for $R \to 0$	$E_d \overset{R \to 0}{=} d/R \to \infty$	$P_d = +\frac{E_d}{D-1} \to \infty$	radiation
(iii) stable for $R \to \infty$	$E_s =$ constant	$P_s = 0$	dust (cold matter)

TABLE 2. The string energy density and pressure
for a gas of strings can be summarized by the formulas
below which become exact for $R \to 0$ and for $R \to \infty$.

STRING ENERGY DENSITY AND PRESSURE FOR ARBITRARY $R(T)$

	Energy density: $\rho \equiv E/R^{D-1}$	Pressure
Qualitatively correct formulas for all R and D	$\rho = \left(u\,R + \frac{d}{R} + s\right)\frac{1}{R^{D-1}}$	$p = \frac{1}{D-1}\left(\frac{d}{R} - u\,R\right)\frac{1}{R^{D-1}}$

Finally, notice that strings continuously evolve from one type of behaviour to the other
two. This is explicitly seen from the string solutions in refs. [5] - [7] . For example the string
described by $q_-(\sigma, \tau)$ for $\tau > 0$ shows unstable behaviour for $\tau \to 0$, dual behaviour for
$\tau \to \tau_0 = 1.246450...$ and stable behaviour for $\tau \to \infty$.

TABLE 3. The **self-consistent** cosmological solution
of the Einstein equations in General Relativity
with the string gas as source.

STRING COSMOLOGY IN GENERAL RELATIVITY

Einstein equations (no dilaton field)	Expansion factor $R(T)$	Temperature $T(R)$	
$T \to 0$	$\frac{D}{2}\left[\frac{2d}{(D-1)(D-2)}\right]^{\frac{1}{D}} T^{\frac{2}{D}}$	$\frac{dD}{S(D-1)}\, 1/R$	
$T \to \infty$	$\left[\frac{(D-1)s}{2(D-2)}\right]^{\frac{1}{D-1}} T^{\frac{2}{D-1}}$	usual matter dominated behaviour	

V. SELF-CONSISTENT STRING COSMOLOGY

In the previous section we investigated the propagation of test strings in cosmological space-times. Let us now investigate how the Einstein equations in General Relativity and the effective equations of string theory (beta functions) can be verified **self-consistently** with our string solutions as sources.

We shall assume a gas of classical strings neglecting interactions as string splitting and coalescing. We will look for cosmological solutions described by metrics of the type (4.1). It is natural to assume that the background will have the same symmetry as the sources. That is, we assume that the string gas is homogeneous, described by a density energy $\rho = \rho(T)$ and a pressure $p = p(T)$. In the effective equations of string theory we consider a space independent dilaton field. Antisymmetric tensor fields wil be ignored.

A. String Dominated Universes in General Relativity (no dilaton field)

The Einstein equations for the geometry (4.1) take the form

$$\frac{1}{2}(D-1)(D-2)\,H^2 = \rho \ ,$$
$$(D-2)\dot{H} + p + \rho = 0 \ . \tag{5.1}$$

where $H \equiv \frac{dR}{dT}/R$. We know p and ρ as functions of R in asymptotic cases. For large R, the unstable strings dominate [eq.(4.35)] and we have for inflationary spacetimes

$$\rho = u\, R^{2-D} \ , \quad p = -\frac{\rho}{D-1} \quad \text{for } R \to \infty \tag{5.2}$$

For small R, the dual regime dominates with

$$\rho = d\, R^{-D} \quad , \quad p = +\frac{\rho}{D-1} \quad \text{for } R \to 0 \tag{5.3}$$

We also know that stable solutions may be present with a contribution $\simeq R^{1-D}$ to ρ and with zero pressure. For intermediate values of R the form of ρ is clearly more complicated but a formula of the type

$$\rho = \left(u_R\, R + \frac{d}{R} + s \right) \frac{1}{R^{D-1}} \tag{5.4}$$

where

$$\lim_{R \to \infty} u_R = \begin{cases} 0 & \text{FRW} \\ u_\infty \neq 0 & \text{Inflationary} \end{cases} \tag{5.5}$$

This equation of state is qualitatively correct for all R and becomes exact for $R \to 0$ and $R \to \infty$. The parameters u_R, d and s are positive constants and the u_R varies smoothly with R.

The pressure associated to the energy density (5.4) takes then the form

$$p = \frac{1}{D-1} \left(\frac{d}{R} - u_R\, R \right) \frac{1}{R^{D-1}} \tag{5.6}$$

Inserting eq.(5.4) into the Einstein-Friedmann equations [eq.(5.1)] we find

$$\frac{1}{2}(D-1)(D-2) \left(\frac{dR}{dT} \right)^2 = \left(u_R\, R + \frac{d}{R} + s \right) \frac{1}{R^{D-3}} \tag{5.7}$$

We see that R is a monotonic function of the cosmic time T. Eq.(5.7) yields

$$T = \sqrt{\frac{(D-1)(D-2)}{2}} \int_0^R dR \, \frac{R^{D/2-1}}{\sqrt{u_R\, R^2 + d + s\, R}} \tag{5.8}$$

where we set $R(0) = 0$.

It is easy to derive the behavior of R for $T \to 0$ and for $T \to \infty$.

For $T \to 0$, $R \to 0$, the term d/R dominates in eq.(5.7) and

$$R(T) \stackrel{T \to 0}{\simeq} \frac{D}{2} \left[\frac{2d}{(D-1)(D-2)} \right]^{\frac{1}{D}} T^{\frac{2}{D}} \tag{5.9}$$

For $T \to \infty$, $R \to \infty$ and the term $u_R\, R$ dominates in eq.(5.7). Hence,

$$R(T) \stackrel{T \to \infty}{\simeq} \left[\frac{(D-2)u_\infty}{2(D-1)} \right]^{\frac{1}{D-2}} T^{\frac{2}{D-2}} \tag{5.10}$$

[u_R tends to a constant u_∞ for $R \to \infty$]. This expansion is faster than (cold) matter dominated universes where $R \simeq T^{\frac{2}{D-1}}$. For example, for $D = 4$, R grows linearly with T whereas for matter dominated universes $R \simeq T^{2/3}$. However, eq.(5.10) **is not** a self-consistent solution. Assuming that the term $u_R\, R$ dominates for large R we find a scale factor $R(T) \simeq T^{\frac{2}{D-2}} \simeq \eta^{\frac{2}{D-4}}$

for $D \neq 4$ and $R(T) \simeq T \simeq e^\eta$ at $D = 4$. This **is not an inflationary universe** but a FRW universe. The term $u_R R$ is absent for large R in FRW universes as explained before. Therefore, we must instead use for large R

$$\rho = \left(\frac{d}{R} + s\right) \frac{1}{R^{D-1}} \tag{5.11}$$

Now, for $T \to \infty, R \to \infty$ and we find a matter dominated regime:

$$R(T) \overset{T \to \infty}{\simeq} \left[\frac{(D-1)s}{2(D-2)}\right]^{\frac{1}{D-1}} T^{\frac{2}{D-1}} \tag{5.12}$$

For intermediate values of T , $R(T)$ is a continuous and monotonically increasing function of T.

In summary, the universe starts at $T = 0$ with a singularity of the type dominated by radiation. (The string behaviour for $R \to 0$ is like usual radiation). Then, the universe expands monotonically, growing for large T as $R \simeq T^{\frac{2}{D-1}}$. In particular, this gives $R \simeq T^{2/3}$ for $D = 4$.

It must be noticed that the qualitative form of the solution $R(T)$ does not depend on the particular positive values of u_R, d and s.

We want to stress that we achieve a **self-consistent** solution of the Einstein equations with string sources since the behaviour of the string pressure and density given by eqs.(5.4)-(5.6) precisely holds in universes with power like $R(T)$.

In ref. [3] similar results were derived using arguments based on the splitting of long strings.

B. Thermodynamics of strings in cosmological spacetimes

Let us consider a comoving volume R^{D-1} filled by a gas of strings. The entropy change for this system is given by:

$$\mathcal{T}dS = d(\rho R^{D-1}) + p \, d(R^{D-1}) \tag{5.13}$$

The continuity equation (4.10) and (5.13) implies that dS/dT vanishes. That is, the entropy per comoving volume stays constant in time. Using now the thermodynamic relation [25]

$$\frac{dp}{d\mathcal{T}} = \frac{p + \rho}{\mathcal{T}} \tag{5.14}$$

it follows [26] that

$$S = \frac{R^{D-1}}{\mathcal{T}}(p + \rho) + \text{constant} \tag{5.15}$$

Eq.(5.15) together with eqs.(5.4) and (5.6) yields the temperature as a function of the expansion factor R. That is,

$$\mathcal{T} = \frac{1}{S}\left\{s + \frac{1}{D-1}\left[\frac{D \, d}{R} + (D-2) \, u_R \, R\right]\right\} \tag{5.16}$$

where S stands for the (constant) value of the entropy.

Eq.(5.16) shows that for small R, $\mathcal{T}$ scales as $1/R$ whereas for large R it scales as R. The small R behaviour of $\mathcal{T}$ is the usual exhibited by radiation.

For large R, in FRW universes $u_R \to 0$ and the constant term in s dominates. We just find a cold matter behaviour for large R.

For large R in inflationary universes, $u_R \to u_\infty$ and eq.(5.16) would indicate a temperature that **grows** proportionally to R. However, as stressed in ref. [3], the decay of long strings (through splitting) makes u_R exponentially decreasing with R.

VI. EFFECTIVE STRING EQUATIONS WITH THE STRING SOURCES INCLUDED

Let us consider now the cosmological equations obtained from the low energy string effective action including the string matter as a classical source. In D spacetime dimensions, this action can be written as

$$S = S_1 + S_2$$
$$S_1 = \frac{1}{2} \int d^D x \sqrt{-G}\, e^{-\Phi} \left[R + G_{AB}\, \partial^A \Phi\, \partial^B \Phi + 2\, U(G, \Phi) - c \right]$$
$$S_2 = -\frac{1}{4\pi\alpha'} \sum_{strings} \int d\sigma d\tau\, G_{AB}(X)\, \partial_\mu X^A\, \partial^\mu X^B \quad , \tag{6.1}$$

Here $A, B = 0, \ldots, D - 1$. This action is written in the so called 'Brans-Dicke frame' (BD) or 'string frame', in which matter couples to the metric tensor in the standard way. The BD frame metric coincides with the sigma model metric to which test strings are coupled.

Eq.(6.1) includes the dilaton field (Φ) with a potential $U(G, \Phi)$ depending on the dilaton and graviton backgrounds; c stands for the central charge deficit or cosmological constant term. The antisymmetric tensor field was not included, in fact it is irrelevant for the results obtained here. Extremizing the action (6.1) with respect to G_{AB} and Φ yields the equations of motion

$$R_{AB} + \nabla_{AB}\Phi + 2\, \frac{\partial U}{\partial G_{AB}} - \frac{G_{AB}}{2} \left[R + 2\nabla^2\Phi - (\nabla\Phi)^2 - c + 2\, U \right] = e^\phi\, T_{AB}$$

$$R + 2\nabla^2\Phi - (\nabla\Phi)^2 - c + 2U - \frac{\partial U}{\partial \Phi} = 0 , \tag{6.2}$$

which can be more simply combined as

$$R_{AB} + \nabla_{AB}\Phi + 2\, \frac{\partial U}{\partial G_{AB}} - G_{AB}\, \frac{\partial U}{\partial \Phi} = e^\Phi\, T_{AB}$$

$$R + 2\, \nabla^2\Phi - (\nabla\Phi)^2 - c + 2\, U - \frac{\partial U}{\partial \Phi} = 0 \tag{6.3}$$

Here T_{AB} stands for the energy momentum tensor of the strings as defined by eq.(2.27). It is also convenient to write these equations as

$$R_{AB} - \frac{G_{AB}}{2}\, R = T_{AB} + \tau_{AB} \tag{6.4}$$

where τ_{AB} is the dilaton energy momentum tensor :

$$\tau_{AB} = -\nabla_{AB}\Phi + \frac{G_{AB}}{2}\left[2\frac{\partial U}{\partial \Phi} - R\right]$$

The Bianchi identity

$$\nabla^A\left(R_{AB} - \frac{G_{AB}}{2}R\right) = 0$$

yields, as it must be, the conservation equation,

$$\nabla^A\left(T_{AB} + \tau_{AB}\right) = 0 \tag{6.5}$$

It must be noticed that eqs.(6.3) do not reduce to the Einstein equations of General Relativity even when $\Phi = U = 0$. Eqs. (6.3) yields in that case the Einstein equations *plus* the condition $R = 0$.

A. Effective String Equations in Cosmological Universes

For the homogeneous isotropic spacetime geometries described by eq.(4.1) we have

$$\begin{aligned} R_0^0 &= -(D-1)(\dot{H} + H^2) \\ R_i^k &= -\delta_i^k\left[\dot{H} + (D-1)H^2\right] \\ R &= -(D-1)(2\,\dot{H} + D\,H^2). \end{aligned} \tag{6.6}$$

where $H \equiv \frac{1}{R}\frac{dR}{dT}$.

The equations of motion (6.3) read

$$\ddot{\Phi} - (D-1)(\dot{H} + H^2) - \frac{\partial U}{\partial \Phi} = e^{\Phi}\,\rho$$

$$\dot{H} + (D-1)H^2 - H\dot{\Phi} + \frac{\partial U}{\partial \Phi} + \frac{R}{D-1}\frac{\partial U}{\partial R} = e^{\Phi}\,p$$

$$2\,\ddot{\Phi} + 2(D-1)\,H\dot{\Phi} - \dot{\Phi}^2 - (D-1)(2\,\dot{H} + D\,H^2) - 2\frac{\partial U}{\partial \Phi} - c + 2U = 0 \tag{6.7}$$

where dot $\cdot$ stands for $\frac{d}{dT}$, and

$$\rho = T_0^0 \qquad , \qquad -\delta_i^k\,p = T_i^k\ . \tag{6.8}$$

The conservation equation takes the form of eq.(4.10)

$$\dot{\rho} + (D-1)\,H\,(p+\rho) = 0\ . \tag{6.9}$$

By defining,

$$\Psi \equiv \Phi - \log\sqrt{-G} = \Phi - (D-1)\log R$$

$$\bar{\rho} = e^{\Phi} \rho \quad , \quad \bar{p} = e^{\Phi} p \, , \tag{6.10}$$

equations (6.7) can be expressed in a more compact form as

$$\ddot{\Psi} - (D-1) H^2 - \left.\frac{\partial U}{\partial \Psi}\right|_R = \bar{\rho}$$

$$\dot{H} - H \dot{\Psi} + \frac{R}{D-1} \left.\frac{\partial U}{\partial R}\right|_{\Psi} = \bar{p}$$

$$\dot{\Psi}^2 - (D-1) H^2 - 2\bar{\rho} - 2U + c = 0 \, , \tag{6.11}$$

The conservation equation reads

$$\dot{\bar{\rho}} - \dot{\Psi}\,\bar{\rho} + (D-1)\,H\bar{p} = 0 \tag{6.12}$$

As is known, under the duality transformation $R \longrightarrow R^{-1}$, the dilaton transforms as $\Phi \longrightarrow \Phi + (D-1)\log R$. The shifted dilaton Ψ defined by eq.(6.10) is invariant under duality. The transformation

$$R' \equiv R^{-1} \quad , \tag{6.13}$$

implies

$$\Psi' = \Psi \quad , \quad H' = -H \quad , \quad \bar{p}' = -p \quad , \quad \bar{\rho}' = \bar{\rho} \tag{6.14}$$

provided $u = d$, that is, a duality invariant string source. This is the duality invariance transformation of eqs.(6.11).

Solutions to the effective string equations have been extensively treated in the literature [28] and they are not our main purpose. For the sake of completeness, we briefly analyze the limiting behaviour of these equations for $R \to \infty$ and $R \to 0$.

It is difficult to make a complete analysis of the effective string equations (6.11) since the knowledge about the potential U is rather incomplete. For weak coupling (e^{Φ} small) the supersymmetry breaking produces an effective potential that decreases very fast (as the exponential of an exponential of Φ) for $\Phi \to -\infty$.

Let us analyze the asymptotic behavior of eqs.(6.11) for $R \to \infty$ and $R \to 0$ assuming that the potential U can be ignored. It is easy to see that a power behaviour Ansatz both for R and for e^{Ψ} as functions of T is consistent with these equations. It turns out that the string sources do not contribute to the leading behaviour here, and we find for $R \to 0$

$$R_{\mp} = C_1 \, T^{\pm 1/\sqrt{D-1}} \to 0 \quad ,$$

$$e^{\Psi_{\mp}} = \quad C_2 \, T^{-1} \to \begin{cases} \infty \\ 0 \end{cases} \tag{6.15}$$

Where C_1 and C_2 are constants. Here the branches $(-)$ and $(+)$ correspond to $T \to 0$ and to $T \to \infty$ respectively. In both regimes $R_{\mp} \to 0$ and $e^{\Phi_{\mp}} \to 0$.

The potential $U(\Phi)$ is hence negligible in these regimes. In terms of the conformal time η , the behaviours (6.15) result

$$R_{\mp} = C_1'\, \eta^{\pm\frac{1}{\sqrt{D-1}\mp 1}} \to 0$$
$$e^{\Psi_{\mp}} = C_2'\, \eta^{-\frac{\sqrt{D-1}}{\sqrt{D-1}\mp 1}} \to \begin{cases} \infty \\ 0 \end{cases} \tag{6.16}$$

Where C_1' and C_2' are constants. The branch $(-)$ would describe an expanding non-inflationary behaviour near the initial singularity $T = 0$, while the branch $(+)$ describes a 'big crunch' situation and is rather unphysical.

Similarly, for $R \to \infty$ and $e^\Phi \to \infty$, we find

$$R_{\mp} = D_1\, T^{\mp 1/\sqrt{D-1}} \to \infty \quad,$$
$$e^{\Psi_{\mp}} = D_2\, T^{-1} \to \begin{cases} \infty \\ 0 \end{cases} \tag{6.17}$$

Where D_1 and D_2 are constants. Here again, the branches $(-)$ and $(+)$ correspond to $T \to 0$ and to $T \to \infty$ respectively, but now in both regimes $R_{\mp} \to \infty$ and $e^{\Phi_{\mp}} \to \infty$. (In this limit, one is not guaranteed that U can be consistently neglected). In terms of the conformal time, eqs.(6.17) read

$$R_{\mp} = D_1'\, \eta^{\mp\frac{1}{\sqrt{D-1}\pm 1}} \to \infty$$
$$e^{\Psi_{\mp}} = D_2'\, \eta^{-\frac{\sqrt{D-1}}{\sqrt{D-1}\pm 1}} \to \begin{cases} \infty \\ 0 \end{cases} \tag{6.18}$$

The branch $(+)$ describes a noninflationary expanding behaviour for $T \to \infty$ faster than the standard matter dominated expansion, while the branch $(-)$ describes a super-inflationary behaviour $\eta^{-\alpha}$, since $0 < \alpha < 1$, for all D.

The behaviours (6.15) for $R_{\mp} \to 0$ and (6.17) for $R_{\mp} \to \infty$ are related by duality $R \leftrightarrow 1/R$.

B. String driven inflation?

Let us consider now the question of whether de Sitter spacetime may be a self-consistent solution of the effective string equations (6.7) with the string sources included. The strings in cosmological universes like de Sitter spacetime have the equation of state (5.4)-(5.6). Since $e^\Psi = e^\Phi\, R^{1-D}$:

$$\bar\rho = e^\Psi \left(u\, R + \frac{d}{R} + s \right) \tag{6.19}$$

$$\bar p = \frac{e^\Psi}{D-1} \left(\frac{d}{R} - u\, R \right) \tag{6.20}$$

In the absence of dilaton potential and cosmological constant term, the string sources do not generate de Sitter spacetime as discussed in sec. V.A. We see that for $U = c = 0$, and $R = e^{HT}$, eqs.(6.11) yields to a contradiction (unless $D = 0$) for the value of Ψ, required to be $-HT + $ constant.

A self-consistent solution describing asymptotically de Sitter spacetime self-sustained by the string equation of state (6.19)-(6.20) is given by

$$R = e^{HT} \quad , \quad H = \text{constant} > 0 \ ,$$
$$2U - c = D\, H^2 = \text{constant}$$
$$\Psi_\pm = \mp HT \pm i\pi + \log \frac{(D-1)\, H^2}{\rho_\pm}$$
$$\rho_+ \equiv u \quad , \quad \rho_- \equiv d \tag{6.21}$$

The branch Ψ_+ describes the solution for $R \to \infty$ ($T \to +\infty$), while the branch Ψ_- corresponds to $R \to 0$ ($T \to -\infty$). De Sitter spacetime with lorentzian signature self-sustained by the strings necessarily requires a constant imaginary piece $\pm i\pi$ in the dilaton field. This makes $e^\Psi < 0$ telling us that the gravitational constant $G \sim e^\Psi < 0$ here describes antigravity.

Is interesting to notice that in the euclidean signature case, i. e. $(+++\dots++)$, the Ansatz $\dot{H} = 0$, $2U - c =$ constant, yields a constant curvature geometry with a real dilaton, but which is of Anti-de Sitter type. This solution is obtained from eqs.(6.20)-(6.21) through the transformation

$$\hat{X}^0 = iT \quad , \quad \hat{H} = -iH \quad , \quad X^i = X^i \quad , \quad \Psi = \Psi \tag{6.22}$$

which maps the Lorentzian de Sitter metric into the positive definite one

$$d\hat{s}^2 = (d\hat{X}^0)^2 + e^{\hat{H}\hat{X}^0}\, (d\vec{X})^2. \tag{6.23}$$

The equations of motion (6.11) within the constant curvature Ansatz ($\dot{\hat{H}} = \ddot{\Psi} = 0$) are mapped onto the equations

$$(D-1)\, \hat{H}^2 - \left.\frac{\partial U}{\partial \Psi}\right|_R = \bar{\rho}$$

$$\hat{H}\frac{d\Psi}{d\hat{X}^0} + \frac{R}{D-1}\left.\frac{\partial U}{\partial R}\right|_\Psi = \bar{p}$$

$$-(\frac{d\Psi}{d\hat{X}^0})^2 + (D-1)\, \hat{H}^2 - 2\,\bar{p} - 2U + c = 0\ , \tag{6.24}$$

with the solution

$$R = e^{\hat{H}\hat{X}^0} \quad , \quad \hat{H} = \text{constant} > 0 \ ,$$
$$c - 2U = D\, \hat{H}^2 = \text{constant}$$
$$\Psi_\pm = \mp \hat{H}\hat{X}^0 + \log \frac{(D-1)\, \hat{H}^2}{\rho_\pm}$$
$$\rho_+ \equiv u \quad , \quad \rho_- \equiv d \tag{6.25}$$

Both solutions (6.25) and (6.21) are mapped one into another through the transformation (6.22).

It could be recalled that in the context of (point particle) field theory, de Sitter spacetime (as well as anti-de Sitter) emerges as an exact selfconsistent solution of the semiclassical Einstein equations with the back reaction included [34] - [35]. (Semiclassical in this context, means that matter fields including the graviton are quantized to the one-loop level and coupled to the (c-number) gravity background through the expectation value of the energy-momentum tensor T_A^B . This expectation value is given by the trace anomaly: $< T_A^A > = \bar{\gamma} \, R^2)$. On the other hand, the α' expansion of the effective string action admits anti-de Sitter spacetime (but not de Sitter) as a solution when the quadratic curvature corrections (in terms of the Gauss-Bonnet term) to the Einstein action are included [36]. It appears that the corrections to the anti-de Sitter constant curvature are qualitatively similar in the both cases, with α' playing the rôle of the trace anomaly parameter $\bar{\gamma}$ [35].

The fact that de Sitter inflation with true gravity $G \sim e^{\Psi} > 0$ does not emerge as a solution of the effective string equations does not mean that string theory excludes inflation. What means is that the effective string equations are not enough to get inflation. The effective string action is a low energy field theory approximation to string theory containing only the *massless* string modes (*massless* background fields).

The vacuum energy scales to start inflation (physical or true vacuum) are typically of the order of the Planck mass [26] - [27] where the effective string action approximation breaks down. One must consider the massive string modes (which are absent from the effective string action) in order to properly get the cosmological condensate yielding de Sitter inflation. We do not have at present the solution of such problem.

TABLE 4. Asymptotic solution of the string effective equations (including the dilaton).

EFFECTIVE STRING EQUATIONS
SOLUTIONS IN COSMOLOGY

Effective String equations	$R(T) \to 0$ behaviour	$R(T) \to \infty$ behaviour
$T \to 0$	$\simeq T^{+1/\sqrt{D-1}}$	$\simeq T^{-1/\sqrt{D-1}}$
$T \to \infty$	$\simeq T^{-1/\sqrt{D-1}}$	$\simeq T^{+1/\sqrt{D-1}}$

VII. MULTI-STRINGS AND SOLITON METHODS IN DE SITTER UNIVERSE

Among the cosmological backgrounds, de Sitter spacetime occupies a special place. This is, in one hand relevant for inflation and on the other hand string propagation turns to be specially interesting there [2] - [8]. String unstability, in the sense that the string proper length grows indefinitely is particularly present in de Sitter. The string dynamics in de Sitter universe is described by a generalized sinh-Gordon model with a potential unbounded from below [4]. The sinh-Gordon function $\alpha(\sigma, \tau)$ having a clear physical meaning : $H^{-1} e^{\alpha(\sigma,\tau)/2}$ determines the string proper length. Moreover the classical string equations of motion (plus the string constraints) turn to be integrable in de Sitter universe [4,5]. More precisely, they are equivalent to a non-linear sigma model on the grassmannian $SO(D,1)/O(D)$ with periodic boundary conditions (for closed strings). This sigma model has an associated linear system [37] and using it, one can show the presence of an infinite number of conserved quantities [38]. In addition, the string constraints imply a zero energy-momentum tensor and these constraints are compatible with the integrability.

The so-called dressing method [37] in soliton theory allows to construct solutions of non-linear classically integrable models using the associated linear system. In ref. [6] we systematically construct string solutions in three dimensional de Sitter spacetime. We start from a given exactly known solution of the string equations of motion and constraints in de Sitter [5] and then we "dress" it. The string solutions reported there indeed apply to cosmic strings in de Sitter spacetime as well.

The invariant interval in D-dimensional de Sitter space-time is given by

$$ds^2 = dT^2 \ - \ \exp[2HT] \sum_{i=1}^{D-1} (dX^i)^2. \tag{7.1}$$

Here T is the so called cosmic time. In terms of the conformal time η,

$$\eta \equiv -\frac{\exp[-HT]}{H} \ , \ -\infty < \eta \le 0 \ ,$$

the line element becomes

$$ds^2 = \frac{1}{H^2\eta^2}[d\eta^2 \ - \ \sum_{i=1}^{D-1}(dX^i)^2] \ .$$

The de Sitter spacetime can be considered as a D-dimensional hyperboloid embedded in a D+1 dimensional flat Minkowski spacetime with coordinates $(q^0, ..., q^D)$:

$$ds^2 = \frac{1}{H^2}[-(dq^0)^2 \ + \ \sum_{i=1}^{D}(dq^i)^2] \tag{7.2}$$

where

$$q^0 = \ \sinh HT \ + \frac{H^2}{2} \exp[HT] \sum_{i=1}^{D-1}(X^i)^2 \ ,$$

$$q^1 = \ \cosh HT \ - \frac{H^2}{2} \exp[HT] \sum_{i=1}^{D-1}(X^i)^2 \ ,$$

$$q^{i+1} = \ H\exp[HT] \ X^i \ , \ \ 1 \le i \le D-1, \ -\infty < T, \ X^i < +\infty. \tag{7.3}$$

The complete de Sitter manifold is the hyperboloid

$$-(q^0)^2 + \sum_{i=1}^{D}(q^i)^2 = 1.$$

The coordinates (T, X^i) and (η, X^i) cover only the half of the de Sitter manifold $q^0 + q^1 > 0$.

We will consider a string propagating in this D-dimensional space-time. The string equations of motion (2.7) in the metric (7.2) take the form:

$$\partial_{+-}q + (\partial_{+}q.\partial_{-}q)\,q = 0 \quad \text{with} \quad q.q = 1, \tag{7.4}$$

where . stands for the Lorentzian scalar product $a.b \equiv -a_0b_0 + \sum_{i=1}^{D} a_ib_i$, $x_{\pm} \equiv \frac{1}{2}(\tau \pm \sigma)$ and $\partial_{\pm}q = \frac{\partial q}{\partial x_{\pm}}$. The string constraints (2.8) become for de Sitter universe

$$T_{\pm\pm} = \frac{\partial q}{\partial x_{\pm}} \cdot \frac{\partial q}{\partial x_{\pm}} = 0. \tag{7.5}$$

Eqs.(7.4) describe a non compact $O(D,1)$ non-linear sigma model in two dimensions. In addition, the (two dimensional) energy-momentum tensor is required to vanish by the constraints eqs.(7.5) . This system of non-linear partial differential equations can be reduced by choosing an appropriate basis for the string coordinates in the $(D+1)$-dimensional Minkowski space time $(q^0, ..., q^D)$ to a noncompact Toda model [4].

These equations can be rewritten in the form of a chiral field model on the Grassmanian $G_D = SO(D,1)/O(D)$. Indeed, any element $\mathbf{g} \in G_D$ can be parametrized with a real vector $q\rangle$ of the unit pseudolength

$$\mathbf{g} = 1 - 2|q\rangle\langle q|J, \quad \langle q|J|q\rangle = 1. \tag{7.6}$$

In terms of $\mathbf{g}$, the string equations (7.4) have the following form

$$2\,\mathbf{g}_{\xi\eta} - \mathbf{g}_{\xi}\,\mathbf{g}\,\mathbf{g}_{\eta} - \mathbf{g}_{\eta}\,\mathbf{g}\,\mathbf{g}_{\xi} = 0 , \tag{7.7}$$

and the conformal constraints (7.5) become

$$\text{tr}\,\mathbf{g}_{\xi}^2 = 0, \quad \text{tr}\,\mathbf{g}_{\eta}^2 = 0 . \tag{7.8}$$

The fact that $\mathbf{g} \in G_D$ implies that $\mathbf{g}$ is a real matrix with the following properties:

$$\mathbf{g} = J\mathbf{g}^{\text{t}}J, \quad \mathbf{g}^2 = I, \quad tr\,\mathbf{g} = D - 1 , \quad \mathbf{g} \in SL(D+1,R). \tag{7.9}$$

These conditions are equivalent to the existence of the representation (7.6). Equation (7.7) is the compatibility condition for the following overdetermined linear system:

$$\Psi_{\xi} = \frac{U}{1-\lambda}\Psi, \quad \Psi_{\eta} = \frac{V}{1+\lambda}\Psi, \tag{7.10}$$

where

$$U = \mathbf{g}_{\xi}\,\mathbf{g}, \quad V = \mathbf{g}_{\eta}\,\mathbf{g} . \tag{7.11}$$

Or, in terms of the vector $q\rangle$

$$U = 2\,q_{\xi}\rangle\langle q\,J - 2\,q\rangle\langle q_{\xi}\,J,$$
$$V = 2\,q_{\eta}\rangle\langle q\,J - 2\,q\rangle\langle q_{\eta}\,J.$$

Eq. (7.6) can be easily inverted yielding q in terms of the matrix g:

$$q_0 = \sqrt{\frac{g_{00}-1}{2}} \quad , \quad q_i = \sqrt{\frac{1-g_{ii}}{2}} \quad 1 \le i \le D \text{ (no sum over } i) \tag{7.12}$$

The use of overdetermined linear systems to solve non-linear partial differential equations associated to them goes back to refs. [39]. (See refs. [40] - [41] for further references).

In order to fix the freedom in the definition of Ψ we shall identify

$$\Psi(\lambda = 0) = \mathbf{g}. \tag{7.13}$$

This condition is compatible with the above equations since the matrix function Ψ at the point $\lambda = 0$ satisfies the same equations as $\mathbf{g}$. Thus the problem of constructing exact solutions of the string equations is reduced to finding compatible solutions of the linear equations (7.10) such that $\mathbf{g} = \Psi(\lambda = 0)$ satisfies the constraints eqs.(7.8) and (7.9).

We concentrate below on the linear system (7.10) since this is the main tool to derive new string solutions in de Sitter spacetime.

In ref. [6] the dressing method was applied as follows. We started from the exact ring-shaped string solution $q_{(0)}$ [5] and we find the explicit solution $\Psi^{(0)}(\lambda)$ of the associated linear system, where λ stands for the spectral parameter. Then, we propose a new solution $\Psi(\lambda)$ that differs from $\Psi^{(0)}(\lambda)$ by a matrix rational in λ. Notice that $\Psi(\lambda = 0)$ provides in general a new string solution.

We then show that this rational matrix must have at least **four** poles, $\lambda_0, 1/\lambda_0, \lambda_0^*, 1/\lambda_0^*$, as a consequence of the symmetries of the problem. The residues of these poles are shown to be one-dimensional projectors. We then prove that these projectors are formed by vectors which can all be expressed in terms of an arbitrary complex constant vector $|x_0\rangle$ and the complex parameter λ_0. This result holds for arbitrary starting solutions $q_{(0)}$.

Since we consider closed strings, we impose a 2π-periodicity on the string variable σ. This restricts λ_0 to take discrete values that we succeed to express in terms of Pythagorean numbers. In summary, our solutions depend on two arbitrary complex numbers contained in $|x_0\rangle$ and two integers n and m. The counting of degrees of freedom is analogous to 2+1 Minkowski spacetime except that left and right modes are here mixed up in a non-linear and precise way.

The vector $|x_0\rangle$ somehow indicates the polarization of the string. The integers (n, m) determine the string winding. They fix the way in which the string winds around the origin in the spatial dimensions (here S^2). Our starting solution $q_{(0)}(\sigma, \tau)$ is a stable string winded $n^2 + m^2$ times around the origin in de Sitter space.

The matrix multiplications involved in the computation of the final solution were done with the help of the computer program of symbolic calculation "Mathematica". The resulting solution $q(\sigma, \tau) = (q^0, q^1, q^2, q^3)$ is a complicated combination of trigonometric functions of σ and hyperbolic functions of τ. That is, these string solitonic solutions do not oscillate in time. This is a typical feature of string unstability [5] - [11] - [9]. The new feature here is that strings (even stable solutions) do not oscillate neither for $\tau \to 0$, nor for $\tau \to \pm\infty$.

We plot in figs. 1-7 the solutions for significative values of $|x_0\rangle$ and (m, n) in terms of the comoving coordinates (T, X^1, X^2)

$$T = \frac{1}{H} \log(q^0 + q^1) \ , \quad X^1 = \frac{1}{H} \frac{q^2}{q^0 + q^1} \ , \quad X^2 = \frac{1}{H} \frac{q^3}{q^0 + q^1} \tag{7.14}$$

The first feature to point out is that our solitonic solutions describe **multiple** (here five or three) strings, as it can be seen from the fact that for a given time T we find several different values for τ. That is, τ is a **multivalued** function of T for any fixed σ (fig.1-2). Each branch of τ as a function of T corresponds to a different string. This is a entirely new feature for strings in curved spacetime, with no analogue in flat spacetime where the time coordinate can always be chosen proportional to τ. In flat spacetime, multiple string solutions are described by multiple world-sheets. Here, we have a **single** world-sheet describing several independent and simultaneous strings as a consequence of the coupling with the spacetime geometry. Notice that we consider *free* strings. (Interactions among the strings as splitting or merging are not considered). Five is the generic number of strings in our dressed solutions. The value five can be related to the fact that we are dressing a one-string solution ($q_{(0)}$) with *four* poles. Each pole adds here an unstable string.

In order to describe the real physical evolution, we eliminated numerically $\tau = \tau(\sigma, T)$ from the solution and expressed the spatial comoving coordinates X^1 and X^2 in terms of T and σ.

We plot $\tau(\sigma, T)$ as a function of σ for different fixed values of T in fig.3-4. It is a sinusoidal-type function. Besides the customary closed string period 2π, another period appears which varies on τ. For small τ , $\tau = \tau(\sigma, T)$ has a convoluted shape while for larger τ (here $\tau \leq 5$), it becomes a regular sinusoid. These behaviours reflect very clearly in the evolution of the spatial coordinates and shape of the string.

The evolution of the five (and three) strings simultaneously described by our solution as a function of T, for positive T is shown in figs. 5-7. One string is stable (the 5th one). The other four are unstable. For the stable string, (X^1, X^2) contracts in time precisely as e^{-HT}, thus keeping the proper amplitude $(e^{HT}X^1, e^{HT}X^2)$ and proper size constant. For this stable string $(X^1, X^2) \leq \frac{1}{H}$. ($1/H$ = the horizon radius). For the other (unstable) strings, (X^1, X^2) become very fast constant in time, the proper size expanding as the universe itself like e^{HT} . For these strings $(X^1, X^2) \geq \frac{1}{H}$. These exact solutions display remarkably the asymptotic string behaviour found in refs. [4,11].

In terms of the sinh-Gordon description, this means that for the strings outside the horizon the sinh-Gordon function $\alpha(\sigma, \tau)$ is the same as the cosmic time T up to a function of σ. More precisely,

$$\alpha(\sigma, \tau) \overset{T \gg \frac{1}{H}}{=} 2H\,T(\sigma, \tau) + \log\left\{2H^2\left[(A^1(\sigma)')^2 + (A^2(\sigma)')^2\right]\right\} + O(e^{-2HT}). \qquad (7.15)$$

Here $A^1(\sigma)$ and $A^2(\sigma)$ are the X^1 and X^2 coordinates outside the horizon. For $T \to +\infty$ these strings are at the absolute *minimum* $\alpha = +\infty$ of the sinh-Gordon potential with infinite size. The string inside the horizon (stable string) corresponds to the *maximum* of the potential, $\alpha = 0$. $\alpha = 0$ is the only value in which the string can stay without being pushed down by the potential to $\alpha = \pm\infty$ and this also explains why only one stable string appears (is not possible to put more than one string at the maximum of the potential without falling down). These features are *generically* exhibited by our one-soliton multistring solutions, independently of the particular initial state of the string (fixed by $|x^0 >$ and (n, m)). For particular values of $|x^0 >$, the solution describes three strings, with symmetric shapes from $T = 0$, for instance like a rosette or a circle with festoons (fig. 5-7).

The string solutions presented here trivially embedd on D-dimensional de Sitter spacetime ($D \geq 3$). It must be noticed that they exhibit the essential physics of strings in D-dimensional de Sitter universe. Moreover, the construction method used here works in any number of dimensions.

New classes of multistring solutions in curved spacetime has been recently found in [14].

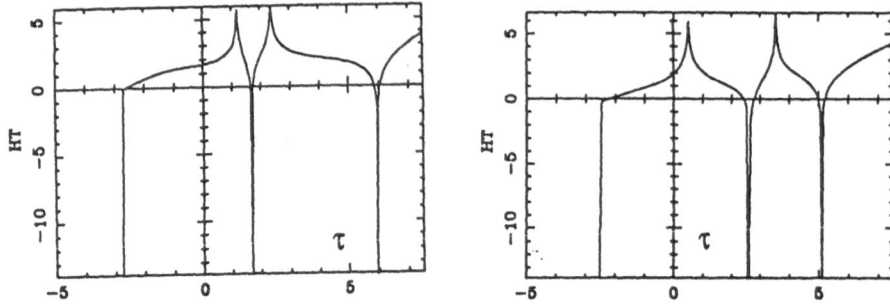

Figure 1: Plot of the function $HT(\tau)$, for two values of σ, for $n = 4, |x^0 \rangle = (1 + i, .6 + .4i, .3 + .5i, .77 + .79i)$. The function $\tau(T)$ is multivalued, revealing the presence of five strings.

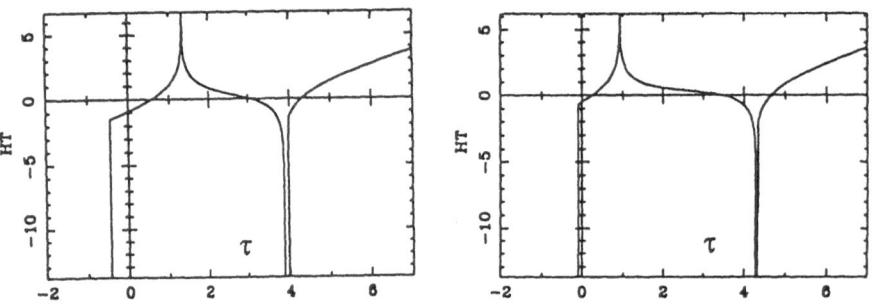

Figure 2: Same as fig.1, for $n = 4, |x^0 \rangle = (1, -1, i, 1)$. Because of a degeneracy, there are now only three strings.

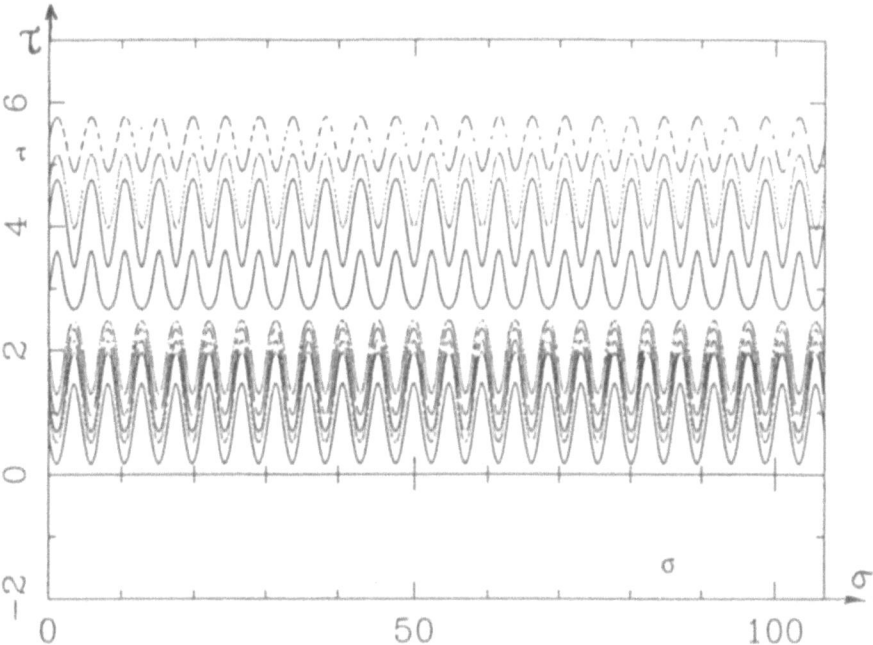

Figure 3: $\tau = \tau(\sigma, T)$ for fixed T for $n = 4, |x^0> = (1, -1, i, 1)$. Three values of HT are displayed, corresponding to HT=0 (full line), 1 (dots), and 2 (dashed line). For each HT, three curves are plotted, which correspond to the three strings. They are ordered with τ increasing.

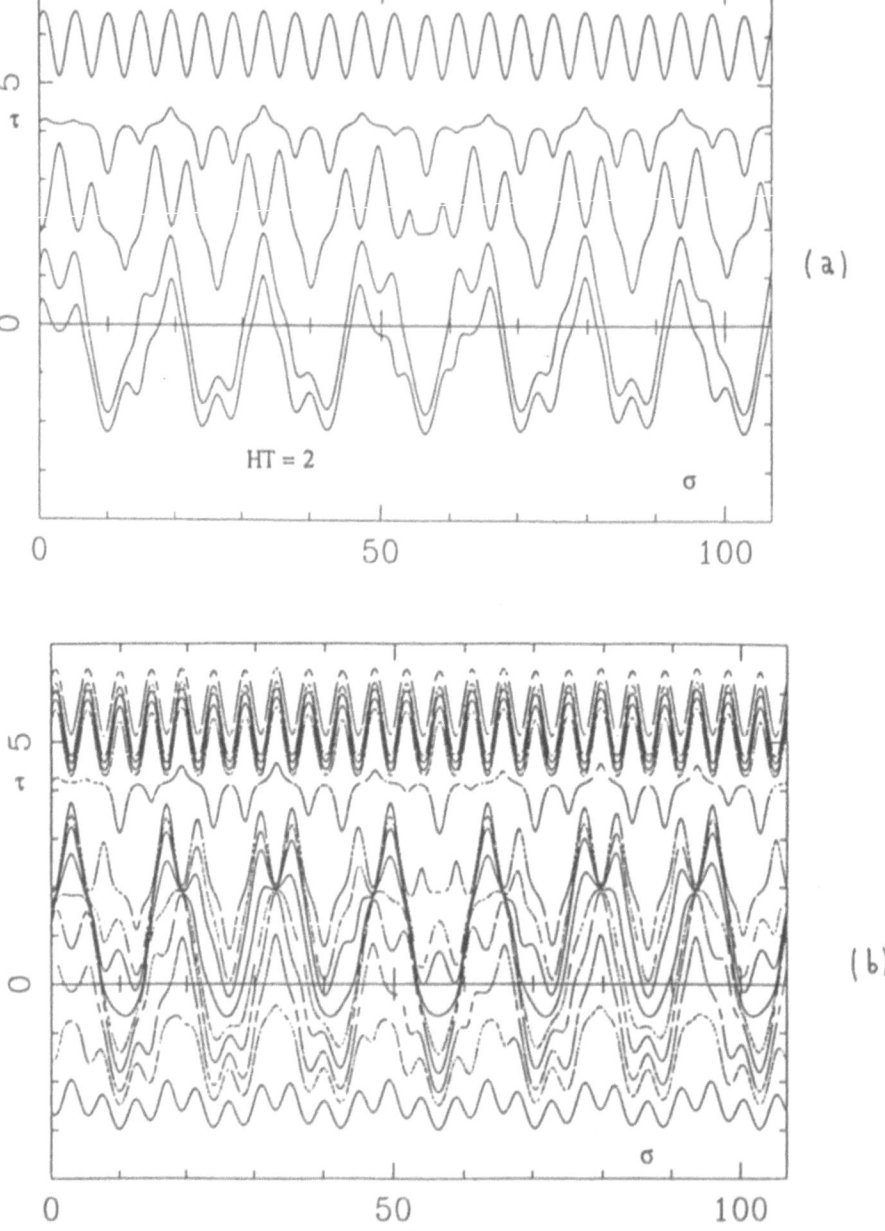

Figure 4: Same as fig. 3 for $n = 4, x^0 >= (1 + i, .6 + .4i, .3 + .5i, .77 + .79i)$. a) The five curves corresponding to the five strings at HT=2. b) The five curves for three values of HT: HT=0 (full line), 1 (dots), and 2 (dashed line).

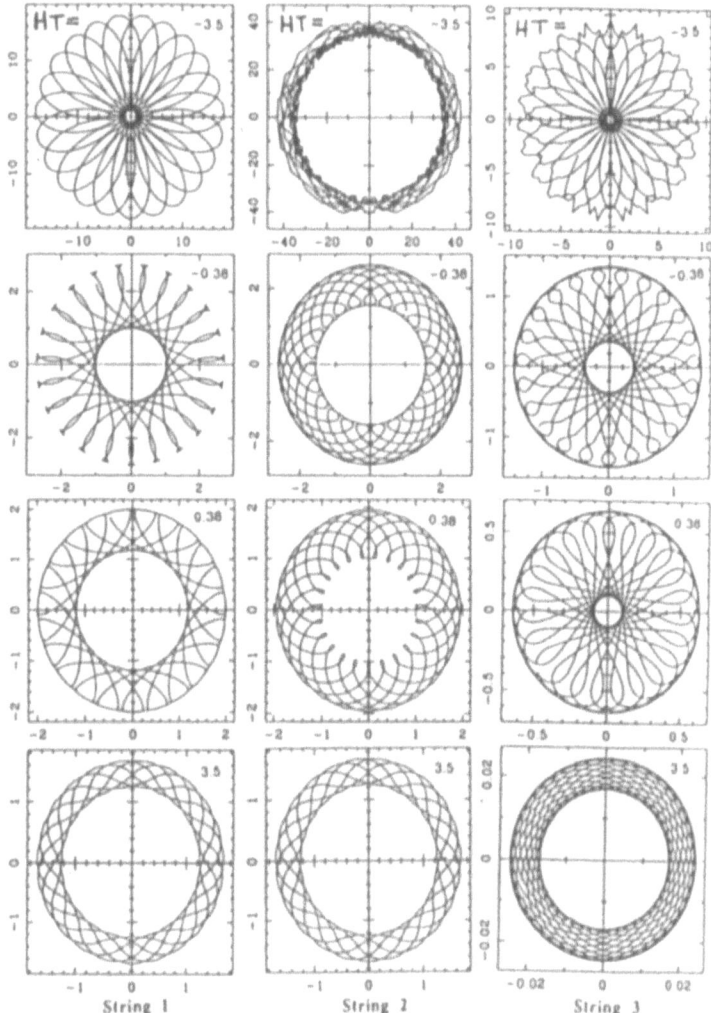

Figure 5: Evolution of the three strings, for $n = 4, |x^0 >= (1, -1, i, 1)$. The comoving size of string (1) stays constant for $HT < -3$, then decreases around $HT = 0$, and stays constant again after $HT = 1$. The invariant size of string (2) is constant for negative HT, then grows as the expansion factor for $HT > 1$, and becomes identical to string (1). The string (3) has a constant comoving size for $HT < -3$, then collapses as e^{-HT} for positive HT.

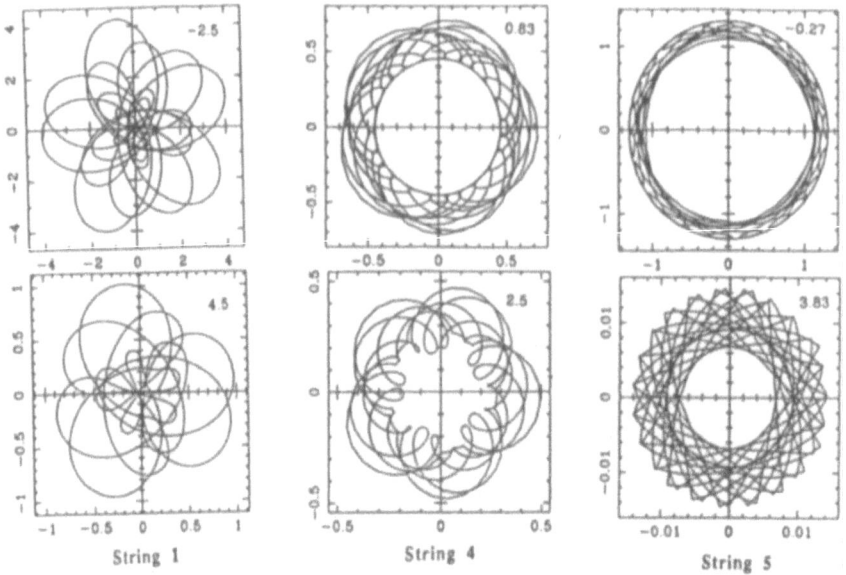

Figure 6: Evolution of three of the five strings for $n = 4, |x^0> = (1 + i, .6 + .4i, .3 + .5i, .77 + .79i)$.

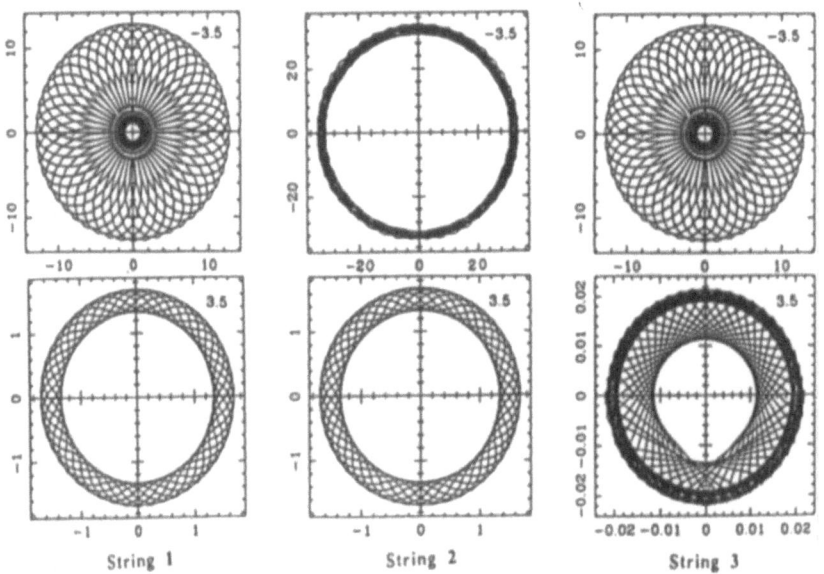

Figure 7: Evolution of the three strings for the degenerate case $n = 6, |x^0> = (1, -1, i, 1)$.

VIII. STRINGS NEXT TO AND INSIDE BLACK HOLES

The classical string equations of motion and constraints were solved near the horizon and near the singularity of a Schwarzschild black hole in ref. [18]. Similar results have been obtained recently in ref. [23] using the null string approach [22].

In a conformal gauge such that $\tau = 0$ (τ = worldsheet time coordinate) corresponds to the horizon ($r = 1$) or to the black hole singularity ($r = 0$), the string coordinates express in power series in τ near the horizon and in power series in $\tau^{1/5}$ around $r = 0$.

In ref. [18] the string invariant size and the string energy-momentum tensor were computed. Near the horizon both are finite and analytic. Near the black hole singularity, the string size, the string energy and the transverse pressures (in the angular directions) tend to infinity as r^{-1}. To leading order near $r = 0$, the string behaves as two dimensional radiation. This two spatial dimensions are describing the S^2 sphere in the Schwarzschild manifold.

A. String Equations of motion in a Schwarzschild Black Hole.

The Schwarzschild metric in Schwarzschild coordinates (t, r, θ, ϕ) takes the following form:

$$ds^2 = \left(1 - \frac{1}{r}\right) dt^2 - \frac{dr^2}{1 - \frac{1}{r}} - r^2(d\theta^2 + \sin^2\theta \, d\phi^2), \tag{8.1}$$

where we choose units where the Schwarzschild radius $R_s = 2m = 1$.

Since we are interested in the whole Schwarzschild manifold and not just in the external part $r > 1$ where the static Schwarzschild coordinates are appropriate, we consider the Kruskal-Szekeres coordinates (u, v, θ, ϕ) defined by

$$u = t_K - r_K \equiv \sqrt{1-r} \, e^{(r-t)/2} \quad , \quad v = t_K + r_K \equiv \sqrt{1-r} \, e^{(r+t)/2}. \tag{8.2}$$

for $v \geq 0, u \geq 0$ and by

$$u = t_K - r_K \equiv -\sqrt{r-1} \, e^{(r-t)/2} \quad , \quad v = t_K + r_K \equiv \sqrt{r-1} \, e^{(r+t)/2}. \tag{8.3}$$

for $v \geq 0, u \leq 0$. For $v \leq 0$ one just flips the sign of v in eq.(8.2) or (8.3) [43].

The coordinate t_K is a time-like coordinate, and r_K spacelike. In Kruskal-Szekeres coordinates the Schwarzschild metric takes the form,

$$ds^2 = -\frac{4}{r} e^{-r} \, du \, dv + r^2 \, (d\theta^2 + \sin^2\theta \, d\phi^2). \tag{8.4}$$

r is a function of the product uv defined by the inverse function of

$$uv = [1 - r] e^r.$$

for $uv \geq 0$. The metric is such coordinates is regular everywhere except at its singularity, $r = 0$.

The string equations of motion in Schwarzschild coordinates and in the conformal gauge, are

$$r_\sigma t_\sigma - r_\tau t_\tau + r(r - 1)(t_{\sigma\sigma} - t_{\tau\tau}) = 0,$$

$$\frac{2r}{1-r}(r_{\tau\tau} - r_{\sigma\sigma}) - \frac{1}{r^2}(t_\tau^2 - t_\sigma^2) + 2r(\theta_\tau^2 - \theta_\sigma^2) +$$
$$2\,r\sin^2\theta\,(\phi_\tau^2 - \phi_\sigma^2) + \frac{1}{(r-1)^2}(r_\tau^2 - r_\sigma^2) = 0. \tag{8.5}$$

$$r\sin\theta\,(\phi_{\tau\tau} - \phi_{\sigma\sigma}) + 2\,r\cos\theta\,(\phi_\tau\theta_\tau - \phi_\sigma\theta_\sigma) + 2\,\sin\theta\,(r_\tau\phi_\tau - r_\sigma\phi_\sigma) = 0,$$
$$r\,(\theta_{\tau\tau} - \theta_{\sigma\sigma}) + 2(r_\tau\theta_\tau - r_\sigma\theta_\sigma) - r\sin\theta\,\cos\theta\,(\phi_\tau^2 - \phi_\sigma^2) \quad = 0. \tag{8.6}$$

The constraints in Schwarzschild coordinates are

$$\frac{1-r}{r}(t_\sigma^2 + t_\tau^2) + \frac{r}{r-1}(r_\tau^2 + r_\sigma^2) + r^2\left[\theta_\tau^2 + \theta_\sigma^2 + \sin^2\theta\,(\phi_\tau^2 + \phi_\sigma^2)\right] = 0,$$
$$\frac{1-r}{r}t_\tau t_\sigma + \frac{r}{r-1}r_\tau r_\sigma + r^2\left(\theta_\tau\theta_\sigma + \sin^2\theta\,\phi_\tau\phi_\sigma\right) = 0. \tag{8.7}$$

The string equations of motion in Kruskal-Szekeres coordinates take the form (always in the conformal gauge),

$$u_{\tau\tau} - u_{\sigma\sigma} + \frac{1}{r}\left(1 + \frac{1}{r}\right)\,e^{-r}\,v\,\left[(u_\tau)^2 - (u_\sigma)^2\right] - \frac{r\,u}{2}\left[\theta_\tau^2 - \theta_\sigma^2 + \sin^2\theta\,(\phi_\tau^2 - \phi_\sigma^2)\right] = 0$$
$$v_{\tau\tau} - v_{\sigma\sigma} + \frac{1}{r}\left(1 + \frac{1}{r}\right)\,e^{-r}\,u\,\left[(v_\tau)^2 - (v_\sigma)^2\right] - \frac{r\,v}{2}\left[\theta_\tau^2 - \theta_\sigma^2 + \sin^2\theta\,(\phi_\tau^2 - \phi_\sigma^2)\right] = 0, \tag{8.8}$$

plus eqs.(8.6) for the angular coordinates.

The constraints in Kruskal-Szekeres coordinates are

$$-\frac{4}{r}\,e^{-r}\,(u_\sigma\,v_\sigma + u_\tau\,v_\tau) + r^2\left[\theta_\tau^2 + \theta_\sigma^2 + \sin^2\theta\,(\phi_\tau^2 + \phi_\sigma^2)\right] = 0,$$
$$-\frac{4}{r}\,e^{-r}\,(u_\tau\,v_\sigma + u_\sigma\,v_\tau) + r^2\left(\theta_\tau\theta_\sigma + \sin^2\theta\,\phi_\tau\phi_\sigma\right) = 0. \tag{8.9}$$

Notice that both the equations of motion and constraints are invariant under the exchange $u \leftrightarrow v$.

Also notice that the equations of motion and constraints in Kruskal-Szekeres coordinates are regular everywhere except at the singularity $r = 0$.

We shall consider closed strings where the string coordinates must be periodic functions of σ:

$$u(\sigma + 2\pi, \tau) = u(\sigma, \tau)\,,\ \ v(\sigma + 2\pi, \tau) = v(\sigma, \tau). \tag{8.10}$$

Therefore, the angular coordinates θ, ϕ may be just quasiperiodic functions of σ:

$$\theta(\sigma + 2\pi, \tau) = \theta(\sigma, \tau) + \text{mod}\,2\pi\,,\ \ \phi(\sigma + 2\pi, \tau) = \phi(\sigma, \tau) + 2n\pi, \tag{8.11}$$

where n is an integer.

B. Strings Near the Singularity $r = 0$

Let us consider the solution of eqs.(8.6,8.8) and constraints (8.9), near $r = 0$. That is to say, near $uv = 1$.

For a generic world-sheet, we choose the gauge such that $\tau = 0$ corresponds to the string at the singularity $uv = 1$. This can be achieved as shown in general in sec. III.A.

Near the singularity $uv = 1$, we propose for $\tau \to 0$ the expansion [18]

$$u(\sigma, \tau) = e^{a(\sigma)} \left[1 - \tau^\alpha \beta(\sigma) + \ldots \right]$$

$$v(\sigma, \tau) = e^{-a(\sigma)} \left[1 - \tau^{\alpha'} \hat{\beta}(\sigma) + \ldots \right]$$

$$\theta(\sigma, \tau) = g(\sigma) + \tau^{\lambda'} \mu(\sigma) + \ldots,$$

$$\phi(\sigma, \tau) = f(\sigma) + \tau^\lambda \nu(\sigma) + \ldots . \tag{8.12}$$

Inserting eqs.(8.12) in eqs.(8.6,8.8) and constraints (8.9) yields [18]

$$\alpha = \alpha' = 4/5 \quad , \quad \lambda = \lambda' = 1/5,$$
$$\hat{\beta}(\sigma) = \beta(\sigma) \quad , \quad \beta(\sigma) = \frac{1}{64} \left[\mu(\sigma)^2 + \nu(\sigma)^2 \sin^2 g(\sigma) \right]^2 \tag{8.13}$$

Since the function $\beta(\sigma)$ is clearly positive, we write it as

$$\beta(\sigma) = \frac{1}{4} \gamma(\sigma)^4.$$

The coordinate r then vanishes as

$$r(\sigma, \tau) = \gamma(\sigma)^2 \, \tau^{2/5} + \ldots . \tag{8.14}$$

The string solution is completely fixed once the functions $f(\sigma), g(\sigma), a(\sigma), \mu(\sigma)$ and $\nu(\sigma)$ are given. These five functions are arbitrary and can be expressed in terms of the initial data. Notice that ϕ and θ approach their limiting values with the same exponent $1/5$ in τ.

Both the equations of motion and constraints are invariant under the exchange $u \leftrightarrow v$ but not the boundary conditions at $\tau = 0$. They differ by $a(\sigma) \leftrightarrow -a(\sigma)$ as we see from eqs.(8.12). Therefore one can obtain $u(\sigma, \tau)$ from $v(\sigma, \tau)$ and viceversa just by flipping the sign of $a(\sigma)$.

We can also find the ring solution of ref. [12] setting $f(\sigma) \equiv n\sigma, a(\sigma) \equiv 0, \mu(\sigma) = $ cte. $g(\sigma) = $ cte. and $\nu(\sigma) \equiv 0$.

The corrections to the leading behaviour appear as positive integer powers of $\tau^{2/5}$. The subdominant leading power in $u(\sigma, \tau)$ and $v(\sigma, \tau)$ is again $\tau^{7/5}$. We find with the help of Mathematica [18],

$$u(\sigma, \tau) = e^{a(\sigma)} \left\{ 1 - \gamma(\sigma)^4 \, \tau^{4/5} \left[1 + O(\tau^{2/5}) \right] \right.$$

$$\left. - \gamma(\sigma)^6 \, \frac{f'(\sigma)\nu(\sigma) \sin^2 g(\sigma) + \mu(\sigma)}{28 \, a'(\sigma)} \, \tau^{7/5} \left[1 + O(\tau^{2/5}) \right] \right\},$$

$$v(\sigma, \tau) = e^{-a(\sigma)} \left\{ 1 - \gamma(\sigma)^4 \, \tau^{4/5} \left[1 + O(\tau^{2/5}) \right] \right.$$

$$+ \gamma(\sigma)^6 \frac{f'(\sigma)\nu(\sigma)\sin^2 g(\sigma) + \mu(\sigma)}{28\, a'(\sigma)}\, \tau^{7/5} \left[1 + O(\tau^{2/5})\right]\right\}. \tag{8.15}$$

Notice that u/v is τ independent up to order $\tau^{7/5}$. Since $u/v = e^{-t}$, this imply that the spatial coordinate t is only σ-dependent up to $O(\tau^{7/5})$. More precisely,

$$t(\sigma, \tau) = \log \frac{v}{u} = -2\, a(\sigma) + \gamma(\sigma)^6 \frac{f'(\sigma)\nu(\sigma)\sin^2 g(\sigma) + \mu(\sigma)}{14\, a'(\sigma)}\, \tau^{7/5} + O(\tau^{9/5}). \tag{8.16}$$

In other words, $t(\sigma, \tau)$ varies slower than the other coordinates ϕ and r when the string approaches the black hole singularity $(\tau \to 0)$.

Using the diagonal conformal transformation (3.6), we can fix one of the arbitrary functions among $f(\sigma), g(\sigma), a(\sigma), \mu(\sigma)$ and $\nu(\sigma)$ keeping in mind the periodic boundary conditions:

$$a(\sigma + 2\pi) = a(\sigma)\,,\ \ \nu(\sigma + 2\pi) = \nu(\sigma)\,,\ \ \mu(\sigma + 2\pi) = \mu(\sigma)\,,$$

$$f(\sigma + 2\pi) = f(\sigma) + 2n\pi\,,\ \ g(\sigma + 2\pi) = g(\sigma) \bmod 2\pi\,. \tag{8.17}$$

We are left with **four** arbitrary functions of σ. This is precisely the number of transverse string degrees of freedom.

C. String energy-momentum and invariant size near the singularity

The string size in the Schwarzschild metric takes the form

$$S^2 = G_{AB}(X)\, \dot{X}^A \dot{X}^B = \frac{4}{r} e^{-r} \dot{u}\dot{v} - r^2\, \dot{\theta}^2 - r^2\, \dot{\phi}^2 \sin^2 \theta. \tag{8.18}$$

where we used eqs.(2.32) and (8.1).

We find near the singularity at $r = 0$ using eqs.(8.14-8.15)

$$
\begin{aligned}
S &= \frac{4\, a'(\sigma)^2}{\gamma(\sigma)^2}\, \tau^{-2/5} - \frac{4}{7}\, a'(\sigma)^2 \left(6 + 25\, \frac{a'(\sigma)^2}{\gamma(\sigma)^7}\right) + O(\tau^{2/5}) \\
&= \frac{4\, a'(\sigma)^2}{r} - \frac{4}{7}\, a'(\sigma)^2 \left(6 + 25\, \frac{a'(\sigma)^2}{\gamma(\sigma)^7}\right) + O(r)\,.
\end{aligned} \tag{8.19}
$$

For simplicity we choose here an equatorial solution at $\theta = \pi/2$. The invariant string size tends then to infinite when the string falls into the $r = 0$ singularity [18,23]. This is due to the infinitely growing gravitational forces that act there on the string.

The string stretching near $r = 0$ was first observed in ref. [42] using perturbative methods and in ref. [14] for a family of exact string solutions inside the horizon.

Inside the horizon we can use t, θ, ϕ as spatial coordinates and r as a coordinate time. We find,

$$\sqrt{g_{rr}}\, \Theta^{AB}(r) = \frac{1}{2\pi\alpha'} \int d\sigma d\tau\, \left(\dot{X}^A \dot{X}^B - X'^A X'^B\right) \delta(r - r(\tau, \sigma)). \tag{8.20}$$

where $g_{rr} = r/(1 - r) > 0$.

We have for the black hole case:

$$G_{AB}(X)\, \dot{X}^A \dot{X}^B = \frac{r\dot{r}^2}{1 - r} - r^2\,\dot{\theta}^2 - r^2\,\dot{\phi}^2\,\sin^2\theta - \frac{1-r}{r}\,\dot{t}^2 \,. \tag{8.21}$$

Using eqs.(8.12) and (8.14) for $\tau \to 0$, we find that each of the first three terms grows as $\tau^{-4/5}$ whereas the last term vanishes as $\tau^{2/5}$. Moreover, the sum of the three terms $O(\tau^{-4/5})$ identically vanishes thanks to eq.(8.13). This cancellation in the trace tells us that near $r = 0$, the dominant (and divergent) components T_r^r, T_ϕ^ϕ and T_θ^θ yield a zero trace. This means that the string behaves to leading order as **two**-dimensional massless particles [18]. This is the so-called dual to unstable behaviour [1] (here for two spatial dimension).

For $\tau \to 0$, $r \to 0$ we can use in eq.(8.20) the dominant behaviours:

$$r(\sigma, \tau) = \gamma(\sigma)^2\,\tau^{2/5} + O(\tau^{4/5}),$$

$$\theta(\sigma, \tau) = g(\sigma) + \mu(\sigma)\,\tau^{1/5} + O(\tau^{3/5}),$$

$$\phi(\sigma, \tau) = f(\sigma) + \nu(\sigma)\,\tau^{1/5} + O(\tau^{3/5}),$$

$$t(\sigma, \tau) = -2a(\sigma) + \gamma(\sigma)^6\,\frac{f'(\sigma)\nu(\sigma)\sin^2 g(\sigma) + \mu(\sigma)}{14\,a'(\sigma)}\,\tau^{7/5} + O\left(\tau^{9/5}\right)\,. \tag{8.22}$$

We thus find for $r \to 0$,

$$2\pi\alpha'\,\Theta^{rr}(r) = \frac{2}{5\,r^2}\int_0^{2\pi} d\sigma\,|\gamma(\sigma)|^5 + O(\frac{1}{r}) \to +\infty\,,$$

$$2\pi\alpha'\,\Theta^{\phi\phi}(r) = \frac{1}{10\,r^3}\int_0^{2\pi} d\sigma\,\nu(\sigma)^2\,|\gamma(\sigma)|^3 + O(\frac{1}{r^2}) \to +\infty\,,$$

$$2\pi\alpha'\,\Theta^{\theta\theta}(r) = \frac{1}{10\,r^3}\int_0^{2\pi} d\sigma\,\mu(\sigma)^2\,|\gamma(\sigma)|^3 + O(\frac{1}{r^2}) \to +\infty\,,$$

$$2\pi\alpha'\,\Theta^{tt}(r) = -10\,r\int_0^{2\pi} d\sigma\,\frac{[a'(\sigma)]^2}{|\gamma(\sigma)|^5} + O(r^2) \to 0^-\,. \tag{8.23}$$

We can identify the string energy with the mixed component $-\Theta_r^r$. We define the mixed components $\Theta_A^B(r)$ by integrating $T_A^B(X)$ over the spatial volume.

This yields for $r \to 0$,

$$E \equiv -\Theta_r^r = \frac{1}{2\pi\alpha'}\,\frac{2}{5\,r}\int_0^{2\pi} d\sigma\,|\gamma(\sigma)|^5 + O(1) \to +\infty\,. \tag{8.24}$$

The transverse pressures are defined as the mixed components Θ_ϕ^ϕ and Θ_θ^θ. They diverge for $r \to 0$:

$$P_\phi \equiv \Theta_\phi^\phi = \frac{1}{2\pi\alpha'}\,\frac{2}{5\,r}\int_0^{2\pi} d\sigma\,\nu(\sigma)^2\,\sin^2 g(\sigma)\,|\gamma(\sigma)|^5 \to +\infty\,,$$

$$P_\theta \equiv \Theta_\theta^\theta = \frac{1}{2\pi\alpha'}\,\frac{2}{5\,r}\int_0^{2\pi} d\sigma\,\mu(\sigma)^2\,|\gamma(\sigma)|^5 \to +\infty. \tag{8.25}$$

Thus, to leading order,

$$E = P_\theta + P_\phi \quad \text{for} \quad r \to 0 .$$

exhibiting a two-dimensional ultrarelativistic gas behaviour. The tidal forces infinitely stretch the string near $r = 0$ in effectively only two directions: ϕ and θ.

We find for the off-diagonal components,

$$2\pi\alpha' \, \Theta^{tr}(r) = \frac{r^{1/2}}{10} \int_0^{2\pi} \frac{d\sigma}{a'(\sigma)} \, \gamma(\sigma)^4 \left[f'(\sigma)\nu(\sigma) \sin^2 g(\sigma) + \mu(\sigma) \right] \to 0^+ ,$$

$$2\pi\alpha' \, \Theta^{t\theta}(r) = \frac{2}{5} \int_0^{2\pi} d\sigma \, \frac{\mu(\sigma)}{a'(\sigma)} \, |\gamma(\sigma)|^3 \left[f'(\sigma)\nu(\sigma) \sin^2 g(\sigma) + \mu(\sigma) \right] = O(1) ,$$

$$2\pi\alpha' \, \Theta^{t\phi}(r) = \frac{2}{5} \int_0^{2\pi} d\sigma \, \frac{\nu(\sigma)}{a'(\sigma)} \, |\gamma(\sigma)|^3 \left[f'(\sigma)\nu(\sigma) \sin^2 g(\sigma) + \mu(\sigma) \right] = O(1) ,$$

$$2\pi\alpha' \, \Theta^{r\phi}(r) = \frac{1}{5 \, r^{5/2}} \int_0^{2\pi} d\sigma \, \nu(\sigma) \, \gamma(\sigma)^4 \to \infty,$$

$$2\pi\alpha' \, \Theta^{r\theta}(r) = \frac{1}{5 \, r^{5/2}} \int_0^{2\pi} d\sigma \, \mu(\sigma) \, \gamma(\sigma)^4 \to \infty,$$

$$2\pi\alpha' \, \Theta^{\theta\phi}(r) = \frac{1}{10 \, r^3} \int_0^{2\pi} d\sigma \, \mu(\sigma) \, \nu(\sigma) \, |\gamma(\sigma)|^3 \to \infty . \tag{8.26}$$

Notice that the invariant string size tends to infinity [see eq.(8.19)] with $4 \, a'(\sigma)^2$ as proportionality factor. Since $-2a(\sigma)$ is the leading behaviour of $t(\sigma, \tau)$, this suggests us that the string stretches infinitely in the (spatial) t direction when $r \to 0$.

As a matter of fact, infinitely growing string sizes are not observed in cosmological spacetimes [1,3] for strings exhibiting radiation (dual to unstable) behaviour.

For particular string solutions the energy-momentum tensor and the string size can be less singular than in the generic case. For ring solutions [12], $\mu(\sigma) = \mu = \text{constant}$, $g(\sigma) = g = \text{constant}$, $a(\sigma) = \nu(\sigma) = 0$, there is no stretching and

$$S = r \sin g \to 0$$

$$E = P_\theta = \frac{1}{\alpha'} \frac{\mu^5}{80 \, r} \to +\infty \quad , \quad P_\phi = 0 .$$

There is no string stretching but the string keeps exhibiting dual to unstable behaviour. This is due to the balance of the tidal forces thanks to the special symmetry of the solution. It behaves in this special case as **one**-dimensional massless particles for $r \to 0$.

As is easy to see, setting $\mu(\sigma) = 0, g(\sigma) = \pi/2$ all equatorial string solutions behave as **one**-dimensional massless particles for $r \to 0$.

The resolution method used here for strings in black hole spacetimes is analogous to the expansions for $\tau \to 0$ developped in ref. [10,11] for strings in cosmological spacetimes (see sec. IVA).

REFERENCES

[1] Lectures by H. J. de Vega and N. Sánchez in 'String Quantum Gravity and the Physics at the Planck Scale', Proceedings of the Erice Workshop held in June 1992. Edited by N. Sánchez, World Scientific, 1993. Pages 73-185, and references given therein.
Lectures by H. J. de Vega and N. Sánchez in 'Current Topics in Astrofundamental Physics: The Early Universe', Proceedings of the Nato ASI Third D. Chalonge School, 4-16 September 1994, p. 99-128, edited by N. Sánchez and A. Zichichi, Kluwer, 1995.

[2] H. J. de Vega and N. Sánchez, Phys. Lett. **B 197**, 320 (1987).

[3] H. J. de Vega and N. Sánchez, Phys. Rev. **D50**, 7202 (1994).

[4] H. J. de Vega and N. Sánchez, Phys. Rev. **D47**, 3394 (1993).

[5] H. J. de Vega, A. V. Mikhailov and N. Sánchez, Teor. Mat. Fiz. **94** (1993) 232.

[6] F. Combes, H. J. de Vega, A. V. Mikhailov and N. Sánchez,
Phys. Rev. **D50**, 2754 (1994).

[7] H. J. de Vega, A. L. Larsen and N. Sánchez, Nucl. Phys. **B 427**, 643 (1994).

[8] I. Krichever, Funct. Anal. and Appl. **28**, 21 (1994),
[Funkts. Anal. Prilozhen. **28**, 26 (1994)].

[9] H. J. de Vega and N. Sánchez, Nucl. Phys. **B309**, 552 and 577 (1988).

[10] N. Sánchez and G. Veneziano, Nucl. Phys. **B333**, 253 (1990).

[11] M. Gasperini, N. Sánchez and G. Veneziano,
Int. J. Mod. Phys. **A 6**, 3853 (1991) and Nucl. Phys. **B364**, 365 (1991).

[12] H. J. de Vega and I. L. Egusquiza, Phys. Rev. **D49**, 763 (1994).

[13] A. L. Larsen and N. Sánchez, Phys. Rev. **D50**, 7493 (1994).

[14] A. L. Larsen and N. Sánchez, Phys. Rev. **D51**, 6929 (1995).

[15] H. J. de Vega and I. L. Egusquiza,
hep-th/9505029, submitted to Class. and Quantum Grav.

[16] S. Hawking and G. F. R. Ellis, 'The large scale structure of the spacetime', Cambridge Univ. Press, 1973.

[17] H. J. de Vega and N. Sánchez, Int. J. Mod. Phys. **A 7**, 3043 (1992).

[18] H. J. de Vega and I. L. Egusquiza, 'Strings next and inside black holes', hep-th/9506214, to appear in Phys. Rev. D.

[19] G. F. R. Ellis, Banff Lectures 1990, *in* Gravitation,
eds. R. Mann and P. Wesson , World Scientific 1991.

[20] See for a review, T. W. B. Kibble, Erice Lectures at the Chalonge School in Astrofundamental Physics, N. Sánchez editor, World Scientific, 1992.

[21] A. Vilenkin, Phys. Rev. **D 24**, 2082 (1981) and Phys. Rep. **121**, 263 (1985).
N. Turok and P. Bhattacharjee, Phys. Rev. **D 29**, 1557 (1984).

[22] H. J. de Vega and A. Nicolaidis, Phys. Lett. **B 295**, 214 (1992).
H. J. de Vega, I. Giannakis and A. Nicolaidis, Mod. Phys. Lett. **A 10**, 2432 (1995).

[23] C. Loustó and N. Sánchez, 'String dynamics in cosmological and black hole backgrounds: the null string approach', in preparation.

[24] R. Myers, Phys. Lett. **B199**, 371 (1987).
M. Mueller, Nucl. Phys. **B337**, 37 (1990).
See for a review: A.A. Tseytlin in the Proceedings of the Erice School "String Quantum Gravity and Physics at the Planck Energy Scale",

21-28 June 1992, Edited by N. Sánchez, World Scientific, 1993.

[25] S. Weinberg, 'Gravitation and Cosmology', J. Wiley, 1972.

[26] E. W. Kolb and M. S. Turner, 'The Early Universe', Addison-Wesley, 1990.

[27] A. D. Linde, 'Particle Physics and Inflationary Cosmology', Harwood (1990).

[28] See for example,
I. Antoniadis, C. Bachas, J. Ellis and D. V. Nanopoulos,
Nucl. Phys. **B 328**, 117 (1989) and Phys. Lett. **B 257**, 278 (1991).
B. A. Campbell, A. Linde and K. A. Olive, Nucl. Phys. **B 355**, 146 (1991).
B. A. Campbell, N. Kaloper and K. A. Olive, Phys. Lett. **B 277**, 265 (1992).
A.A. Tseytlin, Mod. Phys. Lett **A 6**, 1721 (1991).
A.A. Tseytlin and C. Vafa, Nucl. Phys. **B 372**, 443 (1992).
R. Brustein and P. J. Steinhardt, Phys. Lett. **B 302**, 196 (1993).
M. Gasperini and G. Veneziano, Mod. Phys. Lett. **A 8**, 370 (1993),
Phys. Lett. **B 277**, 256 (1992) and Astroparticle Physics **1**, 317 (1993).
E. Raiten, Nucl. Phys. **B 416**, 881 (1994).
R. Brustein and G. Veneziano, Phys. Lett. **B 329**, 429 (1994).
V. A. Kostelecký and M. J. Perry, Nucl. Phys. **B 414**, 174 (1994).
See in addition ref. [24].

[29] See for example:
M. Green, J. Schwarz, E. Witten, 'Superstring Theory'.
Cambridge University Press. 1987.

[30] I.S. Gradshteyn and I.M. Ryzhik, Table of Integrals Series and Products,
(Academic Press, New York, fourth edition, 1965).

[31] See for example,
S. Weinberg, 'Gravitation and Cosmology', J. Wiley, 1972.

[32] F. Müller-Holstein, Class. Quant. Grav. **3**, 665 (1986).
K. G. Akdeniz et al. Mod. Phys. Lett. **A 6**, 1543 (1991) and
Phys. Lett. **B 321**, 329 (1994).

[33] M. V. Fischetti, J. B. Hartle and B. L. Hu, Phys. Rev. **D 20**, 1757 (1979).

[34] S. Wada and T. Azuma, Phys. Lett. **B 132**, 313 (1983).
V. Sahni and L. A. Kofman, Phys. Lett. **A 117**, 275 (1986).

[35] M. A. Castagnino, J. P. Paz and N. Sánchez, Phys. Lett. **B 193**, 13 (1987).

[36] D. G. Boulware and S. Deser, Phys. Rev. Lett. **55**, 2656 (1985).

[37] V. E. Zakharov and A. V. Mikhailov, JETP, **75**, 1953 (1978).

[38] H. J. de Vega, Phys. Lett. **B 87**, 233 (1979).

[39] C. S. Gardner, J. M. Greene, M. D. Kruskal and R. M. Miura,
Phys. Rev. Lett. **19**, 1095 (1967).
P. D. Lax, Comm. Pure and Appl. Math. **21** 467 (1968).

[40] M. J. Ablowitz and H. Segur, "Solitons and the Inverse scattering transformation",
SIAM Philadelphia 1981.
V.E. Zakharov, S.V. Manakov, S.P. Novikov and L.P. Pitaevsky,
"Soliton Theory; The Inverse Method", Nauka, Moscow, 1980.

[41] A. C. Scott, F. Y. F. Chu and D. W. MacLaughlin, Proc. IEEE, **61**, 1443 (1973).
G. L. Lamb, Elements of Soliton Theory, J. Wiley, NY (1980).

[42] C. Loustó and N. Sánchez, Phys. Rev. **D47**, 4498 (1993).

[43] See for example,
   C. W. Misner, K. S. Thorne and J. A. Wheeler, 'Gravitation', Freeman, 1973.

[44] H. J. de Vega and N. Sánchez, Nucl. Phys. **B 317**, 706 (1989) .
   D. Amati and K. Klimĉik, Phys. Lett. **B 210** , 92 (1988) ,
   see also: ref. [45].

[45] M. Costa and H. J. de Vega, Ann. Phys. **211**, 223 and 235 (1991).

[46] H. J. de Vega and N. Sánchez, Phys. Rev. D 45 , 2783 (1992).
   H. J. de Vega, M. Ramón Medrano and N. Sánchez,
   Class. and Quantum Grav. **10**, 2007 (1993).
   G. Horowitz and A.R. Steif, Phys. Rev. Lett. **64**, 260 (1990) and
   Phys. Rev. **D 42** , 1950 (1990).

[47] H. J. de Vega and N. Sánchez, Phys. Rev. **D 42**, 3969 (1990) and
   H. J. de Vega, M. Ramón Medrano and N. Sánchez, Nucl. Phys. **B 374**, 405 (1992)

Strings and Multi-Strings in Black Hole and Cosmological Spacetimes

A.L. Larsen and N. Sánchez

Observatoire de Paris, DEMIRM. Laboratoire Associé au CNRS
UA 336, Observatoire de Paris et École Normale Supérieure.
61, Avenue de l'Observatoire, 75014 Paris, France.

Abstract

Recent results on classical and quantum strings in a variety of black hole and cosmological spacetimes, in various dimensions, are presented. The curved backgrounds under consideration include the $2+1$ black hole anti de Sitter spacetime and its dual, the black string, the ordinary $D \geq 4$ black holes with or without a cosmological constant, the de Sitter and anti de Sitter spacetimes and static Robertson-Walker spacetimes. Exact solutions to the string equations of motion and constraints, representing circular strings, stationary open strings and dynamical straight strings, are obtained in these backgrounds and their physical properties (length, energy, pressure) are described. The existence of *multi-string* solutions, describing finitely or infinitely many strings, is shown to be a general feature of spacetimes with a positive or negative cosmological constant. Generic approximative solutions are obtained using the string perturbation series approach, and the question of the stability of the solutions is addressed.

Furthermore, using a canonical quantization procedure, we find the string mass spectrum in de Sitter and anti de Sitter spacetimes. New features as compared to the string spectrum in flat Minkowski spacetime appear, for instance the *fine-structure effect* at all levels beyond the graviton in both de Sitter and anti de Sitter spacetimes, and the *non-existence* of a Hagedorn temperature in anti de Sitter spacetime. We discuss the physical implications of these results. Finally, we consider the effect of spatial curvature on the string dynamics in Robertson-Walker spacetimes.

N. Sánchez and A. Zichichi (eds.), String Gravity and Physics at the Planck Energy Scale, 65–103.
© *1996 Kluwer Academic Publishers*

1 Introduction

The classical and quantum propagation of strings in curved spacetimes has attracted a great deal of interest in recent years. The main complication, as compared to the case of flat Minkowski spacetime, is related to the non-linearity of the equations of motion. It makes it possible to obtain the complete analytic solution only in a very few special cases like conical spacetime [1] and plane-wave/shock-wave backgrounds [2]. There are however also very general results concerning integrability and solvability for maximally symmetric spacetimes [3, 4] and gauged WZW models [5, 6]. These are the exceptional cases; generally the string equations of motion in curved spacetimes are not integrable and even if they are, it is usually an extremely difficult task to actually separate the equations, integrate them and finally write down the complete solution in closed form. Fortunately, there are several different ways to "attack" a system of coupled non-linear partial differential equations.

The systematic study of string dynamics in curved spacetimes and its associated physical phenomena was started in Refs.[7, 8]. Besides numerical methods, which will not be discussed here, approximative [7-10] and exact [11-14] methods for solving the string equations of motion and constraints in curved spacetimes, have been developed. Classical and quantum string dynamics have been investigated in black hole backgrounds [15-18], cosmological spacetimes [7, 11-14, 17-21], cosmic string spacetime [1], gravitational wave backgrounds [2], supergravity backgrounds (which are necessary for fermionic strings) [22], and near spacetime singularities [23]. Physical phenomena like the Hawking-Unruh effect in string theory [8, 24], horizon string stretching [8, 24], particle transmutation [15, 25], string scattering [1, 2, 15], mass spectrum and critical dimension [1, 7, 15, 27], string instability [7, 11-13, 16, 21] and multi-string solutions [11-13, 28] have been found.

In a generic D-dimensional curved spacetime with metric $g_{\mu\nu}$ and coordinates x^μ, ($\mu = 0, 1, ..., D-1$), the string equations of motion and constraints are:

$$\ddot{x}^\mu - x''^\mu + \Gamma^\mu_{\rho\sigma}(\dot{x}^\rho \dot{x}^\sigma - x'^\rho x'^\sigma) = 0, \qquad (1.1)$$

$$g_{\mu\nu}\dot{x}^\mu x'^\nu = g_{\mu\nu}(\dot{x}^\mu \dot{x}^\nu + x'^\mu x'^\nu) = 0, \qquad (1.2)$$

where dot and prime stand for derivative with respect to the world-sheet coordinates τ and σ, respectively and $\Gamma^\mu_{\rho\sigma}$ are the Christoffel symbols with respect to the metric $g_{\mu\nu}$. In the following we present recent results [26-29] on solutions to Eqs.(1.1)-(1.2) in a variety of curved spacetimes from cosmology, gravitation and string theory. We obtain explicit (exact and/or approximate) mathematical solutions, discuss the corresponding physical properties, we quantize the solutions in different ways (canonical quantization, semiclassical quantization) and find the physical content: the mass spectrum.

The presentation is organized as follows: In Section 2, we discuss the classical string dynamics in the $2 + 1$ black hole anti de Sitter (BH-AdS) spacetime, recently found by Bañados et al [30]. We compare with the string dynamics in ordinary cosmological and black hole spacetimes. This clarifies the geometry (as seen by a string) of the BH-AdS spacetime. In Section 3, generalizing results from Section 2, we derive the quantum string mass spectrum in ordinary D-dimensional anti de Sitter spacetime. We discuss, in particular, the sectors of low and very high mass states. New physical phenomena arise like the fine-structure effect at all levels beyond the graviton and the non-existence of a Hagedorn critical temperature. The results are compared with corresponding results obtained in Minkowski and de Sitter spacetimes. Sections 4 and 5 are devoted to the investigation of the more general underlying structure of solutions to Eqs.(1.1)-(1.2). In Section 4, we find new classes of exact string and multi-string solutions in cosmological and black hole spacetimes, while in Section 5, we consider the effects of a non-zero spatial curvature on the classical and quantum string dynamics in Friedmann-Robertson-Walker (FRW) universes.

2 Classical String Dynamics in $2+1$ BH-AdS

Anti de Sitter (AdS) spacetime is often considered to be a spacetime of minor importance in a cosmological context. However, first of all, it *is* a FRW universe and as such should not be neglected, secondly; it serves as a simple and convenient spacetime for comparison and understanding of results obtained in (say) Minkowski or de Sitter spacetimes and third, it has a tendency to show up (in disguise) as a solution in various models of dilaton-gravity and string theory. An example of the latter is represented by the $2 + 1$ BH-AdS spacetime of Bañados et al [30]. This spacetime background has arised much interest recently [31-35]. It describes a two-parameter family (mass M and angular momentum J) of black holes in 2+1 dimensional general relativity with metric:

$$ds^2 = (M - \frac{r^2}{l^2})dt^2 + (\frac{r^2}{l^2} - M + \frac{J^2}{4r^2})^{-1}dr^2 - Jdtd\phi + r^2d\phi^2. \quad (2.1)$$

It has two horizons $r_\pm = \sqrt{\frac{Ml^2}{2} \pm \frac{1}{2}\sqrt{M^2l^2 - J^2}}$ and a static limit $r_{erg} = \sqrt{M}l$, defining an ergosphere, as for ordinary Kerr black holes. The spacetime is not asymptotically flat; it approaches anti de Sitter spacetime asymptotically with cosmological constant $\Lambda = -1/l^2$. The curvature is constant $R_{\mu\nu} = -(2/l^2)g_{\mu\nu}$ everywhere, except probably at $r = 0$, where it has at most a delta-function singularity. Notice the weak nature of the singularity at $r = 0$ in 2+1 dimensions as compared with the power law divergence of curvature scalars in $D > 3$ (We will not discuss here the geometry near $r = 0$. For a discusion, see Refs.[33, 35]). The spacetime, Eq.(2.1), is also a solution of the low energy effective action of string theory with zero dilaton field $\Phi = 0$, anti-symmetric tensor field $H_{\mu\nu\rho} = (2/l^2)\epsilon_{\mu\nu\rho}$ (i.e. $B_{\phi t} = r^2/l^2$) and $k = l^2$ [31]. Moreover, it yields an exact solution of string theory in 2+1 dimensions, obtained by gauging the WZWN sigma model of the group

$SL(2, R) \times R$ at level k [31, 32] (for non-compact groups, k does not need to be an integer, so the central charge $c = 3k/(k-2) = 26$ when k=52/23). This solution is the black string background [36]:

$$d\tilde{s}^2 = -(1 - \tfrac{M}{\tilde{r}})d\tilde{t}^2 + (1 - \tfrac{Q^2}{M\tilde{r}})d\tilde{x}^2 + (1 - \tfrac{M}{\tilde{r}})^{-1}(1 - \tfrac{Q^2}{M\tilde{r}})^{-1}\tfrac{l^2 d\tilde{r}^2}{4\tilde{r}^2},$$

$$\tilde{B}_{\tilde{x}\tilde{t}} = \tfrac{Q}{r}, \quad \tilde{\Phi} = -\tfrac{1}{2}\log \tilde{r}l, \tag{2.2}$$

which is related by duality [31, 37] to the 2+1 BH-AdS spacetime, Eq.(2.1). It has two horizons $\tilde{r}_\pm = r_\pm$, the same as the metric, Eq.(2.1), while the static limit is $\tilde{r}_{erg} = J/(2\sqrt{M})$.

We first investigate the string propagation in the $2 + 1$ BH-AdS background by considering the perturbation series around the exact center of mass of the string:

$$x^\mu(\tau,\sigma) = q^\mu(\tau) + \eta^\mu(\tau,\sigma) + \xi^\mu(\tau,\sigma) + ... \tag{2.3}$$

The original method of Refs.[7, 8] can be conveniently formulated in covariant form. It is useful to introduce $D-1$ normal vectors n_R^μ ($R = 1,.., D-1$), (which can be chosen to be covariantly constant by gauge fixing), and consider comoving perturbations δx_R, i.e. those seen by an observer travelling with the center of mass, thus $\eta^\mu = \delta x^R n_R^\mu$. After Fourier expansion, $\delta x^R(\tau,\sigma) = \sum_n C_n^R(\tau)e^{-in\sigma}$, the first order perturbations satisfy the matrix Schrödinger-type equation in τ:

$$\ddot{C}_{nR} + (n^2\delta_{RS} - R_{\mu\rho\sigma\nu}n_R^\mu n_S^\nu \dot{q}^\rho \dot{q}^\sigma)C_n^S = 0. \tag{2.4}$$

Second order perturbations ξ^μ and constraints are similarly covariantly treated, ξ^μ also satisfying Schrödinger-type equations with source terms, see Ref.[26].

For our purposes here it is enough to consider the non-rotating ($J = 0$) 2+1 BH-AdS background and a radially infalling string. We have solved completely the c.m. motion $q^\mu(\tau)$ and the first and second order perturbations $\eta^\mu(\tau,\sigma)$ and $\xi^\mu(\tau,\sigma)$ in this background. Eqs.(2.4) become:

$$\ddot{C}_{nR} + (n^2 + \frac{m^2}{l^2})C_{nR} = 0; \quad R = \perp, \| \tag{2.5}$$

The first order perturbations are independent of the black hole mass, only the anti de Sitter (AdS) part emerges. All oscillation frequencies $\omega_n = \sqrt{n^2 + m^2/l^2}$ are real; there are no unstable modes in this case. The perturbations:

$$\delta x_R(\tau,\sigma) = \sum_n [A_{nR}e^{-i(n\sigma+\omega_n\tau)} + \tilde{A}_{nR}e^{-i(n\sigma-\omega_n\tau)}] \tag{2.6}$$

are completely finite and regular. This is also true for the second order perturbations which are bounded everywhere, even for $r \to 0$ ($\tau \to 0$).

We have also computed the conformal generators L_n, (see Ref.[26]), and the string mass:

$$m^2 = 2 \sum_n (2n^2 + \frac{m^2}{l^2})[A_{n\parallel}\tilde{A}_{-n\parallel} + A_{n\perp}\tilde{A}_{-n\perp}]. \tag{2.7}$$

The mass formula is modified (by the term m^2/l^2) with respect to the usual flat spacetime expression. This is due to the asymptotic (here AdS) behaviour of the spacetime. In ordinary $D \geq 4$ black hole spacetimes (without cosmological constant), which are asymptotically flat, the mass spectrum is the same as in flat Minkowski spacetime [15].

We compare with the string perturbations in the ordinary $(D \geq 4)$ black hole anti de Sitter spacetime. In this case Eqs.(2.4) become:

$$\ddot{C}_{nS\perp} + (n^2 + m^2 H^2 + \frac{Mm^2}{r^3})C_{nS\perp} = 0, \quad S = 1, 2 \tag{2.8}$$

$$\ddot{C}_{n\parallel} + (n^2 + m^2 H^2 - \frac{2Mm^2}{r^3})C_{n\parallel} = 0. \tag{2.9}$$

The transverse $\perp$-perturbations are oscillating with real frequencies and are bounded even for $r \to 0$. For longitudinal $\parallel$-perturbations, however, imaginary frequencies arise and instabilities develop. The $(\mid n \mid = 1)$-instability sets in at:

$$r_{\text{inst.}} = (\frac{2Mm^2}{1 + m^2 H^2})^{1/3} \tag{2.10}$$

Lower modes become unstable even outside the horizon, while higher modes develop instabilities at smaller r and eventually only for $r \approx 0$. For $r \to 0$ (which implies $\tau \to \tau_0$) we find $r(\tau) \approx (3m\sqrt{M/2})^{2/3}(\tau_0 - \tau)^{2/3}$ and:

$$\ddot{C}_{nS\perp} + \frac{2}{9(\tau - \tau_0)^2}C_{nS\perp} = 0, \quad S = 1, 2 \tag{2.11}$$

$$\ddot{C}_{n\parallel} - \frac{4}{9(\tau - \tau_0)^2}C_{n\parallel} = 0. \tag{2.12}$$

For $\tau \to \tau_0$ the $\parallel$-perturbations blow up while the string ends trapped into the $r = 0$ singularity. We see the important difference between the string evolution in the 2+1 BH-AdS background and the ordinary 3+1 (or higher dimensional) black hole anti de Sitter spacetime.

We also compare with the string propagation in the 2+1 black string background, Eq.(2.2) (with $J = 0$). In this case, Eqs.(2.4) become:

$$\ddot{C}_{n\perp} + n^2 C_{n\perp} = 0, \tag{2.13}$$

$$\ddot{C}_{n\parallel} + (n^2 - \frac{2m^2 M}{lr})C_{n\parallel} = 0. \tag{2.14}$$

The $\perp$-modes are stable, while $C_{n\parallel}$ develop instabilities. For $r \to 0$ (which implies $\tau \to \tau_0$) we find $r(\tau) \approx \frac{m^2 M}{l}(\tau_0 - \tau)^2$ and:

$$\ddot{C}_{n\parallel} - \frac{2}{(\tau_0 - \tau)^2} C_{n\parallel} = 0, \qquad (2.15)$$

with similar conclusions as for the ordinary 3+1 (or higher dimensional) black hole anti de Sitter spacetime.

In order to extract more information about the string evolution in these backgrounds, in particular exact properties, we consider the circular string ansatz:

$$t = t(\tau), \quad r = r(\tau), \quad \phi = \sigma + f(\tau), \qquad (2.16)$$

in the equatorial plane ($\theta = \pi/2$) of the stationary axially symmetric backgrounds:

$$ds^2 = g_{tt}(r)dt^2 + g_{rr}(r)dr^2 + 2g_{t\phi}(r)dtd\phi + g_{\phi\phi}(r)d\phi^2. \qquad (2.17)$$

This includes all the cases of interest here: The 2+1 BH-AdS spacetime, the black string, as well as the equatorial plane of ordinary Einstein black holes. The string dynamics, determined by Eqs.(1.1)-(1.2), is then reduced to a system of second order ordinary differential equations and constraints, also described as a Hamiltonian system:

$$\dot{r}^2 + V(r) = 0; \quad V(r) = g^{rr}(g_{\phi\phi} + E^2 g^{tt}), \qquad (2.18)$$

$$\dot{t} = -Eg^{tt}, \quad \dot{f} = -Eg^{t\phi}; \quad E = -P_t = \text{const.}, \qquad (2.19)$$

which in all backgrounds considered here are solved in terms of either elementary or elliptic functions. The dynamics of the circular strings takes place at the r-axis in the $(r, V(r))$-diagram and from the properties of the potential $V(r)$ (minima, zeroes, asymptotic behaviour for large r and the value $V(0)$), general knowledge about the string motion can be obtained. On the other hand, the line element of the circular string turns out to be:

$$ds^2 = g_{\phi\phi}(d\sigma^2 - d\tau^2), \quad \text{i.e. } S(\tau) = \sqrt{g_{\phi\phi}(r(\tau))}, \qquad (2.20)$$

$S(\tau)$ being the invariant string size. For all the static black hole AdS spacetimes (2+1 and higher dimensional): $S(\tau) = r(\tau)$.

For the rotating 2+1 BH-AdS spacetime:

$$V(r) = r^2(\frac{r^2}{l^2} - M) + \frac{J^2}{4} - E^2, \qquad (2.21)$$

(see Fig.1). $V(r)$ has a global minimum $V_{\min} < 0$ between the two horizons r_+, r_- (for $Ml^2 \geq J^2$, otherwise there are no horizons). The vanishing of $V(r)$ at $r = r_{01,2}$ (see Ref.[26]) determines three different types of solutions: (i) For $J^2 > 4E^2$, there are two positive zeroes $r_{01} < r_{02}$, the string never comes outside the static limit, never falls into $r = 0$ neither (there is a barrier between $r = r_{01}$ and $r = 0$). The mathematical solution oscillates between r_{01} and r_{02} with $0 < r_{01} < r_- < r_+ < r_{02} < r_{erg}$. It may be interpreted as a string travelling between the different universes described by the maximal analytic extension of the manifold. (ii) For $J^2 < 4E^2$, there is only one positive zero r_0 outside the static limit and there is no barrier preventing the string from collapsing into $r = 0$. The string starts at $r = 0$ with maximal size $S_{max}^{(ii)}$ outside the static limit, it then contracts through the ergosphere and the two horizons and eventually collapses into a point $r = 0$. For $J \neq 0$, it may be still possible to continue this solution into another universe like in the case (i). (iii) $J^2 = 4E^2$ is the limiting case where the maximal string size equals the static limit: $S_{max}^{(iii)} = l\sqrt{M}$. In this case $V(0) = 0$, thus the string contracts through the two horizons and eventually collapses into a point $r = 0$.

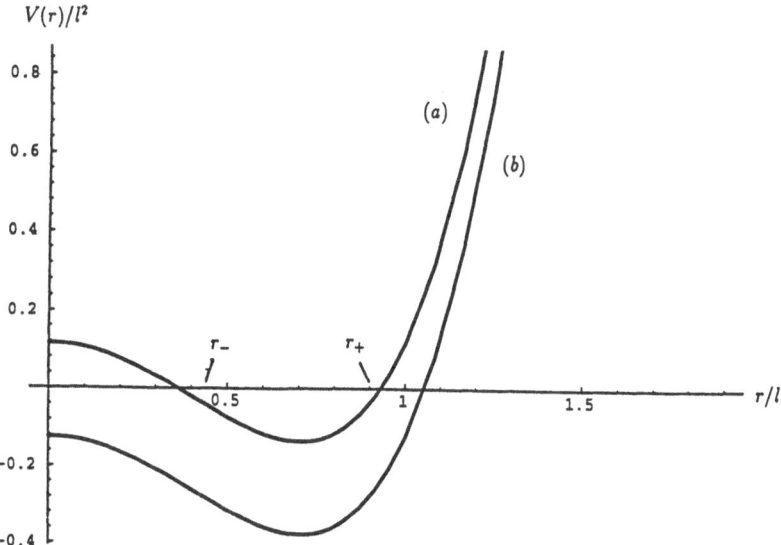

Figure 1. The potential $V(r)$, Eq.(2.21), for a circular string in the $2 + 1$ BH-AdS spacetime. In (a) we have $J^2 > 4E^2$ and a barrier between the inner horizon and $r = 0$, while (b) represents a case where $J^2 < 4E^2$ and a string will always fall into $r = 0$. The static limit is $r_{erg} = l$.

The exact general solution in the three cases (i)-(iii) is given by:

$$r(\tau) = \left| r_m - \frac{1}{c_1 p(\tau - \tau_0) + c_2} \right|, \qquad (2.22)$$

where:

$$r_m = S_{\max} = \sqrt{\frac{Ml^2}{2}} \sqrt{1 + \sqrt{1 - \frac{4V(0)}{Ml^2}}}, \quad V(0) = \frac{J^2}{4} - E^2. \qquad (2.23)$$

c_1, c_2 are constants in terms of (l, M, r_m), given in Ref.[26], and p is the Weierstrass elliptic p-function with invariants (g_2, g_3), discriminant Δ and roots (e_1, e_2, e_3), also given in Ref.[26]. The three cases (i)-(iii) correspond to the cases $\Delta > 0$, $\Delta < 0$ and $\Delta = 0$, respectively. Notice that $S_{\max}^{(ii)} > S_{\max}^{(iii)} = l\sqrt{M} > S_{\max}^{(i)}$. In the case (i), $r(\tau)$ can be written in terms of the Jacobian elliptic function $\mathrm{sn}[\tau^*, k]$, $\tau^* = \sqrt{e_1 - e_3}\,\tau$, $k = \sqrt{(e_2 - e_3)/(e_1 - e_3)}$. It follows that the solution (i) oscillates between the two zeroes r_{01} and r_{02} of $V(r)$, with period 2ω, where ω is the real semi-period of the Weierstrass function: $\omega = K(k)/\sqrt{e_1 - e_3}$, in terms of the complete elliptic integral of the first kind $K(k)$. We have:

$$r(0) = r_m, \quad r(\omega) = \sqrt{Ml^2 - r_m^2}, \quad r(2\omega) = r_m, \ldots \qquad (2.24)$$

In the case (ii) ($\Delta < 0$) two roots (e_1, e_3) become complex, the string collapses into a point $r = 0$ and we have:

$$r(0) = r_m, \quad r(\frac{\omega_2}{2}) = 0, \quad r(\omega_2) = r_m, \ldots \qquad (2.25)$$

where ω_2 is the real semi-period of the Weierstrass function for this case. In the case (iii) ($\Delta = 0$) the elliptic functions reduce to hyperbolic functions:

$$r(\tau) = \frac{\sqrt{M}l}{\cosh(\sqrt{M}\tau)}, \qquad (2.26)$$

so that:

$$r(-\infty) = 0, \quad r(0) = r_m = \sqrt{M}l, \quad r(+\infty) = 0. \qquad (2.27)$$

Here, the string starts as a point, grows until $r = r_m$ (at $\tau = 0$), and then it contracts until it collapses again ($r = 0$) at $\tau = +\infty$. In this case the string makes only one oscillation between $r = 0$ and $r = r_m$.

Notice that for the static background ($J = 0$), the only allowed motion is (ii), i.e. $r_m > r_{\mathrm{hor}} = \sqrt{M}l$ (there is no ergosphere and only one horizon in this case), with:

$$r_m = \sqrt{\frac{Ml^2}{2}} \sqrt{1 + \sqrt{1 + \frac{4E^2}{Ml^2}}}. \qquad (2.28)$$

For $J = 0$, the string collapses into $r = 0$ and stops there. The Penrose diagram of the 2+1 BH-AdS spacetime for $J = 0$ is very similar to the Penrose diagram of the ordinary $(D \geq 4)$ Schwarzschild spacetime, so the string motion outwards from $r = 0$ is unphysical because of the causal structure. The coordinate time $t(\tau)$ can be expressed in terms of the incomplete elliptic integral of the third kind Π, see Ref.[26]. The string has its maximal size r_m at $\tau = 0$, passes the horizon at $\tau = \tau_{\text{hor}}$ (expressed in terms of an incomplete elliptic integral of the first kind) and falls into $r = 0$ for $\tau = \omega_2/2$, ω_2 being the real semi-period of the Weierstrass function. That is, we have:

$$r(0) = r_m, \quad r(\tau_{\text{hor}}) = \sqrt{M}l, \quad r(\omega_2/2) = 0,$$
$$t(0) = 0, \quad t(\tau_{\text{hor}}) = \infty, \tag{2.29}$$

and $t(\omega_2/2)$ can be expressed in terms of the Jacobian zeta function Z, see Ref.[26].

We also study the circular strings in the ordinary $D \geq 4$ spacetimes. In the generic 3+1 Kerr anti de Sitter (or Kerr de Sitter) spacetime, the potential $V(r)$ is however quite complicated, see Ref.[26]. The general circular string solution involves higher genus elliptic functions and it is not necessary to go into details here. We will compare with the non-rotating cases, only.

It is instructive to recall [38] the circular string in Minkowski spacetime (Min), for which $V(r) = r^2 - E^2$ (Fig.2a), the string oscillates between its maximal size $r_m = E$, and $r = 0$ with the solution $r(\tau) = r_m \mid \cos \tau \mid$.

In the Schwarzschild black hole (S) $V(r) = r^2 - 2Mr - E^2$ (Fig.2b), the solution is remarkably simple: $r(\tau) = M + \sqrt{M^2 + E^2} \cos \tau$. The mathematical solution oscillates between $r_m = M + \sqrt{M^2 + E^2}$ and $M - \sqrt{M^2 + E^2} < 0$, but because of the causal structure and the curvature singularity the motion can not be continued after the string has collapsed into $r = 0$.

For anti de Sitter spacetime (AdS), we find: $V(r) = r^2(1 + H^2 r^2) - E^2$ (Fig.2c). The string oscillates between $r_m = \frac{1}{\sqrt{2}H}\sqrt{-1 + \sqrt{1 + 4H^2 E^2}}$ and $r = 0$ with the solution:

$$r(\tau) = r_m \mid \text{cn}[(1 + 4H^2 E^2)^{1/4}, k] \mid, \tag{2.30}$$

which is periodic with period 2ω :

$$\omega = \frac{K(k)}{(1 + 4H^2 E^2)^{1/4}}, \quad k = \sqrt{\frac{\sqrt{1 + 4H^2 E^2} - 1}{2\sqrt{1 + 4H^2 E^2}}}. \tag{2.31}$$

For Schwarzschild anti de Sitter spacetime (S-AdS), we find $V(r) = r^2(1 + H^2 r^2) - 2Mr - E^2$ (Fig.2d) and:

$$r(\tau) = r_m - \frac{1}{d_1 \wp(\tau) + d_2}, \quad r(0) = r_m \tag{2.32}$$

d_1, d_2 are constants given in terms of (M, H, r_m), see Ref.[26] (r_m is the root of the equation $V(r) = 0$, which has in this case exactly one positive solution). The invariants, the discriminant and the roots are also given explicitly in

Ref.[26]. The string starts with $r = r_m$ at $\tau = 0$, it then contracts and eventually collapses into the $r = 0$ singularity. The existence of elliptic function solutions for the string motion is characteristic of the presence of a cosmological constant. For $\Lambda = 0 = -3H^2$ the circular string motion reduces to simple trigonometric functions. From Fig.2 and our analysis we see that the circular string motion is qualitatively very similar in all these backgrounds (Min, S, AdS, S-AdS): the string has a maximal *bounded* size and then it contracts towards $r = 0$. There are however physical and quantitative differences: in Minkowski and pure anti de Sitter spacetimes, the string truly oscillates between r_m and $r = 0$, while in the black hole cases (S, S-AdS), there is only one half oscillation, the string motion stops at $r = 0$. This also holds for the 2+1 BH-AdS spacetime with $J = 0$ (Fig.1b). Notice also that in all these cases, $V(0) = -E^2 < 0$ and $V(r) \sim r^\alpha$; $(\alpha = 2, 4)$ for $r \gg E$.

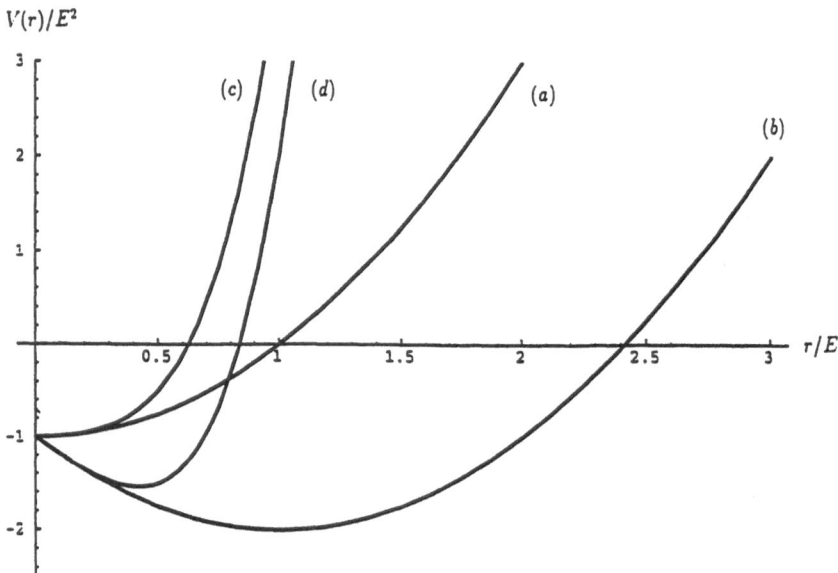

Figure 2. The potential $V(r)$, Eq.(2.18), for a circular string in the equatorial plane of the four $3 + 1$ dimensional spacetimes: (a) Minkowski (Min), (b) Schwarzschild black hole (S), (c) anti de Sitter (AdS), and (d) Schwarzschild anti de Sitter space (S-AdS).

The similarity can be pushed one step further by considering small perturbations around the circular strings. We find [18]:

$$\ddot{C}_n + (n^2 + \frac{r}{2}\frac{da(r)}{dr} + \frac{r^2}{2}\frac{d^2a(r)}{dr^2} - \frac{2E^2}{r^2})C_n = 0, \tag{2.33}$$

determining the Fourier components of the comoving perturbations. For the spacetimes of interest here, $a(r) = 1 - 2M/r + H^2r^2$ (Min, S, AdS, S-AdS), or $a(r) = r^2/l^2 - 1$ (2+1 BH-AdS), the comoving perturbations are regular except near $r = 0$, where we find (for all cases) $r(\tau) \approx -E(\tau - \tau_0)$ and:

$$\ddot{C}_n - \frac{2}{(\tau - \tau_0)^2}C_n = 0. \tag{2.34}$$

It follows that not only the unperturbed circular strings, but also the comoving perturbations around them behave in a similar way in all these non-rotating backgrounds (2+1 and higher dimensional). This should be contrasted with the string perturbations around the center of mass, which behave differently in these backgrounds. It must be noticed that for rotating ($J \neq 0$) spacetimes, the circular string behaviour is qualitatively different from the non-rotating ($J = 0$) spacetimes. For large J, both in the 2+1 BH-AdS as well as in the 3+1 ordinary Kerr-AdS spacetimes, non-collapsing circular string solutions exist. The potential $V(r) \to +\infty$ for $r \to 0$ and no collapse into $r = 0$ is possible.

The dynamics of circular strings in curved spacetimes is determined by the string tension, which tends to contract the string, and by the local gravity (which may be attractive or repulsive). In all the previous non-rotating backgrounds, the local gravity is attractive (i.e. $da(r)/dr > 0$), and it acts together with the string tension in the sense of contraction. But in spacetimes with regions in which repulsion (i.e. $da(r)/dr < 0$) dominates, the strings can expand with unbounded radius (unstable strings [11, 13]). It may also happen that the string tension and the local gravity be of the same order, i.e. the two opposite effects can balance, and the string is stationary. De Sitter spacetime provides an example in which all such type of solutions exist [11, 13]. In de Sitter spacetime, $V(r)$ is unbounded from below for $r \to \infty$ ($V(r) \sim -r^4$) and unbounded expanding circular strings are present. In addition, an interesting new feature appears in the presence of a positive cosmological constant: the existence of *multi-string* solutions [9-11]. The world-sheet time τ turns out to be a multi (finite or infinite) valued function of the physical time. That is, one single world-sheet where $-\infty \leq \tau \leq +\infty$, can describe many (even infinitely many [13]) different and independent strings (in flat spacetime, one single world-sheet describes only one string). In the S, AdS and S-AdS spacetimes, the multi-string feature for circular strings is absent.

The main conclusions of this section are given in Tables 1, 2. Further details can be found in Ref.[26]

Table 1. String motion described by the string perturbation series approach in the 2+1 BH-AdS, ordinary black hole-AdS, de Sitter (dS) and black string backgrounds.

String Perturbation Series Approach
$x^\mu(\tau,\sigma) = q^\mu(\tau) + \eta^\mu(\tau,\sigma) + \xi^\mu(\tau,\sigma) + \ldots$ $\eta^\mu = n^\mu_R \delta x^R, \quad \delta x^R(\tau,\sigma) = \sum_n C^R_n(\tau)e^{-in\sigma}; \quad R = 1,\ldots,(D-1)$ $\ddot{C}_{nR} + (n^2\delta_{RS} - R_{\mu\rho\sigma\nu}n^\mu_R n^\nu_S \dot{q}^\rho \dot{q}^\sigma)C^S_n = 0$

2+1 Black Hole-ADS	Ordinary ($D \geq 4$) Black Hole-ADS
$\ddot{C}_{n\perp} + (n^2 + \frac{m^2}{l^2})C_{n\perp} = 0$ $\ddot{C}_{n\parallel} + (n^2 + \frac{m^2}{l^2})C_{n\parallel} = 0$	$\ddot{C}_{n\perp} + (n^2 + m^2 H^2 + \frac{Mm^2}{r^3})C_{n\perp} = 0$ $\ddot{C}_{n\parallel} + (n^2 + m^2 H^2 - \frac{2Mm^2}{r^3})C_{n\parallel} = 0$
No instability, $\omega_n = \sqrt{n^2 + \frac{m^2}{l^2}}$	Instability, $r_{\text{inst}} = (\frac{2Mm^2}{1+m^2H^2})^{1/3}$
$\delta x^R, \xi^\mu$ bounded everywhere. $\delta x_R = \sum_n[A_{nR}e^{-i(n\sigma+\omega_n\tau)} + \tilde{A}_{nR}e^{-i(n\sigma-\omega_n\tau)}]$	Unbounded perturbations $C_{n\parallel}$ for $r(\tau) \approx (3m\sqrt{M/2})^{2/3}(\tau_0-\tau)^{2/3} \to 0$: $\ddot{C}_{n\parallel} - \frac{4}{9(\tau_0-\tau)^2}C_{n\parallel} = 0$
$m^2 = 2\sum_n(2n^2 + \frac{m^2}{l^2})\sum_R A_{nR}\tilde{A}_{-nR}$	$m^2(H=0) = 4\sum_n n^2 \sum_R A_{nR}\tilde{A}_{-nR}$

Ordinary ($D \geq 4$) de Sitter	Black String
$\ddot{C}_{nR} + (n^2 - m^2 H^2)C_{nR} = 0$ Unbounded modes for $\mid n \mid < mH$	$\ddot{C}_{n\perp} + n^2 C_{n\perp} = 0$ $\ddot{C}_{n\parallel} + (n^2 - \frac{2Mm^2}{lr})C_{n\parallel} = 0$
Instability, $\omega_n = \sqrt{n^2 - m^2 H^2}$	Instability, $r_{\text{inst}} = \frac{2Mm^2}{l}$
	Unbounded perturbations $C_{n\parallel}$ for $r(\tau) \approx \frac{Mm^2}{l}(\tau_0-\tau)^2 \to 0$: $\ddot{C}_{n\parallel} - \frac{2}{(\tau_0-\tau)^2}C_{n\parallel} = 0$
$\delta x_R = \sum_n[A_{nR}e^{-i(n\sigma+\omega_n\tau)} + \tilde{A}_{nR}e^{-i(n\sigma-\omega_n\tau)}]$	
$m^2 = 2\sum_n(2n^2 - m^2 H^2)\sum_R A_{nR}\tilde{A}_{-nR}$	$m^2 = 4\sum_n n^2 \sum_R A_{nR}\tilde{A}_{-nR}$

Table 2. Circular exact string solutions in the indicated backgrounds. $S(\tau)$ is the invariant string size.

Exact Circular Strings
$ds^2 = g_{tt}(r)dt^2 + g_{rr}(r)dr^2 + 2g_{t\phi}(r)dtd\phi + g_{\phi\phi}d\phi^2$ $t = t(\tau), \quad r = r(\tau), \quad \phi = \sigma + f(\tau), \quad (\theta = \pi/2)$ $\dot{r}^2 + V(r) = 0; \quad V(r) = g^{rr}(E^2 g^{tt} + g_{\phi\phi}), \quad \dot{t} = -E g^{tt}, \quad \dot{f} = -E g^{t\phi}$ $ds^2 = g_{\phi\phi}(-d\tau^2 + d\sigma^2), \quad \text{i.e.} \quad S(\tau) = \sqrt{g_{\phi\phi}(r(\tau))}$

2+1 Black Hole ADS	Ordinary $(D \geq 4)$ Black Hole ADS
$V(r) = r^2(r^2/l^2 - M) + J^2/4 - E^2$ $V(0) = J^2/4 - E^2, \quad V(r \to \infty) \propto r^4 \to \infty$	$V(r) \doteq r^2(1 + H^2 r^2) - 2Mr - E^2$ $V(0) = -E^2, \quad V(r \to \infty) \propto r^4 \to \infty$
$S(\tau) = r(\tau) = r_m - (c_1 p(\tau - \tau_0) + c_2)^{-1}$ $S_{\max} = r_m = \sqrt{\frac{Ml^2}{2}} \sqrt{1 + \sqrt{1 - \frac{4V(0)}{Ml^2}}}$	$S(\tau) = r(\tau) = r_m - (c_1 p(\tau - \tau_0) + c_2)^{-1}$ $V(r_m) \equiv 0; \quad S_{\max} = r_m = r_m(M, H, E)$
For $J^2 > 4E^2$, $S_{\min} > 0$, no collapse. For $J^2 \leq 4E^2$, $S_{\min} = 0$, collapse.	String contracts from $r(0) = r_m$ until it collapses into the $r = 0$ singularity.
No unbounded string size. No multi-string solutions.	No unbounded string size. No multi-string solutions.

Ordinary $(D \geq 4)$ Black Hole DS	Black String
$V(r) = r^2(1 - H^2 r^2) - 2Mr - E^2$ $V(0) = -E^2, \quad V(r \to \infty) \propto -r^4 \to -\infty$	$\bar{V}(r) = J^2/(4r^2) - M/r^2 + 1/l^2 - E^2$ $\bar{V}(0) = \infty \ (J \neq 0), \quad \bar{V}(0) = -\infty \ (J = 0)$ $\bar{S}(\tau) = 1/r(\tau), \quad \bar{V}(\infty) = 1/l^2 - E^2$
Contracting, expanding or stationary solutions, depending on the balance between the string tension and the local gravity.	For $J \neq 0$, all solutions are bounded (finite $\bar{S}$). For $J = 0$ unbounded size solutions exist as well.
$M = 0$: $S^2(\tau) = H^{-2} p(\tau - \tau_0) + H^{-2}/3$ Contracting, expanding or oscillating.	$J = 0, El = 1$: $\bar{S}(\tau) = (|2\sqrt{M}\tau|)^{-1/2}$ $\bar{S}(-\infty) = 0, \ \bar{S}(0) = \infty, \ \bar{S}(\infty) = 0$
Unbounded string size. Multi-string solutions.	Unbounded string size. Multi-string solutions.

3   Quantum String Spectrum in Ordinary AdS

In the first part of the previous section, the classical string dynamics was studied in the $2+1$ BH-AdS spacetime. In this section we go further in the investigation of the physical properties of strings in AdS spacetime, by performing string quantization. Since the $2+1$ BH-AdS spacetime is locally AdS, the results of Section 2 can be extended to classical strings in D-dimensional AdS spacetime as well. In AdS spacetime, the string motion is oscillatory in time and is *stable*; all fluctuations around the string center of mass are well behaved and bounded. Local gravity of AdS spacetime is always negative and string instabilities do not develop. The string perturbation series approach, considering fluctuations around the center of mass, is particularly appropriate in AdS spacetime, the natural dimensionless expansion parameter being $\lambda = \alpha'/l^2 > 0$, where α' is the string tension and the 'Hubble constant' $H = 1/l$. The negative cosmological constant of AdS spacetime is related to the 'Hubble constant' H by $\Lambda = -(D-1)(D-2)H^2/2$.

All the spatial ($\mu = 1, ..., D-1$) modes in D-dimensional AdS, oscillate with frequency $\omega_n = \sqrt{n^2 + m^2\alpha'\lambda}$, which are real for all n (m being the string mass). In this section, we perform a canonical quantization procedure. From the conformal generators L_n, $\tilde{L}_n$ and the constraints $L_0 = \tilde{L}_0 = 0$, imposed at the quantum level, we obtain the mass formula (the quantum version of Eq.(2.7) written in units where α' appears explicitly):

$$m^2\alpha' = (D-1)\sum_{n>0}\Omega_n(\lambda) + \sum_{n>0}\Omega_n(\lambda)\sum_{R=1}^{D-1}[(\alpha_n^R)^\dagger\alpha_n^R + (\tilde{\alpha}_n^R)^\dagger\tilde{\alpha}_n^R], \quad (3.1)$$

where:

$$\Omega_n(\lambda) = \frac{2n^2 + m^2\alpha'\lambda}{\sqrt{n^2 + m^2\alpha'\lambda}}, \quad (3.2)$$

and we have applied symmetric ordering of the operators. The operators α_n^R, $\tilde{\alpha}_n^R$ satisfy:

$$[\alpha_n^R, (\alpha_n^R)^\dagger] = [\tilde{\alpha}_n^R, (\tilde{\alpha}_n^R)^\dagger] = 1, \quad \text{for all } n > 0. \quad (3.3)$$

For $\lambda << 1$, which is clearly fulfilled in most interesting cases, we have found the lower mass states $m^2\alpha'\lambda << 1$ and the quantum mass spectrum [27]. Physical states are characterized by the eigenvalue of the number operator:

$$N = \frac{1}{2}\sum_{n>0} n \sum_{R=1}^{D-1}[(\alpha_n^R)^\dagger\alpha_n^R + (\tilde{\alpha}_n^R)^\dagger\tilde{\alpha}_n^R], \quad (3.4)$$

and the ground state is defined by:

$$\alpha_n^R \,|\, 0 >= \tilde{\alpha}_n^R \,|\, 0 >= 0, \quad \text{for all } n > 0. \quad (3.5)$$

We find that $m^2\alpha' = 0$ is an *exact* solution of the mass formula in $D = 25$ and that there is a graviton at $D = 25$, which indicates, as in de Sitter space [7], that the critical dimension of AdS is 25 (although it should be stressed that the question whether these spacetimes are solutions to the full β-function equations remains open). As in Minkowski spacetime, the ground state is a tachyon. Remarkably enough, for $N \geq 2$ we find that a generic feature of all excited states beyond the graviton, is the presence of a *fine structure* effect: for a given eigenvalue $N \geq 2$, the corresponding states have different masses. For the lower mass states the expectation value of the mass operator in the corresponding states (generically labelled $\mid j >$) turns out to have the form (see Ref.[27]):

$$< j \mid m^2\alpha' \mid j >_{\text{AdS}} = a_j + b_j\lambda^2 + c_j\lambda^3 + \mathcal{O}(\lambda^4). \tag{3.6}$$

The collective index "j" generically labels the state $\mid j >$ and the coefficients a_j, b_j, c_j are all well computed numbers, different for each state, their precise values are given in Ref.[27]. The corrections to the mass in Minkowski spacetime appear to order λ^2. Therefore, the leading Regge trajectory for the lower mass states is:

$$J = 2 + \frac{1}{2}m^2\alpha' + \mathcal{O}(\lambda^2). \tag{3.7}$$

In Minkowski spacetime the mass and number operator of the string are related by $m^2\alpha' = -4 + 4N$. In AdS (as well as in de Sitter (dS)), there is no such simple relation between the mass and the number operators; the splitting of levels increases considerably for very large N. The *fine structure* effect we find here is also present in de Sitter space. Up to order λ^2, the lower mass states in dS and in AdS are the same, the differences appear to the order λ^3. The lower mass states in de Sitter spacetime are given by Eq.(3.6) but with the λ^3-term getting an opposite sign $(-c_j\lambda^3)$.

For the very high mass spectrum, we find more drastic effects. States with very large eigenvalue N, namely $N >> 1/\lambda$, have masses:

$$< j \mid m^2\alpha' \mid j >_{\text{AdS}} \approx d_j\lambda N^2 \tag{3.8}$$

and angular momentum:

$$J^2 \approx \frac{1}{\lambda}m^2\alpha', \tag{3.9}$$

where d_j are well computed numbers different for each state, their precise values are given in Ref.[27]. Since $\lambda = \alpha'/l^2$, we see from Eq.(3.8) that the masses of the high mass states are *independent* of α'. In Minkowski spacetime, very large N states all have the same mass $m^2\alpha' \approx 4N$, but here in AdS the masses of the high mass states with the same eigenvalue N are *different* by factors d_j. In addition, because of the fine structure effect, states with different N can get mixed up. For high mass states, the level spacing *grows* with N (instead of being constant as in Minkowski spacetime). As a consequence, the density of states $\rho(m)$ as a function of mass grows like $\text{Exp}[(m/\sqrt{\mid \Lambda \mid})^{1/2}]$ (instead of $\text{Exp}[m\sqrt{\alpha'}]$ as in Minkowski spacetime), and *independently* of α'. The partition function for a gas of strings at a temperature β^{-1} in AdS spacetime is well defined for all finite temperatures β^{-1}, discarding the existence of the Hagedorn temperature.

For the low mass states ($m^2\alpha'\lambda \ll 1$) in anti de Sitter spacetime, our results can be written as [27]:

$$< j \mid m^2(\alpha',l) \mid j >= \frac{4(N-1)}{\alpha'} + \frac{1}{l^2}\sum_{n=0}^{\infty} a_{jn}(\alpha'/l^2)^n, \qquad (3.10)$$

where "j" is a collective index labelling the state $\mid j >$. It is now important to notice that $a_{j0} = 0$ for *all* the low mass states [27], i.e. there is no "constant" term on the right hand side of Eq.(3.10). A non-zero a_{j0}-term would give rise to a α'-independent contribution to the string mass. Its absense, on the other hand, means that the first term on the right hand side of Eq.(3.10) is super-dominant (since, in all cases, $\alpha'/l^2 = \lambda \ll 1$) and that the string scale is therefore set by $1/\alpha'$.

For the high mass states ($m^2\alpha'\lambda \gg 1$) we found instead [27]:

$$< j \mid m^2(\alpha',l) \mid j >\approx \frac{d_j}{l^2}N^2, \qquad \text{for } N \gg l^2/\alpha' \qquad (3.11)$$

where the number d_j depends on the state. The right hand side of Eq.(3.11) is exactly like a non-zero dominant a_{j0}-term in Eq.(3.10). For the high mass states the scale is therefore set by $1/l^2$ which is equal to the absolute value of the cosmological constant Λ (up to a geometrical factor) and *independent* of α'. This suggests that for $\lambda \ll 1$, the masses of *all* string states can be represented by a formula of the form (3.10). For the low mass states $a_{j0} = 0$, while for the high mass states a_{j0} becomes a large positive number.

In the black string background, Eq.(2.2), we have calculated explicitly the first and second order string fluctuations around the center of mass [27]. We then determined the world-sheet energy-momentum tensor and we derived the mass formula in the asymptotic region. The mass spectrum is equal to the mass spectrum in flat Minkowski spacetime. Therefore, for a gas of strings at temperature β^{-1} in the asymptotic region of the black string background, the partition function goes like:

$$Z(\beta) \sim \int^{\infty} dm \; e^{-m(\beta-\sqrt{\alpha'})}, \qquad (3.12)$$

which is only defined for $\beta > \sqrt{\alpha'}$, i.e. there is a Hagedorn temperature:

$$T_{\text{Hg}} = (\alpha')^{-1/2}. \qquad (3.13)$$

In higher dimensional ($D \geq 4$) black hole spacetimes the next step now would be to set up a scattering formalism, where a string from an asymptotic in-state interacts with the gravitational field of the black hole and reappears in an asymptotic out-state [15]. However, this is not possible in the black string background. In the uncharged black string background under consideration here, *all* null and timelike geodesics incoming from spatial infinity pass through the horizon and fall into the physical singularity [36]. No "angular momentum", as in the case of scattering off the ordinary Schwarzschild black hole, can prevent a point particle from falling into the singularity. The string solutions considered here are based on perturbations around the string center of mass which follows, at least approximately, a point particle geodesic. Therefore, a string incoming from spatial infinity inevitably falls into the singularity in the black string background.

Table 3. Characteristic features of the quantum string mass spectrum in anti de Sitter (AdS) and de Sitter (dS) spacetimes.

Quantum Strings	
Anti de Sitter spacetime (AdS)	de Sitter spacetime (dS)
Classical motion is stable and oscillatory in time with real frequencies $\omega_n = \sqrt{n^2 + m^2\alpha'^2H^2}$	Classical motion is unstable with frequencies $\omega_n = \sqrt{n^2 - m^2\alpha'^2H^2}$ Unbounded string size and energy for large de Sitter radius, $R \to \infty$.
The mass formula is well defined for all m. There is an infinite number of states with arbitrary high masses. $m^2\alpha' = 0$ is an exact solution at $D = 25$.	Real mass solutions only for $m < 1/(\alpha'H)$. Finite number of states, $N_{\max} \approx 0.15/(\alpha'H^2)$. $m^2\alpha' = 0$ is an exact solution at $D = 25$.
The coupling to the gravitational background produces a Fine structure effect at all levels in the mass spectrum. The number of levels considerably increases with respect to flat space.	Fine structure effect appears in low mass spectrum. Is similar to AdS; the differences appear to order $(\alpha'H^2)^3$
For the high mass states: $< m^2 > \sim \mid \Lambda \mid N^2, \quad J^2 \sim m^2/\mid \Lambda \mid$ Both are *independent* of α' ! The level spacing grows with N. $\rho(m) \sim \mathrm{Exp}[m/\sqrt{\mid \Lambda \mid})^{1/2}]$, high m. No Hagedorn temperature exists.	The similar region of high mass states does not exist in the de Sitter spacetime.

4 New Classes of Exact String and Multi-String Solutions in Curved Spacetimes

In this section we return to the string equations of motion and constraints, Eqs.(1.1)-(1.2), to look for new classes of exact solutions. In most spacetimes, quite general families of exact solutions can be found by making an appropriate ansatz, which exploits the symmetries of the underlying curved spacetime. In axially symmetric spacetimes, a convenient ansatz corresponds to circular strings, as we saw in Section 2. Such an ansatz effectively decouples the dependence on the spatial world-sheet coordinate σ, and the string equations of motion and constraints reduce to non-linear coupled ordinary differential equations, Eqs.(2.18)-(2.19). In this section we will make instead an ansatz which effectively decouples the dependence on the temporal world-sheet coordinate τ. This ansatz, which we call the "stationary string ansatz" is dual to the "circular string ansatz" in the sense that it corresponds to a formal interchange of the world-sheet coordinates (τ, σ), as well as of the azimuthal angle ϕ and the stationary time t in the target space:

$$\tau \leftrightarrow \sigma, \qquad t \leftrightarrow \phi. \tag{4.1}$$

The stationary string ansatz will describe stationary strings when t (and τ) are timelike, for instance in anti de Sitter spacetime (in static coordinates) and outside the horizon of a Schwarzschild black hole. On the other hand, if t (and τ) are spacelike, for instance inside the horizon of a Schwarzschild black hole or outside the horizon of de Sitter spacetime (in static coordinates), the stationary string ansatz will describe dynamical propagating strings. Considering for simplicity a static line element in the form:

$$ds^2 = -a(r)dt^2 + \frac{dr^2}{a(r)} + r^2(d\theta^2 + \sin^2\theta d\phi^2), \tag{4.2}$$

the stationary string ansatz reads explicitly:

$$t = \tau, \quad r = r(\sigma), \quad \phi = \phi(\sigma), \quad \theta = \pi/2. \tag{4.3}$$

The string equations of motion and constraints, Eqs.(1.1)-(1.2), reduce to two separated first order ordinary differential equations:

$$\phi' = \frac{L}{r^2}, \quad r'^2 + V(r) = 0; \quad V(r) = -a(r)[a(r) - \frac{L^2}{r^2}], \tag{4.4}$$

where L is an integration constant. The qualitative features of the possible string configurations can be read off directly from the shape of the potential $V(r)$. Thereafter, the detailed analysis of the quantitative features can be performed by explicitly solving the (integrable) system of equations (4.4). The induced line element on the world-sheet is given by:

$$ds^2 = a(r)(-d\tau^2 + d\sigma^2). \tag{4.5}$$

Thus, if $a(\tau)$ is negative, the world-sheet coordinate τ becomes spacelike while σ becomes timelike and the stationary string ansatz (4.3) describes a dynamical string. If $a(\tau)$ is positive, the ansatz describes a stationary string.

In this section we solve explicitly Eqs.(4.4) and we analyze in detail the solutions and their physical interpretation in Minkowski, de Sitter, anti de Sitter, Schwarzschild and $2+1$ black hole anti de Sitter spacetimes. In all these cases, the solutions are expressed in terms of elliptic (or elementary) functions. We furthermore analyze the physical properties, energy and pressure, of these solutions. The energy and pressure densities of the strings can be obtained from the $3+1$ dimensional spacetime energy-momentum tensor:

$$\sqrt{-g}T^{\mu\nu} = \frac{1}{2\pi\alpha'} \int d\tau d\sigma (\dot{X}^\mu \dot{X}^\nu - X'^\mu X'^\nu) \delta^{(4)}(X - X(\tau, \sigma)). \qquad (4.6)$$

After integration over a spatial volume that completely encloses the string [39], the energy-momentum tensor takes the form of a fluid:

$$T^\mu{}_\nu = \text{diag.}(-E, P_1, P_2, P_3). \qquad (4.7)$$

In Minkowski spacetime (Min), the potential defined in Eqs.(4.4) is given by:

$$V_{\text{M}}(r) = \frac{L^2}{r^2} - 1, \qquad (4.8)$$

and the solution of Eqs.(4.4) describes one infinitely long straight string with "impact-parameter" L. The equation of state, relating energy and pressure densities, takes the well-known form Ref.[28]:

$$dE = -dP_2 = \text{const.}, \quad P_1 = P_3 = 0. \qquad (4.9)$$

In anti de Sitter spacetime (AdS), the potential is given by:

$$V_{\text{AdS}}(r) = -(1 + H^2 r^2)[1 + H^2 r^2 - \frac{L^2}{r^2}]. \qquad (4.10)$$

The radial coordinate $r(\sigma)$ is periodic with finite period T_σ, which can be expressed in terms of a complete elliptic integral, see Ref.[28]. For $\sigma \in [0, T_\sigma]$, the solution describes an infinitely long stationary string in the wedge $\phi \in]0, \Delta\phi[$, where:

$$\Delta\phi = 2k\sqrt{\frac{1 - 2k^2}{1 - k^2}}[\Pi(1 - k^2, k) - K(k)] \in]0, \pi[\qquad (4.11)$$

The elliptic modulus k, which is a function of HL [28], parametrizes the solutions, $k \in]0, 1/\sqrt{2}[$. The azimuthal angle is generally not a periodic function of σ, thus when the spacelike world-sheet coordinate σ runs through the range $] - \infty, +\infty[$, the solution describes an *infinite number* of infinitely long stationary open strings. The general solution is therefore a *multi-string* solution. Until now multi-string solutions were only found in de Sitter space-time [11-13]. Our results show that multi-string solutions are a general feature of spacetimes with a cosmological constant (positive or negative). The solution in anti de Sitter spacetime describes a *finite* number of strings if the following relation holds:

$$N\Delta\phi = 2\pi M. \tag{4.12}$$

Here N and M are integers, determining the number of strings and the winding in azimuthal angle, respectively, for the multi-string solution, see Fig.3. The equation of state for a full multi-string solution takes the form ($P_3 = 0$):

$$dP_1 = dP_2 = -\frac{1}{2}dE, \quad \text{for } r \to \infty \tag{4.13}$$

corresponding to extremely unstable strings [21].

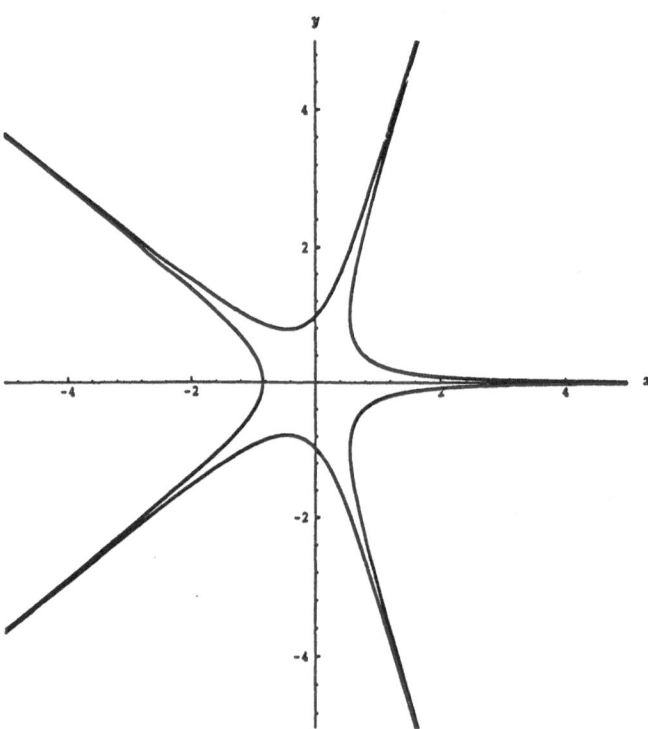

Figure 3. The $(N, M) = (5, 1)$ multi-string solution in anti de Sitter space-time. The (N, M) multi-string solutions describe N stationary strings with M windings in the Azimuthal angle ϕ, in anti de Sitter spacetime.

In de Sitter spacetime (dS), the potential is given by:

$$V_{dS}(r) = -(1 - H^2 r^2)[1 - H^2 r^2 - \frac{L^2}{r^2}]. \qquad (4.14)$$

In this case we have to distinguish between solutions inside the horizon (where τ is timelike) and solutions outside the horizon (where τ is spacelike). Inside the horizon, the generic solution describes one infinitely long open stationary string winding around $r = 0$. For special values of the constants of motion, corresponding to a relation, which formally takes the same form as Eq.(4.12), the solution describes a closed string of finite length $l = N\pi/H$. The integer N in this case determines the number of "leaves", see Fig.4. The energy is positive and finite and grows with N. The pressure turns out to vanish identically, thus the equation of state corresponds to cold matter.

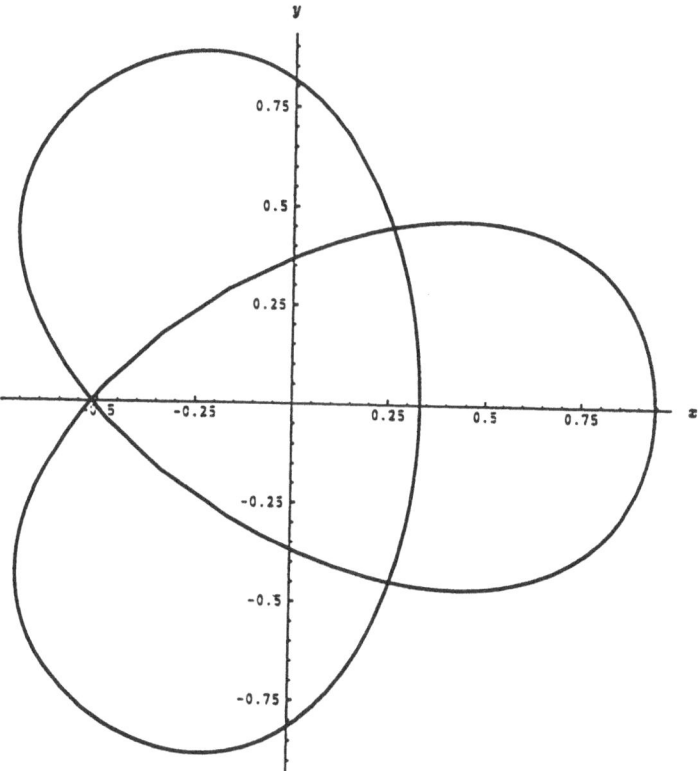

Figure 4. The $(N, M) = (3, 2)$ stationary string solution inside the horizon of de Sitter spacetime. Besides the circular string, this is the simplest stationary closed string configuration in de Sitter spacetime.

Outside the horizon, the world-sheet coordinate τ becomes spacelike while σ becomes timelike, thus we define:

$$\bar{\tau} \equiv \sigma, \qquad \bar{\sigma} \equiv \tau, \tag{4.15}$$

and the string solution is conveniently expressed in hyperboloid coordinates. The radial coordinate $r(\bar{\tau})$ is periodic with a finite period $T_{\bar{\tau}}$. For $\bar{\tau} \in [0, T_{\bar{\tau}}]$, the solution describes a straight string incoming non-radially from spatial infinity, scattering at the horizon and escaping towards infinity again, Fig.5.

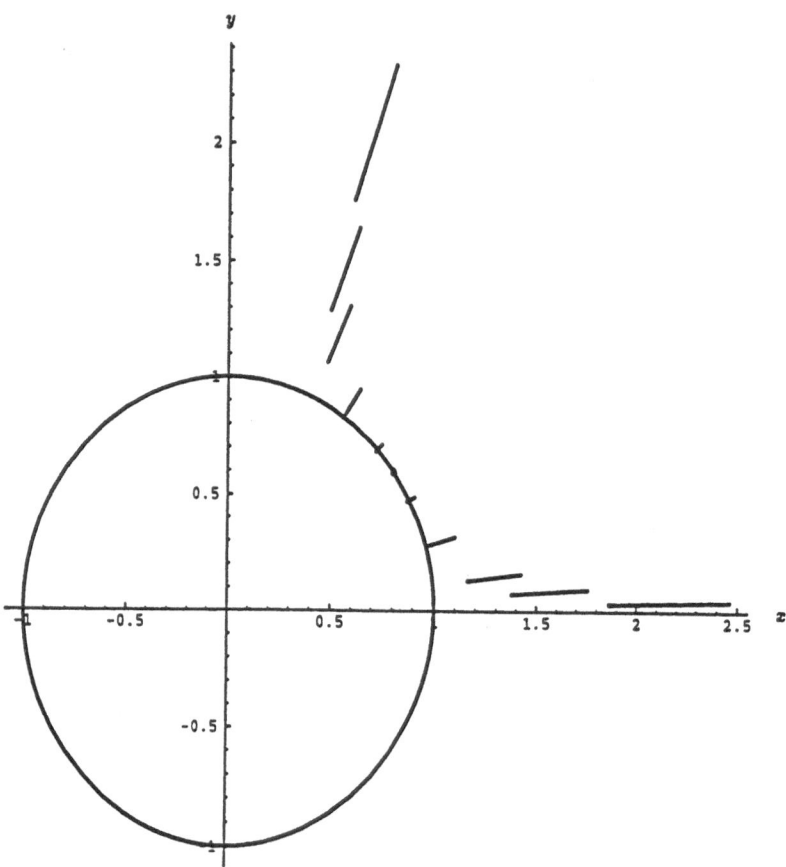

Figure 5. Schematic representation of the time evolution of the $(N, M) = (5, 1)$ dynamical multi-string solution, outside the horizon of de Sitter space-time. Only one of the 5 strings is shown; the others are obtained by rotating the figure by the angles $2\pi/5$, $4\pi/5$, $6\pi/5$ and $8\pi/5$. During the "scattering" at the horizon, the string collapses to a point and re-expands.

The string length is zero at the horizon and grows indefinetely in the asymptotic regions. As in the case of anti de Sitter spacetime, the azimuthal angle is generally not a periodic function, thus when the timelike world-sheet coordinate $\bar{\tau}$ runs through the range $]-\infty, +\infty[$, the solution describes an *infinite* number of dynamical straight strings scattering at the horizon at different angles. The general solution is therefore a multi-string solution. In particular, a multi-string solution describing a *finite* number of strings is obtained if a relation of the form (4.12) is fulfilled. It turns out that the solution describes *at least* three strings. The energy and pressures of a full multi-string solution have also been computed. In the asymptotic region they fulfill an equation of state corresponding to extremely unstable strings, i.e. like Eq.(4.13).

In the Schwarzschild black hole background (S), the potential is given by:

$$V_S(r) = -(1 - 2m/r)[1 - 2m/r - \frac{L^2}{r^2}]. \qquad (4.16)$$

No multi-string solutions are found in this case. Outside the horizon the solution of Eqs.(4.4) describes one infinitely long stationary open string. This solution was already derived in Ref.[40], and we shall not discuss it here. Inside the horizon, where τ becomes spacelike while σ becomes timelike, we make the redefinitions (4.15) and the solution is conveniently expressed in terms of Kruskal coordinates, see Ref.[28]. The solution describes one straight string infalling *non-radially* towards the singularity, see Fig.6. At the horizon, the string length is zero and it grows *indefinitely* when the string approaches the spacetime singularity. Thereafter, the solution can not be continued, see Ref.[28].

In the $2+1$ black hole anti de Sitter spacetime (BH-AdS) [30], the potential is given by:

$$V_{\text{BH-AdS}}(r) = -(\frac{r^2}{l^2} - 1)[\frac{r^2}{l^2} - 1 - \frac{L^2}{r^2}]. \qquad (4.17)$$

Outside the horizon, the solutions "interpolate" between the solutions found in anti de Sitter spacetime and outside the horizon of the Schwarzschild black hole. The solutions thus describe infinitely long stationary open strings. As in anti de Sitter spacetime, the general solution is a *multi-string* describing *infinitely* many strings. In particular, for certain values of the constants of motion, corresponding to the condition of the form (4.12), the solution describes a finite number of strings. In the simplest version of the $2+1$ BH-AdS background ($M = 1$, $J = 0$), it turns out that the solution describes *at least* seven strings, see Ref.[28]. Inside the horizon, we make the redefinitions (4.15) and the solution is conveniently expressed in terms of Kruskal-like coordinates, see Ref.[28]. The solution is similar to the solution found inside the horizon of the Schwarzschild black hole, but there is one important difference at $r = 0$. As in the Schwarzschild black hole background, the solution describes one straight string infalling non-radially towards $r = 0$, and beyond this point the solution can not be continued because of the global structure of the spacetime. At the horizon the string size is zero and during the fall towards $r = 0$, the string size *grows* but stays *finite*. This should be compared with the straight string inside the horizon of the Schwarzschild black hole, where the string size grows indefinitely. The physical reason for this difference is that the point $r = 0$ is not a strong curvature singularity in the $2 + 1$ BH-AdS spacetime.

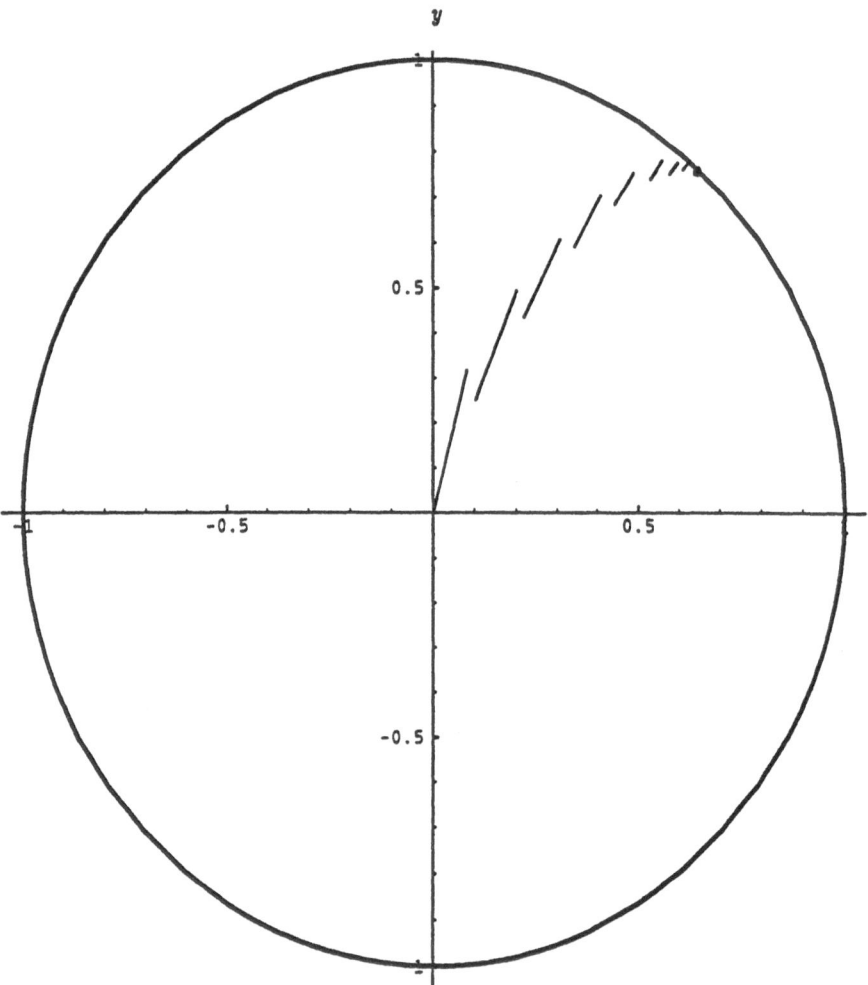

Figure 6. The dynamical straight string inside the horizon of the Schwarzschild black hole. The string infalls non-radially towards the singularity with indefinetely growing length.

We have considered also the Schwarzschild de Sitter and Schwarzschild anti de Sitter spacetimes. These spacetimes contain all the features of the spacetimes already discussed: singularities, horizons, positive or negative cosmological constants. All the various types of string solutions found in the other spacetimes (open, closed, straight, finitely and infinitely long, multi-strings) are therefore present in the different regions of the Schwarzschild-de Sitter and Schwarzschild-anti de Sitter spacetimes. The details are given in Ref.[28].

We close this section with a few remarks on the stability of the solutions. Generally, the question of stability must be addressed by considering small perturbations around the exact solutions. In Ref.[10], a covariant formalism describing physical perturbations propagating along an arbitrary string configuration embedded in an arbitrary curved spacetime, was developed. The resulting equations determining the evolution of the perturbations are however very complicated in the general case, although partial (analytical) results have been obtained in special cases for de Sitter [18, 43] and Schwarzschild black hole [10, 43, 44] spacetimes. The exact solutions found in this section fall essentially into two classes: dynamical and stationary. The dynamical string solutions outside the horizon of de Sitter (or S-dS) and inside the horizon of Schwarzschild (or S-dS, S-AdS) spacetimes, are already unstable at the zeroth order approximation (i.e. without including small perturbations), in the sense that their physical length grows indefinetely. For the stationary string solutions the situation is more delicate. The existence of the stationary configurations is based on an exact balance between the string tension and the local attractive or repulsive gravity. For that reason, it can be expected that the configurations are actually unstable for certain modes of perturbation, especially in strong curvature regions.

The main conclusions of this section are given in Table 4. Further details can be found in Ref.[28].

Table 4. Short summary of the features of the string solutions found in this section. In anti de Sitter spacetime and outside the horizon of de Sitter and $2+1$ BH-AdS spacetimes, the solutions describe a *finite* number of strings provided a condition of the form $N\Delta\phi = 2\pi M$ is fulfilled, where $\Delta\phi$ is the angle between the "arms" of the string and (N, M) are integers.

Line element:	$ds^2 = -a(r)dt^2 + \frac{dr^2}{a(r)} + r^2(d\theta^2 + \sin^2\theta d\phi^2)$
Ansatz:	$t = \tau, \quad r = r(\sigma) \quad \phi = \phi(\sigma), \quad \theta = \pi/2$
String solution:	$\phi' = \frac{L}{r^2} \quad r'^2 + V(r) = 0; \quad V(r) = -a(r)[a(r) - \frac{L^2}{r^2}]$
String length element:	$dl = \sqrt{a(r(\sigma))}\, d\sigma$

MINKOWSKI, $a(r) = 1, \quad V(r) = \frac{L^2}{r^2} - 1$

The solution describes one infinitely long stationary straight string.

$\frac{dE}{dl} = -\frac{dP_1}{dl} = \text{const.}, \qquad P_1 = P_3 = 0$

ANTI DE SITTER, $a(r) = 1 + H^2 r^2, \quad V(r) = -(1 + H^2 r^2)[1 + H^2 r^2 - \frac{L^2}{r^2}]$

The solution describes a finite or infinite number of infinitely long stationary strings.

$\frac{dE}{dl} = -2\frac{dP_1}{dl} = -2\frac{dP_3}{dl}, \qquad P_3 = 0, \quad \text{(asymptotically for } r \to \infty)$

DE SITTER, $a(r) = 1 - H^2 r^2, \quad V(r) = -(1 - H^2 r^2)[1 - H^2 r^2 - \frac{L^2}{r^2}]$

Inside the horizon, the solution describes one finitely or infinitely long stationary string winding around $r = 0$.

$P_1 = P_2 = P_3 = 0, \quad$ as cold matter, $\quad E$ expressed in terms of elliptic functions.

Outside the horizon, the solution describes a finite or infinite number of dynamical straight strings "scattering at the horizon". The string length vanishes at the horizon, but stretches indefinetely at spatial infinity.

$\frac{dE}{dl} = -2\frac{dP_1}{dl} = -2\frac{dP_3}{dl}, \qquad P_3 = 0, \quad \text{(asymptotically for } r \to \infty)$

SCHWARZSCHILD, $a(r) = 1 - 2m/r, \quad V(r) = -(1 - 2m/r)[1 - 2m/r - \frac{L^2}{r^2}]$

Outside the horizon, the solution describes one infinitely long stationary string.

Inside the horizon, the solution describes one dynamical straight string. The string size is zero at the horizon, and grows *indefinetely* as the string falls towards $r = 0$.

2+1 BLACK HOLE-AdS, $a(r) = \frac{r^2}{l^2} - 1, \quad V(r) = -(\frac{r^2}{l^2} - 1)[\frac{r^2}{l^2} - 1 - \frac{L^2}{r^2}]$

Outside the horizon, the solution describes a finite or infinite number of infinitely long stationary string.

Inside the horizon, the solution describes one dynamical straight string. The string size is zero at the horizon, and grows *finitely* as the string falls towards $r = 0$.

5 Spatial Curvature Effects on the String Dynamics

The propagation of strings in Friedmann-Robertson-Walker (FRW) cosmologies has been investigated using both exact and approximative methods, see for example Refs.[7, 11-14, 17-20] (as well as numerical methods, which shall not be discussed here). Except for anti de Sitter spacetime, which has negative spatial curvature, the cosmologies that have been considered until now, have all been spatially flat. In this section we will consider the physical effects of a non-zero (positive or negative) curvature index on the classical and quantum strings. The non-vanishing components of the Riemann tensor for the generic D-dimensional FRW line element, in comoving coordinates:

$$ds^2 = -dt^2 + a^2(t)\frac{d\vec{x}d\vec{x}}{(1 + \frac{K}{4}\vec{x}\vec{x})^2},\tag{5.1}$$

are given by:

$$R_{itit} = \frac{-aa_{tt}}{(1 + \frac{K}{4}\vec{x}\vec{x})^2}, \qquad R_{ijij} = \frac{a^2(K + a_t^2)}{(1 + \frac{K}{4}\vec{x}\vec{x})^4}; \quad i \neq j \tag{5.2}$$

where $a = a(t)$ is the scale factor and K is the curvature index. Clearly, a non-zero curvature index introduces a non-zero spacetime curvature; the exceptional case provided by $K = -a_t^2 = \text{const.}$, corresponds to the Milne-Universe. From Eqs.(5.2), it is also seen that the curvature index has to compete with the first derivative of the scale factor. The effects of the curvature index are therefore most conveniently discussed in the family of FRW-universes with constant scale factor, the so-called static Robertson-Walker spacetimes. This is the point of view we take in the present section.

We consider both the closed ($K > 0$) and the hyperbolic ($K < 0$) static Robertson-Walker spacetimes, and all our results are compared with the already known results in the flat ($K = 0$) Minkowski spacetime. We determine the evolution of circular strings, derive the corresponding equations of state (using Eq.(4.6)), discuss the question of strings as self-consistent solutions to the Einstein equations [20], and we perform a semi-classical quantization. We also find all the stationary string configurations in these spacetimes.

The radius of a classical circular string in the spacetime (5.1), for $a = 1$, is determined by:

$$\dot{r}^2 + V(r) = 0; \qquad V(r) = (1 - Kr^2)(r^2 - b\alpha'^2),\tag{5.3}$$

where b is an integration constant ($b\alpha'^2 = E$, in the notation of Eqs.(2.18)-(2.19)). This equation is solved in terms of elliptic functions and all solutions describe oscillating strings (Fig.7 shows the potential $V(r)$ for $K > 0$, $K = 0$, $K < 0$).

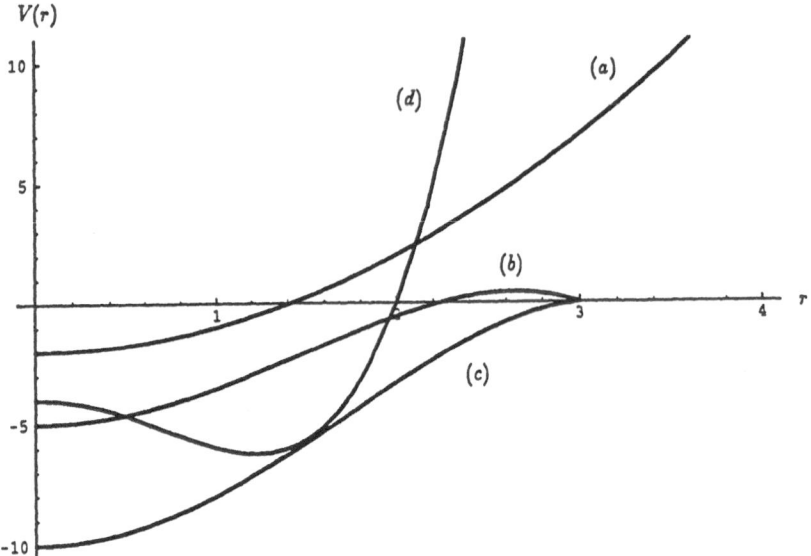

Figure 7. The potential $V(r)$ introduced in Eqs.(5.3) for a circular string in the static Robertson-Walker spacetimes: (a) flat ($K = 0$), (b) closed ($K > 0$ and $bK\alpha'^2 \leq 1$), (c) closed ($K > 0$ and $bK\alpha'^2 > 1$), (d) hyperbolic ($K < 0$).

For $K > 0$, when the spatial section is a hypersphere, the string either oscillates on one hemisphere or on the full hypersphere. The energy is always positive while the average pressure can be positive, negative or zero, depending on the precise value of an elliptic modulus; the equation of state is given explicitly in Table 5, in the different cases. Interestingly enough, we find that the circular strings provide a self-consistent solution to the Einstein equations with a selected value of the curvature index. Self-consistent solutions to the Einstein equations with string sources have been found previously in the form of power law expanding universes [20].

Table 5. Classical circular strings in the static Robertson-Walker spacetimes. Notice that a self-consistent solution to the Einstein equations, with the string back-reaction included, can be obtained only for $K > 0$.

CLASSICAL CIRCULAR STRING SOLUTIONS

Spacetime: $\quad\quad\quad\quad ds^2 = -dt^2 + \frac{dr^2}{1-Kr^2} + r^2 d\phi^2$

String solution: $\quad \phi = \sigma, \ t(\tau) = \sqrt{b}\alpha'\tau, \ \dot{r}^2 + V(r) = 0; \ V(r) = (1 - Kr^2)(r^2 - b\alpha'^2)$

String length, energy and pressure: $\quad l(\tau) = 2\pi|r(\tau)|, \quad T^\mu{}_\nu(\text{string}) = \text{diag.}(-\rho, P, P)$

MINKOWSKI ($K = 0$):

$r(\tau) = \sqrt{b}\alpha' \cos\tau$

$\rho = \sqrt{b}, \quad P = -\frac{\sqrt{b}}{2}\cos 2\tau, \quad <P> = 0, \quad \text{as cold matter.}$

Is not a self-consistent solution to the Einstein equations.

CLOSED ($K > 0$):

$r(\tau) = \begin{cases} \sqrt{b}\alpha' \text{sn}[\tau, k_-] & \text{for } \sqrt{bK}\alpha' \le 1, \quad k_- \equiv \sqrt{bK}\alpha' \in [0,1] \\ \frac{1}{\sqrt{K}}\text{sn}[\tau/k_+, k_+] & \text{for } \sqrt{bK}\alpha' > 1, \quad k_+ \equiv 1/(\sqrt{bK}\alpha') \in\ ]0,1[\end{cases}$

$\rho = \sqrt{b}, \quad P = \begin{cases} \frac{k_-}{2\sqrt{K}\alpha'}(1 - 2\text{sn}^2[\tau, k_-]) & \text{for } \sqrt{bK}\alpha' \le 1 \\ \frac{1}{2k_+\sqrt{K}\alpha'}(1 - 2k_+^2\text{sn}^2[\tau/k_+, k_+]) & \text{for } \sqrt{bK}\alpha' > 1 \end{cases}$

$<P> = \begin{cases} \frac{1}{2\sqrt{K}\alpha'}[\frac{k_-^2-2}{k_-} + \frac{2}{k_-}\frac{E(k_-)}{K(k_-)}] & \text{for } \sqrt{bK}\alpha' \le 1, \quad \text{negative.} \\ \frac{1}{2k_+\sqrt{K}\alpha'}[\frac{2E(k_+)}{K(k_+)} - 1] & \text{for } \sqrt{bK}\alpha' > 1, \quad \text{positive or negative.} \end{cases}$

A self-consistent solution to the Einstein equations is obtained for $k_+ = 0.9089...$

HYPERBOLIC ($K < 0$):

$r(\tau) = \frac{k}{\sqrt{-K}}\text{sd}[\mu\tau, k], \quad \mu \equiv \sqrt{1 - Kb\alpha'^2}, \quad k \equiv \sqrt{-Kb\alpha'^2}/\mu \in [0,1[$

$\rho = \sqrt{b}, \quad P = \frac{k-2k(1-k^2)\text{sd}^2[\mu\tau,k]}{2\sqrt{1-k^2}\sqrt{-K}\alpha'}$

$<P> = \frac{1}{2\sqrt{1-k^2}\sqrt{-K}\alpha'}[-k + \frac{2}{k}(1 - \frac{E(k)}{K(k)})], \quad \text{positive.}$

Is not a self-consistent solution to the Einstein equations.

We have semi-classically quantized the circular string solutions of Eq.(5.3). We used an approach developed in field theory by Dashen et. al. [41], based on the stationary phase approximation of the partition function. The result of the stationary phase integration is expressed in terms of the function:

$$W(m) \equiv S_{\text{cl}}(T(m)) + m\, T(m), \qquad (5.4)$$

where S_{cl} is the action of the classical solution, m is the mass and the period $T(m)$ is implicitly given by:

$$\frac{dS_{\text{cl}}}{dT} = -m. \qquad (5.5)$$

The bound state quantization condition then becomes [41]:

$$W(m) = 2\pi n, \quad n \in N_0 \qquad (5.6)$$

for n 'large'. Parametric plots of $K\alpha' W$ as a function of $K\alpha'^2 m^2$ in the different cases, are shown in Fig.8. The strings oscillating on one hemisphere give rise to a finite number N_- of states with the following mass-formula Ref.[29]:

$$m_-^2 \alpha' \approx \pi\, n, \qquad N_- \approx \frac{4}{\pi K \alpha'}. \qquad (5.7)$$

As in flat Minkowski spacetime, the scale of these string states is set by α'. The strings oscillating on the full hypersphere give rise to an infinity of more and more massive states with the asymptotic mass-formula Ref.[29]:

$$m_+^2 \approx K\, n^2. \qquad (5.8)$$

The masses of these states are independent of α', the scale is set by the curvature index K. Notice also that the level spacing grows with n. A similar result was found recently for strings in anti de Sitter spacetime [42].

For $K < 0$, when the spatial section is a hyperboloid, both the energy and the average pressure of the oscillating strings are positive. The equation of state is given in Table 5. In this case, the strings can not provide a self-consistent solution to the Einstein equations. After semi-classical quantization, we find an infinity of more and more massive states. The mass-formula is given by (see Ref.[29]):

$$\sqrt{-Km^2\alpha'^2}\, \log \sqrt{-Km^2\alpha'^2} \approx -\frac{\pi}{2} K\alpha'\, n \qquad (5.9)$$

Notice that the level spacing grows faster than in Minkowski spacetime but slower than in the closed static Robertson-Walker spacetime.

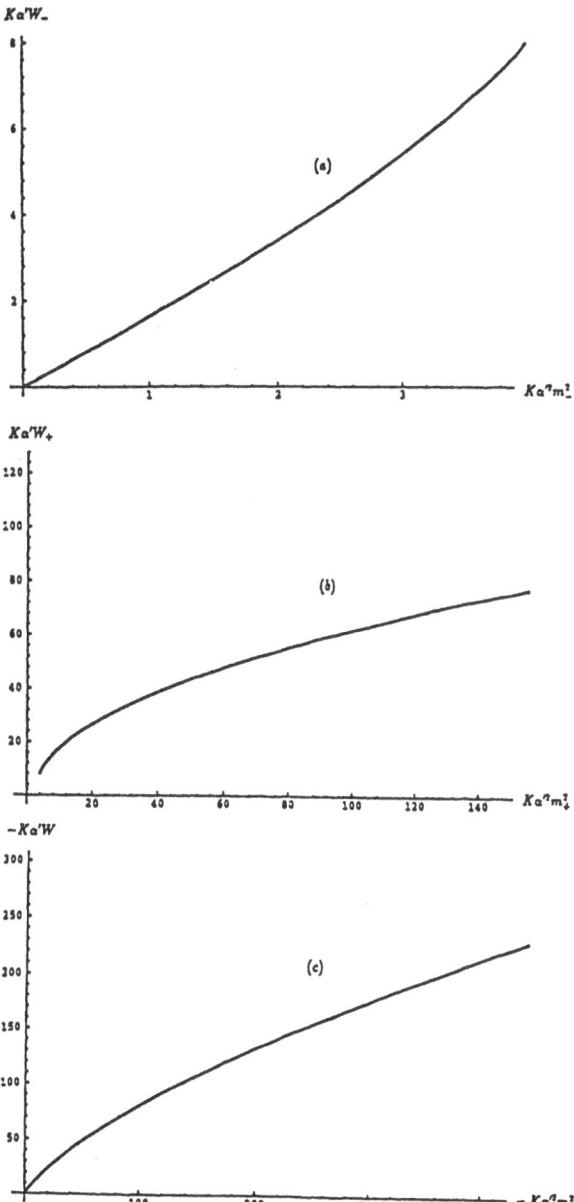

Figure 8. Parametric plot of $K\alpha'W$ as a function of $K\alpha'^2 m^2$ in the three cases: (a) $K > 0$, $bK\alpha'^2 \leq 1$ (strings oscillating on one hemisphere, only finitely many states), (b) $K > 0$, $bK\alpha'^2 > 1$ (strings oscillating on the full hypersphere, infinitely many states), (c) $K < 0$ (strings oscillating on the hyperboloid, infinitely many states).

96

Table 6. Semi-classical quantization of the circular strings in the static Robertson-Walker spacetimes. Notice in particular the different behaviour of the high mass spectrum of strings in the three cases.

SEMICLASSICAL QUANTIZATION OF CIRCULAR STRINGS

Stationary phase: $\quad W(m) = S^{\text{cl}}(T(m)) + m\, T(m)$

Stationary phase point: $\quad \frac{\partial S^{\text{cl}}}{\partial T} = -m, \quad$ determines the period T

Quantization condition: $\quad W(m) = 2\pi\, n, \quad n \in N_0$

MINKOWSKI ($K = 0$):

Mass: $\qquad m = 2\sqrt{b}$

Stationary phase: $\quad W = 2\pi\sqrt{b}\alpha'$

Quantization condition: $\quad \alpha' m^2 = 4\, n, \quad n \in N_0$

There are infinitely many states; $\quad "m^2 \propto n"$

CLOSED ($K > 0$):

Mass: $\quad m = \begin{cases} \frac{2k_-}{\sqrt{K}\alpha'} & \text{for } \sqrt{bK}\alpha' \le 1, \quad k_- \equiv \sqrt{bK}\alpha' \in [0,1] \\ \frac{2}{\sqrt{K}\alpha' k_+} & \text{for } \sqrt{bK}\alpha' > 1, \quad k_+ \equiv 1/(\sqrt{bK}\alpha') \in]0,1[\end{cases}$

Stationary phase: $\quad W = \begin{cases} \frac{8}{K\alpha'}[E(k_-) - (1-k_-^2)K(k_-)] & \text{for } \sqrt{bK}\alpha' \le 1 \\ \frac{8}{K\alpha'}\frac{E(k_+)}{k_+} & \text{for } \sqrt{bK}\alpha' > 1 \end{cases}$

Quantization condition: $\quad \begin{cases} m_-^2\alpha' \approx \pi\, n & \text{for } \sqrt{bK}\alpha' \le 1, \cdot \text{ on one hemisphere.} \\ m_+^2 \approx K\, n^2 & \text{for } \sqrt{bK}\alpha' > 1, \text{ on full hypersphere.} \end{cases}$

Altogether there are infinitely many states; $\quad "m \propto n", \quad$ for high states

---

HYPERBOLIC ($K < 0$):

Mass: $\quad m = \frac{2}{\sqrt{-K}\alpha'} \frac{k}{\sqrt{1-k^2}}, \quad k \equiv \sqrt{\frac{-Kb\alpha'^2}{1-Kb\alpha'^2}}$

Stationary phase: $\quad W = \frac{8}{K\alpha'}\frac{E(k)-K(k)}{\sqrt{1-k^2}}$

Quantization condition: $\quad \sqrt{-Km^2\alpha'^2} \log\sqrt{-Km^2\alpha'^2} \approx -\frac{\pi}{2}K\alpha'\, n$

There are infinitely many states; $\quad "m \log m \propto n", \quad$ for high states

On the other hand, the stationary strings are determined by Eqs.(4.4):

$$\phi' = \frac{L}{r^2}, \qquad r'^2 + V(r) = 0; \quad V(r) = (1 - Kr^2)(\frac{L^2}{r^2} - 1), \qquad (5.10)$$

where L is an integration constant (Fig.9 shows the potential $V(r)$ for $K > 0$, $K = 0$, $K < 0$).

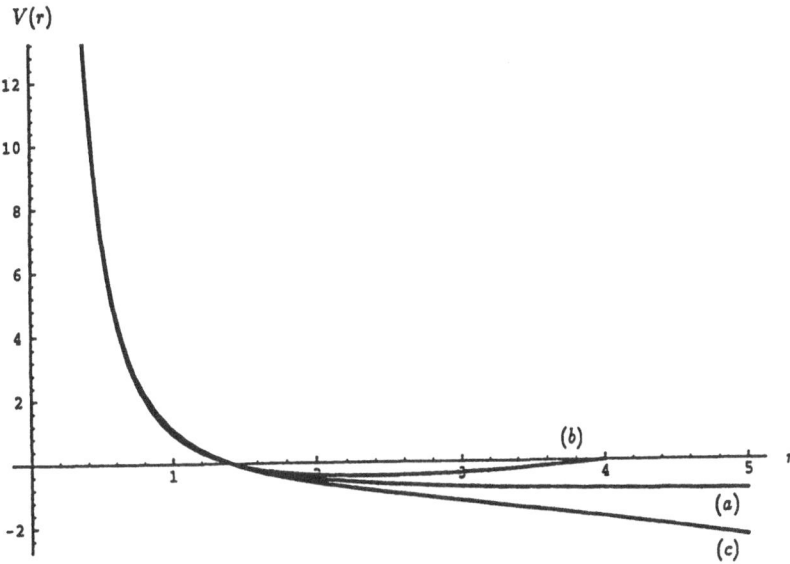

Figure 9. The potential $V(r)$ introduced in Eqs.(5.10) for a stationary string in the static Robertson-Walker spacetimes: (a) flat $(K = 0)$, (b) closed $(K > 0 \ and \ KL^2 \leq 1)$, (c) hyperbolic $(K < 0)$.

For $K > 0$, all the stationary string solutions describe circular strings winding around the hypersphere. The equation of state is of the extremely unstable string type [21]. For $K < 0$, the stationary strings are represented by infinitely long open configurations with an angle between the two "arms" given by:

$$\Delta\phi = \pi - 2\arctan(\sqrt{-K}L). \qquad (5.11)$$

The energy density is positive while the pressure densities are negative. No simple equation of state is found for these solutions.

We also computed the first and second order fluctuations around a static string center of mass, using the string perturbation series approach [7] and its covariant versions [9, 26]. Up to second order, the mass-formula for arbitrary values of the curvature index (positive or negative) is identical to the well-known flat spacetime mass-formula; all dependence on K cancels out.

Table 7. Stationary strings in the static Robertson-Walker spacetimes. Notice that the pressure densities are always negative in all three cases.

STATIONARY STRING SOLUTIONS

Spacetime: $\qquad\qquad ds^2 = -dt^2 + \frac{dr^2}{1-Kr^2} + r^2 d\phi^2$

String solution: $\quad t = r, \quad \phi' = \frac{L}{r^2}, \quad r'^2 + V(r) = 0; \quad V(r) = (1 - Kr^2)(\frac{L^2}{r^2} - 1)$

String length, energy and pressure: $\quad dl = d\sigma, \quad T^\mu{}_\nu(\text{string}) = \text{diag.}(-\rho, P_i)$

MINKOWSKI ($K = 0$):

$r(\sigma) = \sqrt{\sigma^2 + L^2}, \quad \phi(\sigma) = \arctan(\sigma/L), \quad$ straight strings.

$\frac{d\rho}{d\sigma} = \frac{1}{2\pi\alpha'}, \quad \frac{dP_x}{d\sigma} = -\frac{1}{2\pi\alpha'}$

The equation of state corresponds to extremely unstable strings in 1+1 dimensions.

CLOSED ($K > 0$):

$r^2(\sigma) = \frac{1}{K} \frac{KL^2 + \tan^2(\sqrt{K}\sigma)}{1 + \tan^2(\sqrt{K}\sigma)}, \quad \phi(\sigma) = \arctan[\frac{\tan(\sqrt{K}\sigma)}{\sqrt{K}L}], \quad$ circular strings.

$\rho = \frac{1}{\sqrt{K}\alpha'}, \quad P_x = P_z = -\frac{1}{2\sqrt{K}\alpha'}$

The equation of · : corresponds to extremely unstable strings in 2+1 dimensions.

HYPERBOI ($K < 0$):

$r^2(\sigma) = \frac{1}{K} \frac{KL}{1} \frac{\sqrt{-K}\sigma}{\sqrt{-K}\sigma}, \quad \phi(\sigma) = \pm\left\{\frac{\pi}{2} - \arctan[\sqrt{-K}L\tanh^{-1}(\sqrt{-K}\sigma)]\right\}$

Infinitely long ᴄ. ᴧ strings, $\quad \Delta\phi = \pi - 2\arctan(\sqrt{-K}L) \in]0, \pi]$

$\frac{d\rho}{d\sigma} = \frac{1}{2\pi\alpha'}, \quad \frac{dP_x}{d\sigma} = \frac{8KL^2}{\pi\alpha'(1-KL^2)^3}e^{-\sqrt{-K}|\sigma|}, \quad \frac{dP_x}{d\sigma} = \frac{-8}{\pi\alpha'(1-KL^2)^3}e^{-\sqrt{-K}|\sigma|}$

The pressure densities are negative but asymptotically go to zero exponentially. No simple equation of state in the general case.

6 Conclusion

We first studied the string propagation in the 2+1 BH-AdS background. We found the first and second order perturbations around the string center of mass as well as the mass formula, and compared with the ordinary black hole AdS spacetime and with the black string background as well. These results were then generalized to ordinary D-dimensional anti de Sitter spacetime. The classical string motion in anti de Sitter spacetime is stable in the sense that it is oscillatory in time with real frequencies and the string size and energy are bounded. Quantum mechanically, this reflects in the mass operator, which is well defined for any value of the wave number n, and arbitrary high mass states (and therefore an infinite number of states) can be constructed. This is to be contrasted with de Sitter spacetime, where string instabilities develop, in the sense that the string size and energy become unbounded for large de Sitter radius. For low mass states (the stable regime), the mass operator in de Sitter spacetime is given by Eq.(3.1) but with

$$\Omega_n(\lambda)_{\mathrm{dS}} = \frac{2n^2 - m^2 \alpha' \lambda}{\sqrt{n^2 - m^2 \alpha' \lambda}}.$$

Real mass solutions can be defined only up to some *maximal mass* of the order $m^2 \alpha' \approx 1/\lambda$ [7]. For $\lambda \ll 1$, real mass solutions can be defined only for $N \leq N_{\mathrm{max}} \sim 0.15/\lambda$ (where N is the eigenvalue of the number operator) and therefore there exists a *finite* number of states only. These features of strings in de Sitter spacetime have been recently confirmed within a different (semi-classical) quantization approach based on *exact* circular string solutions [42].

The presence of a cosmological constant Λ (positive or negative) increases considerably the number of levels of different eigenvalue of the mass operator (there is a splitting of levels) with respect to flat spacetime. That is, a non-zero cosmological constant *decreases* (although does not remove) the degeneracy of the string mass states, introducing a *fine structure effect*. For the low mass states the level spacing is approximately constant (up to corrections of the order λ^2). For the high mass states, the changes are more drastic and they depend crucially on the sign of Λ. A value $\Lambda < 0$ causes the *growing* of the level spacing linearly with N instead of being constant as in Minkowski space. Consequently, the density of states $\rho(m)$ grows with the exponential of $\sqrt{m}$ (instead of m as in Minkowski space) discarding the existence of a Hagedorn temperature in AdS spacetime, and the possibility of a phase transition. In addition, another important feature of the high mass string spectrum in AdS spacetime is that it becomes independent of α'. The string scale for the high mass states is given by $\mid \Lambda \mid$, instead of $1/\alpha'$ for the low mass states.

We have found new classes of exact string solutions obtained either by the circular string ansatz, Eq.(2.16), or by the stationary string ansatz, Eq.(4.3), in a variety of curved backgrounds including Schwarzschild, de Sitter and anti de Sitter spacetimes. Many different types of solutions have been found: oscillating circular strings, closed stationary strings, infinitely long stationary strings, dynamical straight strings and multi-string solutions describing finitely or infinitely many stationary or dynamical strings. In all the cases we have obtained the exact solutions in terms of either elementary or elliptic functions. Furthermore, we have analyzed the physical properties (length, energy, pressure) of the string solutions.

Finally, we solved the equations of motion and constraints for circular strings in static Robertson-Walker spacetimes. We computed the equations of state and found that there exists a self-consistent solution to the Einstein equations in the case of positive spatial curvature. The solutions were quantized semi-classically using the stationary phase approximation method, and the resulting spectra were analyzed and discussed. We have also found all stationary string configurations in these spacetimes and we computed the corresponding physical quantities, string length, energy and pressure.

Acknowledgements:
A.L. Larsen is supported by the Danish Natural Science Research Council under grant No. 11-1231-1SE

References

[1] H.J. de Vega and N. Sánchez, Phys. Rev. D42 (1990) 3969.
H.J. de Vega, M. Ramón-Medrano and N. Sánchez, Nucl. Phys. B374 (1992) 405.

[2] H.J. de Vega and N. Sánchez, Phys. Lett. B244 (1990) 215, Phys. Rev. Lett. 65C (1990) 1517, IJMP A7 (1992) 3043, Nucl. Phys. B317 (1989) 706 and ibid 731.
D. Amati and C. Klimcik, Phys. Lett. B210 (1988) 92.
M. Costa and H.J. de Vega, Ann. Phys. 211 (1991) 223 and ibid 235.
C. Loustó and N. Sánchez, Phys. Rev. D46 (1992) 4520.

[3] V.E. Zakharov and A.V. Mikhailov, JETP 47 (1979) 1017.
H. Eichenherr, in "Integrable Quantum Field Theories", ed. J. Hietarinta and C. Montonen (Springer, Berlin, 1982).

[4] H.J. de Vega and N. Sánchez, Phys. Rev. D47 (1993) 3394.

[5] I. Bars and K. Sfetsos, Mod. Phys. Lett. A7 (1992) 1091.

[6] H.J. de Vega, J.R. Mittelbrunn, M.R. Medrano and N. Sánchez, Phys. Lett. B232 (1994) 133 and "The Two-Dimensional Stringy Black Hole: a new Approach and a Pathology", PAR-LPTHE 93/14.

[7] H.J. de Vega and N. Sánchez, Phys. Lett. B197 (1987) 320.

[8] H.J. de Vega and N. Sánchez, Nucl. Phys. B299 (1988) 818.

[9] P.F. Mende, in "String Quantum Gravity and Physics at the Planck Energy Scale". Proceedings of the Erice Workshop held in June 1992, ed. N. Sánchez (World Scientific, 1993).

[10] A.L. Larsen and V.P. Frolov, Nucl. Phys. B414 (1994) 129.

[11] H.J. de Vega, A.V. Mikhailov and N. Sánchez, Teor. Mat. Fiz. 94 (1993) 232, Mod. Phys. Lett. A29 (1994) 2745.

[12] F. Combes, H.J. de Vega, A.V. Mikhailov and N. Sánchez, Phys. Rev. D50 (1994) 2754.

[13] H.J. de Vega, A.L. Larsen and N. Sánchez, Nucl. Phys. B427 (1994) 643.

[14] I. Krichever, "Two-Dimensional Algebraic-Geometrical Operators with Self-Consistent Potentials", Landau Institute Preprint, March, 1994, to appear in Funct. Analisis and Appl.

[15] H.J. de Vega and N. Sánchez, Nucl. Phys. B309 (1988) 552 and ibid, 577.

[16] C.O. Loustó and N. Sánchez, Phys. Rev. D47 (1993) 4498.

[17] H.J. de Vega and I.L. Egusquiza, Phys. Rev. D50 (1994) 763.

[18] A.L. Larsen, Phys. Rev. D50 (1994) 2623.

[19] A.L. Larsen and M. Axenides, Phys. Lett. B318 (1993) 47.

[20] H.J. de Vega and N. Sánchez, Phys. Rev. D50 (1994) 7202.

[21] N. Sánchez and G. Veneziano, Nucl. Phys. B333 (1990) 253.
M. Gasperini, N. Sánchez and G. Veneziano, IJMP A6 (1991) 3853,
Nucl. Phys. B364 (1991) 365.

[22] H.J. de Vega, M. Ramón-Medrano and N. Sánchez, Nucl. Phys. B374
(1992) 425.

[23] H.J. de Vega and N. Sánchez, Phys. Rev. D45 (1992) 2783.
M. Ramón-Medrano and N. Sánchez, Class. Quantum Grav. 10 (1993)
2007.

[24] N. Sánchez, Phys. Lett. B195 (1987) 160.

[25] H.J. de Vega, M. Ramón-Medrano and N. Sánchez, Nucl. Phys. B351
(1991) 277.

[26] A.L. Larsen and N. Sánchez, Phys. Rev. D50 (1994) 7493.

[27] A.L. Larsen and N. Sánchez, "Mass Spectrum of Strings in Anti de Sitter
Spacetime", DEMIRM-Paris-94048, to appear in Phys. Rev. D.

[28] A.L. Larsen and N. Sánchez, "New Classes of Exact Multi-String Solutions in Curved Spacetimes", DEMIRM-Paris-95003, to appear in Phys.
Rev. D.

[29] A.L. Larsen and N. Sánchez, "The Effect of Spatial Curvature on the
Classical and Quantum Strings", DEMIRM-Paris-95004.

[30] M. Bañados, C. Teitelboim and J. Zanelli, Phys. Rev. Lett. 69 (1992)
1849.

[31] G.T. Horowitz and D.L. Welch, Phys. Rev. Lett. 71 (1993) 328.

[32] N. Kaloper, Phys. Rev. D48 (1993) 2598.

[33] M. Bañados, M. Henneaux, C. Teitelboim and J. Zanelli, Phys. Rev.
D48 (1993) 1506.

[34] C. Farina, J. Gamboa and A.J. Segui-Santonja, Class. Quantum Grav.
10 (1993) L193.
N. Cruz, C. Martínez and L. Peña, Class. Quantum Grav. 11 (1994)
2731.

[35] A. Achúcarro and M.E. Ortiz, Phys. Rev. D48 (1993) 3600.
D. Cangemi, M. Leblanc and R.B. Mann, Phys. Rev. D48 (1993) 3606.

[36] J. Horne and G.T. Horowitz, Nucl. Phys. B368 (1992) 444.

[37] T. Buscher, Phys. Lett. B201 (1988) 466, B194 (1987) 59.

[38] A. Vilenkin, Phys. Rev. D24 (1981) 2082.

[39] H.J. de Vega and N. Sánchez, Int. J. Mod. Phys. A7 (1992) 3043.

[40] V.P. Frolov, V.D. Skarzhinsky, A.I. Zelnikov and O. Heinrich, Phys. Lett. B224 (1989) 255.

[41] R. Dashen, B. Hasslacher and A. Neveu, Phys. Rev. D11 (1975) 3424.

[42] H.J. de Vega, A.L. Larsen and N. Sánchez, "Semi-Classical Quantization of Circular Strings in de Sitter and anti de Sitter Spacetimes", DEMIRM-Paris-94049, submitted to Phys. Rev. D.

[43] A.L. Larsen, "Stable and Unstable Circular Strings in Inflationary Universes", NORDITA-94/14-P, to appear in Phys. Rev. D.

[44] A.L. Larsen, Nucl. Phys. B412 (1994) 372.

RENORMALIZATION CONSTANTS
FROM STRING THEORY

PAOLO DI VECCHIA AND LORENZO MAGNEA
NORDITA
Blegdamsvej 17, DK-2100 Copenhagen Ø, Denmark

ALBERTO LERDA AND RODOLFO RUSSO
Dipartimento di Fisica Teorica, Università di Torino
Via P.Giuria 1, I-10125 Turin, Italy
and I.N.F.N., Sezione di Torino

AND

RAFFAELE MAROTTA
Dipartimento di Scienze Fisiche, Università di Napoli
Mostra D'Oltremare, Pad. 19, I-80125 Napoli, Italy

Abstract.
We review some recent results on the calculation of renormalization constants in Yang-Mills theory using open bosonic strings. The technology of string amplitudes, supplemented with an appropriate continuation off the mass shell, can be used to compute the ultraviolet divergences of dimensionally regularized gauge theories. The results show that the infinite tension limit of string amplitudes corresponds to the background field method in field theory.

1. Introduction

In the past few years it has become clear that string theory is not only a good candidate for a unified theory of all interactions, but also a useful tool to understand the structure of perturbative field theories. Field-theoretical results can be recovered from string theory by decoupling the infinite tower of massive string states, that is by taking the limit of infinite string tension, or equivalently of vanishing Regge slope α'.

N. Sánchez and A. Zichichi (eds.), String Gravity and Physics at the Planck Energy Scale, 105–119.

Since in the limit $\alpha' \to 0$ string theories reduce to non-abelian gauge theories, unified with gravity, order by order in perturbation theory, in this limit we may expect to reproduce, order by order, scattering amplitudes, ultraviolet divergences, and other physical quantities that one computes perturbatively in non-abelian gauge theories.

A very useful feature of string theory for this purpose is the fact that, at each order of string perturbation theory, one does not get the large number of diagrams characteristic of field theories, which makes it very difficult to perform high order calculations. Using closed strings, one gets only one diagram at each order, while with open strings the number of diagrams remains limited. Furthermore, compact expressions for these diagrams are known explicitly for an arbitrary perturbative order [1], in contrast with the situation in field theory, where no such all-loop formula is known. Finally, string amplitudes are naturally written in a way that takes maximal advantage of gauge invariance: the color decomposition is automatically performed, and so are integrations over loop momenta, so that the helicity formalism is readily implemented.

The combination of these different features of string theory has led several authors [2, 3, 4, 5, 6, 7] to use string theory as an efficient conceptual and computational tool in different areas of perturbative field theory. In particular, because of the compactness of the multiloop string expression, it is in some cases easier to calculate non-abelian gauge theory amplitudes by starting from a string theory, and performing the zero slope limit, rather than using traditional techniques. In this way the one-loop amplitude involving four external gluons has been computed, reproducing the known field-theoretical result with much less computational cost [8]. Following the same approach, also the one-loop five gluon amplitude has been computed for the first time [9].

The aim of this talk is to summarize the results obtained in Refs. [10] and [11]. There it was shown that, provided a simple off-shell continuation is performed, string theory also contains information on the ultraviolet divergences of Yang-Mills theory, and the information can be consistently extracted in the language of dimensional regularization. In particular, starting from the one-loop two, three and four-gluon amplitudes in the open bosonic string, we performed the field theory limit, and we showed that in this limit the renormalization constants Z_A, Z_3 and Z_4 of non-abelian gauge theories can be consistently recovered. It is interesting to note that string theory leads unambiguously to the background field method.

Before going into the details of the calculation, let us first recall how field theory amplitudes are obtained from string theory, and how we expect those amplitudes to be renormalized.

In field theory one normally computes either connected Green functions, denoted here by $W_M(p_1 \ldots p_M)$, or one-particle irreducible (1PI) Green functions, $\Gamma_M(p_1 \ldots p_M)$. In both cases, in general, an off-shell continuation is performed, in order to avoid possible infrared divergences.

In string theory, on the other hand, one computes S-matrix elements involving gluon states, which are connected, via the reduction formulas, to on-shell connected Green functions, truncated with free propagators.

Taking the field theory limit, the natural ultraviolet regulator of string theory, $1/\alpha'$, is removed, so that the usual divergences are recovered. The Green functions one computes are thus unrenormalized, and a new regulator must be introduced, in our case dimensional continuation. We will see that also in this case an off-shell extrapolation is necessary in order to avoid infrared problems.

Once the field theory limit is taken, it is possible to isolate 1PI contributions, which lead to the 1PI Green functions Γ_M, or to compute the full amplitudes, which lead to the Green functions W_M. From the knowledge on how they renormalize we can then extract the renormalization constants. For example,

$$\Gamma_2(g) = Z_A^{-1}\Gamma_2^{(R)}(g) \qquad \Gamma_3(g) = Z_3^{-1}\Gamma_3^{(R)}(g) \qquad \Gamma_4(g) = Z_4^{-1}\Gamma_4^{(R)}(g) \ , \tag{1.1}$$

while

$$W_3(g) = Z_3^{-1}Z_A^3 W_3^{(R)}(g) \ , \tag{1.2}$$

where g is the renormalized coupling constant.

The talk is organized as follows. In Section 2 we consider the open bosonic string, and we write the explicit expression of the M gluon amplitude at h loops, including the overall normalization. In Section 3 we give the relevant amplitudes for the tree and one-loop diagrams. In Section 4 we sketch the calculation of the one-loop two gluon amplitude, already presented in [10], and we extract the gluon wave function renormalization constant Z_A. In Section 5 we present an alternative method, that allows one to exactly integrate over the punctures, and we use it to extract the renormalization constants Z_A, Z_3 and Z_4. Finally, in Section 6 we extend the calculation of Section 4 to the one-loop three gluon amplitude, and we discuss how to extract the contribution of the one-particle reducible diagrams, that were neglected in Section 5. Section 7 contains concluding remarks.

2. The M-gluon h-loop amplitude

In string theory the M-gluon scattering amplitude can be computed perturbatively and is given by

$$\begin{aligned} A(p_1, \ldots, p_M) &= \sum_{h=0}^{\infty} A^{(h)}(p_1, \ldots, p_M) \\ &= \sum_{h=0}^{\infty} g_s^{2h-2} \hat{A}^{(h)}(p_1, \ldots, p_M) \ , \end{aligned} \tag{2.1}$$

where g_s is a dimensionless string coupling constant, which is introduced to formally control the perturbative expansion. In Eq. (2.1), $A^{(h)}$ represents the h-loop contribution. For the closed string $A^{(h)}$ is given by only one diagram, while for the open string the number of diagrams is small in comparison with the large number of diagrams encountered in field theory.

Let us consider the open bosonic string, and let us restrict ourselves only to planar diagrams. For such diagrams the M-gluon h-loop amplitude, including the appropriate Chan-Paton factor, is given by

$$
\begin{aligned}
A_P^{(h)}(p_1,\ldots,p_M) \;=\;& N^h \, \mathrm{Tr}(\lambda^{a_1}\cdots\lambda^{a_M}) \, C_h \mathcal{N}_0^M \\
&\times \int [dm]_h^M \left\{ \prod_{i<j} \left[\frac{\exp\left(\mathcal{G}^{(h)}(z_i,z_j)\right)}{\sqrt{V_i'(0)\,V_j'(0)}} \right]^{2\alpha' p_i \cdot p_j} \right. \\
&\times \; \exp\left[\sum_{i\neq j} \sqrt{2\alpha'} p_j \cdot \varepsilon_i \, \partial_{z_i} \mathcal{G}^{(h)}(z_i,z_j) \right. \tag{2.2} \\
&\left.\left. + \frac{1}{2} \sum_{i\neq j} \varepsilon_i \cdot \varepsilon_j \, \partial_{z_i}\partial_{z_j} \mathcal{G}^{(h)}(z_i,z_j) \right] \right\}_{\mathrm{m.l.}} \; ,
\end{aligned}
$$

where the subscript "m.l." stands for multilinear, meaning that only terms linear in each polarization should be kept. Eq. (2.2) is written for transverse gluons, satisfying the condition $\varepsilon_i \cdot p_i = 0$, whereas the mass-shell condition $p_i^2 = 0$, though necessary for conformal invariance of the amplitude, has not been enforced yet.

The main ingredient in Eq. (2.2) is the h-loop world-sheet bosonic Green function $\mathcal{G}^{(h)}(z_i,z_j)$, which plays a key role in the field theory limit. $[dm]_h^M$ is the measure of integration on moduli space for an open Riemann surface of genus h with M operator insertions on the boundary [1]. The Green function $\mathcal{G}^{(h)}(z_i,z_j)$ can be expressed as

$$
\mathcal{G}^{(h)}(z_i,z_j) = \log E^{(h)}(z_i,z_j) - \frac{1}{2} \int_{z_i}^{z_j} \omega^\mu \, (2\pi \mathrm{Im}\tau_{\mu\nu})^{-1} \int_{z_i}^{z_j} \omega^\nu \; , \tag{2.3}
$$

where $E^{(h)}(z_i,z_j)$ is the prime-form, ω^μ ($\mu = 1,\ldots,h$) the abelian differentials and $\tau_{\mu\nu}$ the period matrix of an open Riemann surface of genus h. All these objects, as well as the measure on moduli space $[dm]_h^M$, can be explicitly written down in the Schottky parametrization of the Riemann surface, and their expressions for arbitrary h can be found for example in Ref. [12]. Here we will only write the explicit expression for the measure, to give a flavor of the ingredients that enter the full string theoretic calculations. It is

$$
[dm]_h^M \;=\; \frac{\prod_{i=1}^M dz_i}{dV_{abc}} \prod_{\mu=1}^h \left[\frac{dk_\mu d\xi_\mu d\eta_\mu}{k_\mu^2 (\xi_\mu - \eta_\mu)^2} (1 - k_\mu)^2 \right] \tag{2.4}
$$

$$
\times \; [\det(-i\tau_{\mu\nu})]^{-d/2} \prod_\alpha{}' \left[\prod_{n=1}^\infty (1 - k_\alpha^n)^{-d} \prod_{n=2}^\infty (1 - k_\alpha^n)^2 \right] \; .
$$

Here $\tau_{\mu\nu}$ is the period matrix, while k_μ are the multipliers and ξ_μ and η_μ the fixed points of the generators of the Schottky group; dV_{abc} is the projective invariant volume element

$$
dV_{abc} = \frac{d\rho_a \, d\rho_b \, d\rho_c}{(\rho_a - \rho_b)\,(\rho_a - \rho_c)\,(\rho_b - \rho_c)} \; , \tag{2.5}
$$

where ρ_a, ρ_b, ρ_c are any three of the M Koba-Nielsen variables, or of the $2h$ fixed points of the generators of the Schottky group, which can be fixed at will; finally, the primed product over α denotes a product over classes of elements of the Schottky group [12].

Notice that in the open string the Koba-Nielsen variables must be cyclically ordered, for example according to

$$z_1 \geq z_2 \cdots \geq z_M \quad , \tag{2.6}$$

and the ordering of Koba-Nielsen variables automatically prescribes the ordering of color indices.

The amplitude in Eq. (2.2) contains two normalization constants which were calculated in Ref. [11], and are given by

$$C_h = \frac{1}{(2\pi)^{dh}} g_s^{2h-2} \frac{1}{(2\alpha')^{d/2}} \qquad \mathcal{N}_0 = g_d \sqrt{2\alpha'} \quad , \tag{2.7}$$

where the string coupling g_s and the d-dimensional gauge coupling g_d are related by

$$g_s = \frac{g_d}{2} (2\alpha')^{1-d/4} \quad . \tag{2.8}$$

An efficient way to explicitly obtain $A^{(h)}(p_1, \ldots, p_M)$ is to use the M-point h-loop vertex $V_{M;h}$ of the operator formalism. The explicit expression of $V_{M;h}$ for the planar diagrams of the open bosonic string can be found in Ref. [1]. The vertex $V_{M;h}$ depends on M real Koba-Nielsen variables z_i through M projective transformations $V_i(z)$, which define local coordinate systems vanishing around each z_i, i.e. such that

$$V_i^{-1}(z_i) = 0 \quad . \tag{2.9}$$

When $V_{M;h}$ is saturated with M physical string states satisfying the mass-shell condition, the corresponding amplitude does not depend on the V_i's. However, as we discussed in Ref. [10], to extract informations about the ultraviolet divergences that arise when the field theory limit is taken, it is necessary to relax the mass-shell condition, so that also the amplitudes $A^{(h)}$ will depend on the choice of projective transformations V_i's, just like the vertex $V_{M;h}$. This is the reason of the appearence of V_i in Eq. (2.2).

3. Tree and one-loop diagrams

For tree-level amplitudes, corresponding to $h = 0$, the situation is particularly simple. The Green function in Eq. (2.3) reduces to

$$\mathcal{G}^{(0)}(z_i, z_j) = \log(z_i - z_j) \quad , \tag{3.1}$$

while the measure $[dm]_0^M$ is simply

$$[dm]_0^M = \frac{\prod_{i=1}^{M} dz_i}{dV_{abc}} \quad . \tag{3.2}$$

Inserting Eqs. (3.1) and (3.2) into Eq. (2.2), and writing explicitly all the normalization coefficients, we obtain the color ordered, planar, on-shell M gluon amplitude at tree level

$$A_P^{(0)}(p_1, \ldots, p_M) = 4 \, \mathrm{Tr}(\lambda^{a_1} \cdots \lambda^{a_M}) \, g_d^{M-2} \, (2\alpha')^{M/2-2}$$

$$\times \int_{\Gamma_0} \frac{\prod_{i=1}^{M} dz_i}{dV_{abc}} \left\{ \prod_{i<j} (z_i - z_j)^{2\alpha' p_i \cdot p_j} \right. \tag{3.3}$$

$$\times \left. \exp \left[\sum_{i<j} \left(\sqrt{2\alpha'} \frac{p_j \cdot \varepsilon_i - p_i \cdot \varepsilon_j}{(z_i - z_j)} + \frac{\varepsilon_i \cdot \varepsilon_j}{(z_i - z_j)^2} \right) \right] \right\}_{\mathrm{m.l.}},$$

where Γ_0 is the region identified by Eq. (2.6). Notice that any dependence on the local coordinates $V_i(z)$ drops out in the amplitude after enforcing the mass-shell condition. Notice also that Eq. (3.3) is valid only for $M \geq 3$, since the measure given by Eq. (3.2) is ill-defined for $M \leq 2$.

We readily derive the three-gluon amplitude

$$A^{(0)}(p_1, p_2, p_3) = -4 \, g_d \, \mathrm{Tr}(\lambda^a \lambda^b \lambda^c) \left(\varepsilon_1 \cdot \varepsilon_2 \, p_2 \cdot \varepsilon_3 \right.$$

$$\left. + \varepsilon_2 \cdot \varepsilon_3 \, p_3 \cdot \varepsilon_1 + \varepsilon_3 \cdot \varepsilon_1 \, p_1 \cdot \varepsilon_2 + O(\alpha') \right), \tag{3.4}$$

and the four-gluon amplitude

$$A_4^{(0)}(p_1, p_2, p_3, p_4) = 4 g_d^2 \, Tr(\lambda^{a_1} \lambda^{a_2} \lambda^{a_3} \lambda^{a_4}) \frac{\Gamma(1 - \alpha's)\Gamma(1 - \alpha't)}{\Gamma(1 + \alpha'u) \, s \, t} \tag{3.5}$$

$$\times \, [(\varepsilon_1 \cdot \varepsilon_2)(\varepsilon_3 \cdot \varepsilon_4) \, t \, u + (\varepsilon_1 \cdot \varepsilon_3)(\varepsilon_2 \cdot \varepsilon_4) \, t \, s + (\varepsilon_1 \cdot \varepsilon_4)(\varepsilon_2 \cdot \varepsilon_3) \, s \, u + \ldots],$$

where we have not written explicitly terms of the form $(\varepsilon \cdot \varepsilon)(\varepsilon \cdot p)(\varepsilon \cdot p)$ and higher orders in α'.

At one loop ($h = 1$) we keep the gluons off the mass shell, and Eq. (2.2) gives, for $M \geq 2$ transverse gluons,

$$A_P^{(1)}(p_1, \ldots, p_M) = N \, \mathrm{Tr}(\lambda^{a_1} \cdots \lambda^{a_M}) \frac{g_d^M}{(4\pi)^{d/2}} (2\alpha')^{(M-d)/2} (-1)^M$$

$$\times \int_0^\infty d\tau e^{2\tau} \tau^{-d/2} \prod_{n=1}^\infty \left(1 - e^{-2n\tau}\right)^{2-d} \int_0^\tau d\nu_M \int_0^{\nu_M} d\nu_{M-1} \ldots \int_0^{\nu_3} d\nu_2$$

$$\times \left\{ \prod_{i<j} \left[\sqrt{\frac{z_i \, z_j}{V_i'(0) \, V_j'(0)}} \exp\left(G(\nu_{ij})\right) \right]^{2\alpha' p_i \cdot p_j} \right. \tag{3.6}$$

$$\times \left. \exp\left[\sum_{i \neq j} \left(\sqrt{2\alpha'} p_j \cdot \varepsilon_i \, \partial_i G(\nu_{ij}) + \frac{1}{2} \varepsilon_i \cdot \varepsilon_j \, \partial_i \partial_j G(\nu_{ij}) \right) \right] \right\}_{\mathrm{m.l.}},$$

where $\nu_{ij} \equiv \nu_j - \nu_i$, $\partial_i \equiv \partial/\partial\nu_i$ and τ is related to the period $\bar{\tau}$ of the annulus by the relation

$$\tau = -i\pi\bar{\tau} \quad . \tag{3.7}$$

Instead of the Koba-Nielsen variables z_i, we have used the real variables

$$\nu_i = -\frac{1}{2}\log z_i \quad , \tag{3.8}$$

while the Green function $G(\nu_{ij})$ is given by

$$G(\nu_{ji}) = \log\left[-2\pi i \frac{\theta_1\left(\frac{i}{\pi}(\nu_j - \nu_i)|\frac{i}{\pi}\tau\right)}{\theta_1'\left(0|\frac{i}{\pi}\tau\right)}\right] - \frac{(\nu_j - \nu_i)^2}{\tau} \quad , \tag{3.9}$$

where θ_1 is the first Jacobi θ function.

If we enforce the mass-shell condition $p_i^2 = 0$, any dependence on the local coordinates V_i's drops out. However, in order to avoid infrared divergences, we will continue the gluon momenta off shell, in an appropriate way to be discussed later. Then, following Ref. [11], we will regard the freedom of choosing V_i as a gauge freedom. We make the simple choice

$$V_i'(0) = z_i \quad , \tag{3.10}$$

which will lead, in the field theory limit, to the background field Feynman gauge. The conditions (2.9) and (3.10) are easily satisfied by choosing for example

$$V_i(z) = z_i z + z_i \quad . \tag{3.11}$$

4. The two-gluon amplitude

The one-loop two-gluon amplitude is given by

$$A^{(1)}(p_1, p_2) = N\,\text{Tr}(\lambda^a\lambda^b)\,\frac{g_d^2}{(4\pi)^{d/2}}(2\alpha')^{2-d/2}\varepsilon_1\cdot\varepsilon_2 p_1\cdot p_2\,R(p_1\cdot p_2)\ , \tag{4.1}$$

where

$$R(s) = \int_0^\infty d\tau\ e^{2\tau}\,\tau^{-d/2}\prod_{n=1}^\infty\left(1 - e^{-2n\tau}\right)^{2-d}\int_0^\tau d\nu e^{2\alpha's G(\nu)}\,[\partial_\nu G(\nu)]^2\ . \tag{4.2}$$

Notice that if the two gluons are on mass shell, the two-gluon amplitude becomes ill defined, as the kinematical prefactor vanishes, while the integral diverges. In order to avoid this problem we keep the two gluons off shell.

To take the field theory limit, we must remember that the modular parameter τ and the coordinate ν are related to proper-time Schwinger parameters for the Feynman diagrams contributing to the two point function. In particular, $t \sim \alpha'\tau$ and $t_1 \sim \alpha'\nu$, where t_1 is the proper time associated with one of the two internal gluon propagators, while t is the total proper time around the loop. In the field theory limit these proper times have to remain finite, and thus the limit $\alpha' \to 0$ must correspond to the limit $\{\tau, \nu\} \to \infty$ in the integrand. The field theory limit is then determined by the asymptotic behavior of the Green function for large τ, namely

$$G(\nu, \tau) = -\frac{\nu^2}{\tau} + \log\left(2\sinh(\nu)\right) - 4\,e^{-2\tau}\sinh^2(\nu) + 0(e^{-4\tau})\ , \tag{4.3}$$

where ν must also be taken to be large, so that $\hat{\nu}$ remains finite; in this region, we may use

$$G(\nu, \tau) \sim (\hat{\nu} - \hat{\nu}^2)\tau - e^{-2\hat{\nu}\tau} - e^{-2\tau(1-\hat{\nu})} + 2e^{-2\tau} \quad , \qquad (4.4)$$

so that

$$\frac{\partial G}{\partial \nu} \sim 1 - 2\hat{\nu} + 2e^{-2\hat{\nu}\tau} - 2e^{-2\tau(1-\hat{\nu})} \quad . \qquad (4.5)$$

We now substitute these results into Eq. (4.1), and keep only terms that remain finite when $k = e^{-2\tau} \to 0$. Divergent terms must be discarded by hand, since they correspond to the propagation of the tachyon in the loop. The next-to-leading term corresponds to gluon exchange, and while it is also divergent in the field theory limit, the corresponding divergence is regularized by dimensional regularization. Finally, higher order terms $e^{-2n\tau}$ with $n > 0$ are vanishing in the field theory limit.

Notice that by taking the large τ and ν limit we have discarded two singular regions of integration that potentially contribute in the field theory limit, namely $\nu \to 0$ and $\nu \to \tau$. In these regions (often referred to as "pinching" regions) the Green function has a logarithmic singularity corresponding to the insertion of the two external states very close to each other, and this singularity in general gives non-vanishing contributions in the field theory limit. However, in the case of the two gluon amplitude, these regions correspond to Feynman diagrams with a loop consisting of a single propagator, *i. e.* a "tadpole". Massless tadpoles are defined to vanish in dimensional regularization, and thus we are justified in discarding these contributions as well.

Replacing the variable ν with $\hat{\nu} \equiv \nu/\tau$, Eq. (4.2) becomes

$$R(s) = \int_0^\infty d\tau \int_0^1 d\hat{\nu} \, \tau^{1-d/2} \, e^{2\alpha' s (\hat{\nu} - \hat{\nu}^2)\tau} \left[(1 - 2\hat{\nu})^2(d - 2) - 8 \right] \quad , \quad (4.6)$$

so that the integral is now elementary, and yields

$$R(s) = -\Gamma\left(2 - \frac{d}{2}\right)(-2\alpha' s)^{d/2-2} \frac{6 - 7d}{1 - d} B\left(\frac{d}{2} - 1, \frac{d}{2} - 1\right) \quad , \qquad (4.7)$$

where B is the Euler beta function.

If we substitute Eq. (4.7) into Eq. (4.1), we see that the α' dependence cancels, as it must. The ultraviolet finite string amplitude, Eq. (4.1), has been replaced by a field theory amplitude which diverges in four space-time dimensions, because of the pole in the Γ function in Eq. (4.7). Defining as usual a dimensionless coupling constant $g_d = g\,\mu^\epsilon$, with μ an arbitrary mass scale, and having set $d = 4 - 2\epsilon$, we find

$$A^{(1)}(p_1, p_2) = -N\frac{g^2}{(4\pi)^2}\left(\frac{4\pi\,\mu^2}{-p_1 \cdot p_2}\right)^\epsilon \Gamma(\epsilon)\frac{11 - 7\epsilon}{3 - 2\epsilon} B(1-\epsilon, 1-\epsilon)A^{(0)}(p_1, p_2)$$

$$(4.8)$$

Eq. (4.8)) is exactly equal to the gluon vacuum polarization of the $SU(N)$ gauge field theory that one computes with the background field method, in

Feynman gauge, with dimensional regularization, provided we use for the tree-level two-gluon amplitude the expression

$$A^{(0)}(p_1, p_2) = \delta^{ab} [\varepsilon_1 \cdot \varepsilon_2 \, p_1 \cdot p_2 - \varepsilon_1 \cdot p_2 \, \varepsilon_2 \cdot p_2] \qquad (4.9)$$

Comparing Eq. (4.8) with the equation for Γ_2 in Eq. (1.1) we can extract the minimal subtraction wave function renormalization constant

$$Z_A = 1 + N \frac{g^2}{(4\pi)^2} \frac{11}{3} \frac{1}{\epsilon} \quad . \qquad (4.10)$$

While this result is what we expected, it relies on our prescription to continue the string amplitude off shell, and on our choice of the projective transformations V_i. To make sure that our prescription is consistent we need to compute the three and four point renormalizations as well, and verify that gauge invariance is preserved.

5. An alternative computation of proper vertices

In the previous section we have computed the 1PI two-gluon amplitude and we have extracted the wave function renormalization constant. In this section we present an alternative method, introduced by Metsaev and Tseytlin [2]. This method isolates the 1PI part of the amplitude, and is thus particularly suited to the evaluation of renormalization constants. It is based on the following alternative expression for the bosonic Green function [13]

$$G(\nu_i, \nu_j) = - \sum_{n=1}^{\infty} \frac{1 + q^{2n}}{n(1 - q^{2n})} \cos 2\pi n \left(\frac{\nu_j - \nu_i}{\tau} \right) + \ldots \quad , \qquad (5.1)$$

where $q = e^{-\pi^2/\tau}$ and the dots stand for terms independent of ν_i and ν_j, that will not be important in our discussion.

An important advantage of this approach is that, at least at one loop, it allows to integrate exactly over the punctures before the field theory limit is taken. The result does not present pinching singularities, that are regularized directly in the Green function. As a consequence, for the two gluon amplitude, we will get the same expression that we derived in Section 4, while for the three and four gluon amplitudes we will get only the contributions that do not include pinchings and are therefore one-particle irreducible.

As a first step, we rewrite the one-loop M-gluon planar amplitude as

$$A_P^{(1)}(p_1, \ldots, p_M) = N \, \mathrm{Tr}(\lambda^{a_1} \cdots \lambda^{a_M}) \frac{g_d^M}{(4\pi)^{d/2}} (2\alpha')^{2-d/2} \qquad (5.2)$$

$$\times \; (-1)^M \int_0^{\infty} d\tau \, e^{2\tau} \, \tau^{-d/2} \prod_{n=1}^{\infty} \left(1 - e^{-2n\tau}\right)^{2-d} I_M^{(1)}(\tau) ,$$

where $I_M^{(1)}(\tau)$ is the integral over the punctures ν_i, and can be read off from Eq. (3.6).

For $M = 2$, after a partial integration with vanishing surface term, we get

$$I_2^{(1)}(\tau) = p_1 \cdot p_2 \, \varepsilon_1 \cdot \varepsilon_2 \int_0^\tau d\nu \, (\partial_\nu G(\nu))^2 \left(e^{G(\nu)}\right)^{2\alpha' p_1 \cdot p_2} \quad . \qquad (5.3)$$

Since we are only interested in divergent renormalizations, and since the overall power of α' is already appropriate to the field theory limit, as it vanishes when $d \to 4$, we can now neglect the exponential, which would contribute $1 + O(\alpha')$. Using the expression in Eq. (5.1) for the Green function, we can easily perform exactly the integral over the puncture, and we get

$$I_2^{(1)}(\tau) = \frac{2\pi^2}{\tau} p_1 \cdot p_2 \varepsilon_1 \cdot \varepsilon_2 \sum_{n=1}^\infty \left(\frac{1 + q^{2n}}{1 - q^{2n}}\right)^2 \quad , \qquad (5.4)$$

so that we can write

$$
\begin{aligned}
A^{(1)}(p_1, p_2) &= \frac{N}{2} \mathrm{Tr}(\lambda^{a_1}\lambda^{a_2}) \frac{g_d^2}{(4\pi)^{d/2}} (2\alpha')^{2-d/2} p_1 \cdot p_2 \varepsilon_1 \cdot \varepsilon_2 \, Z(d) \\
&= \frac{N}{4} \frac{g_d^2}{(4\pi)^{d/2}} (2\alpha')^{2-d/2} Z(d) \, A^{(0)}(p_1, p_2) \quad .
\end{aligned}
\qquad (5.5)
$$

Here

$$Z(d) \equiv (2\pi)^2 \int_0^\infty d\tau \, e^{2\tau} \, \tau^{-1-d/2} \prod_{n=1}^\infty \left(1 - e^{-2n\tau}\right)^{2-d} \sum_{m=1}^\infty \left(\frac{1 + q^{2m}}{1 - q^{2m}}\right)^2 \qquad (5.6)$$

is the string integral that generates the renormalization constants as $\alpha' \to 0$.

With three gluons we get

$$
\begin{aligned}
I_3^{(1)}(\tau) = &\int_0^\tau d\nu_3 \int_0^{\nu_3} d\nu_2 \, \{\varepsilon_1 \cdot \varepsilon_2 \, \partial_1 \partial_2 G(\nu_{21}) \\
&\times \, [p_1 \cdot \varepsilon_3 \, \partial_3 G(\nu_{31}) + p_2 \cdot \varepsilon_3 \, \partial_3 G(\nu_{32})] + \ldots \} \quad ,
\end{aligned}
\qquad (5.7)
$$

where terms needed for cyclic symmetry and terms of order α' are not written explicitly, and we discarded the exponentials of the Green functions, that are not contributing since the external gluons are on shell.

The integrals over ν_2 and ν_3 can be done by using the expression in Eq. (5.1) for the Green function. The result is

$$
\begin{aligned}
I_3^{(1)}(\tau) = &\frac{(2\pi)^2}{\tau} [\varepsilon_1 \cdot \varepsilon_2 p_2 \cdot \varepsilon_3 + \varepsilon_2 \cdot \varepsilon_3 p_3 \cdot \varepsilon_1 + \varepsilon_1 \cdot \varepsilon_3 p_1 \cdot \varepsilon_2] \\
&\times \sum_{n=1}^\infty \left(\frac{1 + q^{2n}}{1 - q^{2n}}\right)^2 + 0(\alpha') \quad ,
\end{aligned}
\qquad (5.8)
$$

so that the three gluon amplitude is given by

$$A^{(1)}(p_1, p_2, p_3) = \frac{N}{4} \frac{g_d^2}{(4\pi)^{d/2}} (2\alpha')^{2-d/2} Z(d) A^{(0)}(p_1, p_2, p_3) + O(\alpha') \quad . \quad (5.9)$$

Finally, the same calculation can be done for the four-gluon amplitude, where we can concentrate on the terms whose kinematical prefactor has no powers of the external momenta (and thus is of the form $\varepsilon_i \cdot \varepsilon_j \, \varepsilon_h \cdot \varepsilon_k$). Other terms are suppressed by powers of α'. Then we need to consider the expression

$$
\begin{aligned}
I_4^{(1)}(\tau) \;=\; & \int_0^\tau d\nu_4 \int_0^{\nu_4} d\nu_3 \int_0^{\nu_3} d\nu_2 \Big[\varepsilon_1 \cdot \varepsilon_2 \, \varepsilon_3 \cdot \varepsilon_4 \, \partial_1 \partial_2 G(\nu_{21}) \, \partial_3 \partial_4 G(\nu_{43}) \\
& + \; \varepsilon_1 \cdot \varepsilon_3 \, \varepsilon_2 \cdot \varepsilon_4 \, \partial_1 \partial_3 G(\nu_{31}) \, \partial_2 \partial_4 G(\nu_{42}) \\
& + \; \varepsilon_1 \cdot \varepsilon_4 \, \varepsilon_3 \cdot \varepsilon_2 \, \partial_1 \partial_4 G(\nu_{41}) \, \partial_3 \partial_2 G(\nu_{32}) \Big] \quad.
\end{aligned}
\tag{5.10}
$$

Using again Eq. (5.1), we can perform the integrals over the punctures, and we get

$$
\begin{aligned}
I_4^{(1)}(\tau) = & \; \frac{(2\pi)^2}{\tau} \sum_{n=1}^\infty \left(\frac{1+q^{2n}}{1-q^{2n}} \right)^2 \\
& \times \left[\varepsilon_1 \cdot \varepsilon_3 \, \varepsilon_2 \cdot \varepsilon_4 - \frac{1}{2} \varepsilon_1 \cdot \varepsilon_2 \, \varepsilon_3 \cdot \varepsilon_4 - \frac{1}{2} \varepsilon_2 \cdot \varepsilon_3 \, \varepsilon_1 \cdot \varepsilon_4 \right] \quad.
\end{aligned}
\tag{5.11}
$$

The amplitude becomes then

$$
A^{(1)}(p_1, p_2, p_3, p_4) = \frac{N}{4} \frac{g_d^2}{(4\pi)^{d/2}} (2\alpha')^{2-d/2} Z(d) A^{(0)}(p_1, p_2, p_3, p_4) + O(\alpha') \quad,
\tag{5.12}
$$

where the 1PI part of the four-gluon amplitude at tree level is given by

$$
\begin{aligned}
A^{(0)}(p_1, p_2, p_3, p_4) \;=\; & \; 4 g_d^2 \, Tr(\lambda^{a_1} \lambda^{a_2} \lambda^{a_3} \lambda^{a_4}) \\
& \times \left[\epsilon_1 \cdot \epsilon_3 \, \epsilon_2 \cdot \epsilon_4 - \frac{1}{2} \epsilon_1 \cdot \epsilon_2 \epsilon_3 \cdot \epsilon_4 - \frac{1}{2} \epsilon_2 \cdot \epsilon_3 \, \epsilon_1 \cdot \epsilon_4 \right] \quad.
\end{aligned}
\tag{5.13}
$$

Defining the factor

$$
K(d) = \frac{N}{4} \frac{g_d^2}{(4\pi)^{d/2}} (2\alpha')^{2-d/2} Z(d) \quad,
\tag{5.14}
$$

we can now perform the limit $\alpha' \to 0$, keeping the ultraviolet cutoff $\epsilon \equiv 2 - d/2$ small but positive, and eliminating by hand the tachyon contribution. The calculation of the integral $Z(d)$ in this limit is described in detail in Ref. [11]. The result is

$$
K(4 - 2\epsilon) \to -\frac{11}{3} N \frac{g^2}{(4\pi)^2} \frac{1}{\epsilon} + O(\epsilon^0) \quad.
\tag{5.15}
$$

If we finally compare Eqs. (1.1) with Eqs. (5.5), (5.9) and (5.12) we can determine the renormalization constants. They are given by

$$
Z_A = Z_3 = Z_4 = 1 + \frac{11}{3} N \frac{g^2}{(4\pi)^2} \frac{1}{\epsilon} \quad,
\tag{5.16}
$$

in agreement with the result of the previous section for Z_A, and as dictated by the background field method Ward identities.

6. The three-gluon amplitude

The methods described in the previous two sections are both adequate to compute one-particle irreducible contributions to the Green functions. Reducible diagrams, on the other hand, correspond to regions in moduli space where the gluons are inserted on the string world sheet very close to each other (pinching regions). These regions were excluded by hand in Section 5, since the corresponding logarithmic singularity in the world-sheet Green function was regularized by a ζ-function regularization [11]. If we wish to include them along the lines of Section 4, we have to perform the field theory limit in a slightly different way. To see this, let us consider the simplest case in which these contributions arise, namely the three-gluon amplitude.

The one-loop correction to Eq. (3.4) can be written as

$$A^{(1)}(p_1, p_2, p_3) = -N \operatorname{Tr}(\lambda^a \lambda^b \lambda^c) \frac{g_d^3}{(4\pi)^{d/2}} (2\alpha')^{2-d/2}$$
$$\times \int_0^\infty \mathcal{D}\tau \int_0^\tau d\nu_3 \int_0^{\nu_3} d\nu_2 \, f_3(\nu_2, \nu_3, \tau) \quad , \qquad (6.1)$$

where

$$f_3(\nu_2, \nu_3, \tau) \equiv e^{2\alpha' p_1 \cdot p_2 \, G(\nu_2)} \, e^{2\alpha' p_2 \cdot p_3 \, G(\nu_{32})} \, e^{2\alpha' p_3 \cdot p_1 \, G(\nu_3)}$$
$$\times \left\{ \left[-\varepsilon_1 \cdot \varepsilon_2 \, \partial_2^2 G(\nu_2) \, (p_1 \cdot \varepsilon_3 \, \partial_3 G(\nu_3) + p_2 \cdot \varepsilon_3 \, \partial_3 G(\nu_{32})) \right. \right.$$
$$+ \varepsilon_2 \cdot \varepsilon_3 \, \partial_3^2 G(\nu_{32}) \, (p_2 \cdot \varepsilon_1 \, \partial_2 G(\nu_2) + p_3 \cdot \varepsilon_1 \, \partial_3 G(\nu_3))$$
$$\left. + \varepsilon_1 \cdot \varepsilon_3 \, \partial_3^2 G(\nu_3) \, (p_3 \cdot \varepsilon_2 \, \partial_3 G(\nu_{32}) - p_1 \cdot \varepsilon_2 \, \partial_2 G(\nu_2)) \right]$$
$$\left. + O(\alpha') \right\} \quad , \qquad (6.2)$$

and

$$\mathcal{D}\tau \equiv d\tau \, e^{2\tau} \, \tau^{-d/2} \prod_{n=1}^\infty \left(1 - e^{-2n\tau}\right)^{2-d} \quad . \qquad (6.3)$$

One-particle irreducible contributions can be calculated along the lines of Section 4, expanding the bosonic world-sheet Green function for large values of τ as in Eq. (4.3). The calculation is described in some detail in Ref. [11], and gives

$$A^{(1)}(p_1, p_2, p_3)\Big|_{\text{1PI}} = \left(-\frac{11}{3}\right) N \left(\frac{g}{4\pi}\right)^2 \frac{1}{\epsilon} A^{(0)}(p_1, p_2, p_3) + O(\epsilon^0) \quad , \qquad (6.4)$$

which agrees with the results of Section 5.

Next, we turn to the analysis of the pinching regions. There are clearly three such regions, corresponding to $\nu_2 \to 0$, $\nu_2 \to \nu_3$ and $\nu_3 \to \tau$, as dictated by cyclic symmetry and periodicity on the annulus.

Let us consider, for example, the first region, $\nu_2 \to 0$. Since this pinching contribution is localized in a neighbourhood of 0, we can replace the integral $\int_0^{\nu_3} d\nu_2$ with an integral $\int_0^\eta d\nu_2$, where η is an arbitrary small number.

Further, we can use for the bosonic Green function the approximation

$$G(\nu) \sim \log(2\nu) \quad . \tag{6.5}$$

After this is done, in $f_3(\nu_2, \nu_3, \tau)$ we can expand $G(\nu_{32})$ in powers of ν_2, which turns the amplitude $A^{(1)}(p_1, p_2, p_3)$ into an infinite series. The n-th term of this series is proportional to an integral of the form

$$C_n \equiv \int_0^\eta d\nu_2 \, \nu_2^{n-2+2\alpha' p_1 \cdot p_2} \quad , \tag{6.6}$$

with $n \geq 0$. After a suitable analytic continuation in the momenta to insure convergence, we get

$$C_n = \frac{\eta^{n-1+2\alpha' p_1 \cdot p_2}}{n-1+2\alpha' p_1 \cdot p_2} \quad . \tag{6.7}$$

We see that, when the pinching $\nu_2 \to 0$ is performed, the amplitude becomes an infinite sum over all possible string states that are exchanged in the (12)-channel, $n = 0$ corresponding to the tachyon, $n = 1$ to the gluon and so on. In the case of the three-gluon amplitude, one can verify that the exchange of a tachyon does not give any contribution: in fact the coefficient of the quadratic divergence $\frac{1}{\nu_2^2}$ is zero because of the trasversality of the externals states. The gluon contribution, on the other hand, survives in the field theory limit, and contributes to the ultraviolet divergence, as expected: the single pole in ν_2 in fact generates, through the change of variable to $\hat{\nu}_2$, the negative power of τ needed for the integral to diverge. All other terms in the series, corresponding to $n \geq 2$, and to states whose mass becomes infinite as $\alpha' \to 0$, vanish in the field theory limit. Notice also that for $n = 1$ the dependence on the cutoff η in Eq. (6.7) disappears as $\alpha' \to 0$.

Keeping this in mind, and collecting all relevant factors, we find that the pinching contribution to the three gluon amplitude that we are considering is

$$A^{(1)}(p_1, p_2, p_3)\Big|_{\nu_2 \to 0} = -N \operatorname{Tr}(\lambda^a \lambda^b \lambda^c) \frac{g_d^3}{(4\pi)^{d/2}} (2\alpha')^{2-d/2} \tag{6.8}$$

$$\times \frac{(p_1 + p_2) \cdot p_3}{p_1 \cdot p_2} R\left[(p_1 + p_2) \cdot p_3\right] \varepsilon_1 \cdot \varepsilon_2 \, p_2 \cdot \varepsilon_3 \quad ,$$

where R is the integral defined in Eq. (4.2). Notice that Eq. (6.8) contains a ratio of momentum invariants which are vanishing on shell. The appearance of such ratios in string amplitudes, in the corners of moduli space corresponding to loops isolated on external legs, is a well-known fact, which for example motivated the work of Ref. [14]. As we already remarked in Ref. [10], this "0/0" ambiguity is similar to the one that appears in the unrenormalized connected Green functions of a massless field theory, if the external legs are kept on the mass-shell and divergences are regularized with dimensional regularization. Our prescription to deal with this ambiguity is to continue off shell the momentum of the gluon attached to the loop, according to

$$p_3^2 = (p_1 + p_2)^2 = m^2 \quad . \tag{6.9}$$

The other gluon momenta, p_1 and p_2, on the other hand, are kept on shell. Here we rely on the assumption, subtantiated by the results obtained so far, that string amplitudes lead to field theory amplitudes calculated with the background field method. As was shown in Ref. [15], S-matrix elements are obtained in this method by first calculating one-particle irreducible vertices to the desired order, and then gluing them together with propagators that are defined by fixing the gauge for the background field. This leads us to interpret Eq. (6.8) as a one-loop, one-particle irreducible two point function, whose momentum must be continued off shell according to Eq. (6.9), glued to a tree-level three point vertex, for which no off-shell continuation is necessary. We thus keep $p_1^2 = p_2^2 = 0$, which, using momentum conservation, implies

$$p_1 \cdot p_2 = \frac{m^2}{2} \quad . \tag{6.10}$$

Then, comparing Eq. (6.8) with Eq. (3.4), and including a factor of three to account for the three pinching regions, we can write

$$A^{(1)}(p_1, p_2, p_3)\Big|_{pinch.} = -\frac{3}{2} N \frac{g_d^2}{(4\pi)^{d/2}} (2\alpha')^{2-d/2} R(-\tilde{m}^2) A^{(0)}(p_1, p_2, p_3) \quad . \tag{6.11}$$

Extracting the ultraviolet divergence of Eq. (6.11), and adding Eq. (6.4), we find the total divergence of the unrenormalized, connected, three gluon Green function,

$$A^{(1)}(p_1, p_2, p_3)\Big|_{div} = 2 \left(\frac{11}{3}\right) N \left(\frac{g}{4\pi}\right)^2 \frac{1}{\epsilon} A^{(0)}(p_1, p_2, p_3) \quad , \tag{6.12}$$

which leads again to the background field Ward identity (see Eq. (1.2))

$$Z_3 = Z_A = 1 + N \left(\frac{g}{4\pi}\right)^2 \frac{11}{3}(\frac{1}{\epsilon}) \quad . \tag{6.13}$$

The same analysis can be carried out for the four-point amplitude, as described in Ref. [11], and no surprises arise.

7. Concluding remarks

We have shown that it is possible to calculate renormalization constants in Yang-Mills theories using the simplest of string theories, the open bosonic string. To do so it is necessary to continue off shell some of the external momenta, but this can be done consistently in the field theory limit, and the results concide with the ones obtained using the background field method and dimensional regularization. Since bosonic string amplitudes are well understood at all orders in perturbation theory, this technique may be useful beyond one loop.

References

1. See, for example, P. Di Vecchia, *"Multiloop amplitudes in string theory"* in Erice, *Theor. Phys.* (1992), and references therein.
2. R. R. Metsaev and A. A. Tseytlin, *Nucl. Phys.* B **298** (1988) p. 109.
3. J. A. Minahan, *Nucl. Phys.* B **298** (1988) p. 36.
4. T. R. Taylor and G. Veneziano, *Phys. Lett.* **212** B (1988) p. 147.
5. V. S. Kaplunovsky, *Nucl. Phys.* B **307** (1988) p. 145, and *Nucl. Phys.* B **382** (1992) p. 436.
6. Z. Bern and D. A. Kosower, *Phys. Rev.* D **38** (1988) p. 1888.
7. Z. Bern and D. A. Kosower, *Nucl. Phys.* B **321** (1989) p. 605.
8. Z. Bern and D. A. Kosower, *Nucl. Phys.* B **379** (1992) p. 451.
9. Z. Bern, L. Dixon and D. A. Kosower, *Phys. Rev. Lett.* **70** (1993) p. 2677.
10. P. Di Vecchia, A. Lerda, L. Magnea and R. Marotta, *Phys. Lett.* B **351** (1995) p. 445.
11. P. Di Vecchia, A. Lerda, L. Magnea, R. Marotta and R. Russo, to appear.
12. P. Di Vecchia, F. Pezzella, M. Frau, K. Hornfeck, A. Lerda and S. Sciuto, *Nucl. Phys.* B **322** (1989) p. 317.
13. E. S. Fradkin and A. A. Tseytlin, *Phys. Lett.* **163** B (1985) p. 123.
14. Z. Bern, D. A. Kosower and K. Roland, *Nucl. Phys.* B **334** (1990) p. 309.
15. L. F. Abbott, M. T. Grisaru and R. K. Schaefer, *Nucl. Phys.* B **229** (1983) p. 372.

Closed Superstrings in Magnetic Field: Instabilities and Supersymmetry Breaking

A.A. TSEYTLIN*[†]

Theoretical Physics Group, Blackett Laboratory,
Imperial College, London SW7 2BZ, U.K.

Abstract

We consider a 2-parameter class of solvable closed superstring models which 'interpolate' between Kaluza-Klein and dilatonic Melvin magnetic flux tube backgrounds. The spectrum of string states has similarities with the Landau spectrum for a charged particle in a uniform magnetic field. The presence of spin-dependent 'gyromagnetic' interaction implies breaking of supersymmetry and possible existence of tachyonic instabilities (for certain values of magnetic parameters). We study in detail the simplest example of the Kaluza-Klein Melvin model describing superstring moving in a flat but non-trivial 10-dimensional space containing a 3d factor which is a 'twisted' product of a 2-plane and an internal circle. We also discuss the compact version of this model constructed by 'twisting' the product of $SU(2)$ and $U(1)$ in $SU(2) \times U(1)$ WZNW theory without changing the local group-space geometry (and the central charge). We explain how the supersymmetry is broken by continuous 'magnetic' twist parameters and comment on possible implications for internal space compactification models.

1. Introduction

An important problem in string theory is to study how quantised strings propagate in non-trivial backgrounds described by conformal 2d models. This may help to understand the structure of the space of exact string solutions as well as certain generic properties of string theory like the existence 'critical' or 'maximal' values of fields, instabilities, mechanisms of supersymmetry breaking, residual symmetries of string spectrum, etc.

Magnetic field is one of the simplest probes of the spectrum and critical properties of a physical system. In the context of a gravitational theory like closed string theory, a

---

* e-mail address: tseytlin@ic.ac.uk

† On leave from Lebedev Physics Institute, Moscow, Russia.

N. Sánchez and A. Zichichi (eds.), String Gravity and Physics at the Planck Energy Scale, 121–149.
© *1996 Kluwer Academic Publishers.*

magnetic background is accompanied by a curvature of space-time. In spite of that, certain closed string magnetic models can be solved exactly. In particular, these are static flux tube type configurations with approximately uniform magnetic field which generalize the Melvin solution of the Einstein-Maxwell theory [1]. Such backgrounds are exact solutions of (super)string theory [2,3] and, moreover, the spectrum of the corresponding unitary conformal string models can be explicitly determined [3,4].

Below we shall first review the solution [4] of this class of magnetic models. In the last section we shall discuss related compact models which may be used for string compactifications and the issue of supersymmetry breaking induced by 'magnetic' twist parameters.

Before turning to closed string theory let us first recall the solution of similar uniform magnetic field problems in particle (field) theory and open string theory.

1.1. Point particles and open strings in magnetic field

The reason why the quantum-mechanical or field-theoretical problem of a particle in a uniform abelian (electro)magnetic field is exactly solvable is that the action $I = \int d\tau [\dot{x}^\mu \dot{x}^\mu + ie\dot{x}^\mu A_\mu(x)]$ (which determines the Hamiltonian in quantum mechanics and the heat kernel in field theory) becomes gaussian if the field strength is constant, $A_\mu = -\frac{1}{2} F_{\mu\nu} x^\nu$, $F_{\mu\nu} = const$. Assuming the magnetic field is directed along x^3-axis (so that $F_{ij} = H\epsilon_{ij}$, $i,j = 1,2$) and introducing $x = x_1 + ix_2$, $a = -i(\partial_x^* + \frac{1}{4} eHx)$, $a^\dagger = -i(\partial_x - \frac{1}{4} eHx^*)$, $[a, a^\dagger] = \frac{1}{2} eH$, one can reduce the corresponding quantum-mechanical problem to a free oscillator one. The resulting energy spectrum is the special ($S = 0$) case of the Landau spectrum for a particle of charge e, mass M_0 and third component of the spin S (we assume $eH > 0$)

$$E^2 = M_0^2 + p_3^2 - 2eHJ , \quad J \equiv -l - \tfrac{1}{2} + \tfrac{1}{2} gS , \quad l = 0, 1, 2, \dots . \tag{1.1}$$

Here l is the Landau level number (which replaces the continuous momenta p_1, p_2) and g is the gyromagnetic ratio which is $1/S$ for minimally coupled particles but can be equal to 2 for non-minimally coupled ones. Thus E^2 can become negative for large enough values of H, e.g., $H > H_{crit} = M_0^2/e$ for spin 1 charged states. That applies, for example, to W-bosons in the context of electroweak theory [5] suggesting the presence of a transition to a phase with a W-condensate. In the case of unbroken gauge theory with massless charged vector particles the instability is present for any (e.g., infinitesimal) value of the magnetic field [5]. This infra-red instability of a magnetic background is not cured by supersymmetry, i.e. it remains also in supersymmetric gauge theories (e.g., in ultra-violet finite $N = 4$ supersymmetric Yang-Mills theory) since the small fluctuation operator for the gauge field $-\delta_{\mu\nu} D^2 - 2F_{\mu\nu}$ still has negative modes due to the 'anomalous magnetic moment' term. This is not surprising given that the magnetic spin-dependent coupling breaks supersymmetry.

The meaning of this instability is that the expansion is carried out near a classical solution of the Yang-Mills equations (abelian $A_i = -\frac{1}{2}H\epsilon_{ij}x^j$ belonging to the Cartan subalgebra) which is *not* a vacuum one: the energy is proportional to H^2 and thus is minimal for $H = 0$. As a result, a non-zero magnetic field will tend to dissipate. The presence of tachyonic modes in the magnetic background with infinitesimal (but non-vanishing H) does not indicate an instability of the trivial vacuum but only that of a configuration with $H \neq 0$ (this corresponds to a resummation of the expansion in small perturbations near the trivial vacuum).

The solution of the uniform magnetic field problem in the open string theory is similar to the one in the particle case. Indeed, the open string is coupled to the (abelian) vector field only at its ends, $I = \frac{1}{4\pi\alpha'}\int d^2\sigma \, \partial_a x^\mu \partial^a x^\mu + ie\int d\tau \, A_\mu(x)\dot{x}^\mu$, and thus for the abelian background $A_\mu = -\frac{1}{2}F_{\mu\nu}x^\nu$ the resulting gaussian path integral can be computed exactly [6]. This is a consistent 'on-shell' problem since $F_{\mu\nu} = $ const is an exact solution of the effective field equations of the open string theory. The corresponding 2d world-sheet theory is a conformal field theory [7] which can be solved explicitly in terms of free oscillators (thus representing a string generalization of the Landau problem in quantum particle mechanics). As a result, one is able to determine the spectrum of an open string moving in a constant magnetic field [7,8,9]. The Hamiltonian (L_0) of the open (super)string in a constant magnetic field is given by [7,9]

$$\hat{\mathcal{H}} = \tfrac{1}{2}\alpha'\left(-E^2 + p_\alpha^2\right) + \hat{N} - \gamma\hat{J},\tag{1.2}$$

$$\hat{\gamma} \equiv \frac{2}{\pi}\left|\arctan(\tfrac{1}{2}\alpha'\pi e_1 H) + \arctan(\tfrac{1}{2}\alpha'\pi e_2 H)\right|, \quad 0 \le \hat{\gamma} < 1.$$

Here e_1, e_2 are charges at the two ends of the string and H is the magnetic field, $F_{ij} = H\epsilon_{ij}$. $\hat{N} = 0, 1, 2...$ is the number of states operator and $\hat{J} = -l - \frac{1}{2} + S$ is the angular momentum operator of the open string in the $(1,2)$ plane. The energy spectrum is found from the constraint $\hat{\mathcal{H}} = 0$ and for small $e_i H$ is in agreement with (1.1) with $g = 2$. A novel feature of this spectrum as compared to the free (super)string one is the presence of tachyonic states above certain critical values of the magnetic field ($H_{crit} \sim \alpha'^{-1}\hat{N}/S \sim \alpha'^{-1}$). Thus the constant magnetic field background is unstable in the open string theory (as it is in the non-abelian gauge theory). A qualitative reason for this instability is that the free open string spectrum contains electrically charged higher spin massive particle states and the latter have (approximately) the Landau spectrum (1.1) ($\hat{\gamma} \approx \alpha'(e_1 + e_2)H$ for a weak field). The tachyonic states appear only in the bosonic (Neveu-Schwarz) sector and only on the leading Regge trajectory ($l = 0$, maximal S for a given mass level $\hat{N}$, $S = \hat{N} + 1$).

1.2. Closed strings and Melvin-type magnetic flux tube backgrounds

Similar background magnetic field problem is also exactly solvable in *closed* (super)string theories [10,3,4]. This may be unexpected since the abelian vector field is now coupled to the internal points of the string, and such interaction terms, e.g., $\Delta L = \partial y \bar{\partial} y + A_\mu(x) \partial x^\mu \bar{\partial} y + \tilde{A}_\mu(x) \bar{\partial} x^\mu \partial y + ...$, do not become gaussian for $A_\mu = -\frac{1}{2} F_{\mu\nu} x^\nu$. Here $y \in (0, 2\pi R)$ is a compact internal Kaluza-Klein coordinate that 'charges' the string. The spectrum of the closed string compactified on a circle contains states with arbitrarily large masses, spins *and* charges $Q_{L,R} = mR^{-1} \pm w\alpha'^{-1}R$, $(m, w = 0, \pm 1, ...)$. One difference compared to the open string case is that there are infinitely many possible values of charges. Another is that both Q_L and Q_R are, in general, non-vanishing and coupled to (a combination of) two abelian vector fields ($G_{5\mu}$ and $B_{5\mu}$).[1]

The important observation is that in contrast to the tree level abelian open string case, the $F_{\mu\nu} = $ const background in flat space does not represent a solution of a *closed* string theory, i.e. the above interaction terms added to the free string Lagrangians do not lead to conformally invariant 2d σ-models. Since the closed string theory contains gravity, a uniform magnetic field, which has a non-vanishing energy, must curve the space (as well as possibly induce other 'massless' background fields). One should thus first find a consistent conformal model which is a closed string analogue of the uniform magnetic field background in the flat space field (or open string) theory and then address the question of its solvability. Remarkably, it turns out that extra terms which should be added to the closed string action in order to satisfy the conformal invariance condition (i.e. to satisfy the closed string effective field equations) produce solvable 2d models. Like the particle and open string models, these models are exactly solvable in terms of free oscillators (one is able to find explicitly the spectrum, partition function, etc).

A simple example of an approximately uniform magnetic field background in the Einstein-Maxwell theory is the static cylindrically symmetric Melvin 'magnetic universe' or 'magnetic flux tube' solution [1]. It has R^4 topology and can be considered [12] as a gravitational analogue of the abelian Higgs model vortex [13] with the magnetic pressure (due to repulsion of Faraday's flux lines) being balanced not by the Higgs field but by the gravitational attraction. The magnetic field is approximately constant inside the tube and decays to zero at infinity in the direction orthogonal to x_3-axis. The space is not, however, asymptotically euclidean: the (ρ, φ) 2-plane orthogonal to x_3 asymptotically closes at large ρ (so the solution should be interpreted not as a flux tube embedded in an approximately flat space, but rather as a 'magnetic universe'). Several interesting features of the Melvin solution in the context of Kaluza-Klein (super)gravity (e.g., instability against monopole or magnetic black hole pair creation) were discussed in [12,14,15].

---

[1] As a result, the gyromagnetic ratio of closed string states is $g \le 2$, e.g., $g = 1$ for standard non-winding Kaluza-Klein states [11,3].

This Einstein-Maxwell ('$a = 0$') solution has two direct analogues among solutions of the low-energy closed string theory (superstring toroidally compactified to $D = 4$). Assuming $x^5 = y$ is a compact internal coordinate, the $D = 5$ string effective action can be expressed in terms of $D = 4$ fields: metric $G_{\mu\nu}$, dilaton ϕ, antisymmetric tensor $B_{\mu\nu}$, two vector fields $\mathcal{A}_\mu$ and $\mathcal{B}_\mu$ (related to $G_{5\mu}$ and $B_{5\mu}$) and the 'modulus' σ. The dilatonic ('$a = 1$') and Kaluza-Klein ('$a = \sqrt{3}$') Melvin solutions have zero $B_{\mu\nu}$ but ϕ or σ being non-constant. Starting with the $D = 5$ bosonic string effective action and assuming that one spatial dimension x^5 is compactified on a small circle, one finds (ignoring massive Kaluza-Klein modes) the following dimensionally reduced $D = 4$ action

$$S_4 = \int d^4x \sqrt{G}\, e^{-2\Phi}\, \left[\, R\ + 4(\partial_\mu \Phi)^2 - (\partial_\mu \sigma)^2 \right. \tag{1.3}$$

$$\left. - \tfrac{1}{12}(\hat{H}_{\mu\nu\lambda})^2 - \tfrac{1}{4}e^{2\sigma}(F_{\mu\nu}(\mathcal{A}))^2 - \tfrac{1}{4}e^{-2\sigma}(F_{\mu\nu}(\mathcal{B}))^2 + O(\alpha') \right] \ ,$$

where $\hat{G}_{\mu\nu} = G_{\mu\nu} - G_{55}\mathcal{A}_\mu\mathcal{A}_\nu$, $F_{\mu\nu}(\mathcal{A}) = 2\partial_{[\mu}\mathcal{A}_{\nu]}$, $F_{\mu\nu}(\mathcal{B}) = 2\partial_{[\mu}\mathcal{B}_{\nu]}$, $\hat{H}_{\lambda\mu\nu} = 3\partial_{[\lambda}B_{\mu\nu]} - 3\mathcal{A}_{[\lambda}F_{\mu\nu]}(\mathcal{B})$, $\Phi = \phi - \tfrac{1}{2}\sigma$. The effective equations following from (1.3) have a 3-parameter (α, β, q) class of stationary axisymmetric (electro)magnetic flux tube solutions [3]. The most interesting subclass of these solutions ($\alpha = \beta$) describes *static* magnetic flux tube backgrounds (β and q are magnetic field strength parameters). It contains the $a = \sqrt{3}$ and $a = 1$ Melvin solutions [16] as the special cases of $\beta = 0$ and $\beta = q$. The four-dimensional geometry is given by (in terms of the string metric in (1.3); $x^\mu = (t, \rho, \varphi, x^3)$)

$$ds_4^2 = -dt^2 + d\rho^2 + F(\rho)\tilde{F}(\rho)\rho^2 d\varphi^2 + dx_3^2 \ , \tag{1.4}$$

$$\mathcal{A}_\varphi = qF(\rho)\rho^2 \ , \qquad \mathcal{B}_\varphi = -\beta\tilde{F}(\rho)\rho^2 \ , \tag{1.5}$$

$$e^{2(\phi - \phi_0)} = \tilde{F}(\rho) \ , \quad e^{2\sigma} = \tilde{F}(\rho)F^{-1}(\rho) \ , \quad F \equiv \frac{1}{1 + q^2\rho^2} \ , \quad \tilde{F} \equiv \frac{1}{1 + \beta^2\rho^2} \ .$$

There are two magnetic fields with the effective strengths q and β (when both are non-vanishing the radius of φ-circle goes to zero at large ρ). In particular,[2] the '$a = \sqrt{3}$ Melvin' background is ($\beta = 0$, $q \neq 0$):

$$ds_4^2 = -dt^2 + d\rho^2 + F(\rho)\rho^2 d\varphi^2 + dx_3^2 \ , \tag{1.6}$$

$$\mathcal{A} = qF(\rho)\rho^2 d\varphi \ , \quad e^{2\sigma} = F^{-1} = 1 + q^2\rho^2 \ , \quad \mathcal{B} = B = 0 \ , \quad \phi = \phi_0 \ ,$$

and the '$a = 1$ Melvin' background is ($\beta = q \neq 0$):

$$ds_4^2 = -dt^2 + d\rho^2 + \tilde{F}^2(\rho)\rho^2 d\varphi^2 + dx_3^2 \ , \tag{1.7}$$

---

[2] In these special cases there is effectively just one non-trivial vector and one scalar so that the Einstein-frame action can be put into the form $\int d^4x \sqrt{G'}\left[R' - \tfrac{1}{2}(\partial_\mu \psi)^2 - \tfrac{1}{4}e^{-a\psi}F_{\mu\nu}^2\right]$.

$$\mathcal{A} = -\mathcal{B} = \beta \tilde{F}(\rho) \rho^2 d\varphi , \quad e^{2(\phi - \phi_0)} = \tilde{F} = (1 + \beta^2 \rho^2)^{-1} , \quad B = \sigma = 0 .$$

The string model corresponding to the $a = 1$ Melvin background [16] was constructed in [2] and solved in [3]. The general case of arbitrary (β, q) was studied in [3]. The model describing $a = \sqrt{3}$ Melvin background is the simplest possible special case. The reason is that when viewed from higher dimensions the background (1.6) corresponds to a *flat* (but globally non-trivial) 5-dimensional space-time [14,15].[3] This is why the associated string model is explicitly solvable.

The remarkable simplicity of the $a = \sqrt{3}$ Melvin string model makes it a good pedagogical example which we shall discuss first (Section 2), before turning to more general (β, q) models (which no longer correspond to flat higher dimensional spaces but are still solvable). The quantum Hamiltonian of the corresponding type II superstring model will be equal to the free superstring one plus terms linear and quadratic in angular momentum operators. As a result, the mass spectrum can be explicitly determined. We will show that supersymmetry is broken and that there exist intervals of values of moduli parameters (Kaluza-Klein radius and magnetic field strength) for which the model is unstable. The string partition function on the torus is IR finite or infinite depending on the values of the parameters. We shall consider both the Ramond-Neveu-Schwarz and the light-cone Green-Schwarz formulation of the theory (in the latter the breaking of supersymmetry is related to the absence of Killing spinors in the Melvin background).

In Section 3 the results obtained for the $a = \sqrt{3}$ Melvin model will be generalized to the (β, q) class of static magnetic flux tube models and, in particular, to the $a = 1$ Melvin model. We shall explain the reason for solvability of these models and clarify the nature of perturbative instabilities that appear for generic values of the magnetic field parameters.

In Section 4 we shall first explain the relation between the $a = \sqrt{3}$ Melvin model and superstring compactifications on twisted tori where supersymmetry is broken by discrete rotation angles. Then we shall consider the compact version of the $a = \sqrt{3}$ Melvin model which is obtained by 'twisting' $SU(2) \times U(1)$ WZNW model and discuss the issue of supersymmetry breaking by the twist parameters.

---

[3] This is an interesting example of a Kaluza-Klein background which looks non-trivial (curved) from lower-dimensional point of view but is actually flat as 5-dimensional space. That means that the total contribution of the three 4-dimensional fields (metric, vector and scalar) in any local observable will vanish. However, the global properties of the corresponding string theory will be non-trivial. Note also that since the effective Kaluza-Klein radius $e^\sigma R$ grows with ρ, the Kaluza-Klein interpretation may not apply for large q (see [15]).

2. Superstring model for $a = \sqrt{3}$ Melvin background

2.1. Bosonic string model

Let us consider the closed bosonic string propagating in the space $M^D = M^5 \times T^{D-5}$, where T^n is a torus and $M^5 = R_t \times R_{x_3} \times \mathcal{M}^3$. $\mathcal{M}^3$ is flat but globally non-euclidean space which can be represented as a twisted product (symbolically, $\mathcal{M}^3 =: R^2 * S^1$) of the 2-plane R^2 (ρ, φ) with Kaluza-Klein circle S^1 $(y \in (0, 2\pi R))$ (or as a bundle with S^1 as a base and R^2 as a fibre). It can also be obtained [15] by factorizing R^3 over the group generated by translations in two angular directions: in the coordinates where $ds^2 = d\rho^2 + \rho^2 d\theta^2 + dy^2$ one should identify the points $(\rho, \theta, y) = (\rho, \theta + 2\pi n + 2\pi q R m, y + 2\pi R m)$ $(n, m =$integers$)$, i.e. combine the shift by $2\pi R$ in y with a rotation by an arbitrary angle $2\pi q R$ in the 2-plane. In terms of the globally defined 2π-periodic coordinate φ the metric of $\mathcal{M}^3$ is

$$ds^2 = d\rho^2 + \rho^2 (d\varphi + q\,dy)^2 + dy^2 \ . \tag{2.1}$$

It is flat since locally one may introduce the coordinate $\theta = \varphi + qy$ and decouple y from ρ, φ. The global structure of this 3-space is non-trivial: the fixed ρ section is a 2-torus (with ρ-dependent conformal factor and complex modulus) which degenerates into a circle at $\rho = 0$ (the space is actually regular everywhere, including $\rho = 0$).

The Lagrangian describing string propagation in such flat but globally non-trivial M^D is

$$L = L_0 + L_1 \ , \qquad L_0 = -\partial_a t \partial^a t + \partial_a x_\alpha \partial^a x^\alpha \ , \tag{2.2}$$

$$L_1(\mathcal{M}^3) = \partial_a \rho \partial^a \rho + \rho^2 (\partial_a \varphi + q\partial_a y)(\partial^a \varphi + q\partial^a y) + \partial_a y \partial^a y \ . \tag{2.3}$$

Here $\rho \geq 0$ and $0 < \varphi \leq 2\pi$ correspond to the cylindrical coordinates on a (x_1, x_2)-plane, y is a 'Kaluza-Klein' coordinate with period $2\pi R$, and x^α include the flat x^3-coordinate of $D = 4$ space-time and, e.g., 21 (or 5 in the superstring case) internal coordinates compactified on a torus. To give the 4-dimensional interpretation to this model L_1 should be represented in the 'Kaluza-Klein' form

$$L_1 = \partial_a \rho \partial^a \rho + F(\rho)\rho^2 \partial_a \varphi \partial^a \varphi + e^{2\sigma}(\partial_a y + \mathcal{A}_\varphi \partial_a \varphi)(\partial^a y + \mathcal{A}_\varphi \partial^a \varphi) \ , \tag{2.4}$$

where $F^{-1} = e^{2\sigma} = 1 + q^2\rho^2$, $\mathcal{A}_\varphi = q\rho^2 F$. The resulting $D = 4$ background (metric, Abelian vector field $\mathcal{A}_\mu$ and scalar σ) is indeed the $a = \sqrt{3}$ Melvin geometry (1.6). The parameter q thus has the interpretation of the magnetic field strength at the core of the flux tube. From the 4-dimensional point of view this model describes the motion of charged string states in the $a = \sqrt{3}$ Melvin magnetic flux tube background.[4]

Although the 5-space is flat, the string theory (2.3) will be non-trivial already at the classical level (due to the existence of winding string states) and also at the quantum level in the non-winding sector (where there will be a 'magnetic' coupling to the total angular momentum in the 2-plane). This represents an example of a gravitational 5d (space-time) Aharonov-Bohm-type phenomenon: the value of the magnetic field strength parameter q does not influence the (zero) curvature of the space but affects the global properties like masses of string states.

Since $\mathcal{M}^3$ is flat, the model (2.3) is conformal for arbitrary values of the two parameters q and R. Certain values of these moduli are special: if $qR = n$, $n = 0, \pm 1, ...$, the coordinate θ is globally defined (2π periodic) and so (2.4) is equivalent to a free bosonic string theory compactified on a circle.[5] Models with $n < qR < n + 1$ are equivalent to models with $0 < qR < 1$. This periodicity condition in qR will be modified in the superstring theory: because of the presence of fermions of half-integer spin n will be replaced by $2n$, i.e. only models with $qR = 2n$ will be trivial (more generally, superstring theories with (R, q) and $(R, q + 2nR^{-1})$ will be equivalent).

The Lagrangian (2.4) has the following useful form ($x = x_1 + ix_2 = \rho e^{i\varphi}$) :

$$L_1 = (\partial_a x_i - q\epsilon_{ij} x_j \partial_a y)(\partial^a x_i - q\epsilon_{ij} x_j \partial^a y) + \partial_a y \partial^a y \tag{2.5}$$

$$= D^a x D_a^* x^* + \partial_a y \partial^a y , \qquad D_a \equiv \partial_a + iA_a , \qquad A_a \equiv q\partial_a y ,$$

---

[4] We assume that q is a continuous parameter. There is no reason for its quantization at the level of string model (2.3). Given the magnetic flux tube (1.6) interpretation of this model, one may, however, wonder how this is reconciled with the flux quantization in similar magnetic backgrounds like the Higgs scalar vortex or magnetic monopole on 2-sphere. Though the magnetic flux through the (ρ, φ) 2-plane is finite in the case of the Melvin background, the topological argument for its quantization (cf. [12]) does not apply since the 2-space is non-compact (the $\rho = \infty$ point is not part of the space). Given a magnetic field configuration with a finite flux through a 2-plane, the flux may be quantised once charges are added since this may corresponds to a state of minimal energy (cf. the case of the Cooper pairs in superconductors; the minimal energy condition leads also to the asymptotic condition $D_\mu \phi = 0$, $ieA_\mu = \partial_\mu \phi$, etc., in the case of the scalar vortex).

[5] The trivial models with $qR = n$ may still look non-trivial from lower dimensional Kaluza-Klein point of view. The equivalence between higher dimensional and lower dimensional descriptions is, of course, established once the contribution of the whole tower of higher massive Kaluza-Klein states is taken into account.

where the 2d gauge potential A_a is flat (locally pure gauge). Since y is compact, the effect of this gauge potential will be non-trivial if the world sheet will have non-trivial holonomy (2d or 'world sheet' Aharanov-Bohm effect).

The flatness of the potential $A_a = q\partial_a y$ in (2.5) implies that x can be formally 'rotated' to decouple it from y. Then y satisfies the free-field equation and x is also expressed in terms of free fields. The only interaction which effectively survives in the final expressions is the coupling of x to the derivative of the zero mode part of y (e.g., $y_* = y_0 + 2\alpha' p\tau + 2Rw\sigma$ in the case of the cylinder as a world sheet). It is then straightforward to carry out the canonical quantization procedure, expressing all observables in terms of free oscillators. The resulting Hamiltonian will be given by the sum of the free string Hamiltonian plus $O(q)$ and $O(q^2)$ terms depending on the left and right components of the free string angular momentum operators $\hat{J}_L$ and $\hat{J}_R$ [3,4].

This bosonic string model is stable in the non-winding sector, where there are no new instabilities in addition to the usual flat space tachyon [3]. This means, in particular, that the Kaluza-Klein field theory corresponding to the Melvin background is perturbatively stable with respect to the 'massless' (graviton, vector, scalar) *and* massive perturbations (the theory may still be unstable at a non-perturbative level [15]). At the same time, there exists a range of parameters q and R for which there are tachyonic states in the *winding* sector. This instability (whose origin is essentially in the gyromagnetic coupling term $wqR(\hat{J}_R - \hat{J}_L)$ which may have a negative sign, see below and cf. (1.1),(1.2)) is not related to the presence of the flat bosonic string tachyon and survives also in the superstring case [4].

2.2. Solution of the superstring model

Let us now consider the type II superstring version of (2.2) using first the RNS formulation of the model. The $(1,1)$ world-sheet supersymmetric extension of the model (2.3),(2.5) has the form ($x^\mu \equiv (x^i, y)$)

$$L_{RNS} = G_{\mu\nu}(x)\partial_+ x^\mu \partial_- x^\nu + \lambda_{Rm}D_+\lambda_R^m + \lambda_{Lm}D_-\lambda_L^m \ , \tag{2.6}$$

$$D_{\pm n\mu}^m \equiv \delta_n^m \partial_\pm + \omega_{n\mu}^m \partial_\pm x^\mu \ .$$

$\lambda^m = e_\mu^m \lambda^\mu$ are vierbein components of the 2d Majorana-Weyl spinors and $\omega_{n\mu}^m$ is flat spin connection (so that the quartic fermionic terms are absent).[6] In terms of the left and right Weyl spinors $\lambda = \lambda_1 + i\lambda_2$ corresponding to $x = x_1 + ix_2$ and $\lambda^y \equiv \psi$, we get (cf. (2.5))

$$L_{RNS} = \tfrac{1}{2}(D_+xD_-^*x^* + c.c.) + \partial_+ y\partial_- y + \lambda_R^* D_+\lambda_R + \lambda_L^* D_-\lambda_L \tag{2.7}$$

---

[6] In the basis $e^i = dx^i - q\epsilon^{ij}x_j dy$, $e^y = dy$, the spin connection 1-form has the following components: $\omega^{ij} = -q\epsilon^{ij}dy$, $\omega^{iy} = 0$.

$$+ \psi_R \partial_+ \psi_R + \psi_L \partial_- \psi_L , \qquad D_\pm \equiv \partial_\pm + iq\partial_\pm y ,$$

where the covariant derivative $D_\pm$ is the same as in (2.5), i.e. it contains the flat 2d $U(1)$ potential. This means that, as in the bosonic case, it is possible to redefine the fields x, λ so that the only non-trivial coupling will be to the zero mode of y. Although it may seem that, as in the bosonic case, the model with $qR = n$ should be equivalent to the free superstring theory compactified on a circle (since for $qR = n$ one can, in principle, eliminate the coupling terms in (2.7) by rotating the fields) this will not actually be true unless the integer n is even, $n = 2k$. The non-triviality for $n = 2k+1$ is directly related to the presence of space-time fermions in the spectrum, which change sign under 2π spatial rotation accompanying the periodic shift in y. This will be obvious in the GS formulation (see below).

Taking the world sheet to be a cylinder ($0 < \tau < \infty, 0 < \sigma \le \pi$) so that x, y, λ obey the usual closed-string boundary conditions

$$x(\tau, \sigma + \pi) = x(\tau, \sigma) , \qquad y(\tau, \sigma + \pi) = y(\tau, \sigma) + 2\pi R w , \qquad w = 0, \pm 1, \ldots , \qquad (2.8)$$

$$\lambda_{R,L}(\tau, \sigma + \pi) = \pm \lambda_{R,L}(\tau, \sigma) , \qquad (2.9)$$

we can solve the classical equations corresponding to (2.7) by introducing the fields X and $\Lambda_{R,L}$, which satisfy the free string equations but have 'twisted' boundary conditions $(\sigma_\pm \equiv \tau \pm \sigma)$[7]

$$x(\tau, \sigma) = e^{-iqy(\tau, \sigma)} X(\tau, \sigma) , \qquad \partial_+ \partial_- X = 0 , \qquad X = X_+(\sigma_+) + X_-(\sigma_-) , \qquad (2.10)$$

$$X(\tau, \sigma + \pi) = e^{2\pi i \gamma} X(\tau, \sigma) , \qquad \gamma \equiv qRw , \qquad (2.11)$$

$$\lambda_{R,L}(\tau, \sigma) = e^{-iqy(\tau, \sigma)} \Lambda_{R,L}(\tau, \sigma) , \qquad \partial_\pm \Lambda_{R,L} = 0 , \qquad \Lambda_{R,L} = \Lambda_{R,L}(\sigma_\mp) , \qquad (2.12)$$

$$\Lambda_{R,L}(\tau, \sigma + \pi) = \pm e^{2\pi i \gamma} \Lambda_{R,L}(\tau, \sigma) , \qquad (2.13)$$

with the signs '$\pm$' in (2.13) corresponding to the Ramond (R) and Neveu-Schwarz (NS) sectors. The crucial observation is that if x, λ are on shell, y still satisfies the free-field equation:

$$\partial_+ \partial_- y = 0 , \qquad y = y_* + y' , \qquad y_* = y_0 + 2\alpha' p\tau + 2Rw\sigma . \qquad (2.14)$$

The explicit expressions for the fields $X = X_+ + X_-$ and $\Lambda_{L,R}$ are then

$$X_\pm(\sigma_\pm) = e^{\pm 2i\gamma\sigma_\pm} \mathcal{X}_\pm(\sigma_\pm) , \qquad \mathcal{X}_\pm(\sigma_\pm \pm \pi) = \mathcal{X}_\pm(\sigma_\pm) , \qquad (2.15)$$

$$\Lambda_{L,R}(\sigma_\pm) = e^{\pm 2i\gamma\sigma_\pm} \eta_{L,R}(\sigma_\pm) , \qquad (2.16)$$

---

[7] The twist parameter γ can be interpreted as a flux corresponding to the 2d field $A_a = q\partial_a y$ on the cylinder, $\int A = 2qRw \int d\sigma = 2\pi\gamma$.

where $\mathcal{X}_\pm$ and $\eta_{L,R}$ are the free fields with the standard free closed string boundary conditions, i.e. having the standard oscillator expansions, e.g.,

$$\mathcal{X}_+ = i\sum_{n\in\mathbb{Z}} \tilde{a}_n e^{-2in\sigma_+} , \quad \mathcal{X}_- = i\sum_{n\in\mathbb{Z}} a_n e^{-2in\sigma_-} , \tag{2.17}$$

$$\eta_R^{(NS)} = \sum_{r\in\mathbb{Z}+\frac{1}{2}} c_r \, e^{-2ir\sigma_-} , \quad \eta_R^{(R)} = \sum_{n\in\mathbb{Z}} d_n \, e^{-2in\sigma_-} ,$$

and similar expressions for the left fermions with oscillators having extra tildes. One can then proceed with the canonical quantization of the model expressing the observables in terms of the above free oscillators. It is convenient to choose the light-cone gauge, eliminating the oscillator part of $u = y - t$ (see [3,17,4] for details). Then the string states are parametrised by the following global quantum numbers: the total energy E, the Kaluza-Klein linear momentum number m ($p_y = mR^{-1}$) and the winding number w, the orbital momenta l_R and l_L in the 2-plane (analogues of the Landau level which replace the linear momenta p_1, p_2) and by the continuous momentum p_3 (as well as by discrete momenta corresponding to extra 5 toroidal directions).

The left and right angular momentum operators in the 2-plane contain the orbital momentum parts plus the spin parts

$$\hat{J}_{L,R} = \pm(l_{L,R} + \tfrac{1}{2}) + S_{L,R} , \quad \hat{J} \equiv \hat{J}_L + \hat{J}_R = l_L - l_R + S_L + S_R , \tag{2.18}$$

where the orbital momenta $l_{L,R} = 0, 1, 2, \ldots$ and $S_{L,R}$ have the standard free superstring expressions [18] in terms of free oscillators, e.g., $S_R = \sum_{n=1}^{\infty} \left(b_{n+}^\dagger b_{n+} - b_{n-}^\dagger b_{n-}\right)$+fermionic terms ($b_n$ are bosonic oscillators a_n in (2.17) rescaled by factors of $(n \pm \gamma)^{1/2}$). In the case when $\gamma = 0$ (or, more generally, $\gamma = n$) the zero-mode structure changes in that the translational invariance in the 2-plane is restored (the zero mode oscillators are replaced by the standard zero-mode operators $x_{1,2}, p_{1,2}$ [10,3]). The number of states operators $\hat{N}_R$ and $\hat{N}_L$ have the standard expressions [18] in terms of the oscillators in (2.17) so that $\hat{N}_{R,L} = N_{R,L} - a$, $a^{(R)} = 0$, $a^{(NS)} = \tfrac{1}{2}$, $N_R = \sum_{n=1}^{\infty} n(b_{n+}^\dagger b_{n+} + b_{n-}^\dagger b_{n-} + \ldots)$, etc. Under the usual GSO projection (which is implied by the GS formulation and is necessary for correspondence with the free RNS superstring theory in the limit $q = 0$ but will not automatically lead to space-time supersymmetry for generic q) $\hat{N}_R$ and $\hat{N}_L$ can take only non-negative integer values They satisfy the standard constraint $\hat{N}_R - \hat{N}_L = mw$.

Computing the stress tensor one finds the resulting expression for the light-cone gauge Hamiltonian[8]

$$\hat{\mathcal{H}} = \hat{\mathcal{H}}_0 - \alpha' q(Q_L \hat{J}_R + Q_R \hat{J}_L) + \tfrac{1}{2}\alpha' q^2 \hat{J}^2 , \tag{2.19}$$

---

[8] Here p_α^2 includes p_3^2 as well as the contributions of the linear and winding momenta in other 5 free compactified dimensions (for simplicity we shall sometimes set them equal to zero).

$$\hat{\mathcal{H}}_0 \equiv \tfrac{1}{2}\alpha'\big(-E^2 + p_\alpha^2 + \tfrac{1}{2}Q_L^2 + \tfrac{1}{2}Q_R^2\big) + \hat{N}_R + \hat{N}_L, \qquad Q_{L,R} = mR^{-1} \pm \alpha'^{-1}Rw.$$

It can be interpreted as describing charged states of closed superstring compactified on S^1 moving in the Melvin flux tube background. $\hat{\mathcal{H}}$ is different from the Hamiltonian of the free string on a circle $\hat{\mathcal{H}}_0$ by $O(q)$ ('gyromagnetic' interaction, cf. (1.1),(1.2)) and $O(q^2)$ (charge-independent 'gravitational' interaction) terms. Charged string states are 'trapped' by the magnetic field (they cannot move freely in the 2-plane having discrete orbital momentum numbers l_L, l_R instead of continuous linear momenta p_1, p_2).

$\hat{\mathcal{H}}$ can be represented also in the following ('free superstring compactified on a circle') form

$$\hat{\mathcal{H}} = \tfrac{1}{2}\alpha'\big[-E^2 + p_\alpha^2 + (m - qR\hat{J})^2 R^{-2} + \alpha'^{-2}w^2 R^2\big] + \hat{N}_R + \hat{N}_L - qRw(\hat{J}_R - \hat{J}_L)$$
$$= \tfrac{1}{2}\alpha'\big(-E^2 + p_\alpha^2 + m'^2 R^{-2} + \alpha'^{-2}w^2 R^2\big) + \hat{N}_R' + \hat{N}_L' , \tag{2.20}$$

where[9] $m' \equiv m - qR\hat{J}$, $\hat{N}_R' \equiv \hat{N}_R - \gamma\hat{J}_R$, $\hat{N}_L' \equiv \hat{N}_L + \gamma\hat{J}_L$, $\gamma \equiv qRw$. The Virasoro condition $\hat{\mathcal{H}} = 0$ then leads to the expression for the mass spectrum $M^2 \equiv E^2 - p_\alpha^2$. The mass spectrum is invariant under

$$q \to q + 2nR^{-1} , \qquad n = 0, \pm 1, \dots , \tag{2.21}$$

since (for $w = 0$) this transformation can be compensated by $m \to m - 2n\hat{J} =$ integer. Note that because $\hat{J}$ can take both integer (NS-NS, R-R sectors) and half-integer (NS-R, R-NS sectors) values, the symmetry of the bosonic part of the spectrum $q \to q + nR^{-1}$ is not a symmetry of its fermionic part, i.e. the full superstring spectrum is invariant only under (2.21).

The same conclusion about the periodicity in q is true in general for $w \neq 0$. In the form given above, eq. (2.19) is valid for $0 \leq w < (qR)^{-1}$, i.e. for $0 \leq \gamma < 1$. The generalization to other values of γ is straightforward [3,4]: one is to replace $\gamma = qRw$ in (2.20) by $\hat{\gamma} \equiv \gamma - [\gamma]$, where $[\gamma]$ denotes the integer part of γ ($0 \leq \hat{\gamma} < 1$, cf. (1.2)). For fixed radius R the mass spectrum is thus periodic in q, i.e. it is mapped into itself under (2.21) (combined with $m \to m - 2n\hat{J}$). In the case of $qR = 2n$ (i.e. $\gamma = 2nw = 2k$) the spectrum is thus equivalent to that of the free superstring compactified on a circle. For $qR = 2n + 1$ (i.e. $\gamma = (2n + 1)w = 2k + 1$ if w is odd) the spectrum is the same as that of the free superstring compactified on a circle with antiperiodic boundary conditions for space-time fermions [19] (see also [20,21]). This relation will become clear in the GS formulation discussed below.[10]

---

[9] Up to the orbital momentum terms, $\hat{N}_{R,L}'$ can be put into the same form as free operators $\hat{N}_{R,L}$ with the factor n replaced by $n \pm \gamma$. This is related to the fact that the model we are solving is 'locally trivial', i.e. the q-dependence could be eliminated by a rotation of coordinates if not for the global effects.

[10] In particular, it will be apparent that the interaction term in the superstring action can be eliminated by a globally defined field transformation only if $qR = 2n$, while for $qR = 2n + 1$ this can be done at the expense of imposing antiperiodic boundary conditions (in σ or in the y-direction) on fermions (under the rotation by the angle $2\pi qR = 2\pi$ in the 2-plane, which is associated with a periodic shift in y, the bosons remain invariant but spinors change sign).

2.3. Green-Schwarz formulation

Given a generic curved bosonic background, the corresponding Green-Schwarz (GS) superstring action [22] defines a complicated non-linear 2d theory. When one is able to fix a light-cone gauge and, moreover, the background geometry is flat as in the case of the Melvin model (2.3) (so that conformal invariance and κ-supersymmetry are guaranteed) the action becomes very simple (cf. (2.6))[11]

$$L_{GS} = G_{\mu\nu}(x)\partial_+ x^\mu \partial_- x^\nu + iS_R \mathcal{D}_+ S_R + iS_L \mathcal{D}_- S_L , \tag{2.22}$$

$$\mathcal{D}_a \equiv \partial_a + \tfrac{1}{4}\gamma_{mn}\omega_\mu^{mn}\partial_a x^\mu .$$

Here $S^p_{R,L}$ $(p = 1, ..., 8)$ are the right and left real spinors of $SO(8)$ (we consider type IIA theory). In the case of (2.3) we get (cf. (2.5))

$$L_{GS} = (\partial_+ + iq\partial_+ y)x(\partial_- - iq\partial_- y)x^* + \partial_+ y\partial_- y \tag{2.23}$$

$$+ iS_R(\partial_+ - \tfrac{1}{4}qe^{ij}\gamma_{ij}\partial_+ y)S_R + iS_L(\partial_- - \tfrac{1}{4}qe^{ij}\gamma_{ij}\partial_- y)S_L .$$

It is natural to decompose the $SO(8)$ spinors according to $SO(8) \to SU(4) \times U(1)$, i.e. $S^p_L \to (S^r_L, \bar{S}^r_L)$, $S^p_R \to (S^r_R, \bar{S}^r_R)$, $r = 1, .., 4$. Then the fermionic terms in (2.23) become

$$L_{GS}(S) = i\bar{S}^r_R(\partial_+ + \tfrac{1}{2}iq\partial_+ y)S^r_R + i\bar{S}^r_L(\partial_- - \tfrac{1}{2}iq\partial_- y)S^r_L . \tag{2.24}$$

The connection terms in the covariant derivatives in the fermionic part of the GS action (2.24) have extra coefficients $\tfrac{1}{2}$ with respect to the ones in the RNS action (2.7). This immediately implies that the full theory is periodic under $qR \to qR + 2n$.

The condition that the GS action (2.22),(2.23) has residual supersymmetry invariance $S \to S + \epsilon(x)$ is equivalent to $\mathcal{D}_a \epsilon(x(\tau,\sigma)) = \partial_a x^\mu(\partial_\mu + \tfrac{1}{4}\gamma_{mn}\omega_\mu^{mn})\epsilon(x) = 0$. The absence of supersymmetry invariance is the consequence of the absence of zero modes of the above covariant derivative operators, or, equivalently, of the non-existence of solutions of the Killing spinor equation

$$(\partial_\mu + \tfrac{1}{4}\gamma_{mn}\omega^{mn}_{\mu})\epsilon = 0 . \tag{2.25}$$

In the $D = 3$ background corresponding to (2.5) $\epsilon = \epsilon(x^i, y)$ is a space-time spinor and ω^{mn}_{μ} is the same flat spin connection as in (2.6),(2.7) so that (2.25) reduces (after $SU(4) \times U(1)$ split of ϵ as in (2.24)) to

$$(\partial_y \mp \tfrac{1}{2}iq)\epsilon = 0 . \tag{2.26}$$

---

[11] Its form [23,24,25] can be explicitly determined, e.g., by comparing [24] with the known light-cone superstring vertex operators [18].

The formal solution of (2.26) $\epsilon(y) = \exp(\pm\frac{1}{2}iq) \epsilon(0)$ does not, however, satisfy the periodic boundary condition in y, $\epsilon(y + 2\pi R) = \epsilon(y)$ (unless $qR = 2n$ when the Killing spinor does exist, in agreement with the fact that in this case the theory is equivalent to the free superstring). The conclusion is that for $qR \neq 2n$ there is no residual space-time supersymmetry in the higher-dimensional (e.g., $D = 5$ supergravity) counterpart of the $a = \sqrt{3}$ Melvin background.[12]

As in the bosonic and RNS cases, one can explicitly solve the classical string equations corresponding to (2.23) (cf. (2.12),(2.13))

$$S_{R,L}(\tau,\sigma) = e^{-\frac{i}{2}qy(\tau,\sigma)}\Sigma_{R,L}(\sigma_{\mp}) , \quad \Sigma_{R,L}(\tau,\sigma + \pi) = e^{i\pi\gamma}\Sigma_{R,L}(\tau,\sigma) , \qquad (2.27)$$

with the final result that the only essential difference, as compared to the free superstring case, is the coupling of bosons and fermions to the zero-mode part of the flat $U(1)$ connection $\partial_a y_*$. The expressions for the superstring Hamiltonian and mass spectrum are effectively the same as in the RNS approach (2.19).[13]

In general, the model with $qR = 2n + 1$ ($\gamma = 2k + 1$ for odd w) is equivalent to the free superstring compactified on a twisted 3-torus (in the limit when the 2-torus part is replaced by 2-plane), or on a circle with antiperiodic boundary conditions for the fermions [19] (in particular, the theory with $qR = 1$ and $R < \sqrt{2\alpha'}$ will have tachyons).

The fundamental world-sheet fermions S that appear in GS action (2.22) are always *periodic* in σ (this is necessary for supersymmetry of the model in the $q = 0$ limit). This implies that the 'redefined' fermions Σ in (2.27) must change phase under a shift in y-direction. For $qR = 2n + 1$ this results in antiperiodic boundary conditions for *space-time* fermions as functions of y (the space-time fields can be represented, e.g., as coefficients in expansion of a super string field $\Phi(y, S, ...)$ in powers of world-sheet fermions). As a result, there exists a 1-parameter family of models interpolating between the standard supersymmetric $qR = 0$ model with fermions which are periodic in y and a non-supersymmetric $qR = 1$ model with fermions which are antiperiodic in y.

---

[12] Even though the $D = 5$ supergravity background M^5 is flat, it is the presence of the flat but non-trivial spin connection that leads to the breaking of supersymmetry. Let us note also that the absence of Killing spinors in the case of the $a = 0$ Melvin solution of the Einstein-Maxwell theory was pointed out in [12].

[13] For $2k \leq \gamma < 2k+1$ the operators $\hat{N}_{L,R}$, $\hat{J}_{L,R}$ have the usual free GS superstring form, which is similar to their form in the R-sector of the RNS formalism with vanishing zero-point energy. For $2k - 1 \leq \gamma < 2k$ the operators $\hat{N}_{L,R}$ have the 'NS-sector' form, i.e. they take half-integer eigenvalues starting from $-\frac{1}{2}$.

2.4. Mass spectrum: supersymmetry breaking and (in)stability

The expression for the mass spectrum that follows from (2.19),(2.20) is

$$M^2 \equiv E^2 - p_\alpha^2 = M_0^2 - 2qR^{-1}m\hat{J} - 2\alpha'^{-1}qRw(\hat{J}_R - \hat{J}_L) + q^2\hat{J}^2 \tag{2.28}$$

$$= 2\alpha'^{-1}(\hat{N}_L + \hat{N}_R) + (m - qR\hat{J})^2 R^{-2} + \alpha'^{-2}w^2R^2 - qRw(\hat{J}_R - \hat{J}_L) , \tag{2.29}$$

where $M_0^2 = 2\alpha'^{-1}(\hat{N}_L + \hat{N}_R) + m^2 R^{-2} + \alpha'^{-2}w^2R^2$ is the mass operator of the free superstring compactified on a circle. It is easy to see that in general M^2 is not positive definite in the winding ($w \neq 0$) sector because of the last $O(qRw)$ gyromagnetic interaction term in (2.29). Thus one should expect the presence of instabilities, in agreement with the magnetic interpretation of the model and the existence of charged higher spin states in the spectrum.

Indeed, it follows from (2.28) that (i) the space-time supersymmetry is broken for $qR \neq 2n$, and (ii) there exists a range of values of parameters q and R for which there are tachyonic states in the spectrum.

The breaking of supersymmetry is of course expected in view of the magnetic interpretation of the model (the coupling is spin-dependent). Suppose that we start with the free superstring compactified on a circle y and study what happens with the spectrum when we switch on the magnetic field, $q \neq 0$. Since the mass shift in (2.28) involves *both* components $\hat{J}_L$ and $\hat{J}_R$ of the angular momentum the masses of bosons and fermions that were equal for $q = 0$ will become different for $q \neq 0$ (it is impossible to have both $\hat{J}_L$ and $\hat{J}_R$ equal for bosons and fermions). Supersymmetry is absent already in the non-winding sector (where the coupling is to the total angular momentum $\hat{J}$).

For example, the free superstring massless ground states ($\hat{N}_{L,R} = 0 = m = w$) will, according to (2.28), get masses $M = |q\hat{J}|$ proportional to their total angular momenta, which must be integer for bosons and half-integer for fermions (cf. (2.18)). Note that these states are neutral, so that from the 4-dimensional point of view the shift in the masses can be interpreted as a gravitational effect. This shift implies, in particular, that supersymmetry is broken already at the field-theory ($D = 5$ or $D = 4$ supergravity) level, in agreement with the absence of Killing spinors in the $D = 4$ Melvin background discussed above.

In the absence of supersymmetry some instabilities of the bosonic string model may survive also in the superstring case. The mass operator (2.29) is positive in the non-winding sector, but, as in the bosonic case, tachyonic states may appear in the winding sector. Consider, for example, the NS-NS winding states with zero Kaluza-Klein momentum and zero orbital momentum quantum numbers and with maximal absolute values of the spins $S_{R,L}$ at given levels (leading Regge trajectory)

$$w > 0 , \quad m = 0 , \quad l_R = l_L = 0 , \quad S_R = \hat{N}_R + 1 , \quad S_L = -\hat{N}_L - 1 . \tag{2.30}$$

We shall assume that $0 < qRw < 1$ (states with $w > (qR)^{-1}$ can be analysed in a similar way). Then $\hat{N}_R = \hat{N}_L \equiv N$, $\hat{J} = 0$, $\hat{J}_R - \hat{J}_L = 2N + 1$, and

$$\alpha' M^2 = 4N + \alpha'^{-1} w^2 R^2 - 2qRw(2N + 1) . \qquad (2.31)$$

The state with given N and w will be tachyonic for $q > q_{crit}$, $q_{crit} = \frac{4N + \alpha'^{-1} w^2 R^2}{2(2N+1)wR}$. For $N = 0$ we get $\alpha' q_{crit} = \frac{1}{2}wR$. The condition $qRw < 1$ is satisfied provided $wR < \sqrt{2\alpha'}$.

In general, states with $M^2 < 0$ can be present only for $R < \sqrt{2\alpha'}$, i.e. the full spectrum is *tachyon-free* if $R > \sqrt{2\alpha'}$. For fixed $R < \sqrt{2\alpha'}$ the minimal value of the magnetic field strength parameter at which tachyons first appear is $\alpha' q_{crit} = \frac{1}{2}R$, corresponding to the $N = 0, w = 1$ case discussed above.

All other sectors (R-R, R-NS, NS-R) are tachyon-free. The absence of tachyons in the fermionic sectors is a direct consequence of unitarity. Since a unitary tree-level S-matrix should correspond to a string field theory with a hermitian action, the 'square' of hermitian fermionic kinetic operator should be positive in any background. This translates into the positivity of M^2 for the fermionic states in the case of static backgrounds. One implication is that similar models should have another general property of the spectrum: the states which become tachyonic should originate only from the states of the free superstring spectrum which belong to the leading Regge trajectory [4]. If there were bosonic tachyons not only on the leading Regge trajectory, but also on the subleading one, then a fermionic state with an 'intermediate' value of the spin (but otherwise the same quantum numbers) would have $M^2 < 0$. Since this is not allowed by unitarity, in any unitary superstring model corresponding to a static background tachyonic states can only appear on the first (bosonic) Regge trajectory. This is indeed true in the open superstring case (2.9) and in the present and more general closed superstring models discussed in Section 4 [4].

Let us note also that the fact that some higher spin winding states may become tachyonic means that that there are new *massless* states at the critical values of the magnetic field. This suggests a possibility of symmetry enhancement in similar models. The magnetic perturbations may also reveal certain hidden symmetries of the superstring spectrum.[14]

One can consider also the heterotic version of the above model (where the magnetic field is embedded in the Kaluza-Klein sector) by combining the 'left' or 'right' part of the superstring model with the free internal part. Then [4] (cf. (2.29))

$$\alpha' M^2 = 2(\hat{N}_R + \hat{N}_L) + p_I^2 + \alpha'(mR^{-1} - q\hat{J})^2 + \alpha'^{-1} w^2 R^2 - 2qRw(\hat{J}_R - \hat{J}_L), \qquad (2.32)$$

where $\hat{N}_R - \hat{N}_L = mw + \frac{1}{2}p_I^2$, $\hat{N}_R = 0, 1, 2, ...$, $\hat{N}_L = N_L - 1 = -1, 0, 1, ...$. In addition to the instabilities discussed above there are also new ones, which (for the 'self-dual' value of the radius $R = \sqrt{\alpha'}$) appear for infinitesimal values of the magnetic field. These are the usual Yang-Mills-type magnetic instabilities, associated with the gauge bosons ($m = w = \pm 1$, $p_I^2 = l_R = l_L = 0$, $\hat{N}_R = N_L = 0$, $S_R = 1$, $S_L = 0$) of the $SU(2)_L$ group.

---

[14] Let us note in this connection that a special symmetry of a general class of tachyon-free string models with finite 1-loop cosmological constant was discussed in [26].

2.5. Partition function

The basic properties of the spectrum are reflected in the 1-loop (torus) partition function Z of the model which will be non-vanishing for $q \neq 2nR^{-1}$ due to the absence of the GS fermionic zero modes, i.e. the absence of supersymmetry. Z is straightforward to compute by computing the path integral in the GS formulation [4]. The first step is to expand y in eigen-values of the Laplacian on the 2-torus and redefine the fields x, x^* and $S_{L,R}, \bar{S}_{L,R}$ in (2.23),(2.24) to eliminate the non-zero-mode part of y from the $U(1)$ connection. The zero-mode part of y on the torus $(ds^2 = |d\sigma_1 + \tau d\sigma_2|^2, \ \tau = \tau_1 + i\tau_2, \ 0 < \sigma_a \leq 1)$ is $y_* = y_0 + 2\pi R(w\sigma_1 + w'\sigma_2)$, where w, w' are integer winding numbers. Integrating over the fields x, x^* and $S^r_{L,R}, \bar{S}^r_{L,R}$, we get a ratio of determinants of scalar operators of the type $\partial + iA, \ \bar{\partial} - i\bar{A}$ with constant connection $A = q\partial y_* = \pi\chi, \ \chi \equiv qR(w' - \tau w)$. The partition function has the simple form [4]

$$Z(R, q) = cV_7 R \int \frac{d^2\tau}{\tau_2^2} \sum_{w,w'=-\infty}^{\infty} \exp\left(-\pi(\alpha'\tau_2)^{-1} R^2 |w' - \tau w|^2\right) \tag{2.33}$$

$$\times \ \mathcal{Z}_0(\tau, \bar{\tau}; \chi, \bar{\chi}) \ \frac{Y^4(\tau, \bar{\tau}; \frac{1}{2}\chi, \frac{1}{2}\bar{\chi})}{Y(\tau, \bar{\tau}; \chi, \bar{\chi})} \ ,$$

where

$$Y(\tau, \bar{\tau}; \chi, \bar{\chi}) = \exp[\frac{\pi(\chi - \bar{\chi})^2}{2\tau_2}] \left|\frac{\theta_1(\chi|\tau)}{\chi\theta_1'(0|\tau)}\right|^2 . \tag{2.34}$$

The factor $\mathcal{Z}_0$ in (2.33) stands for the contributions of the integrals over the constant fields $x, x^*, S_{L,R}, \bar{S}_{L,R}$ which become zero modes in the free-theory $(q = 0)$ limit

$$\mathcal{Z}_0 = \frac{(\frac{1}{2}\chi\tau_2^{-1/2})^4 \ (\frac{1}{2}\bar{\chi}\tau_2^{-1/2})^4}{\chi\bar{\chi}\tau_2^{-1}} = 2^{-8}q^6 R^6 |w' - \tau w|^6 \tau_2^{-3} . \tag{2.35}$$

$\mathcal{Z}_0$ vanishes for $q \to 0$ in agreement with the restoration of supersymmetry in this limit.[15]

The partition function vanishes at all supersymmetric points $qR = 2n$ where the fermionic determinants have zero modes (θ_1-functions in Y-factors in (2.33) have zeros for any w, w'). More generally, Z is periodic in q (see (2.21))

$$Z(R, q) = Z(R, q + 2nR^{-1}) , \qquad n = 0, \pm 1, \dots . \tag{2.36}$$

For $qR = 2n + 1$ the partition function is the same as that of the free superstring compactified on a circle with antiperiodic boundary conditions for space-time fermions [19] (the

---

[15] The $q \to 0$ divergence of the bosonic 'constant mode' factor $\sim q^{-2}$ corresponds to the restoration of the translational invariance in the x_1, x_2-plane in the zero magnetic field limit (this infrared divergence reproduces the factor of the area of the 2-plane).

dependence on odd qR can be eliminated from (2.24) at the expense of making $\mathcal{S}_{R,L}, \bar{\mathcal{S}}_{R,L}$ to satisfy antiperiodic boundary conditions in σ or y, cf. (2.27)).

Z is infrared-divergent for those values of the moduli q and R for which there are tachyonic states in the spectrum and is finite for all other values. In particular, it is finite for $R > \sqrt{2\alpha'}$ and arbitrary q. Z is usually interpreted as a cosmological term or 1-loop effective potential for the moduli (so that one may study its extrema, etc., cf. [27]). Since (2.33) is non-negative and periodic in qR, its minima are at supersymmetric points. It is not clear, however, that this interpretation of Z as an effective potential applies in the present case of a space-time (i.e. not internal space) string model since the $D = 4$ background is curved (Z is rather the full 1-loop effective action evaluated on a classical solution).

3. Superstring models for $a = 1$ Melvin and more general static magnetic flux tube backgrounds

In the previous section we have discussed the simplest possible static magnetic flux tube model. Now we shall consider more general models [3,4] corresponding to the 2-parameter flux tube backgrounds (1.4). It turns out that the superstring versions of these models (which depend on compactification radius, vector and axial magnetic field parameters R, q and β) have properties analogous to those of the $\beta = 0$ ($a = \sqrt{3}$ Melvin) model. In particular, supersymmetry is broken for generic values of (β, q) and these models reduce to the free superstring theory when both qR and $\alpha'\beta R^{-1}$ are even integers. The spectrum contains tachyonic states which (in agreement with the general remarks in Section 2.3) appear only in the bosonic NS-NS sector and belong to the leading Regge trajectory.

The string models corresponding to (1.4) with $\beta > q$ are related to the models with $\beta < q$ by the duality transformation in the Kaluza-Klein coordinate y.[16] More precisely, the (R, β, q) model is y-dual to $(\alpha'R^{-1}, q, \beta)$ model so that the $q = \beta$ ($a = 1$ Melvin) is the 'self-dual' point. For fixed q these models thus fill the interval $0 \leq \beta \leq q$ parametrized by β, with $a = \sqrt{3}$ and $a = 1$ Melvin models being the boundary points. The non-trivial $\mathcal{M}^3 = (\rho, \varphi, y)$ part of the corresponding bosonic string Lagrangian is [3] (cf. (2.3))

$$L = \partial_+\rho\partial_-\rho + F(\rho)\rho^2(\partial_+\varphi + q_+\partial_+y)\,(\partial_-\varphi + q_-\partial_-y) \tag{3.1}$$

$$+ \partial_+y\partial_-y + \mathcal{R}[\phi_0 + \tfrac{1}{2}\ln F(\rho)]\,, \qquad F^{-1} = 1 + \beta^2\rho^2\,, \qquad q_\pm \equiv q \pm \beta\,.$$

Note that in addition to the metric of $\mathcal{M}^3$ (which is no longer flat) there is also the antisymmetric tensor and dilaton backgrounds ($\mathcal{R}$ is 2d curvature). This model is related

---

[16] At the level of the effective action (1.3) this transformation is the usual T-duality map, $\mathcal{A} \leftrightarrow \pm\mathcal{B}$, $\sigma \to -\sigma$, $\Phi \to \Phi$, $\hat{G}_{\mu\nu} \to \hat{G}_{\mu\nu}$, $\hat{H}_{\mu\nu\lambda} \to \hat{H}_{\mu\nu\lambda}$, which is obviously a symmetry of (1.3).

to (2.3) by the formal $O(2,2;R)$ duality rotation (combination of a shift of φ by y and duality in y). Indeed, it can be obtained from the model which is y-dual to (2.3) by first changing $q \to \beta$, $\tilde{y} \to y$ and then shifting $\varphi \to \varphi + qy$. This explains why this bosonic model is solvable [3,4] even though the 10-dimensional target space geometry is no longer flat. The equivalent form of (3.1) is

$$L = \partial_+\rho\partial_-\rho + F(\rho)(\partial_+y - \beta\rho^2\partial_+\varphi')(\partial_-y + \beta\rho^2\partial_-\varphi') \tag{3.2}$$

$$+ \rho^2\partial_+\varphi'\partial_-\varphi' + \mathcal{R}[\phi_0 + \tfrac{1}{2}\ln F(\rho)] \,,$$

where we have used the formal notation $\varphi' = \varphi + qy$. Introducing an auxiliary 2d vector field with components V_+, V_- we can represent (3.2) as follows, cf. (2.5) (this corresponds to 'undoing' the duality transformation mentioned above)

$$L = \tfrac{1}{2}(D_+x \ D_-^*x^* + c.c.) + V_+V_- - V_-\partial_+y + V_+\partial_-y \,, \tag{3.3}$$

$$D_\pm \equiv \partial_\pm + i\beta V_\pm + iq\partial_\pm y \,.$$

Now it is easy to understand why the classical equations of this model are explicitly solvable in terms of free fields and the partition function is computable. In spite of the y-dependence in the first term, the equation of motion for y still imposes the constraint that V_a has zero field strength, $\mathcal{F}(V) = \partial_-V_+ - \partial_+V_- = 0$: the variation over y of the first term vanishes once one uses the equation for x (as follows from the fact that qy-terms can be formally absorbed into a phase of x). Then $V_+ = C_+ + \partial_+\tilde{y}$, $V_- = C_- + \partial_-\tilde{y}$, $C_\pm = $ const. In the equations for V_+, V_- one can again ignore the variation of the first term in (3.3) since it vanishes under $\mathcal{F}(V) = 0$. We find that $V_+ = C_+ + \partial_+\tilde{y} = \partial_+y$, $V_- = C_- + \partial_-\tilde{y} = -\partial_-y$. The solution of the model then effectively reduces to that of the model (2.3), the only extra non-trivial contribution being the zero-mode parts of the two dual fields y and $\tilde{y}$. Interchanging of q and β is essentially equivalent (after solving for C_+, C_-) to interchanging of y and $\tilde{y}$ and thus of momentum and winding modes. Eliminating C_+, C_- one gets terms quartic in the angular momentum operators in the final Hamiltonian. Similar approach applies to the computation of the partition function Z. Once x, x^* have been integrated out, the integrals over the constant parts of V_+, V_- cannot be easily computed for $q\beta \neq 0$ and thus remain in the final expression [3,4].

This discussion has a straightforward generalization to superstring case. The corresponding RNS action now contains the quartic fermionic terms which reflect the presence of a non-trivial (generalized) curvature of the space $\mathcal{M}^3$. The direct analogue of the 'first-order' Lagrangian (3.3) is (cf. (2.7))[17]

$$L_{\text{RNS}} = \tfrac{1}{2}(D_+x \ D_-^*x^* + c.c.) + \lambda_R^*D_+\lambda_R + \lambda_L^*D_-\lambda_L \tag{3.4}$$

---

[17] The fermionic part of this Lagrangian is reminiscent of the fermionic models studied in [28].

$$+ V_+ V_- - V_- \partial_+ y + V_+ \partial_- y \ .$$

The final expressions for the Hamiltonian and partition function then look very similar to the bosonic ones (the role of fermions is just to supersymmetrize the corresponding free superstring number of states and angular momentum operators and to cancel certain normal ordering terms). One finds (cf. (2.19))

$$\hat{\mathcal{H}} = \tfrac{1}{2}\alpha'(-E^2 + p_\alpha^2) + \hat{N}_R + \hat{N}_L \tag{3.5}$$

$$+ \tfrac{1}{2}\alpha' R^{-2}(m - qR\hat{J})^2 + \tfrac{1}{2}\alpha'^{-1} R^2 (w - \alpha'\beta R^{-1}\hat{J})^2 - \hat{\gamma}(\hat{J}_R - \hat{J}_L) \ ,$$

where $\hat{N}_R - \hat{N}_L = mw$, $\hat{\gamma} \equiv \gamma - [\gamma]$, $\gamma \equiv qRw + \alpha'\beta R^{-1}m - \alpha'q\beta\hat{J}$. $[\gamma]$ denotes the integer part of γ and the operators $\hat{N}_{R,L}$, $\hat{J}_{R,L}$ are the same as in (2.19).

The duality symmetry in the compact Kaluza-Klein direction y (which interchanges the axial and vector magnetic field parameters β and q) is now manifest: (3.5) is invariant under $R \leftrightarrow \alpha'R^{-1}$, $\beta \leftrightarrow q$, $m \leftrightarrow w$. The resulting expression for the mass spectrum can be written in terms of the 'left' and 'right' magnetic field parameters and charges, $q_\pm \equiv q \pm \beta$, $Q_{L,R} = mR^{-1} \pm \alpha'^{-1}Rw$ (it reduces to (2.28) when $\beta = 0$)

$$M^2 = M_0^2 - 2(q_+ Q_L \hat{J}_R + q_- Q_R \hat{J}_L) + (q_+^2 \hat{J}_R + q_-^2 \hat{J}_L)\hat{J} \ . \tag{3.6}$$

The only states which can be tachyonic are bosonic states on the first Regge trajectory with the maximal value for S_R, minimal value for S_L, and zero orbital momentum, i.e. $\hat{J}_R = S_R - \tfrac{1}{2} = \hat{N}_R + \tfrac{1}{2}$, $\hat{J}_L = S_L + \tfrac{1}{2} = -\hat{N}_L - \tfrac{1}{2}$. Then

$$\alpha' M^2 = 2(\hat{N}_R + \hat{N}_L)(1 - \hat{\gamma}) + \alpha' R^{-2}(m - qR\hat{J})^2 + \alpha'^{-1} R^2(w - \alpha'\beta R^{-1}\hat{J})^2 - 2\hat{\gamma}, \tag{3.7}$$

which is not positive definite due to the last term $-2\hat{\gamma}$. One finds [4] that for generic values of (q, β) there are instabilities (associated with states with high spin and charge) for arbitrarily small values of the magnetic field parameters. The special case of $\beta = 0$ (or $q = 0$), corresponding to the $a = \sqrt{3}$ Melvin model discussed in Section 2, is the only exception: in this (type II) model there are no tachyons below some *finite* value of q. The example which illustrates the generic pattern is the $a = 1$ Melvin model where $q = \beta$ ($q_- = 0, q_+ = 2\beta$) and

$$\alpha' M^2 = 4\hat{N}_R + \alpha' Q_R^2 - 4\hat{\gamma}\hat{J}_R \ , \qquad \gamma = \alpha'\beta Q_L - \alpha'\beta^2 \hat{J} \ . \tag{3.8}$$

If we choose for $R = \sqrt{\alpha'}$ then the states with $w = m$, $\hat{N}_L = 0$, $\hat{J}_R = \hat{N}_R + \tfrac{1}{2}$ and $\hat{J}_L = -\tfrac{1}{2}$ become tachyonic for β in the interval $\beta_1 < \beta < \beta_2$, $\beta_{1,2} = m^{-1}(1 \mp \sqrt{1 - \gamma_{crit}})$, $\gamma_{crit} = m^2/(m^2 + \tfrac{1}{2})$. For large $m = \hat{N}_R$ these magnetic field parameters are very small. These infinitesimal instabilities appear because of the presence of states with arbitrarily large charges and thus are a special property of the *closed* string theory. Unlike the usual Yang-Mills-type magnetic instabilities, they (being associated with higher level states) remain even after the massless-level states get small masses. This suggests that a configuration with generic values of (q, β) will 'decay' to become a stable one with special values of q, β.

The supersymmetry is broken for generic values of the magnetic field parameters β, q (the two magnetic fields couple to both L, R- components of the spin which cannot simultaneously be the same for bosons and fermions). When $qR = 2n_1$ and $\alpha'\beta R^{-1} = 2n_2$, $n_{1,2} = 0, \pm 1, ...$, the theory is equivalent to the free superstring compactified on a circle (in this case $\hat{\gamma} = 0$ and, after appropriate shifts of m, w by integers, (3.5) reduces to the free superstring Hamiltonian). If $qR = 2n_1 + 1$ or $\alpha'\beta R^{-1} = 2n_2 + 1$, then the necessary shift in m or w in the fermionic sector involves half-integer numbers. In these cases the theory can be interpreted as a free superstring on a circle with antiperiodic boundary conditions for space-time fermions.

The corresponding partition function is [3,4] (cf. (2.33))

$$Z(R,q,\beta) = cV_7 R \int \frac{d^2\tau}{\tau_2^2} \int dC d\bar{C} \, (\alpha'\tau_2)^{-1} \sum_{w,w'=-\infty}^{\infty} \tag{3.9}$$

$$\times \exp\left(-\pi(\alpha'\beta^2\tau_2)^{-1}[\chi\bar{\chi} - R(q+\beta)(w' - \tau w)\bar{\chi} - R(q-\beta)(w' - \bar{\tau}w)\chi\right.$$

$$\left. + R^2 q^2 (w' - \tau w)(w' - \bar{\tau}w)]\right) \times Z_0(\tau, \bar{\tau}; \chi, \bar{\chi}) \frac{Y^4(\tau, \bar{\tau}; \frac{1}{2}\chi, \frac{1}{2}\bar{\chi})}{Y(\tau, \bar{\tau}; \chi, \bar{\chi})},$$

where $\chi \equiv 2\beta C + qR(w' - \tau w)$, $\bar{\chi} \equiv 2\beta\bar{C} + qR(w' - \bar{\tau}w)$, and $Y(\tau, \bar{\tau}; \chi, \bar{\chi})$ and $Z_0(\tau, \bar{\tau}; \chi, \bar{\chi})$ were defined in (2.34) and (2.35). The auxiliary parameters $C, \bar{C}$ are proportional to the constant parts of $V_\pm$ in (3.4). In the limit $\beta \to 0$ we recover the partition function (2.33) of the model discussed in the previous section.

The partition function (3.9) has the following symmetries (cf. (2.36))

$$Z(R,q,\beta) = Z(\alpha'R^{-1}, \beta, q), \quad Z(R,q,\beta) = Z(R, q + 2n_1 R^{-1}, \beta + 2n_2 \alpha'^{-1} R). \tag{3.10}$$

These are symmetries of the full conformal field theory (as can be seen directly from the string action in the Green-Schwarz formulation). If $qR \neq n_1$ and $\alpha'\beta R^{-1} \neq n_2$, there are tachyons for any value of the radius R, and the partition function contains infrared divergences. As follows from (3.10), when $\alpha'\beta R^{-1}$ (or qR) is an even integer, the partition function reduces to that of the $a = \sqrt{3}$ Melvin model (2.33). In particular, in the special case when both qR and $\alpha'\beta R^{-1}$ are even, the partition function is identically zero (then the theory is equivalent to the free superstring). When either $\alpha'\beta/R$ or qR is an odd integer, the partition function is finite in a certain range of values of the radius.

4. 'Twisted' $SU(2) \times U(1)$ WZNW model as compact analogue of magnetic flux tube model and supersymmetry breaking

The non-trivial part of the 10-dimensional space-time corresponding to the $a = \sqrt{3}$ Melvin model (2.3),(2.7) is a flat non-compact space $\mathcal{M}^3$. The breaking of supersymmetry in the model (2.7),(2.22) is a consequence of an incompatibility between periodicity of space-time spinors in the Kaluza-Klein direction y and the presence of mixing between y and the angular coordinate of 2-plane which produces a flat but globally non-trivial spin connection. Replacing the 2-plane by a *compact* space with a non-trivial isometry and mixing the isometric coordinate with another compact internal coordinate y, one may try to construct similar models in which supersymmetry is broken while the Lorentz symmetry in the remaining flat non-compact directions is preserved. Below we shall discuss such a model where the 4-space $R_{x_3} \times \mathcal{M}^3 = (x_3, \rho, \varphi, y)$ is replaced by $\mathcal{M}^4 = SU(2) * U(1)$, i.e. the 'twisted' product of the $SU(2)$ WZNW model and a circle. $\mathcal{M}^4$ (plus the corresponding torsion) is locally the group space $SU(2) \times U(1)$ and thus defines a conformal model.

We shall consider $\mathcal{M}^4$ as (part of) the *internal* space and discuss how the twist parameters lead to supersymmetry breaking.[18] This will not be in contradiction with the 'no-go' theorem on impossibility of spontaneous supersymmetry breaking by continuous parameters [30,31] since the corresponding (heterotic) string vacuum will not be supersymmetric already in the absence of the twist (the central charge condition will not be satisfied unless one adds a linear dilaton background or considers special values of the level k). In what follows we shall ignore this problem, concentrating on 'additional' supersymmetry breaking induced by continuous twist parameters. In view of a relation to the magnetic flux tube model, this supersymmetry breaking can be given a 'magnetic' interpretation.[19]

Simplest examples of models with spontaneous supersymmetry breaking are string compactifications on 'twisted' tori (or string analogues of the 'Scherk-Schwarz' [33] compactifications) [19,34,21]. Consider, e.g., the 3-torus $(x_1, x_2, y) \equiv (x_1 + 2\pi R' n_1, x_2 + 2\pi R' n_2, y + 2\pi Rm)$ and twist it by imposing the condition that the shift by period in y should be accompanied by a rotation in the (x_1, x_2)-plane. For a *finite* R' the only possible rotations are by angles $\frac{1}{2}\pi m$, i.e. one may identify the points $(\theta, y) = (\theta + 2\pi n + \frac{1}{2}\pi m, y + 2\pi Rm)$, $\cot \theta = x_1/x_2$. The superstring theory with this flat but non-trivial 3-space $\mathcal{M}_0^3$ as part of the internal space was considered in [19] (see also [34,21]) where it was found that such twist of the torus breaks supersymmetry and leads to the

---

[18] A special case of this model was considered in [29] where it was interpreted as describing a space-time magnetic background (with the curvature of S^3 being small, i.e. the level of $SU(2)$ being large).

[19] In that sense this model is similar to other models where supersymmetry is broken (in a discrete way) by magnetic fields in internal dimensions, see [32] and refs. there.

existence of tachyons for $R^2 < 2\alpha'$ and non-vanishing (and finite for $R^2 > 2\alpha'$) partition function. The $R' \to \infty$ limit of this model is actually equivalent to the special case $qR = \frac{1}{4}m$ of the $a = \sqrt{3}$ Melvin model (2.5) (the case of $m = 4$ explicitly considered in [19] is equivalent to the superstring compactified on a circle with antiperiodic boundary conditions for the fermions). Since in the model (2.5) the 2-plane is non-compact and thus the twisting angle $2\pi qR$ is arbitrary, this model continuously connects large R' limits of the models of [19] with different values of the integer m.

Such models with compact *flat* internal spaces always have *discrete* allowed values of the twisting parameter (a symmetry group of a lattice which generates a torus from R^N is discrete). As a result, the supersymmetry breaking mass scale μ is directly proportional to the compactification mass scale R'^{-1} [35]. Given that μ should be of TeV order, this implies the large value for R' and thus the existence of a tower of 'light' $(M \sim TeV)$ Kaluza-Klein states [36,37]. This leads to fast growth of coupling with energy making perturbation theory unapplicable [36].

It could happen that analogous 'twistings' of models with compact *curved* internal spaces with isometries lead to vacua where the supersymmetry breaking scale could be continuously adjusted and thus decoupled from the Kaluza-Klein scale. Such possibility, however, is ruled out by the results of [30,31].[20] Even though this will not resolve the problem of large internal dimension (since the 'discrete' part of the supersymmetry breaking will already relate the supersymmetry breaking and compactification scales) it may still be of interest to look for other continuous mechanisms of supersymmetry breaking which may complement the 'discrete' one. The idea is to separate the issue of discrete supersymmetry breaking due to non-compensation of the central charge from the *additional* continuous supersymmetry breaking induced by the twists.

To construct a generalisation of the model (2.2),(2.3) which will be compact and, at the same time, remain to be conformal, let us consider the $SU(2) \times U(1)$ WZNW theory and 'twist' the product by shifting the two isometric Euler angles θ_L and θ_R of $SU(2)$ $(g = \exp(\frac{i}{2}\theta_L\sigma_3)\exp(\frac{i}{2}\psi\sigma_2)\exp(\frac{i}{2}\theta_R\sigma_3))$ by the periodic coordinate $y \in (0, 2\pi R)$ corresponding to $U(1)$

$$\theta'_L = \theta_L + q_1 y , \qquad \theta'_R = \theta_R + q_2 y . \tag{4.1}$$

---

[20] According to [30] it is not possible to break supersymmetry in a continuous way by adding a marginal perturbation to an $N = 2$ supersymmetric world sheet conformal model describing a space-time supersymmetric vacuum of heterotic string theory. This 'no-go' theorem does not apply to the non-compact model (2.3),(2.7) which can be considered as a marginal perturbation of the free non-compact model: the left-right symmetric perturbation $q\epsilon_{ij}(x_i\partial x_j\bar{\partial}y + x_i\bar{\partial}x_j\partial y)$ in (2.5) is marginal and integrable (since the space $\mathcal{M}^3$ is flat so that $R_{\mu\nu} = 0$ order by order in q) but is not well-defined as a CFT operator. Note that according to (2.19),(2.28) the supersymmetry is broken there already at $O(q)$ order.

Here q_1, q_2 are continuous twist parameters.[21] The special case of such model with $q_2 = 0$ was considered in [29]. To make contact with the model (2.3) it is necessary to keep both q_1 and q_2 non-vanishing [4].

The resulting $SU(2) \times U(1)$ WZNW Lagrangian

$$L'(q_1, q_2) = L_{SU(2)}(\psi, \theta'_L, \theta'_R) + \partial y \bar{\partial} y , \qquad (4.2)$$

$$L_{SU(2)}(\psi, \theta_L, \theta_R) = k(\partial \psi \bar{\partial} \psi + \partial \theta_L \bar{\partial} \theta_L + \partial \theta_R \bar{\partial} \theta_R + 2 \cos \psi \partial \theta_R \bar{\partial} \theta_L) , \qquad (4.3)$$

defines a conformal theory since locally $\mathcal{M}^4 = (\psi, \theta_L, \theta_R, y)$ is still the same $SU(2) \times U(1)$ group manifold. In particular, the corresponding central charge is unchanged, $c = \frac{3k}{k+2} + 1$. Its independence of q_i makes it clear that 'trivial' discrete breaking of supersymmetry due to non-zero central charge deficit will be unrelated to 'non-trivial' continuous one induced by non-vanishing 'magnetic' twists q_i.

The case of $q_1 = -q_2 = q$ is a compact analogue of the model (2.3). Let us first note that $L_{SU(2)}$ in (4.3) can be written as

$$L_{SU(2)}(\psi, \theta, \tilde{\theta}) = k \left[\partial \psi \bar{\partial} \psi + 4 \sin^2 \frac{\psi}{2} \, \partial \theta \bar{\partial} \theta + 4 \cos^2 \frac{\psi}{2} \, \partial \tilde{\theta} \bar{\partial} \tilde{\theta} \right. \qquad (4.4)$$

$$\left. + 2 \cos \psi (\partial \tilde{\theta} \bar{\partial} \theta - \partial \theta \bar{\partial} \tilde{\theta}) \right] , \qquad \theta = \tfrac{1}{2}(\theta_L - \theta_R) , \quad \tilde{\theta} = \tfrac{1}{2}(\theta_L + \theta_R) .$$

For small ψ (large k) (4.4) reduces to $k(\partial \psi \bar{\partial} \psi + \psi^2 \partial \theta \bar{\partial} \theta + 4 \partial \tilde{\theta} \bar{\partial} \tilde{\theta} + ...)$, i.e. describes a product of a 2-disc (ψ, θ) and a line $\tilde{\theta}$. Observing that for $q_1 = -q_2 = q$ the shift (4.1) implies $\theta' = \theta + qy$, $\tilde{\theta}' = \tilde{\theta}$, we can establish a relation between $L'(q, -q)$ (4.2) and (2.3) by identifying the coordinates in the following way: $\sqrt{k}\psi \to \rho$, $\theta \to \varphi$, $\sqrt{k}\tilde{\theta} \to x_3$.

The Lagrangian (4.2) can be represented in the form of a perturbation of the $SU(2) \times U(1)$ WZNW model

$$L'(q_1, q_2) = L_{SU(2)}(\theta_L, \theta_R, \psi) + 2q_1 \bar{J}_3 J_y + 2q_2 J_3 \bar{J}_y \qquad (4.5)$$

$$+ (1 + kq_1^2 + kq_2^2 + 2kq_1 q_2 \cos \psi) \partial y \bar{\partial} y ,$$

where $J_3 = -ik\mathrm{Tr}(\sigma_3 g^{-1} \partial g)$, $\bar{J}_3 = -ik\mathrm{Tr}(\sigma_3 \bar{\partial} g g^{-1})$, J_y, $\bar{J}_y$ are the Cartan currents of the $SU(2) \times U(1)$ model[22]

$$J_3 = k(\partial \theta_L + \cos \psi \partial \theta_R), \quad \bar{J}_3 = k(\bar{\partial} \theta_R + \cos \psi \bar{\partial} \theta_L), \quad J_y = \partial y, \quad \bar{J}_y = \bar{\partial} y. \qquad (4.6)$$

---

[21] The discussion that follows can be generalised to the case of other WZNW models (and their cosets) using the parametrisation $g = \exp(i\theta_L^s H_s) \exp(i\psi^\alpha E_\alpha) \exp(i\theta_R^s H_s)$ (where H_s are generators of the Cartan subalgebra) and mixing θ_L, θ_R with coordinates of an extra torus.

[22] Another way to see why (4.2),(4.5) with $q_1 = -q_2 = q$ is related to (2.3),(2.5) is to use the parametrization $g = \exp(\frac{i}{2\sqrt{k}} x_n \sigma_n)$ in which (for small x_n or large k) $J_3 = \sqrt{k}\partial x_3 + \epsilon_{ij} x_i \partial x_j + ...$, $\bar{J}_3 = \sqrt{k}\bar{\partial} x_3 - \epsilon_{ij} x_i \bar{\partial} x_j + ...$, $L_{SU(2)} = \partial x_i \bar{\partial} x_i + \partial x_3 \bar{\partial} x_3 + ...$, $i, j = 1, 2$. Then $O(q)$ terms in (4.5) coincide with $O(q)$ terms in $L_1 + \partial x_3 \bar{\partial} x_3$ in (2.5).

$O(q_1)$ and $O(q_2)$ terms in (4.5) are thus integrable marginal perturbations of the $SU(2) \times U(1)$ WZNW model. This follows directly from conformal invariance of (4.2) and is also in agreement with the fact that marginal $J\bar{J}$-perturbations by Cartan currents are integrable [38]. Note that it is only in the 'chiral' case of $q_1 = 0$ or $q_2 = 0$ considered in [29] that L' (4.2) can be represented as the original $SU(2) \times U(1)$ CFT (with rescaled radius of y) plus $J\bar{J}$-term.

The $SU(2) \times U(1)$ group space preserves half of maximal space-time supersymmetry ($N = 4, D = 4$); in particular, there is the corresponding number of Killing spinors. The heterotic string compactification on $S^3 \times S^1$ was discussed, e.g., in [25] where it was pointed out that in the context of the effective field theory approach the extra condition [23] on the Killing spinor $\gamma^{mnk} \hat{H}_{mnk} = 0$ coming from the gaugino transformation law (under the assumption that the dilaton is constant) is not satisfied. This is related to the issue of cancellation of the central charge, i.e. the presence or absence of the tree-level potential term for the dilaton ('cosmological constant'). For special values of k the central charge can be cancelled by combining this model with a minimal model [39] or with 'untwisted' $N = 2$ coset model [40]. A central charge deficit can be compensated by a linear dilaton background with resulting 'discrete' supersymmetry breaking [36].[23] As mentioned above, we shall concentrate on additional supersymmetry breaking induced by the continuous twist parameters q_i.

Let us show that the background associated with the twisted model (4.2) does not admit Killing spinors so that the corresponding light-cone gauge Green-Schwarz superstring action does not have residual supersymmetry. In the case of the WZNW model the fermionic part of GS action is given by [24,25] (cf. (2.22))

$$L_{GS}(\mathcal{S}) = i\mathcal{S}_R \mathcal{D}_+ \mathcal{S}_R + i\mathcal{S}_L \mathcal{D}_- \mathcal{S}_L , \qquad (4.7)$$

$$\mathcal{D}_\pm \equiv \partial_\pm + \tfrac{1}{4}\gamma_{mn}\omega_{\pm\mu}^{mn}\partial_\pm x^\mu , \qquad \omega_{\pm\mu}^{mn} = \omega_\mu^{mn} \pm \tfrac{1}{2}H^{mn}{}_\mu .$$

The quartic fermionic terms are absent since the generalised curvature vanishes.[24] If one chooses the left-invariant vierbein basis ($e^m = e_\mu^m dx^\mu = -i\mathrm{Tr}(\sigma^m g^{-1} dg)$) then one of the two generalised connections vanishes ($H_{\mu\nu\lambda} = -f_{mnk}e_\mu^m e_\nu^n e_\lambda^k$): $\omega_{-\mu}^{mn} = 0$, $\omega_{+\mu}^{mn} = -\tfrac{3}{2}f^{mn}{}_\mu$ (in the right-invariant basis $\omega_{+\mu}^{mn} = 0$). As a result, half of the maximal supersymmetry is preserved since $\mathcal{S}_R$ fermions remain free (the corresponding Killing spinor equation (2.25) with $\omega \to \omega_+$ has ϵ=const as a solution). It may seem that the same conclusion should be true also in the 'shifted' theory (4.2) since locally (4.1) can be considered as a coordinate

---

[23] For a discussion of supersymmetric $R_Q \times SU(2)_k$-type models with linear dilaton see [41,29].

[24] The corresponding supersymmetric σ-model (RNS) action was discussed in [42,43]. Let us note that the twist (4.1) *preserves* the extended world-sheet supersymmetry the σ-model action since the existence of the supersymmetry is determined by local conditions on a background.

transformation and thus the transformed $\omega^{mn}_{-\mu}$ should also vanish. However, the left-invariant vierbein basis e^m_μ which is independent of θ_L explicitly depends on $\cos\theta_R$, $\sin\theta_R$ and thus its direct analogue obtained by making the shift (4.1) is not defined unless $q_i R$ are integers. If one uses the original left-invariant basis, one needs to make an extra local Lorentz transformation to make the full metric diagonal. Then $\mathcal{D}_-$ gets a flat but non-trivial q_i-dependent connection term. As a result, for generic $q_i R \neq 2n_i$ the Killing spinor equation $\mathcal{D}_{-\mu}\epsilon = 0$ has no solutions consistent with periodic boundary conditions in y (cf. (2.23),(2.24),(2.26)), i.e. supersymmetry is completely broken.

This can be seen more explicitly by choosing another ('isometry-adapted') basis corresponding to the diagonal form of the WZNW action (4.4): $e^1 = d\psi$, $e^2 = 2\sin\frac{\psi}{2}d\theta$, $e^3 = 2\cos\frac{\psi}{2}d\tilde{\theta}$, $e^y = dy$. Then $\omega^{\theta\psi} = -\cos\frac{\psi}{2}d\theta$, $\omega^{\tilde{\theta}\psi} = \sin\frac{\psi}{2}d\theta$, $\omega^{\theta\tilde{\theta}} = 0$, $H = \frac{1}{2}e^\psi \wedge e^\theta \wedge e^{\tilde{\theta}}$, $\omega^{\theta\psi}_\pm = -\cos\frac{\psi}{2}d(\theta \pm \tilde{\theta})$, $\omega^{\tilde{\theta}\psi}_\pm = \sin\frac{\psi}{2}d(\tilde{\theta} \pm \theta)$). The solution of the corresponding Killing equations is $\epsilon_{R,L}(\psi,\theta,\tilde{\theta}) = \exp(\frac{i}{2}\psi\sigma_3)\exp(\pm\frac{i}{2}\theta_{R,L}\sigma_1)\epsilon_{R,L}(0)$, where $\theta_{L,R} = \tilde{\theta} \pm \theta$ and $\epsilon_{L,R}(0)$=const. Making the transformation (4.1) we find that the formal solutions of the Killing spinor equations corresponding to the 'twisted' model are (cf. (2.24),(2.26)):

$$\epsilon_L = e^{\frac{i}{2}\psi\sigma_3}e^{-\frac{i}{2}\theta_L\sigma_1}e^{-\frac{i}{2}q_1 y\sigma_1}\epsilon_L(0) , \qquad \epsilon_R = e^{\frac{i}{2}\psi\sigma_3}e^{\frac{i}{2}\theta_R\sigma_1}e^{\frac{i}{2}q_2 y\sigma_1}\epsilon_L(0) . \qquad (4.8)$$

As in (2.26) the periodic boundary conditions in y imply that supersymmetry is broken unless $q_i R = 2n_i$.

The dependence of the Killing spinors on θ_L and θ_R is related to the existence of the fixed points ($\psi = 0, 2\pi$) in the action of the isometries corresponding to shifts along θ_L and θ_R. The same is true in the 2-plane case in the polar coordinate basis. The breaking of supersymmetry by the twist (4.1) may be attributed to this dependence. Similar breaking will thus happen in general when an isometry which has fixed points is 'mixed' with another circular dimension. This dependence of Killing spinors on angular coordinates leads also to an apparent 'breakdown' of supersymmetry [44] after the duality transformations in these coordinates (it is only the local realisation of supersymmetry that is actually broken by the duality [45,46]). Note that the shift (4.1) becomes part of the $O(3,3;Z)$ duality transformation group of the $SU(2) \times U(1)$ model [47] only when $q_i R = 2n_i$, i.e. when supersymmetry is unbroken.

It would be interesting to determine the spectrum of the model (4.2) to see the supersymmetry breaking explicitly. This can probably be done by generalizing the approach of [29] where the special case of $q_2 = 0$ was solved. It is clear from the spectrum given in [29] that q_1 plays the role of the the supersymmetry breaking parameter.

To try to construct 'realistic' models which include this 'magnetic' supersymmetry breaking one needs to address the question of saturation of the central charge condition. One possible suggestion is to relax this condition, assuming that the dilaton equation should eventually be satisfied with loop and non-perturbative corrections included [36,32] (note, however, that $\delta c \sim 1/k$ is small only if k is large and that leads back to the problem of large compactification scale). Another is to consider special values of k for which the total c can be balanced by combining this model, e.g., with $N = 2$ coset one. An interesting aspect of the supersymmetry breaking induced by the 'magnetic' twists is that the resulting contribution to the cosmological constant is likely to be very small. Assuming that the compactification radius is $R \sim \sqrt{\alpha'} \sim M_{Pl}^{-1}$ while the 'magnetic' supersymmetry breaking scale $\mu^2 \sim q$ is of order of TeV^2 or M_W^2, and that the partition function of a 'realistic' model will be similar to (2.33) (which is proportional to q^6 for small q), one may expect to find $\Lambda_4 \sim M_W^{12}/M_{Pl}^8$ which is very small indeed.

Acknowledgements

I would like to thank T. Banks, M. Green, E. Kiritsis, K. Kounnas, D. Lüst, F. Quevedo and M. Tsypin for useful discussions and remarks. I am grateful to J. Russo for collaboration and discussions. I acknowledge also the support of PPARC and of ECC grant SC1*-CT92-0789.

References

[1] M.A. Melvin, Phys. Lett. 8 (1964) 65.

[2] A.A. Tseytlin, Phys. Lett. B346 (1995) 55.

[3] J.G. Russo and A.A. Tseytlin, Nucl. Phys. B449 (1995) 91, hep-th/9502038.

[4] J.G. Russo and A.A. Tseytlin, "Magnetic flux tube models in superstring theory", hep-th/9508068.

[5] N.K. Nielsen and P. Olesen, Nucl. Phys. B144 (1978) 376.

[6] E.S. Fradkin and A.A. Tseytlin, Phys. Lett. B163 (1985) 123.

[7] A. Abouelsaood, C. Callan, C. Nappi and S. Yost, Nucl. Phys. B280 (1987) 599.

[8] C.P. Burgess, Nucl. Phys. B294 (1987) 427; V.V. Nesterenko, Int. J. Mod. Phys. A4 (1989) 2627.

[9] S. Ferrara and M. Porrati, Mod. Phys. Lett. A8 (1993) 2497.

[10] J.G. Russo and A.A. Tseytlin, Nucl. Phys. B448 (1995) 293, hep-th/9411099.

[11] J.G. Russo and L. Susskind, Nucl. Phys. B437 (1995) 611; A. Sen, Nucl. Phys. B440 (1995) 421.

[12] G.W. Gibbons, in: *Fields and Geometry*, Proceedings of the 22nd Karpacz Winter School of Theoretical Physics, ed. A. Jadczyk (World Scientific, Singapore, 1986).

[13] H.B. Nielsen and P. Olesen, Nucl. Phys. B61 (1973) 45.

[14] F. Dowker, J.P. Gauntlett, D.A. Kastor and J. Traschen, Phys. Rev. D49 (1994) 2909; F. Dowker, J.P. Gauntlett, S.B. Giddings and G.T. Horowitz, Phys. Rev. D50 (1994) 2662.

[15] F. Dowker, J.P. Gauntlett, G.W. Gibbons and G.T. Horowitz, "The decay of magnetic fields in Kaluza-Klein theory", hep-th/9507143.

[16] G.W. Gibbons and K. Maeda, Nucl. Phys. B298 (1988) 741.

[17] J.G. Russo and A.A. Tseytlin, "Heterotic strings in a uniform magnetic field", hep-th/9506071.

[18] M.B. Green, J.H. Schwarz and E. Witten, *Superstring Theory* (Cambridge U.P., 1987).

[19] R. Rohm, Nucl. Phys. B237 (1984) 553.

[20] J.J. Atick and E. Witten, Nucl. Phys. B310 (1988) 291.

[21] C. Kounnas and B. Rostant, Nucl. Phys. B341 (1990) 641.

[22] M.B. Green and J.H. Schwarz, B136 (1984) 307; Nucl. Phys. B243 (1984) 285.

[23] P. Candelas, G. Horowitz, A. Strominger and E. Witten, Nucl. Phys. B258 (1985) 46.

[24] E.S. Fradkin and A.A. Tseytlin, Phys. Lett. B158 (1985) 316; Phys. Lett. B160 (1985) 69.

[25] I. Bars, D. Nemeschansky and S. Yankielowicz, Nucl. Phys. B278 (1986) 632.

[26] K. R. Dienes, Nucl. Phys. B429 (1994) 533; hep-th/9409114; hep-th/9505194.

[27] P. Ginsparg and C. Vafa, Nucl. Phys. B289 (1987) 414.

[28] M. Porrati and E.T. Tomboulis, Nucl. Phys. B315 (1989) 615; E.T. Tomboulis, Phys. Lett. B198 (1987) 165.

[29] E. Kiritsis and C. Kounnas, "Infrared behavior of closed superstrings in strong magnetic and gravitational fields", hep-th/9508078.

[30] T. Banks and L. Dixon, Nucl. Phys. B307 (1988) 93.

[31] M. Dine and N. Seiberg, Nucl. Phys. B301 (1988) 357.

[32] C. Bachas, "A way to break supersymmetry", hep-th/9503030.

[33] J. Scherk and J.H. Schwarz, Phys. Lett. B82 (1979) 60; Nucl. Phys. B153 (1979) 61.

[34] S. Ferrara, C. Kounnas and M. Porrati, Nucl. Phys. B304 (1988) 500; Phys. Lett. B206 (1988) 25; C. Kounnas and M. Porrati, Nucl. Phys. B310 (1988) 355.

[35] I. Antoniadis, C. Bachas, D. Lewellen and T. Tomaras, Phys. Lett. B207 (1988) 441.

[36] S. de Alwis, J. Polchinski and R. Schimmrigk, Phys. Lett. B218 (1989) 449.

[37] I. Antoniadis, Phys. Lett. B246 (1990) 377; I. Antoniadis, C. Munoz and M. Quiros, Nucl. Phys. B397 (1993) 515; K. Benakli, "Perturbative supersymmetry breaking in orbifolds with Wilson line backgrounds", hep-th/9509115.

[38] S. Chaudhuri and J.A. Schwarz, Phys. Lett. B219 (1989) 291.

[39] D. Gepner, Nucl. Phys. B296 (1987) 757.

[40] Y. Kazama and H. Suzuki, Nucl. Phys. B321 (1989) 232.

[41] C. Kounnas, Phys. Lett. B321 (1994) 26.

[42] S.J. Gates, C.M. Hull and M. Roček, Nucl. Phys. B248 (1984) 157; P. Howe and G. Sierra, Phys. Lett. B148 (1984) 451.

[43] R. Rohm, Phys. Rev. D32 (1985) 2849.

[44] E. Bergshoeff, R. Kallosh and T. Ortín, Phys. Rev. D51 (1995) 3009; I. Bakas, Phys. Lett. B343 (1995) 103.

[45] I. Bakas and K. Sfetsos, Phys. Lett. B349 (1995) 448.

[46] E. Álvarez, L. Álvarez-Gaumé and I. Bakas, "T-duality and space-time supersymmetry", hep-th/9507112; K. Sfetsos, "Duality and restoration of manifest supersymmetry", hep-th/9510034.

[47] S.F. Hassan and A. Sen, Nucl. Phys. B405 (1993) 143; M. Henningson and C. Nappi, Phys. Rev. D48 (1993) 861; E. Kiritsis, Nucl. Phys. B405 (1993) 109.

SOLUTION OF THE SL(2,R) STRING

IN CURVED SPACETIME[1]

ITZHAK BARS

Department of Physics and Astronomy
University of Southern California
Los Angeles, CA 90089-0484, USA
E-mail: bars@physics.usc.edu

1. Abstract

The SL(2,R) WZW model, one of the simplest models for strings propagating in curved space time, was believed to be non-unitary in the algebraic treatment involving affine current algebra. It is shown that this was an error that resulted from neglecting a zero mode that must be included to describe the correct physics of non-compact WZW models. In the presence of the zero mode the mass-shell condition is altered and unitarity is restored. The correct currents, including the zero mode, have logarithmic cuts on the worldsheet. This has physical consequences for the spectrum because a combination of zero modes must be quantized in order to impose periodic boundary conditions on mass shell in the physical sector of the theory. To arrive at these results and to solve the model completely, the SL(2,R) WZW model is quantized in a new free field formalism that differs from previous ones in that the fields and the currents are Hermitean, there are cuts, and there is a new term that could be present more generally, but is excluded in the WZW model.

2. Introduction

One of the main reasons to study string theory in curved spacetime is to develop the appropriate methods to investigate physical phenomena in the presence of quantum gravity in a mathematically consistent theory. Quantum gravity is important during the early universe and this must have an impact on the symmetries and matter content (gauge bosons, families of quarks and leptons) observed at accelerator energies. In addition, in order to develop an understanding of gravitational singularities such as

---

[1] Based on lectures delivered at the Strings '95 conference, USC, March 1995, and at the Strings, Gravity and Physics at the Planck Scale conference, Erice, August 1995.

N. Sánchez and A. Zichichi (eds.), String Gravity and Physics at the Planck Energy Scale, 151–169.
© *1996 Kluwer Academic Publishers.*

black holes or the big bang, including quantum aspects of gravity, string theory in curved spacetime must be investigated.

A string propagating in curved spacetime with one time and $(d-1)$ space coordinates is described by a string action that has the following form in the conformal gauge

$$S = \int d\tau d\sigma \left[\partial_+ X^\mu \partial_- X^\nu G_{\mu\nu}(X) + \cdots \right] \tag{1}$$

where $G_{\mu\nu}(X)$ is a background metric in d-dimensions with signature $(-1, 1, 1, \cdots)$. The terms in the action denoted by $\cdots$ may contain additional background fields such as an antisymmetric tensor $B_{\mu\nu}(X)$ a dilaton $\Phi(X)$ etc.. Conformal invariance must be imposed on these background fields at the quantum level, otherwise reparametrization invariance (Virasoro constraints) cannot be used to remove ghosts from the theory.

In one time plus $(d-1)$ space dimensions many models that are exactly conformally invariant at the quantum level have been constructed by now (1)(2)(3). A first example of string propagation in curved spacetime was the SL(2,R) WZW model that potentially could be solved through algebraic methods. However, one immediately came across an unexpected inconsistency problem involving the unitarity of the theory (1)(4). It was noticed a long time ago that the Virasoro constraints were insufficient to remove all the negative norm states encountered in the affine current algebra treatment of the model. There seemed to be more negative norm states than those introduced by the time-like string coordinates. This problem persited with all other non-compact affine current algebra coset models, including the much studied SL(2,R)/R two dimensional black hole (1)(5). This observation casted doubt on the physical validity of the models and prevented the development of physical ideas based on the algebraic properties of these (in principle) completely solvable models.

In this talk, and in a related paper (6), I explain the solution of the unitarity problem. I show that the WZW model has more degrees of freedom than those described by the affine current algebra. The extra degrees of freedom are zero modes (in particular, in non-compact directions). In the old treatment one inadvertently set the zero mode to zero values, and hence missed important properties of the model. One of the important effects of the non-trivial zero mode is that the mass shell condition is altered. In the absence of the zero mode the string cannot be put correctly on mass shell and this begins to explain why the non-unitary states emerged.

To actually show that the model has no ghosts a second step is needed. In particular one must argue that the discrete series representation of SL(2,R) does not appear at the base, since in the purely algebraic approach (including the new zero mode) there would still be negative norm states in this representation, according to the old approach. To show that the WZW model excludes the discrete series, I formulate the quantum theory in terms of a suitable parametrization of SL(2,R) that corresponds to free fields, and then show that only the unitary principal series is allowed.

In this formulation the spectrum is completely solved and the Virasoro constraints implemented. The no ghost theorem is then shown to be valid.

In the presence of non trivial values of the zero mode (which is needed for the correct physics) the conserved currents of the theory have logarithmic singularities on the world sheet. *A priori* one may think that this would imply that the closed string boundary conditions are not satisfied. However, this is not the case, because periodicity is required only on mass shell. In the presence of the new zero modes the Hilbert space is larger than before. Imposing periodicity gives quantized values for a combination of the zero modes, thus satisfying the correct closed (or open) string boundary conditions for the physical on-mass-shell states.

To arrive at these results, and also to solve the model completely, a new free field formalism is introduced as mentioned above. The structure of the currents in terms of the free fields is reminiscent of the one found by Wakimoto (7) and Gerasimov et. al. (8), but it differs from theirs in three important aspects: (1) The free fields as well as the currents are Hermitean; this is important for the discussion of unitarity. (2) The currents have logaritmic cuts that are associated naturally with the zero modes of the free fields; this affects the spectrum and monodromy. (3) A new term, that can be present in the free field formulation to reproduce the effects of the most general SL(2,R) currents, is introduced in order to obtain all unitary representations at the base. The source of all the unwanted ghosts is traced to the new term. But, for the specific model at hand, i.e. the WZW model, the extra term is shown to be absent, thus elucidating the mechanism by which the model becomes unitary.

3. The unitarity problem

The SL(2,R) WZW model has a timelike coordinate that introduces negative norm oscillators. These create negative norm states, but this is not the source of the problem. On the basis of naïve counting one may hope that the Virasoro constraints will remove these ghosts from the theory (this expectation is born out in our final result). A similar situation occurs also in the flat theory. As is well known, in the flat case one can indeed prove the no ghost theorem (9) which implies that the theory is unitary. However, in the case of curved spacetime current algebra models, one finds that even after imposing the Virasoro constraints there remains negative norm states that render the theory non-unitary. This has been the main stumbling block that discouraged the application of these ideas to model building for the past five years.

Negative norm states that satisfy the Virasoro constraints can be displayed ex-

154

plicitly. An example is (1)

$$|\phi, l> = \left(J_{-1}^1 - iJ_{-1}^2\right)^l |j, m = j + 1>,$$

$$\tilde{L}_n |\phi, l> = 0, \qquad n \geq 1, \tag{2}$$

$$< \phi, l |\phi, l> = N_{j(l)} \, (l!) \, \prod_{r=0}^{l-1} (k - 2j(l) - 2 + r).$$

where $N_{j(l)} = < j, m = j + 1 | j, m = j + 1 >$ is the norm of the state at the base. The base is in the discrete series representation of SL(2,R), with $m \geq (j + 1)$. It is required to be in the discrete series[a] by the mass-shell condition $L_0 = 1$, which gives

$$-\frac{j(j+1)}{k-2} + l = 1. \tag{3}$$

Evidently, for sufficiently large values of the excitation number l the norm switches between positive and negative values. Hence, despite the Virasoro constraints this model is not unitary and cannot describe a physical string.

4. Solution of the unitarity problem

Until now a solution to this problem, and the related SL(2,R)/R black hole problem, has not been found despite many attempts (10)(11). Suggestions included: (1) Restrict (artificially) $j(l) + 1 < k/2$ so that the norm never becomes negative; (2) Allow large values of $j(l)$ as needed by the excited level l, but also permit the base state to have negative norm $N_{j(l)}$ in such a way as to make the norm of the excited state $< \phi, l |\phi, l >$ positive; (3) Hope that modular invariants will fix the problem. All of these suggestions are rejected (6).

The resolution of the problem lies in understanding that wrong assumptions have been made about the algebraic structure of the WZW model. In particular, the assumption that the SL(2,R) WZW model is described by affine SL(2,R) is not entirely correct. There is an additional zero mode that is present in the local conserved SL(2,R) currents of the WZW model, whose presence is crucial for the resolution of the unitarity issues. This zero mode is missed by the assumption of affine currents that are written in the form of a Laurent series $\tilde{J}^a(z) = \sum J_n^a z^{-n-1}$. When the additional zero mode is included, the true currents $J^a(z)$ have a logarithmic term $\ln z$ in addition to the usual powers z^n. Hence, manipulations such as (2,3) based on the old affine currents $\tilde{J}^a(z)$ do not fully reflect the correct theory, and this is why we find inadmissible unphysical results.

In general one can include such $\ln z$ parts, but still have the correct local commutation rules or operator products with only poles. To see this, let $\tilde{J}^a(z)$ be the usual

---

[a] When $j(j+1) > 0$ only the discrete series can occur among the unitary representations of SL(2,R). By contrast, the principal series occurs only when $j(j+1) < -1/4$.

Laurent series (with the usual operator products) that have modes J_n^a as above, and in addition introduce a new zero mode α_0^- that commutes with all the other modes J_n^a. The new currents are

$$J^0(z) + J^1(z) = \left[\tilde{J}^0(z) + \tilde{J}^1(z)\right]$$

$$J^0(z) - J^1(z) = \left[\tilde{J}^0(z) - \tilde{J}^1(z)\right] - 2i\alpha_0^- \ln z \; \tilde{J}^2(z)$$

$$- \frac{k}{z}\alpha_0^- + \left(-i\alpha_0^- \ln z\right)^2 \left[\tilde{J}^0(z) + \tilde{J}^1(z)\right] \qquad (4)$$

$$J^2(z) = \tilde{J}^2(z) - i\alpha_0^- \ln z \left[\tilde{J}^0(z) + \tilde{J}^1(z)\right]$$

It can be shown that they have the usual correct operator products, with only poles, for any value of the zero mode α_0^- (6).

The $\ln z$ terms arise naturally in the canonical formulation of the WZW model. There are left/right moving string coordinates $X_{L,R}$ that parametrize the group element $g(X_L(z), X_R(\bar{z}))$. As usual, string coordinates have a "momentum" zero mode $p \ln z$. The currents in this model depend both on $X_{L,R}$ as well as on their derivatives. Therefore, the currents are expected to include $\ln z$ pieces proportional to the zero modes. This is unlike the flat theory, where the currents depend only on the derivatives of the $X_{L,R}$. If one sets the $p \ln z$ parts equal to zero, as is inadvertently done by assuming affine currents in the form of Laurent series, one forces the string to lie in a sector of fixed zero mode $p = 0$. This may be harmless in compact directions, but it is fatal in non-compact directions. For example, for the flat string, if one requires the lightcone momentum $p^- = 0$, then the mass-shell condition

$$p^+ p^- - p_i^2 = l - 1$$

cannot be satisfied in that sector. In the SL(2,R) string exactly this situation arises when $\alpha_0^- = 0$. It turns out that, in SL(2,R), in the absence of the zero mode, which is analogous to p^-, it is still possible to satisfy a mass shell condition, but only in the discrete series representation. As seen above the discrete series gives rise to ghosts. On the other hand, when the zero mode $\alpha_0^- \neq 0$ is included (and hence $\ln z$ is present in the currents), the mass shell condition is satisfied in the principal series representation, which is free of ghosts at the excited levels.

5. SL(2,R) Currents and free fields

Consider the free fields $X^-(z)$, $P^+(z)$, $S(z)$, $T'(z)$. They have naïve dimensions 0,1,1,2 respectively and they are defined as follows

$$X^-(z) \;=\; q^- - i\alpha_0^- \ln z + i \sum_{n \neq 0} \frac{1}{n}\alpha_n^- z^{-n}, \qquad \left(\alpha_n^-\right)^\dagger = \alpha_{-n}^-$$

$$P^+(z) \;=\; \sum_{n=-\infty}^{\infty} \alpha_n^+ \, z^{-n-1}, \quad \left(\alpha_n^+\right)^\dagger = \alpha_{-n}^+$$

$$S(z) \;=\; \sum_{n=-\infty}^{\infty} s_n \, z^{-n-1}, \quad (s_n)^\dagger = s_{-n} \tag{5}$$

$$T'(z) \;=\; \sum_{n=-\infty}^{\infty} L_n' \, z^{-n-2}, \quad (L_n')^\dagger = L_{-n}'$$

These fields are Hermitean. The currents are

$$
\begin{aligned}
J_0(z) + J_1(z) \;&=\; P^+(z) \\
J_0(z) - J_1(z) \;&=\; \; :X^-(z)\, P^+(z)\, X^-(z): \, +2S(z)\, X^-(z) \\
&\quad - ik\partial_z X^-(z) - \frac{(k-2)T'(z)}{P^+(z)} \\
J_2(z) \;&=\; \; :X^-(z)\, P^+(z): \, +S(z)
\end{aligned}
\tag{6}
$$

The currents have a Wakimoto type structure, but with the exception that these currents are Hermitean. Furthermore, they contain two other new features: (i) the ln z parts that have the same structure as (4), and (ii) the new terms T'/P^+. It can be shown that the operator products of free fields produce the correct operator products of the currents (6).

The free field commutation rules are

$$[q^-, \alpha_0^+] = i \,,$$

$$[\alpha_n^-, \alpha_m^+] = n\, \delta_{n+m,0} \,,$$

$$[s_n, s_m] = \left(\tfrac{k}{2} - 1\right) n\, \delta_{n+m,0} \,, \tag{7}$$

$$[L_n', L_m'] = (n-m)L_{n+m}' + 0$$

while all other commutators are zero. The $\alpha_n^\pm$ oscillators may be rewritten in terms of light-cone type combinations of one time-like α_n^0 and one space-like α_n^1 oscillator, i.e. $\alpha_n^\pm = (\alpha_n^1 \pm \alpha_n^0)/\sqrt{2}$. This shows one source of negative norms, but by naive counting, they are expected to be removed by the Virasoro constraints. In fact, the usual proof of the no-ghost theorem can be extended to show that these ghosts are indeed removed (6). The zero modes $q^-, p^\pm$ are interpreted as light-cone type canonical variables $q = x^-$, $p = p^+$. We have not introduced a canonical variable corresponding to x^+, hence $\alpha_0^- = p^-$ commutes with all the operators and acts like a constant. Similarly the zero mode s_0 also acts like a constant. $\alpha_0^\pm, s_0, L_0'$ are simultaneously diagonalized in the Hilbert space, and they label the base.

The L_n' operators act like Virasoro operators with zero central charge. It is always possible to construct such an operator in terms of free fields, the simplest being a (negative norm) free boson with a background charge. Another example is any critical conformal field theory including the conformal ghosts with central charge $c = -26$. Both of these examples have ghosts and this is true more generally. Indeed, there is no construction of a zero central charge Virasoro operator that does not contain negative norm states in its Hilbert space. Hence, in the presence of L'_n there is an additional source of negative norms that will not be possible to be removed by the Virasoro constraints. For example a base state with the property $L'_0 |h' >= h'|h' >$ produces a negative norm state $|\phi >= L'_{-1}|h' >$, $< \phi|\phi >= 2h' < h'|h' >$, if $h' < 0$. As we will see below only when L' is present and $h' < 0$, the discrete series is possible. Hence the unwanted negative norm states and the discrete series go hand-in-hand. We will show that L' is absent in the WZW model, hence the WZW model has no additional sources of negative norms.

6. Stress tensor and free fields

The energy momentum tensor is obtained from the normal ordered product of the currents

$$T(z) = \frac{1}{k-2} : \left(-(J_0(z))^2 + (J_1(z))^2 + (J_2(z))^2 \right) :$$
(8)

The result of the computation gives

$$T(z) =: P^+ i\partial_z X^- : +T_S(z) + T'(z) \quad ,$$
(9)

If the computation is repeated with the $\tilde{J}(z)$ currents the only difference is dropping the α_0^- term contained in

$$i\partial_z X^- = \sum_{n=-\infty}^{\infty} \alpha_n^- z^{-n-1}.$$
(10)

In (9) T_S is a Hermitean stress tensor

$$T_S(z) = \frac{1}{k-2} \left[: (S(z))^2 : -\frac{i}{z}\partial_z (zS(z)) + \frac{1}{4z^2} \right]$$
(11)

The structure $\frac{i}{z}\partial_z (zS(z))$ differs from the usual one $i\partial S$, and thus is Hermitean. The operator products of $T_S(z)$ are

$$T_S(z) \times T_S(w) = \frac{c_s/2}{(z-w)^2} + \frac{2T_S(w)}{(z-w)^2} + \frac{\partial_w T_S(w)}{z-w} + \cdots$$
(12)

with the central charge

$$c_s = 1 + \frac{6}{k-2}.$$
(13)

Note that the term $P^+ i\partial_z X^-$ is identical to the energy momentum tensor of flat light-cone coordinates constructed from the oscillators $\alpha_n^{\pm}$. Therefore, that part is mathematically equivalent to a $c = 2$ stress tensor constructed from one time and one space coordinate in flat spacetime. Then the total central charge is

$$
\begin{aligned}
c &= 2 + c_s + c' \\
&= 2 + \left(1 + \tfrac{6}{k-2}\right) + 0 \\
&= \tfrac{3k}{k-2},
\end{aligned}
\tag{14}
$$

which is the right central charge for the $SL(2, R)$ WZW model. Finally, as a further consistency check, by using only the operator products of the elementary fields, one finds that $T(z)$ has the correct operator products with the currents.

The zero mode of the stress tensor takes the form $L_0 = L_0^{\pm} + L_0^S + L_0'$, where each piece has the eigenvalues

$$
\begin{aligned}
L_0^{\pm} &= p^+ p^- + l_{\pm}, \\
L_0^S &= (s_0^2 + 1/4)/(k - 2) + l_s, \\
L_0' &= h' + l'.
\end{aligned}
\tag{15}
$$

where $l_{\pm}, l_s, l'$ are positive integers and h' is the eigenvalue of L_0' at the base (whose possible values depend on the model for T'). The mass shell condition $L_0 = a$ is

$$
p^+ p^- + (s_0^2 + 1/4)/(k - 2) + h' + \text{integer} = a
\tag{16}
$$

where $a \leq 1$. The term $p^+ p^-$ is crucial since it takes negative values, as seen below.

To identify the value of the Casimir operator $j(j + 1)$ associated with the affine currents $\tilde{J}$ in (4) we set $p^- = 0$, and compare the eigenvalue of the resulting L_0 to the standard formula. We then see that the Casimir of the old currents $(\tilde{J}_0)^2 = -j(j+1)$ takes the value

$$
j(j + 1) = -(s_0^2 + 1/4) - h'(k - 2).
\tag{17}
$$

Therefore, if the T' piece is absent in the construction ($h' = 0$), then $j = -1/2 \pm i s_0$ is only in the principal series. The supplementary series could occur for $-1/4 < j(j + 1) < 0$ and the discrete series occurs for $-1/4 < j(j + 1)$. We see that the field T' with a positive h' contributes only to the principal series and with a negative h' it leads to the other representations as well. This construction may find various applications in the future. We will see below that T' is absent in the SL(2,R) WZW model, hence only the special case of our construction ($T' = 0$) finds an application in the WZW model. Then, for excited string states, since the integer in (16) is positive it would not be possible to satisfy the mass shell condition in the absence of the p^-. So, the logarithmic structure plays a role.

7. The SL(2,R) WZW model

We now relate the algebraic structures above to the WZW model for SL(2,R).

The quantum theory for any WZW model at the critical point is conveniently formulated in terms of the left and right moving currents after writing the group element $g(\tau,\sigma) = g_L(\tau+\sigma)\, g_R^{-1}(\tau-\sigma)$

$$J_L(z) = ik\partial_z g_L g_L^{-1}, \quad J_R(\bar{z}) = ik\partial_{\bar{z}} g_R g_R^{-1}, \tag{18}$$

where $z = e^{i(\tau+\sigma)}$, $\bar{z} = e^{i(\tau-\sigma)}$. The quantum rules are most conveniently given in terms of operator products among the currents and the group elements.

$$J^i_{L,R}(z)\, J^j_{L,R}(w) \;\to\; \frac{k/2}{(z-w)^2} + i\epsilon^{ijl}\,\eta_{lk}\frac{J^k_{L,R}(w)}{z-w} + \cdots \tag{19}$$

$$J^i_{L,R}(z)\, g_{L,R}(w) \;\to\; \frac{-t^i}{z-w} g_{L,R}(w) + \cdots$$

The t_i is a basis for the SL(2,R) Lie algebra which is given in terms of Pauli matrices $t_0 = \sigma_2/2$, $t_1 = i\sigma_1/2$, $t_2 = -i\sigma_3/2$. They satisfy $\eta_{ij} = -2tr(t_i t_j) = diag(-1,1,1)$.

Any group element $g_{L,R}$ in SL(2,R) can be rewritten in terms of the Gauss decomposition as follows[b]

$$g_{L,R} = \begin{pmatrix} 1 & 0 \\ X^-_{L,R} & 1 \end{pmatrix} \begin{pmatrix} :e^{-\frac{u_{L,R}}{k-2}}: & 0 \\ 0 & :e^{\frac{u_{L,R}}{k-2}}: \end{pmatrix} \begin{pmatrix} 1 & X^+_{L,R} \\ 0 & 1 \end{pmatrix}. \tag{20}$$

We compute the left/right currents (omitting the L, R indices for simplicity)

$$i(k-2) : \partial_z g g^{-1} := \begin{pmatrix} -J^2(z) & J^0(z) + J^1(z) \\ -J^0(z) + J^1(z) & J^2(z) \end{pmatrix} \tag{21}$$

As compared to the classical currents (18) we have shifted $k \to (k-2)$ in both g (20) and the definition of the current (21), and applied normal ordering. This renormalization is necessary for the commutation rules to work out, and is consistent with similar phenomena concerning the quantization of the WZW model (16). One

---

[b] For any SL(2,R) group element $g = (a, b; c, d)$, the Gauss decomposition is given by $X^+ = b/a$, $X^- = c/a$. Instead of the exponentials in the middle factor $\exp[\mp u/(k-2)]$ one could take more generally $diag\,(a, a^{-1})$, where a can have any sign, unlike the exponentials. One may carry out the quantization in terms of a instead of u. However, in the final analysis the currents depend only on a^2 or on $S \sim a\partial a^{-1}$ which may be rewritten as $S \sim |a|\,\partial\,|a|^{-1}$ even if a changes sign. Therefore, the parametrization used here, $|a| = \exp[-u/(k-2)]$, is adequate for the general case.

finds then

$$J^0(z) + J^1(z) = \; : (k-2)\, i\partial_z X^+ e^{-2u/(k-2)} :$$
$$J^2(z) = (k-2) : i\partial_z X^+ X^- e^{-2u/(k-2)} : +i\partial_z u$$
$$J^0(z) - J^1(z) = (k-2) : i\partial_z X^+ \left(X^-\right)^2 e^{-2u/(k-2)} : \tag{22}$$
$$+2X^- i\partial_z u - ik\partial_z X^-$$

The coefficient of $-ik\partial_z X^-$ is ambiguous because of the normal ordering of the term $: i\partial_z X^+ (X^-)^2 e^{-2u/(k-2)} :$. Again this has to be fixed by requiring that the commutation rules work out. Therefore, instead of having naively $-i(k-2)\partial_z X^-$, we actually must have $-ik\partial_z X^-$. These results are established by applying the canonical formalism and identifying these structures with canonical conjugate variables. Velocities must be replaced by canonical momenta. Note that for left/right movers ∂_z can be related to time derivatives ∂_τ or space derivatives ∂_σ. So, at the quantum level we find that we must identify the canonical pairs (X^-, P^+) and (u, S) as follows

$$P^+(z) = (k-2)\, i\partial_z X^+ e^{-2u/(k-2)} \tag{23}$$
$$S(z) = i\partial_z u$$

and then the currents take the form

$$J^0(z) + J^1(z) = P^+(z)$$
$$J^2(z) = \; : X^- P^+ : +S \tag{24}$$
$$J^0(z) - J^1(z) = \; : X^- P^+ X^- : +2SX^- - ik\partial_z X^-$$

This is the form used in the previous section without the extra field $L'(z)$. Thus, as discussed before, only the principal series will emerge in the WZW model. Using the oscillator form introduced in (5) we can express $u(z)$ and $X^+(z)$ in terms of the basic oscillators s_n, α_n^+ by inverting the formulas in (23), thus

$$u(z) = u_0 - is_0 \ln z + i \sum_{n \neq 0} \frac{1}{n} s_n z^{-n} \tag{25}$$

$$X^+(z) = -i \int^z dz' \frac{P^+(z')}{(k-2)} : \exp\left[\frac{2u(z')}{k-2}\right] : .$$

Then these structures satisfy the operator products

$$< u(z)\, S(w) > = \left(\frac{i}{z-w} + \frac{i}{2w}\right)\left(\frac{k}{2} - 1\right) \tag{26}$$

$$[J^0(z) - J^1(z)] \times X^+(w) \to \frac{-i}{z-w} : e^{2u(w)/(k-2)} :$$

Thus, $u(z)$ is just the canonical conjugate to $S(z)$. Another property of X^+ that follows from the fundamental operator products is that it is a singlet under the action of $J_2(z)$

$$J_2(z) \times X^+(w) \to 0. \tag{27}$$

Actually ∂X^+ is a screening current (see below). Its operator products with all the currents is either zero or a total derivative. Therefore, its zero mode commutes with all the currents.

Inserting the expressions in eq.(25) into (20) we obtain the quantum operator version of the group element g. The operator products may now be evaluated. We find the correct quantum products (19) with the above construction in terms of oscillators. That is,

$$\left[J^0_{L,R}(w) + J^1_{L,R}(w)\right] \times g_{L,R}(w) \rightarrow \frac{-i}{z-w} \begin{pmatrix} 0 & 0 \\ 1 & 0 \end{pmatrix} g_{L,R}(w)$$

$$J^2_{L,R}(z) \times g_{L,R}(w) \rightarrow \frac{i/2}{z-w} \begin{pmatrix} 1 & 0 \\ 0 & -1 \end{pmatrix} g_{L,R}(w) \tag{28}$$

$$\left[J^0_{L,R}(w) - J^1_{L,R}(w)\right] \times g_{L,R}(w) \rightarrow \frac{i}{z-w} \begin{pmatrix} 0 & 1 \\ 0 & 0 \end{pmatrix} g_{L,R}(w)$$

This result, combined with the current $\times$ current operator products that we have proven earlier, is convincing evidence that the free field formalism that we have discussed corresponds to the quantization of the SL(2,R) WZW model.

8. Physical states

8.1 No ghosts

Since we have rewritten the WZW theory in terms of free fields, the space of states consists of the Fock space for the oscillators $\alpha^\pm_n, s_n$ applied on the base $|p^+, p^-, s_0 >$ that diagonalizes the zero mode operators $\alpha^\pm_0, s_0$.

$$\prod_{n=1}^{\infty} \left(\alpha^+_{-n}\right)^{a_n} \prod_{m=1}^{\infty} \left(\alpha^-_{-m}\right)^{b_m} \prod_{k=1}^{\infty} (s_{-k})^{c_k} |p^+, p^-, s_0 > \tag{29}$$

where the powers a_n, b_m, c_k are positive integers or zero. This is the space of states that provide a representation basis for the SL(2,R) currents with only the principal series. The physical states are identified as those linear combinations that are annihilated by the the total Virasoro generators

$$L_n|\psi >= 0, \quad n \geq 1. \tag{30}$$

In the present case the total Virasoro generators include the following terms

$$L_n = L^\pm_n + L^S_n. \tag{31}$$

where

$$L_n^\pm = \sum_m : \alpha_{-m}^- \alpha_{n+m}^+ :$$

$$L_n^S = \frac{1}{k-2} \left(\sum_m : s_{-m} s_{n+m} : + i n s_n + \frac{1}{4} \delta_{n,0} \right)$$

(32)

Note that the $L_n^\pm$ is equivalent to the $c = 2$ Virasoro operator in 2D flat spacetime. The full central charge is

$$c = \frac{3k}{k-2}.$$

(33)

The eigenvalue of the total L_0 is

$$L_0 = p^+ p^- + \frac{1}{k-2} \left[s_0^2 + 1/4 \right] + \text{integer} = a$$

(34)

Thus, the theory has been reduced to a 2D lightcone in flat spacetime plus a Liouville type space-like free field that has positive norm. A small but important difference as compared to the standard Liouville formalism is that the linear term in L_n^S is Hermitean in our case, and does not contribute to L_0^S.

The only negative norm states are the ones produced by the time-like oscillator $\alpha_n^0 = (\alpha_n^+ - \alpha_n^-)/\sqrt{2}$. However, this is no worse than the usual flat spacetime case. The space of physical states is defined by

$$(L_n - a\delta_{n,0}) |\phi >= 0$$

(35)

with $a \leq 1$ fixed. A proof of no ghosts can now be given by following step by step the same arguments that prove the no ghost theorem in flat spacetime (9). There is no need to repeat it here. We only recall that there are no ghosts as long as $a \leq 1$ and $c \leq 26$.

8.2 Monodromy

So far we have not taken into account the physical effects of the $\ln z$ cut in the currents. At first sight, the presence of $\ln z$ in the currents appears to be contrary to the periodicity requirement of closed strings. However, this is not true. The periodicity requirement arises as a boundary condition in the process of minimizing the action. In other words only on mass shell physical string configurations are required to be periodic. In the presence of the extra zero mode the Hilbert space is larger. One must require periodicity in the physical on shell sector. The physical sector is identified as the subspace of states for which the matrix elements of the currents are periodic.

$$< phys | J^i(ze^{i2\pi n}) | phys' >=< phys | J^i(z) | phys' > .$$

(36)

As described below, under this requirement, it turns out that the extra zero mode must have quantized eigenvalues in the physical sector. Quantum mechanically it is possible to impose the monodromy condition simultaneously with the Virasoro constraints since the latter commute with the monodromy operator as seen below. These monodromy conditions are easily taken care of in the free field formalism, thus finally giving a complete physical unitary spectrum of the SL(2,R) WZW model.

To implement the monodromy let us first consider its effect on the currents. From the modified currents in (4) we see that under the monodromy the currents undergo a linear transformation

$$
\begin{aligned}
&[J^0 + J^1]\left(ze^{i2\pi n}\right) = [J^0 + J^1]\left(z\right) \\
&[J^0 - J^1]\left(ze^{i2\pi n}\right) = [J^0 - J^1]\left(z\right) + 4\pi n \alpha_0^- \; J^2(z) \\
&\qquad\qquad + \left(2\pi n \alpha_0^-\right)^2 [J^0 + J^1]\left(z\right) \\
&J^2(ze^{i2\pi n}) = J^2(z) + 2\pi n \alpha_0^- \; [J^0 + J^1]\left(z\right)
\end{aligned}
\tag{37}
$$

Note that on the right hand side one finds J^a, not $\tilde{J}^a$. Therefore we expect that the right hand side can be rewritten as the adjoint action with a global SL(2,R) transformation. Since the current $J^0(z) + J^1(z)$ remains unchanged the generator of this transformation must be the zero mode of this current. Indeed, since α_0^- acts like a number, we can rewrite the monodromy in the form

$$
J^i(ze^{i2\pi n}) = e^{-2i\pi n \alpha_0^- \left(J_0^0 + J_0^1\right)} J^i(z)\, e^{2i\pi n \alpha_0^- \left(J_0^0 + J_0^1\right)}
\tag{38}
$$

Therefore physical states that satisfy (36) are the subset of states that are invariant under the monodromy

$$
e^{2i\pi n \alpha_0^- \left(J_0^0 + J_0^1\right)}|phys> = |phys>\,.
\tag{39}
$$

In the free boson representation this is easy to implement. Using $(J_0^0 + J_0^1) = \alpha_0^+$ this condition is applied on the Fock space of the free bosons in the form

$$
e^{2i\pi n \alpha_0^- \alpha_0^+} \prod_{n,m,k=1}^{\infty} \left(\alpha_{-n}^+\right)^{a_n} \left(\alpha_{-m}^-\right)^{b_m} \left(s_{-k}\right)^{c_k} |p^+ p^- s_0>
\tag{40}
$$

Since $\alpha_0^- \alpha_0^+$ commutes with all oscillators, it can be moved to the right and applied on the base. The result is the quantization condition

$$
\begin{aligned}
&e^{2i\pi n \alpha_0^- \alpha_0^+}|p^+, p^-, s_0> = |p^+, p^-, s_0> \\
&\alpha_0^- \alpha_0^+ = p^- p^+ = -r, \quad r = 0, 1, 2, \cdots
\end{aligned}
\tag{41}
$$

We must take negative integers because according to the mass shell condition $p^- p^+$ is negative. So, the mass shell condition on physical states at excitation level l takes the form

$$-r + \frac{1}{k-2}\left[s_0^2 + 1/4\right] + l = a. \tag{42}$$

It is always possible to satisfy this condition with some value of s_0 which is quantized in terms of the positive integers r, l. In terms of the original Casimir $j(j+1)$ this corresponds to a principal series representation of SL(2,R) with quantized values of j given by $j = -\frac{1}{2} + is_0 = -\frac{1}{2} \pm i\sqrt{(k-2)(r-l+a) - 1/4}$ where r must be chosen so that the square root is real.

Therefore, it is sufficient to require that j (or s_0) has quantized values

$$j_n = -\frac{1}{2} \pm i\sqrt{(k-2)(n+a) - 1/4}. \tag{43}$$

Then the on-shell physical states automatically satisfy the periodicity condition.

8.3   Open and Closed strings

An open string action $S = \int d\tau \int_0^\pi d\sigma\, L(\tau, \sigma)$ is minimized by allowing free variation of the end points. For the WZW model for any group G this produces the boundary terms

$$\delta S = \int d\tau \left\{ \begin{array}{c} Tr\left((\delta g g^{-1})(\partial_\sigma g g^{-1})\right)|_\pi \\ - Tr\left((\delta g g^{-1})(\partial_\sigma g g^{-1})\right)|_0 \end{array} \right\} \tag{44}$$

In addition to the equations of motion, these terms must also vanish at each end of the string. That is,

$$\left. \partial_\sigma g g^{-1}\right|_{\sigma=0} = 0 = \left. \partial_\sigma g g^{-1}\right|_{\sigma=\pi}. \tag{45}$$

At the conformal critical point the equations of motion are satisfied by the general form $g(\tau, \sigma) = g_L(\tau + \sigma)\, g_R^{-1}(\tau - \sigma)$. Then the boundary conditions require that g_L and g_R be related to each other by the constraint

$$g_L^{-1}(\tau)\, \partial_\tau g_L(\tau) + g_R^{-1}(\tau)\, \partial_\tau g_R(\tau) = 0. \tag{46}$$

Furthermore, each term in this equation is required to be periodic. As discussed in the rest of this paper, we impose periodicity on the physical states. The relation (46) between $g_L(\tau)$ and $g_R(\tau)$ is not easy to solve explicitly. However, we may carry out the quantum theory in terms of one current $\hat{J}$

$$\hat{J}(z) = g_L^{-1}(z)\, \partial_z g_L(z) = -g_R^{-1}(z)\, \partial_z g_R(z). \tag{47}$$

This is neither the left moving current $J_L = \partial g_L g_L^{-1}$ nor the right moving one $J_R = \partial g_R g_R^{-1}$, but is related to them by transformations involving g_L or g_R.

$$J_L = g_L \hat{J} g_L^{-1}, \quad J_R = -g_R \hat{J} g_R^{-1}.$$

The current $\hat{J}$ generates transformations on the right side of g_L and the left side of g_R^{-1}, and the meaning of (46) is that the total current on both g_L and g_R vanishes at the end points. The canonical commutation rules for this current are identical to the ones we have already discussed in the rest of the paper. The stress tensor constructed from it is equal to the stress tensor constructed from either the left movers or the right movers

$$Tr(\hat{J}^2) = Tr(J_L^2) = Tr(J_R^2). \tag{48}$$

The quantum spectrum is obtained from the properties of $\hat{J}$, whose mathematical structure is the same as either left movers or right movers as discussed in the previous sections. For an open string we choose to parametrize the $\hat{J}$ current in terms of free fields. Thus, the quantum spectrum of the open string in the SL(2,R) curved spacetime becomes identical to the spectrum discussed above.

For a closed string we have independent left and right moving sectors. The full group element is $g = g_L(z)g_R^{-1}(\bar{z})$ and there are left and right moving currents. Therefore we now need two sets of oscillators, the left movers $\alpha_n^\pm, s_n$, and the right movers $\tilde{\alpha}_n^\pm, \tilde{s}_n$. So, the direct product Hilbert space has a base labelled by $|p^-, p^+, s_0; \tilde{p}^-, \tilde{p}^+, \tilde{s}_0 >$ with $p^- p^+ = -r$ and $\tilde{p}^- \tilde{p}^+ = -\tilde{r}$ to insure that the currents obey the monodromy conditions in the physical sector. We now need to figure out if these are all independent labels or if they must be constrained by physical considerations.

For this purpose we recall that a possible modular invariant is the so called "diagonal invariant" that requires the same unitary representation labelled by the same j for both left and right movers. This may be understood as being related to the representation of the full group element $D^j(g) = D^j(g_L(z))D^j(g_R^{-1}(\bar{z}))$ which requires the same j for both left and right movers. Therefore, we must demand $s_0 = \tilde{s}_0$.

In addition, we examine $g(z, \bar{z})$ in more detail. Keeping the order of operators, it may be written in the form

$$g = g_L(z)g_R^{-1}(\bar{z}) = \begin{pmatrix} u & a \\ -b & v \end{pmatrix} \tag{49}$$

with

$$
\begin{aligned}
u &= e^{\frac{-u_L+u_R}{k-2}} - e^{\frac{-u_L}{k-2}} \left(X_L^+ - X_R^+\right) X_R^- e^{\frac{-u_R}{k-2}} \\
v &= e^{\frac{u_L-u_R}{k-2}} + e^{\frac{-u_L}{k-2}} X_L^- \left(X_L^+ - X_R^+\right) e^{\frac{-u_R}{k-2}} \\
a &= e^{\frac{-u_L}{k-2}} \left(X_L^+ - X_R^+\right) e^{\frac{-u_R}{k-2}} \\
b &= -\left(X_L^- - X_R^-\right) e^{\frac{-u_L+u_R}{k-2}} \\
&\quad + e^{\frac{-u_L}{k-2}} X_L^- \left(X_L^+ - X_R^+\right) X_R^- e^{\frac{-u_R}{k-2}}
\end{aligned}
\tag{50}
$$

We see that g is not periodic under $\sigma \to \sigma + 2\pi n$ since there are logarithms in the expressions for every $X^{\pm}_{L,R}, u_{L,R}$. However, provided we impose $p^+ = -\tilde{p}^+$ on physical states (to cancel the non-periodic behavior in $X^+_L - X^+_R$), we find that we can rewrite this monodromy in the form

$$g(ze^{i2\pi n}, \bar{z}e^{-i2\pi n}) = U\tilde{U}g(z,\bar{z})\tilde{U}^{-1}U^{-1}$$
$$U\tilde{U} = e^{-ip^+p^-2\pi n}e^{-is_0^2 2\pi n}e^{i\tilde{p}^+\tilde{p}^-2\pi n}e^{i\tilde{s}_0^2 2\pi n} \tag{51}$$

where p^+, s_0 are operators which do not commute with q^-, u_0, and similarly for right movers (note that we have never introduced a canonical conjugate to p^- (or $\tilde{p}^-$)). To insure that the matrix elements of the overall g are consistent with monodromy in the physical sector it is sufficient to impose the conditions

$$2p^+p^- + 2s_0^2 - 2\tilde{p}^+\tilde{p}^- - 2\tilde{s}_0^2 = 2m \tag{52}$$

where m is an integer. Since we have already seen that $s_0 = \tilde{s}_0$ we find that this condition reduces to $r - \tilde{r} = m$, and does not impose any additional constraints on $r, \tilde{r}$. Furthermore, for a closed string we should also have $L_0 - \tilde{L}_0 = 0$ on the physical states. According to the mass shell condition (42) this requires $r - l = \tilde{r} - \tilde{l}$.

So, modular invariant physical closed string states are labelled at the base as follows

$$|p^-, \ p^+, s_0(n) > \times |\tilde{p}^-, -p^+, s_0(n) > \tag{53}$$

where the following restrictions are imposed

$$\tilde{p}^+ = -p^+, \quad \tilde{s}_0 = s_0 = s_0(n)$$
$$s_0(n) \equiv [(k-2)(n+a) - 1/4]^{1/2} \tag{54}$$
$$n = 0, 1, 2, \cdots \tag{55}$$

Then the on-shell physical states automatically satisfy the periodicity conditions.

9. Chiral Vertex operators

In string theory there is a vertex operator corresponding to every physical state. In the usual algebraic approach one would try to construct a chiral vertex operator $V_{jm}(z)$ corresponding to the "tachyon" state $|jm >$. In our case we have diagonalized $J_0^0 + J_0^1 \to p^+$ instead of $J_0^0 \to m$. Hence, our vertex operator is labelled as $V_{jp^+}(z)$. In this basis, we need a vertex operator with the following operator product properties

$$J^a(z) V_{jp^+}(w) \sim \frac{1}{z-w} [t^a V_{jp^+}(w)] \tag{56}$$

where t^a is a Hermitean generator of SL(2,R) that acts on the p^+ label in the unitary principal series labelled by $j = -1/2 + is$. This representation is given by

$$
\begin{aligned}
\left(t^0 + t^1\right) V_{jp^+}(w) &= p^+ V_{jp^+}(w) \\
t^2 V_{jp^+}(w) &= \left[\frac{1}{2}\left\{p^+, i\partial_{p^+}\right\} + s\right] V_{jp^+}(w) \\
\left(t^0 - t^1\right) V_{jp^+}(w) &= \left[i\partial_{p^+} p^+ i\partial_{p^+} + 2si\partial_{p^+}\right] V_{jp^+}(w)
\end{aligned}
\tag{57}
$$

A vertex operator with these properties is given by

$$
V_{jp^+}(w) = \exp\left(ip^+ X^-(w)\right) : \exp\left(\frac{1 + 2is}{k - 2} u(w)\right) :
\tag{58}
$$

It is straightforward to verify that it has the correct operator product properties with the currents in (24). This product of exponentials has a rather simple structure which can be readily manipulated in the computation of corellation functions.

There is a $\ln z$ term in the first exponential

$$
\exp\left(p^+ \alpha_0^- \ln w\right) = w^{p^+ \alpha_0^-}
$$

The monodromy condition requires that the power $p^+ \alpha_0^- =$ integer, in agreement with the previous approach.

In the construction of corellation functions in other WZW models it has been found that there are screening charges that play an important role. In the SL(2,R) case we found the following two screening currents (note that $S_1 \sim \partial X^+$)

$$
\begin{aligned}
S_1(z) &= \frac{P^+(z)}{k - 2} : \exp\left(\frac{2u(z)}{k - 2}\right) : \\
S_2(z) &= \left(P^+(z)\right)^{k-2} : \exp(2u(z)) :
\end{aligned}
\tag{59}
$$

Their operator products with the currents are

$$
\begin{aligned}
\left[J^0(z) + J^1(z)\right] S_1(w) &\sim 0 \\
J^2(z) S_1(w) &\sim 0 \\
\left[J^0(z) - J^1(z)\right] S_1(w) &\sim \partial_w\left(\frac{\exp\left[2u(w)/(k-2)\right]}{z - w}\right)
\end{aligned}
\tag{60}
$$

and

$$
\begin{aligned}
\left[J^0(z) + J^1(z)\right] S_2(w) &\sim 0 \\
J^2(z) S_2(w) &\sim 0 \\
\left[J^0(z) - J^1(z)\right] S_2(w) &\sim \partial_w\left(\frac{\left(P^+(w)\right)^{k-3} e^{2u(w)}}{z - w}\right)
\end{aligned}
\tag{61}
$$

Therefore, the zero modes of these screening currents

$$Q_{1,2} \equiv \frac{1}{2\pi i} \oint dz \, S_{1,2}(z) \tag{62}$$

are the screening charges that commute with all the currents. These play an important role in the construction of corellation functions. We hope to discuss these issue further in future work.

10. Comments

We have shown that a unitary string theory in SL(2,R) curved spacetime can be constructed and its spectrum solved exactly. A crucial ingredient was an additional zero mode whose presence introduces many new technical and physical features. This should provide a lesson for other more involved models.

Attention was drawn to some new technical points. The first is that currents are allowed to contain logarithmic cuts provided monodromy conditions are applied on the physical states. The second is a new representation of the currents in terms of free bosons that render the theory completely solvable. Vertex operators and screening currents that should be useful in computations were also suggested. These features have obvious generalizations to higher dimensions as well as to gauged WZW models (coset models).

We have also shown that the free boson methods permit a more general representation of SL(2,R) current algebra when the extra degrees of freedom L'_n are introduced. These were absent in the WZW model, but they may be present in more general models.

As emphasized in the introduction, the main purpose for the present exercise is to develop the appropriate methods to study string theory during the early universe and to understand the impact of string theory on the symmetries and matter content observed at accelerator energies. For this purpose the current methods must be generalized to heterotic strings such as those described in (18). Methods used for other special models of curved spacetimes may also be helpful (19).

REFERENCES

I. Bars and D. Nemeschansky, Nucl. Phys. B348 (1991) 89.

For a recent review, see: I. Bars, "Curved Space-time Geometry for Strings ...", in Perspectives in Mathematical physics" Vol.III, Eds. R. Penner and S.T.Yau, International Press (1994), page 51. See also hep-th 930942.

A.A. Tseytlin, reviews, hep-th/9408040, 9410008. G.T.Horowitz and A.A. Tseytlin, hep-th/9409021.

J. Balog, L. O'Raighfertaigh, P. Forgacs, and A. Wipf, Nucl. Phys. B325 (1989) 225.

E. Witten, Phys. Rev. D44 (1991) 314.

I. Bars, "Ghost free spectrum of a quantum string in SL(2,R) curved spacetime", hep-th/9503205.

M. Wakimoto, Commun. Math. Phys. 104 (1986) 605.

A. Gerasimov, A. Morozov, M. Olshanetsky, A. Marshakov and S. Shatashvili, Int. J. Mod. Phys. A5 (1990) 2495.

C. Thorn, Nucl. Phys. B248 (1974) 551, and lecture in Unified String Theory, Eds. D. Gross and M. Green, page 5.

I. Bars, "Curved Spacetime Strings and Black Holes", in Proc. of Strings-91 conference, Strings and Symmetries 1991, Eds. N. Berkovitz et.al. World Scientific (1992), page 135.

P. Petropoulos, Phys. Lett. 236B (1990) 151. M. Henningson, S. Hwang, P. Roberts, and B. Sundborg, Phys. Lett. 267B (1991) 350. J. Distler and P. Nelson, Nucl. Phys. B366 (1991) 255. Shyamoli Chaudhuri and J.D. Lykken, Nucl.Phys.B396 (1993) 270.

I. Bars and K. Sfetsos, Mod. Phys. Lett A7 (1992) 1091.

I. Bars and J. Schulze, hep-th/9405156, Phys. Rev. D.51 (1995) 1854.

I. Bars, "Folded Strings in Curved Spacetime", hep-th/9411078, to appear in Phys. Rev D.

J.K.Freericks and M.B.Halpern, Ann.of Phys. 188 (1988) 258; Erratum, ibid.190 (1989) 212. M.B. Halpern, E. Kiritsis, N. A. Obers, K. Clubok, "Irrational Conformal Field Theory", hep-th/9501144.

I. Bars and K. Sfetsos, Phys. Rev. D48 (1993) 844. A.A. Tseytlin, Nucl. Phys. B399 (1993) 601. A.A. Tseytlin and K. Sfetsos, Phys. Rev. D49 (1994) 2933.

See e.g. Superstring Theory I, M.B. Green, J.S. Schwarz and E. Witten, pages 111-113.

I. Bars, Phys. Lett. B293 (1992) 315; Nucl. Phys B334 (1990) 125.

I. Antoniadis, S. Ferrara, S. Kounnas, Nucl. Phys. B421 (1994) 343. E. Kiritsis and C.Kounnas, Phys.Lett.B331 (114) 321. J. Russo and A. Tseytlin, hep-th/9502038.

INTERNAL PHYSICS OF BLACK HOLES: RECENT DEVELOPMENTS

WERNER ISRAEL
Canadian Institute for Advanced Research
Cosmology Program,
Theoretical Physics Institute, University of Alberta,
Edmonton,Alberta T6G 2J1, Canada

1. Introduction

Surely there is nothing which more plainly invites ridicule than the study of black hole interiors. No matter what happens inside the hole, we can never observe it, nor will it affect anything we can observe. Penrose's famous 1965 theorem assures the presence of a (classical) singularity internally, signalling a breakdown of all the laws of physics. Finally, "hair" (irregularities) swept into the hole at the time of collapse will surely pile up, producing a state of internal chaos. On the face of it, this hardly seems a topic for serious scientists of sound views.

It is thus fitting to begin by explaining why I believe the outlook is not quite as bleak as this. I have so far neglected to mention a key property of the hole's interior. *Descent into a black hole is fundamentally a progression in time.* (Recall that Schwarzschild's radial co-ordinate r becomes timelike within the event horizon.) This means we are confronted with an evolutionary, not a structural, problem. We do not need to understand the physics of quantum gravity to follow the evolution up to the stage where curvatures begin to approach Planck levels. Simple causality prevents our ignorance of the inner, high-curvature regions from infecting the description afforded by contemporary theory of the outer and *preceding* layers.

171

N. Sánchez and A. Zichichi (eds.), String Gravity and Physics at the Planck Energy Scale, 171–185.
© *1996 Kluwer Academic Publishers.*

Moreover, we know precisely the initial conditions for this evolution (in contrast to the situation in cosmology), thanks to the no-hair theorems and the work of Richard Price. Initially, near the event horizon, we have a Kerr-Newman geometry perturbed by a tail of gravitational waves decaying according to a power law. The task, then, is to evolve these initial data, using the Einstein field equations up to the point where a singularity forms or where curvatures reach Planck levels.

Exploration of the outer, sub-Planckian zone of the hole is thus seen to reduce to a standard Cauchy evolutionary problem, no more speculative or exotic than following the motion of a fluid, using Euler's equations, up to onset of turbulence or a shock. But technically it is a bit harder!

The first studies of this problem [1, 2] used spherical models to simplify the mathematics. The results which emerged were sufficiently unexpected and unorthodox to arouse vague disquiet as well as specific objections. By now the dust seems to have settled enough to permit a claim that at least the spherical case is well understood [3, 4]. Attention has shifted to the development of efficient techniques for dealing with the generic case.

To minimize overlap with my previous "D. Chalonge" lectures [5], I shall concentrate here on these latest developments. For introductory accounts and broader reviews the reader may consult this reference or [6, 7].

2. The Spherical Case: A Retrospective and Update

The effect of a spherical (e.g. scalar-field) wave-tail on the internal structure of a charged spherical hole was examined by Poisson and myself [1] in 1989.

We found it convenient to employ a two-dimensional reduction of the spherisymmetric Einstein equations. Any spherisymmetric 4-metric is expressible as

$$ds^2 = g_{AB}(x^C)dx^A\,dx^B + r^2\,d\Omega^2 \qquad (A, B, \cdots = 0, 1) \qquad (1)$$

where (x^0, x^1) are an arbitrary pair of co-ordinates which identify a privileged 2-sphere of area $4\pi r^2$. The gradient of r defines the

Schwarzschild mass function $M(x^A)$ through

$$(\nabla r)^2 = g^{AB}(\partial_A r)(\partial_B r) = 1 - 2M/r.$$

The Einstein field equations can now be reformulated as two-dimensionally covariant equations involving the scalars $r(x^A)$, $M(x^A)$. The central result is a $(1 + 1)$-dimensional wave equation for M,

$$\Box M = -16\pi^2 r^3 T_{ab} T^{ab} \qquad (2)$$

(modulo some additional, inessential terms) which brings out very clearly the effects of nonlinearity of the Einstein equations.

A charged spherical hole shares with a generic black hole the crucial feature of an inner, Cauchy horizon, a lightlike hypersurface characterized by infinite external advanced time v, but reachable in finite proper time by an infalling observer. Wave-tails and other time-dependent disturbances propagating into the hole experience unbounded blueshift near the Cauchy horizon, causing the source term in (2) to diverge internally as $v \to \infty$.

This results in a "mass-inflationary" singularity at the Cauchy horizon CH (Figs. 1 and 2). For a wave-tail whose flux decays externally like v^{-p}, M diverges internally like

$$M \sim C|uv|^{-p} e^{\kappa_0(u+v)}, \qquad (3)$$

while remaining nearly constant externally. Here, u is an internal retarded time, calibrated so that $u = -\infty$ on the event horizon (EH), and κ_0 is the inner surface gravity of the asymptotic Reissner-Nordström static black hole that an external observer would see.

This divergence hinges on the presence of at least a nominal outflux transverse to CH. (For a pure influx [8], no matter how blueshifted, T_{ab} is lightlike and the right-hand side of (2) vanishes.) In reality, it is of course impossible to avoid the outflux resulting from backscatter of incoming waves.

It is conceivable that this backscatter could lead to problems, since it falls off only as an inverse power $|u|^{-p}$ of retarded time u as $u \to -\infty$ toward the past of CH inside the hole. Conceivably, this might not be fast enough to allow the Cauchy horizon to begin

174

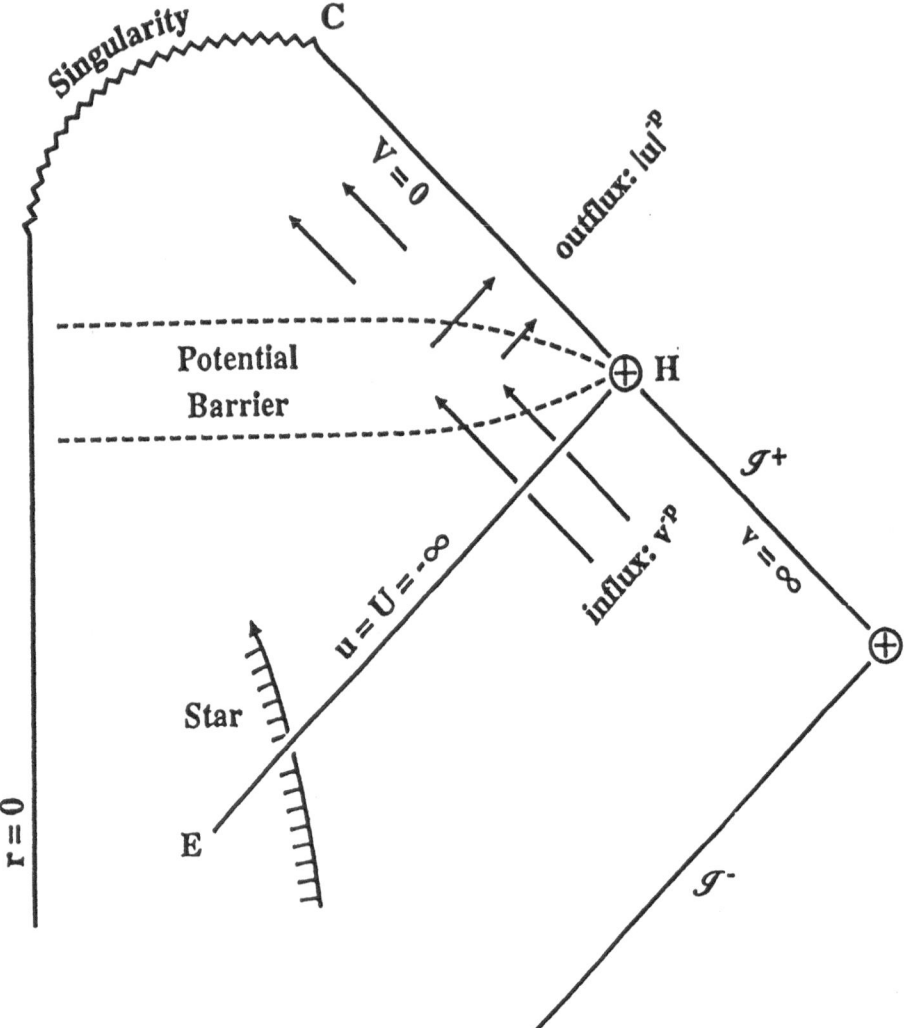

Fig. 1: Waves propagating into a charged spherical hole are scattered off a potential barrier situated between the event horizon EH and the Cauchy horizon CH. This is a Penrose conformal map: tracks of radial photons are straight lines which are inclined at ±45°, and past and future lightlike infinity have been "compactified" to the lines $\mathcal{I}^-$ and $\mathcal{I}^+$ by a conformal transformation. "Points" such as H, marked by circled crosses, are singularities of the conformal mapping. In reality, EH and CH are 3-cylinders of different radii which extend indefinitely without intersecting.

175

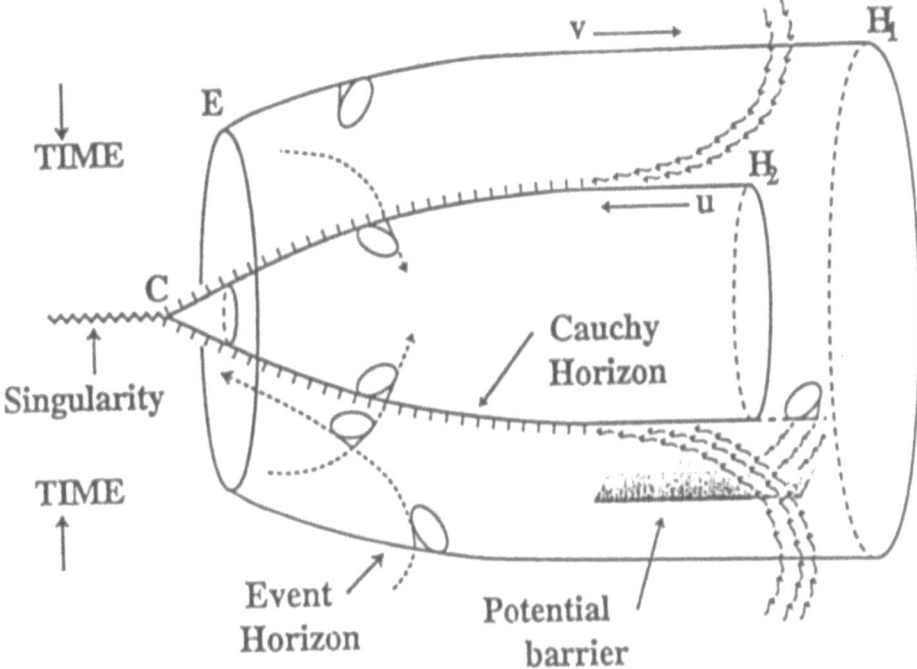

Fig. 2: Geometrically faithful rendering of the scenario of Fig. 1. One angular co-ordinate has been suppressed. Inside the hole, increasing time coincides with diminishing radius. The points H_1 and H_2 of this figure were telescoped into a single "point" H in Fig. 1.

176

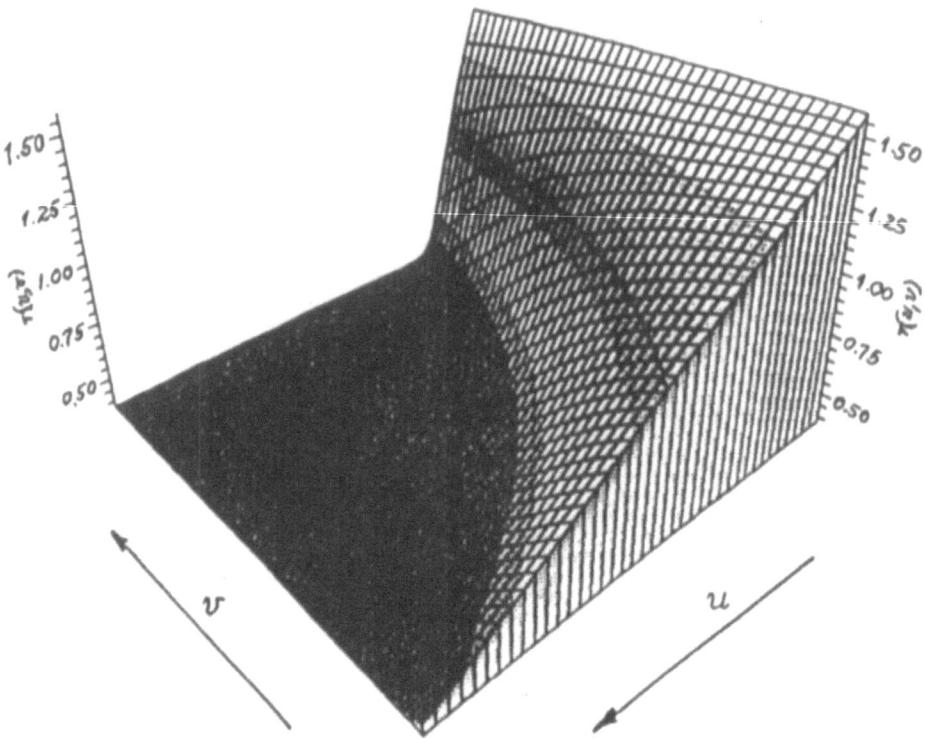

Fig. 3: Effect of scalar wave-tail on the internal geometry of a charged spherical hole. The axes at the rear of the figure parallel to the u- and v-directions respectively are the Cauchy and event horizons. This three-dimensional plot of the radial co-ordinate $r(u, v)$ shows the radius of the Cauchy horizon decreasing continuously to zero with increasing retarded time u.

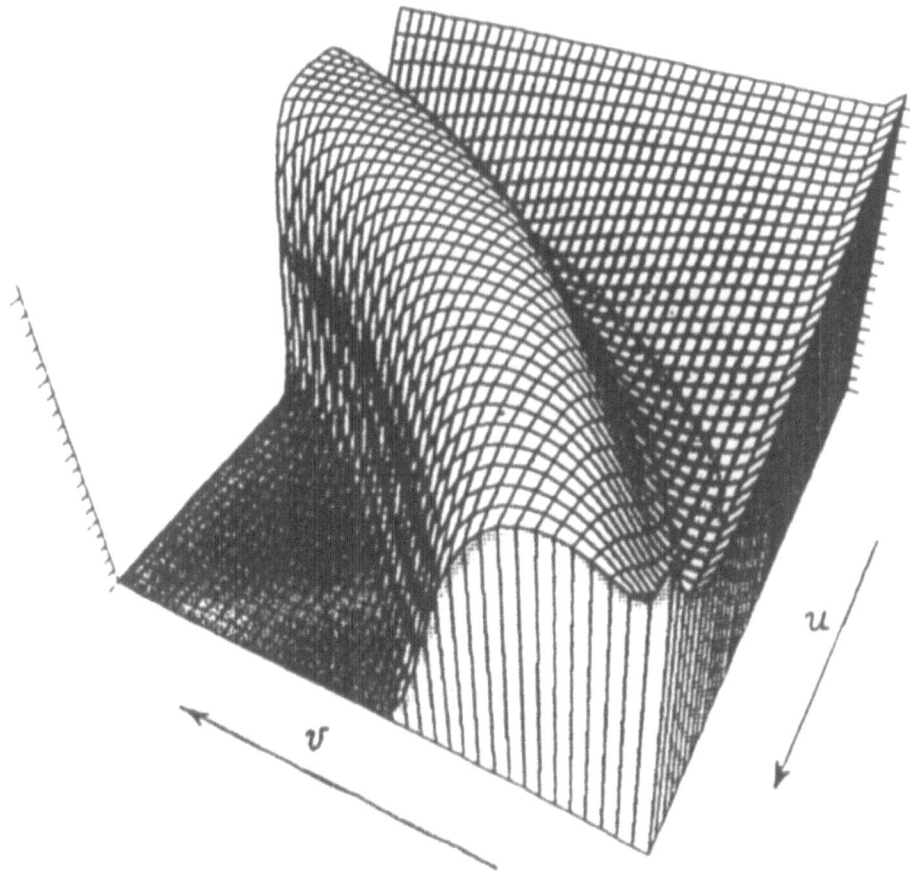

Fig. 4: A plot of $\exp(1 - 2M/r)$, showing the effect of mass inflation near the Cauchy horizon. Figs. 3 and 4 are from numerical integrations by P. R. Brady and J. D. Smith [4], reproduced by kind permission of the authors.

its contraction from an asymptotically static, Reissner-Nordström configuration, and might lead to its pre-emption by a spacelike singularity. A 1993 numerical integration [9] of scalar-wave scattering by a charged spherical hole gave evidence that precisely this would actually happen.

However, analytical investigations [3], undertaken incidentally with a view to examining this conclusion, have not supported it. Instead of entering into the gory details of this, I shall trust that seeing is believing, and present in Figs. 2 and 3 typical results of recent numerical integrations by Brady and Smith [4], which confirm that the Cauchy horizon survives as a lightlike, mass-inflationary singularity.

3. Structure of Aspherical Holes

In the generic case, one would expect the mass-inflationary singularity (more accurately: a divergence of the Newman-Penrose component Ψ_2 of the Weyl tensor in accordance with (3)) to be overlaid by a gravitational shock wave imploding along the Cauchy horizon. A growing body of evidence [5, 7, 10–12] suggests that, apart from this, the spherical picture is qualitatively unchanged.

But, again, there have been doubts. Yurtsever [13] has stressed that Cauchy horizons are generically spacelike rather than lightlike, at least for plane-wave spacetimes. However, the plane-wave analogy is probably misleading [5, 7, 14]. (It has to be borne in mind that the concept of Cauchy horizon is not absolute, but relative to the hypersurface on which initial data are set. In the black hole context, the Cauchy horizon for externally set initial data *precedes* what one would normally consider to be the plane-wave Cauchy horizon in a homogeneous space.)

A full treatment of the aspherical case needs techniques for analyzing the geometry near a generic lightlike singularity. I shall devote the rest of this report to an introductory account of a versatile technique suited to this purpose.

4. Covariant Double-null Dynamics

The classic analysis of Arnowitt, Deser and Misner [15] formulates gravitational dynamics in terms of the evolution of a spatial 3-geometry. The geometrical framework is the imbedding formalism of Gauss and Codazzi for the foliation of spacetime by spacelike hypersurfaces.

Quite often, however, —as in the analysis of Cauchy horizon singularities, but also in other contexts—one encounters circumstances where a lightlike foliation is especially suitable. Because of various degeneracies that arise in the lightlike case, the imbedding relations are quite different and the situation not quite as familiar and under control. To bypass the degeneracies, one is forced to fall back to a foliation of co-dimension 2, by spacelike 2-surfaces.

Several such (2+2)-formalisms are extant, notably Geroch, Held and Penrose [16], d'Inverno and Smallwood [17] and Hayward [18]. Of course, all have basically the same content, but they take very different forms.

The distinguishing feature of the approach I shall present here is that it maintains manifest two-dimensional covariance and operates directly with geometrically significant objects. Two-dimensional covariance permits reduction of the Einstein field equations to an especially concise and geometrically transparent form: the ten Ricci components are embraced in a set of just three compact, two-dimensionally covariant expressions.

(2 + 2)-formalisms are a versatile tool having a range of applications which go far beyond the specific problem of Sec. 3. They are useful in the analysis and numerical integration of the characteristic initial-value problem of general relativity, the dynamics of strings, of real and apparent horizons, the study of Planck-energy collisions and light-cone quantization. The covariant formalism has been applied by our group ([20]) to elucidate the singular geometry of the black hole Cauchy horizon.

5. Basic Metric Notions

I shall briefly sketch the elements of the formalism.

We consider a double foliation of spacetime by a net of two intersecting families of lightlike hypersurfaces Σ^0 (with equations $u^0 =$ const.) and Σ^1 (given by $u^1 =$ const.), where $u^A(x^\alpha)$ ($A, B, \ldots, = 0, 1$; $\alpha, \beta, \ldots, = 1, \ldots, 4$) are a given pair of scalar fields over spacetime, with lightlike gradients:

$$\nabla u^A \cdot \nabla u^B = e^{-\lambda} \eta^{AB},$$

and where the matrix

$$\eta^{AB} = \eta_{AB} = \begin{pmatrix} 0 & -1 \\ -1 & 0 \end{pmatrix}$$

will be used to raise and lower upper-case Latin indices, and $\lambda(x^\alpha)$ is a scalar function. The generators $\ell_\alpha^{(A)}$ of Σ^A are conveniently defined as

$$\ell^{(A)} = e^\lambda \nabla u^A.$$

Two hypersurfaces Σ^0 and Σ^1 intersect in a 2-surface S, with parametric equations

$$x^\alpha = x^\alpha(u^A, \theta^a) \qquad (a, b, \cdots = 2, 3)$$

where (θ^2, θ^3) are intrinsic co-ordinates of S. Both generators $\ell^{(A)}$ are orthogonal to S.

Holonomic basis vectors $e_{(a)}$ and the intrinsic metric of S may now be defined:

$$e_{(a)}^\alpha = \frac{\partial x^\alpha}{\partial \theta^a}, \qquad g_{ab} = e_{(a)} \cdot e_{(b)}.$$

The matrix g_{ab} and its inverse g^{ab} are used to lower and raise lowercase Latin indices, so that $e^{(a)} = g^{ab} e_{(b)}$ are the dual basis vectors tangent to S.

Two-dimensional shift vectors s_A^a are defined by

$$s_A^a = \frac{\partial x^a}{\partial u^A} e_\alpha^{(a)} = -\ell_{(A)}^\alpha \frac{\partial \theta^a}{\partial x^\alpha}.$$

As in the Arnowitt-Deser-Misner formalism, the shift vector s_A^a measures how much one has to deviate from the normal direction $\ell_{(A)}$ to connect points on different 2-surfaces having the same intrinsic co-ordinates θ^a. An infinitesimal four-dimensional displacement dx^α can be decomposed as

$$dx^\alpha = \ell_{(A)}^\alpha du^A + e_{(a)}^\alpha (d\theta^a + s_A^a du^A).$$

Together with the completeness relation

$$g_{\alpha\beta} = e^{-\lambda} \eta_{AB} \ell_\alpha^{(A)} \ell_\beta^{(B)} + g_{ab} e_\alpha^{(a)} e_\beta^{(b)}$$

for the basis $(\ell^{(A)}, e_{(a)})$, this implies that the spacetime metric is decomposable as

$$g_{\alpha\beta} dx^\alpha \, dx^\beta = e^\lambda \eta_{AB} du^A \, du^B + g_{ab}(d\theta^a + s_A^a du^A)(d\theta^b + s_B^b \, du^B).$$

6. Covariant Geometrical Objects Embodying First Derivatives of Metric

Associated with its two normals $\ell_{(A)}$, a 2-surface S has two extrinsic curvatures, defined by

$$K_{Aab} = (\nabla_\beta \ell_{(A)\alpha}) e_{(a)}^\alpha \, e_{(b)}^\beta$$

and easily shown to be symmetric in a, b. (Since we are free to rescale the null vectors $\ell_{(A)}$, a certain scale-arbitrariness is inherent in this definition.)

A further basic geometrical property of the double foliation is given by the Lie bracket of $\ell_{(0)}$ and $\ell_{(1)}$. One finds

$$[\ell_{(B)}, \ell_{(A)}] = \epsilon_{AB} \omega^a e_{(a)} \tag{4}$$

where

$$\omega^a = \epsilon^{AB}(\partial_B s_A^a - s_B^b s_{A;b}^a),$$

the semicolon indicates two-dimensional covariant differentiation associated with metric g_{ab}, and ϵ_{AB} is the two-dimensional permutation symbol.

The geometrical significance of the "twist" ω^a can be read off from (4): the curves tangent to the generators $\ell_{(0)}$, $\ell_{(1)}$ mesh together to form 2-surfaces (orthogonal to the surfaces S) if and only if $\omega^a = 0$. In this case, it would be consistent to allow the co-ordinates θ^a to be dragged along both sets of generators, and thus to gauge both shift vectors to zero.

I shall denote by D_A the two-dimensionally invariant operator associated with differentiation along the normal direction $\ell_{(A)}$. Acting on any two-dimensional geometrical object $X^{a\cdots}_{b\cdots}$, D_A is formally defined by

$$D_A X^{a\cdots}_{b\cdots} = (\partial_A - \mathcal{L}_{s^d_A}) X^{a\cdots}_{b\cdots}.$$

Here, ∂_A is the partial derivative with respect to u^A and $\mathcal{L}_{s^d_A}$ the Lie derivative with respect to the 2-vector s^d_A. As an example:

$$D_A g_{ab} = \partial_A g_{ab} - 2s_{A(a;b)} = 2K_{Aab}.$$

Geometrically, $D_A X^{a\cdots}_{b\cdots}$ is the projection onto S of the Lie derivative with respect to $\ell_{(A)}$ of the equivalent tangential 4-tensor $X^{\alpha\cdots}_{\beta\cdots}$.

The objects K_{Aab}, ω^a and D_A are all simple projections onto S of four-dimensional geometrical objects. Consequently, they transform very simply under two-dimensional co-ordinate transformations. Under the arbitrary reparametrization

$$\theta^a \rightarrow \theta^{a'} = f^a(\theta^b, u^A) \tag{5}$$

(which leaves u^A and hence the surfaces Σ^A and S unchanged), ω_a and K_{Aab} transform cogrediently with

$$e_{(a)} \rightarrow e'_{(a)} = e_{(b)} \partial \theta^b / \partial \theta^{a'}.$$

By contrast, the shift vectors s^a_A undergo a more complicated gauge-like transformation, arising from the u-dependence in (5).

7. Ricci Tensor

This geometrical groundwork is already sufficient to allow me to display the simple form that the Ricci components take in this formalism. (Notation for the tetrad components is typified by $R_{aA} = R_{\alpha\beta}e^{\alpha}_{(a)}\ell^{\beta}_{(A)}$.)

The results are

$$^{(4)}R_{ab} = \frac{1}{2}\,^{(2)}Rg_{ab} - e^{-\lambda}(D_A + K_A)K^A_{ab}$$

$$+ 2e^{-\lambda}K^d_{A(a}K^A_{b)d} - \frac{1}{2}e^{-2\lambda}\omega_a\omega_b - \lambda_{;ab} - \frac{1}{2}\lambda_{,a}\lambda_{,b}$$

$$R_{AB} = -D_{(A}K_{B)} - K_{Aab}K^{ab}_B + K_{(A}D_{B)}\lambda$$

$$- \frac{1}{2}\eta_{AB}\big[(D^E + K^E)D_E\lambda - e^{-\lambda}\omega^a\omega_a + (e^\lambda)^{;a}\,_a\big]$$

$$R_{Aa} = K^b_{Aa;b} - \partial_a K_A - \frac{1}{2}\partial_a D_A\lambda + \frac{1}{2}K_A\partial_a\lambda$$

$$+ \frac{1}{2}\epsilon_{AB}e^{-\lambda}\big[(D^B + K^B)\omega_a - \omega_a D^B\lambda\big],$$

where $^{(2)}R$ is the curvature scalar associated with the 2-metric g_{ab}, and $K_A \equiv K^a_{Aa}$.

It should be clear that this provides a remarkably economical and geometrically transparent generalization of the $(2+2)$ spherical formalism, whose application to spherical black holes was outlined in Sec. 2. For further details and developments, I must refer to the original papers [19, 20].

8. Summary and Conclusion

It appears that the inside of a black hole is, after all, not completely inscrutable. The investigations I have described lead to a rather definite working picture of at least the outer, classical layers of the hole, illustrated in Figs. 1, 2.

The crushing spacelike singularity has a milder, lightlike precursor, situated on the Cauchy horizon CH. In effect, CH serves as a lightlike bridge, linking the quiescent, asymptotically hairless, final phases of the hole's history (at late advanced times v) to the "hairy" crunch near C associated with its formation.

Observers entering the hole long after it is formed fall toward the Cauchy horizon and enter the Planck layer close to CH. What happens then? On this question, contemporary theory is silent. We have only charted a coastline; we do not know whether it heralds a narrow strip of land or a continent. Its further exploration is a task for physics of the millenium.

Acknowledgments

It is a pleasure to record my debt to Alfio Bonanno, Pat Brady, Serge Droz, Sharon Morsink and Eric Poisson, whose work is reported in these pages, for their input and stimulation, and to Pat Brady in particular for permission to reproduce Figs. 3 and 4.

This work was supported by the Canadian Institute for Advanced Research and by NSERC of Canada.

References

1. Poisson, E. and Israel, W. (1990) Phys. Rev. **D41**, 1796.
2. Ori, A. (1991) Phys. Rev. Letters **67**, 789.
3. Bonanno, A., Droz, S., Israel, W. and Morsink, S. M. (1994) Phys. Rev. **D50**, 755;
   (1995) Proc. Roy. Soc. **A450**, 553.
4. Brady, P. R. and Smith, J. D. (1995) Phys. Rev. Letters **75**, 1256.
5. Israel, W. (1993) The internal constitution of black holes, in N. Sanchez and A. Zichichi (eds.), *Current Topics in Astrofundamental Physics*, World Scientific, Singapore, pp. 430–448. [Note typographical omission of factor v^{-p} in first of eqns. (8).]
6. Droz, S., Israel, W. and Morsink, S. M. (1996) Physics World (in press).

7. Bonanno, A., Droz, S., Israel, W. and Morsink, S. M. (1994) Can. J. Phys. **72**, 755; erratum. (1995) Can. J. Phys. **73**, 251.

8. Hiscock, W. A. (1981) Physics Letters **83A**, 110.

9. Gnedin, M. L. and Gnedin, N. Y. (1993) Class Quantum Grav. **10**, 1083.

10. Ori, A. (1992) Phys. Rev. Letters **68**, 2117.

11. Brady, P. R. and Chambers, C. M. (1995) Phys. Rev. **D51**, 4177.

12. Brady, P. R., Droz, S., Israel, W. and Morsink, S. M. (1995), paper in preparation.

13. Yurtsever, U. (1993) Class. Quantum Grav. **10**, 117.

14. Flanagan, E. and Ori, A. (1995), preprint.

15. Arnowitt, R., Deser, S., Misner, C. W. (1962) *Gravitation: an Introduction to Current Research* (L. Witten, ed.) Wiley, New York, Chap. 7.

16. Geroch, R., Held, A. and Penrose, R. (1973), J. Math. Phys. **14**, 874.

17. d'Inverno, R. A. and Smallwood, J. (1980), Phys. Rev. **D22**, 1233.

18. Hayward, S. A. (1994) Class. Quantum Grav. **11**, 3025.

19. Brady, P. R., Droz, S., Israel, W. and Morsink, S. M. (1995) Class. Quantum Grav. (in press).

20. Brady, P. R., Droz, S., Israel, W. and Morsink, S. M. (1995), paper in preparation.

BLACK HOLE ENTROPY AND PHYSICS AT PLANCKIAN SCALES

VALERI FROLOV

*CIAR Cosmology Program; Theoretical Physics Institute,
University of Alberta, Edmonton, Canada T6G 2J1**

1 Black-Hole Entropy

According to the thermodynamical analogy in black hole physics, the entropy of a black hole in the Einstein theory of gravity is

$$S^{BH} = A_H/(4l_P^2), \tag{1.1}$$

where A_H is the area of a black hole surface and $l_P = (\hbar G/c^3)^{1/2}$ is the Planck length [1, 2]. In black hole physics the Bekenstein-Hawking entropy S^{BH} plays essentially the same role as in the usual thermodynamics. In particular it allows to estimate what part of the internal energy of a black hole can be transformed into work. Four laws of black hole physics that form the basis in the thermodynamical analogy were formulated in [3]. According to this analogy the entropy S is defined by the response of the free energy F of the system containing a black hole to the change of its temperature:

$$dF = -SdT. \tag{1.2}$$

The generalized second law [1, 2, 4] (see also [5, 6, 7, 8] and references therein) implies that when a black hole is a part of the thermodynamical system the total entropy (i.e. the sum of the entropy of a black hole and the entropy of the surrounding matter) does not decrease.

The Euclidean approach [9, 10, 11] provides a natural way to derive black hole thermodynamical properties. Doing a Wick's rotation $t \to -i\tau$ in the Schwarzschild metric one gets the metric with the Euclidean signature. The corresponding manifold with the Euclidean metric is regular if and only if the imagine time τ is periodic and the period is $2\pi/\kappa$, where κ is the surface gravity of the black hole. The corresponding

---

*E-mail address: frolov@phys.ualberta.ca

N. Sánchez and A. Zichichi (eds.), String Gravity and Physics at the Planck Energy Scale, 187–207.
© *1996 Kluwer Academic Publishers.*

regular Euclidean metric is known as the Gibbons-Hawking instanton. The period $2\pi/\kappa$ of the imagine time τ is naturally identified with the inverse temperature of the black hole, while the Euclidean action, calculated for the Gibbons-Hawking instanton, is directly related with its free energy. York and collaborators [12, 13] made an important observation that the formal derivation of the black hole entropy using the Euclidean approach requires certain modifications for its consistency. Namely, to get a well defined canonical ensemble, one needs to consider a black hole in a box of finite size and fix the corresponding thermodynamical quantities (temperature) at the boundary. Box filled with thermal radiation and a black hole in it in thermal equilibrium with radiation is thermodynamically stable if the radius of a spherical box r is less than $3M$. If one changes the temperature of the box, the mass of a black hole inside the box also changes, as required by the conditions of the thermal equilibrium. As the result of this process the free energy F of the system is changed. One can use the relation (1.2) and single out the contribution to the entropy due to the black hole. In the classical approximation the so defined thermodynamical entropy coincides with the Bekenstein-Hawking entropy S^{BH}.

The simple relation between the thermodynamical entropy of a black hole and its surface area is characteristic for Einstein theory. The definition of the black hole entropy can be generalized to non-Einstein versions of gravitational theory, provided they allow existence of black holes. Wald [14] showed that in the general case the entropy of a black hole is defined by Nöther charge related with the Killing vector. (For application and developing this idea, see [15, 16, 17].) Recently the interest to the problem of black hole entropy was increased by observation that the entropy of extremal black hole might vanish [18, 19].

The success of the thermodynamical analogy in black hole physics allows one to hope that this analogy may be even deeper and it is possible to develop statistical-mechanical foundation of black hole thermodynamics. It is worthwhile to remind that the thermodynamical and statistical-mechanical definitions of the entropy are logically different. *Thermodynamical entropy* S^{TD} is defined by the response of the free energy F of a system to the change of its temperature:

$$dF = -S^{TD}dT. \tag{1.3}$$

(This definition applied to a black hole determines its Bekenstein-Hawking entropy.)
Statistical-mechanical entropy S^{SM} is defined as

$$S^{SM} = -\text{Tr}(\hat{\rho}\ln\hat{\rho}), \tag{1.4}$$

where $\hat{\rho}$ is the density matrix describing the internal state of the system under consideration. It is also possible to introduce the *informational entropy* S^I by counting different possibilities to prepare a system in a final state with given macroscopical parameters from different initial states

$$S^I = -\sum_n p_n \ln p_n, \tag{1.5}$$

with p_n being the probabilities of different initial states. In standard case all three definitions give the same answer.

Is the analogy between black holes thermodynamics and the 'standard' thermodynamics complete? Are there internal degrees of freedom of a black hole responsible for its entropy? Is it possible to apply the statistical-mechanical and informational definitions of the entropy to black holes and how are they related with the Bekenstein-Hawking entropy? These are the questions that are to be answered.

Historically first attempts of the statistical-mechanical foundation of the entropy of a black hole were connected with the informational approach [2, 20]. According to this approach the black hole entropy is interpreted as "the logarithm of the number of quantum mechanically distinct ways that the hole could have been made"[5, 20]. The so defined informational entropy of a black hole is simply related to the amount of information lost by stretching the horizon, and as was shown by Thorne and Zurek it is equal to the Bekenstein-Hawking entropy [5, 20]. Quite interesting results relating informational and statistical-mechanical entropies can be obtained in a special model proposed by Bekenstein and Mukhanov [21, 22]. According to this model a black hole is identified with a system having discrete internal states, so that an absorption or emission of a particle by a black hole is accompanied by the transition from one state to another. In such model the mass of a black hole is quantized[1]. Unfortunately the physical origin of internal degrees of freedom of a black hole in the model and their discreteness is not derived but postulated.

The dynamical origin of the entropy of a black hole and the relation between the statistical-mechanical and Bekenstein-Hawking entropy have remained unclear. In the present talk I describe some recent results obtained in this direction.

2 Dynamical Degrees of Freedom of a Black Hole

The problem of the dynamical origin of the black hole entropy was intensively discussed recently. The proposed basic idea is to relate the dynamical degrees of freedom of a black hole with its quantum excitations. This idea has different realizations[2].

In the framework of the *membrane paradigm* the dynamical degrees of freedom are identified with different possible states of thermal atmosphere of a black hole, while the entropy of a black hole is identified with the amount of information about the state of thermal atmosphere, which is lost by stretching the horizon[5, 20].

In his *brick wall model* 't Hooft [25] proposed to consider a mirror-like boundary, located outside a black hole at the close distance to its horizon. He assumed that outside the boundary there exist thermal radiation with temperature equal to the Hawking temperature. He has shown that the entropy of such thermal radiation is

---

[1]Another approach to the quantization of the mass of a black hole can be found in [23].

[2]For recent review of the problem of the dynamical origin of the entropy of a black hole, see also [24].

of the same order of magnitude as the Bekenstein-Hawking entropy, provided the distance of the mirror-like boundary from the horizon is chosen to be of order of Planck's length.

Bombelli *et al* [26] attracted attention to the fact that even in a flat spacetime in a Minkowski vacuum state one can obtain a non-vanishing entropy if one restricts his observations to the spatially bounded part of the space. The corresponding *entanglement entropy* arises due to the presence of correlations of those modes of zero-point fluctuations which are propagating in the vicinity of the boundary of chosen spatial region. It was proposed to relate black hole entropy with the entanglement entropy related with the presence of the horizon [26, 27]

We (with Igor Novikov) arrived to the similar idea independently by analyzing the gedanken experiment proposed in our earlier paper [28]. Namely we assumed that there exist a traversable wormhole, and its mouths are freely falling into a black hole. If one of the mouths crosses the gravitational radius earlier than the other, then rays passing through the first mouth can escape from the region lying inside the gravitational radius. Such rays would go through the wormhole and enter the outside region though the second mouth. As the result during the period of time when the first of the mouth is inside and the other mouth is outside the gravitational radius the surface area of the horizon decreases. If we assume that the black hole entropy is related with the surface area of a black hole, then the only possibility to escape contradictions with the second law is to assume that during such process the decrease of entropy is related with the possibility to get access to some new information concerning black hole internal states. At first sight it looks like a puzzle. We know that (at least in the classical General Relativity) a black hole at late time is completely specified by finite number of parameters. For a non-rotating uncharged black hole one need to know only one parameter (mass M) to describe all its properties. It is true not only for the exterior where this property is the consequence of no-hair theorems, but also for the black hole interior [29]. The reason why it is impossible for an isolated black hole at late time to have non-trivial classical states in its interior in the vicinity of the horizon is basically the same as for the exterior regions. These states might be excited only if a collapsing body emits the pulse of fields or particles immediately after it crosses the event horizon. Due to red-shift effect the energy of the emitted pulse must be exponentially large in order to reach the late time region with any reasonable energy. After short time (say $100M$) after the formation of a black hole it is virtually impossible because it requires the energy of emission much greater than the black hole mass M.

In quantum physics the situation is completely different. The above mentioned puzzle can be solved if we remember that any quantum field has zero-point fluctuations. To analyze states of a quantum field it is convenient to use its decomposition into modes. Besides the positive-frequency modes which have positive energy, there exist also positive-frequency modes with negative energy. In a non-rotating uncharged black holes such modes can propagate only inside the horizon where the Killing vector used to define the energy is spacelike. It is possible to show that at late time for any regular initial state of the field these states are thermally excited and the corresponding temperature coincides with the black hole temperature $T_H = (8\pi M)^{-1}$. (For more details see, e.g.[30].) In principle by using a traversable wormhole described above one can get information concerning these internal modes propagating near the horizon and change their states.

In the framework of this *dynamical-black-hole-interior* model we proposed to identify the dynamical degrees of freedom of a black hole with the internal modes of all physical fields. The set of the fields must include the gravitational one. It can be shown that for any chosen field the number of the modes infinitely grows as one considers the regions located closer and closer to the horizon. For this reason the contribution of a field to the statistical-mechanical entropy of a black hole calculated by counting the internal modes of a black hole is formally divergent. In order to make it finite one might restrict himself by considering only those modes which are located at the proper distance from the horizon greater than some chosen value l. For this choice of the cut-off the contribution of a field to the statistical-mechanical entropy of a black hole is

$$S^{SM} = \alpha \frac{A}{l^2}, \qquad (2.1)$$

where A is the surface area of a black hole, and a dimensionless parameter α depends on the type of the field.

3   No-Boundary Wave Function of a Black Hole

The calculation of the black hole entropy in the dynamical-black-hole-interior model is made by counting the number of thermally excited internal modes existing at given moment of time (more accurately on the chosen spacelike surface, crossing the horizon). This calculations can be simplified by using the following trick. Consider an eternal version of a black hole, i.e. an eternal black hole with the same mass M as the original black hole formed as the result of collapse. At late time the geometry of both holes are identical. One can trace back in time all the perturbations, propagating at late time in the geometry of the eternal version of a black hole . As the result one can relate perturbations at late time in a spacetime of a real black hole with initial data on the Einstein-Rosen bridge (spatial slice of $t =$const) of the eternal black hole geometry (see Fig. 1).

We denote the 3-surface of the Einstein-Rosen bridge by Σ. This surface has the topology $S^2 \times R^1$. The 2-surface S of the horizon $r = 2M$ splits it into two isometric parts: 'external' Σ_+ and 'internal' Σ_-. In a spacetime of the eternal black hole the Killing vector ξ which is used to define energy is future directed on Σ_+ and past directed on Σ_-. For this reason initial date having a support located on Σ_+ correspond to the field configurations having positive energy, while the energy of the field configurations with the initial data on Σ_- possess negative energy. The

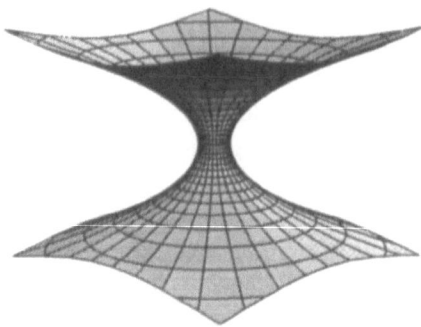

Figure 1: Embedding diagram for a two-dimensional $\theta = $ const section of the Einstein-Rosen bridge.

former describes external degrees of freedom of a black hole, while the latter describes the internal ones. The set of fields representing the degrees of freedom of a black hole contains the gravitational perturbations. For given initial values of fields and gravitational perturbations on Σ the gravitational constraint equations determine the deformation of the 3-geometry of the Einstein-Rosen bridge (see Fig. 2). We shall use the notion 'deformation' in order to describe not only deformed geometry of the Einstein-Rosen bridge, but also the physical fields on it. By using this terminology we can say that the states of a black hole at late time are uniquely related with deformations of the Einstein-Rosen bridge.

It was proposed in [31] to introduce a wave function of a black hole as the functional over the space of deformations of the Einstein-Rosen bridge. In this approach the wave function of a black hole depends on data located on both parts of the Einstein-Rosen bridge: an external Σ_+ (external degrees of freedom) and an inner Σ_- one (internal degrees of freedom).

Certainly there exist infinite number of different wavefunctions of a black hole. Our aim is to get a useful tool for the description of the canonical ensemble of black holes inside the cavity restricted by a spherical boundary of the radius r_B and with fixed inverse temperature β on it. For this reason only very special wavefunctions will be important for us. Here we present a modified version of the no-boundary approach of Ref.[31] which is analogous to the 'no-boundary ansatz' in quantum cosmology [36]. This ansatz singles out a set of no-boundary wavefunctions which is convenient for our purpose.

Instead of the complete Einstein-Rosen bridge we consider its part Σ' lying between two spherical 2-dimensional boundaries $S_\pm$ located from both sides of S at

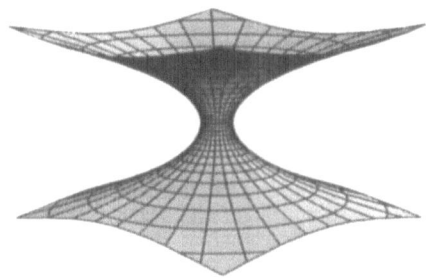

Figure 2: Deformation of the Einstein-Rosen bridge.

the radius $r = r_B$. Σ' has the topology $S^2 \times I$, where I is the unit interval $[0,1]$. We denote by M_β a Euclidean manifold with a boundary ∂M_β, which consists of two parts: Σ' and another 3-surface Σ^B with the same topology $S^2 \times I$, which intersects Σ' at S_+ and S_-, and which represents the Euclidean evolution of the external boundary B.

We define the no-boundary wavefunction depending on one parameter β by the following path integral

$$\Psi_\beta(^3g(\boldsymbol{x}), \varphi(\boldsymbol{x})) = \int \mathcal{D}\,^4g\, \mathcal{D}\phi\, e^{-I[^4g, \phi]}. \tag{3.1}$$

Here $I[^4g, \phi]$ is the Euclidean gravitational action. The integral is taken over Euclidean 4-geometries and matter-field configurations on a spacetime M_β with a boundary $\partial M_\beta \equiv \Sigma' \cup \Sigma^B$. The integration variables are subject to the conditions

$$(^3g(\boldsymbol{x}), \varphi(\boldsymbol{x})), \quad \boldsymbol{x} \in \partial M_\beta,$$

the collection of 3-geometry and boundary matter fields on ∂M_β, which are just the argument of the wavefunction (3.1).

We assume that the 3-metric on the boundary is of the form

$$ds^2_{\Sigma^B} = d\tau^2 + r_B^2 d\omega^2 + \dots, \qquad \tau \in (-\beta/4, \beta/4), \tag{3.2}$$

$$ds^2_{\Sigma'} = (1 - r_+/r)^{-1} dr^2 + r^2 d\omega^2 + \dots, \qquad r \in [r_+, r_B), \tag{3.3}$$

where $r_+ \equiv 2M$, $d\omega^2$ is the line element on a unit sphere, and dots indicate omitted terms describing perturbations of the metric.

If (g_0, ϕ_0) is a point of the extremum of the action I, then we can write

$$g = g_0 + \tilde{g}, \qquad \phi = \phi_0 + \tilde{\phi}, \tag{3.4}$$

$$I[g, \phi] = I_0[g_0, \phi_0] + I_2[\tilde{g}, \tilde{\phi}] + \dots. \tag{3.5}$$

In accordance with this decomposition the no-boundary wavefunction (3.1) in the quasiclassical approximation reads

$$\Psi_\beta(^3g(x), \varphi(x)) = \Psi_\beta^0(^3g_0(x), \varphi_0(x)) \; \Psi_\beta^1(^3\tilde{g}(x), \tilde{\varphi}(x)), \tag{3.6}$$

where

$$\Psi_\beta^0(^3g_0(x), \varphi_0(x)) = e^{-I_0[g_0, \phi_0]}, \tag{3.7}$$

is a classical (tree-level) contribution, and

$$\Psi_\beta^1(^3\tilde{g}(x), \tilde{\varphi}(x)) = \int \mathcal{D}^4\tilde{g} \; \mathcal{D}\tilde{\phi} \; e^{-I_2[^4\tilde{g}, \tilde{\phi}]}. \tag{3.8}$$

is a one-loop part.

We consider a theory for which $\phi_0 = 0$, so that g_0 is a solution of the vacuum Einstein equations. The corresponding Euclidean solution is a part M_β of the Gibbons-Hawking instanton (see Fig. 3), i.e. the Euclidean Schwarzschild solution

$$ds^2 = F d\tau^2 + F^{-1}dr^2 + d\omega^2, \qquad F = 1 - r_+/r, \tag{3.9}$$

with $\tau \in (-\frac{1}{4}\beta_\infty, \frac{1}{4}\beta_\infty)$, where $\beta_\infty = (F(r_B))^{-1/2}\beta$ is the inverse temperature at infinity. (For a special choice of $\beta_\infty = 8\pi M$ this part is a half of the instanton.)

To calculate the Euclidean Einstein-Hilbert action

$$I_0[g_0] = -\frac{1}{16\pi} \int_{M_\beta} R\sqrt{g} d^4x - \frac{1}{8\pi} \int_{\partial M_\beta} K\sqrt{h} d^3x - \frac{1}{8\pi} \int_{\partial M_\beta} K_0\sqrt{h} d^3x \tag{3.10}$$

for M_β we note that for the vacuum solution $R = 0$, so that only the surface terms of the action contribute. The calculation of the trace of the extrinsic curvature K for Σ^B is straightforward and gives

$$K = \frac{1}{2}\frac{r_+}{r_B^2}\frac{1}{\sqrt{F(r_B)}} + \frac{2\sqrt{F(r_B)}}{r_B}. \tag{3.11}$$

By using this expression we get

$$\frac{1}{8\pi} \int_{\Sigma^B} K\sqrt{h} d^3x = \frac{1}{2}\beta r_B\sqrt{F(r_B)} + \frac{\beta r_+}{8\sqrt{F(r_B)}}. \tag{3.12}$$

Figure 3: Part M_β of the Gibbons-Hawking instanton, which gives the main contribution into the no-boundary wavefunction Ψ_β in the quasiclassical approximation.

The extrinsic curvature vanishes identically everywhere on the boundary Σ' except the point $r = r_+$, where it has δ-type singularity. The corresponding contribution is

$$\frac{1}{8\pi} \int_{\Sigma^B} K\sqrt{h}d^3x = \frac{\pi}{2}r_+^2 \left(1 - \frac{\beta}{4r_+\sqrt{F(r_B)}}\right). \tag{3.13}$$

There is a well known ambiguity in the term $\frac{1}{8\pi}\int_{\partial M_\beta} K_0\sqrt{h}d^3x$ which is to be subtracted and which depends on the choice of a reference space. Here we fix this ambiguity simply by subtracting from $\frac{1}{8\pi}\int_{\partial M_\beta} K\sqrt{h}d^3x$ its value for $M = 0$. Finally we have

$$I_0[g_0] = \frac{1}{2}\beta r_B\left(1 - \sqrt{F(r_B)}\right) - \frac{\pi}{2}r_+^2. \tag{3.14}$$

Now we calculate the one-loop contribution to the no-boundary wavefunction of a black hole. First of all we note that each of the fields (including the gravitational perturbations) give independent contribution to I_2. It means that Ψ_β^1 is a product of wavefunctions $\Psi_\beta^1[\varphi]$ depending on only one particular type of field φ. (We remind that $\phi_0 = 0$ and hence the value φ on the boundary coincides with its perturbation $\tilde{\phi}$.) The Gaussian integral (3.8) in the definition of $\Psi_\beta^1[\varphi]$ can be easily calculated. Really, let us denote by $\phi(\varphi)$ a solution of field equations for the action $I_2[\phi]$ obeying the boundary conditions $\phi\big|_{\partial M_\beta} = \varphi$. Then

$$\Psi_\beta^1(\varphi) = Ce^{-I_2[\varphi]}. \tag{3.15}$$

We assume that φ on Σ^B does not depend on τ. In this case a solution $\phi(\varphi)$ obeying boundary conditions

$$\phi(x)\Big|_{\Sigma_\pm} \equiv \phi(\pm\beta_\infty/4, x) = \varphi_\pm(x), \tag{3.16}$$

can be written as a decomposition

$$\phi(\tau, x) = \sum_\lambda \left\{ \varphi_{\lambda,+} u_{\lambda,-}(\tau, x) + \varphi_{\lambda,-} u_{\lambda,+}(\tau, x) \right\} \tag{3.17}$$

in the basis functions of the field equation

$$u_{\lambda,\pm}(\tau, x) = \frac{\sinh(\beta_\infty/4 \mp \tau)\omega}{\sinh(\beta_\infty/2)} R_\lambda(x), \tag{3.18}$$

where $R_\lambda(x)$ is a complete set of spatial harmonics on M_β with a chosen boundary conditions on Σ^B. The coefficients $\varphi_{\lambda,\pm}$ in (3.17) are just the decomposition coefficients of the fields (3.16) in the basis of spatial harmonics

$$\varphi_\pm(x) = \sum_\lambda \varphi_{\lambda,\pm} R_\lambda(x). \tag{3.19}$$

Substituting (3.17) into $I_2[\varphi]$, integrating by parts with respect to the Euclidean time and spatial coordinates and taking into account the equations of motion, one finds that the Euclidean action reduces to the following quadratic form in $\varphi_{\lambda,\pm}$:

$$I_2[\varphi_+, \varphi_-] = \sum_\lambda \left\{ \frac{\omega_\lambda \cosh(\beta_\infty \omega_\lambda/2)}{2 \sinh(\beta_\infty \omega_\lambda/2)} (\varphi_{\lambda,+}^2 + \varphi_{\lambda,-}^2) - \frac{\omega_\lambda}{\sinh(\beta_\infty \omega_\lambda/2)} \varphi_{\lambda,+} \varphi_{\lambda,-} \right\} \tag{3.20}$$

This action is a sum of Euclidean actions for quantum oscillators of frequency ω_λ for the interval β_∞ of the Euclidean time with the initial value of its amplitude φ_- and the final value φ_+.

To summarize we obtain the following expression for the no-boundary wavefunction of a black hole in the semiclassical approximation

$$\Psi_\beta[M, \varphi_+, \varphi_-] = N e^{-1/2\beta r_B [1 - (1 - 2M/r_B)^{1/2}] + 2\pi M^2 - \sum I_2[\varphi_+, \varphi_-]}, \tag{3.21}$$

Here N is a normalization constant. The symbol of summation in the exponent indicates that the additional a summation over all physical fields must be done. The square of this wavefunction gives the probability to find a given configuration in the state determined by the parameter β. For large M ($M \gg m_P$) this probability is a sharp peak with width $\approx m_P$ located near the value of $M = M_\beta \equiv \beta_\infty/8\pi$. For $\beta_\infty = 8\pi M$ and $r_B \to \infty$ this wavefunction coincides with a no-boundary wavefunction obtained in [31].

For fixed M the density matrix for internal variables φ_- of a black hole is defined as

$$\hat{\rho}_\beta[\varphi_-, \varphi_-'] = \int D\varphi_+ \Psi_\beta[\varphi_+, \varphi_-] \Psi_\beta[\varphi_+, \varphi_-'], \tag{3.22}$$

and it is of the form

$$\hat{\rho}[\varphi_-, \varphi_-'] = P' \, e^{-\tilde{I}_2[\varphi_-, \varphi_-']}, \tag{3.23}$$

where $\tilde{I}_2$ is given by the expression (3.20) with β changed by 2β. It is easy to show that

$$\hat{\rho}[\varphi_-, \varphi_-'] = P'' \langle \varphi_- | e^{-\beta \hat{H}} | \varphi_-' \rangle, \tag{3.24}$$

where P'' is a normalization constant and $\hat{H}$ is the Hamiltonian of free fields φ propagating on the Schwarzschild background. The statistical-mechanical entropy S^{SM} of a black hole obtained by using this density matrix coincides with the expression (2.1).

The statistical-mechanical entropy S^{SM} in this as well as other 'dynamical' approaches possesses the following main properties: (1) $S^{SM} \sim A$, where A is the surface area of a black hole; (2) S^{SM} is divergent and requires regularization: $S^{SM} \sim A/l^2$, where l is the cut-off parameter; (3) S^{SM} depends on the number of fields, which exist in nature; (4) $S^{SM} \sim S^{BH}$ for $l \sim l_{\mathrm{P}}$.

The following two problems are of importance: (1) What is the relation between the statistical-mechanical entropy S^{SM} introduced by counting the internal degrees of freedom of a black hole and its thermodynamical entropy S^{TD}? In particular how to explain the universality of the Bekenstein-Hawking entropy S^{BH}, while S^{SM} is not universal and depends on the number of fields? (2) The formal expression for the statistical-mechanical entropy S^{SM} contains the Planck scale cut-off. Does it mean that by studying the thermodynamical properties of black holes we can obtain certain conclusions concerning physics at Planckian scales?

In what follows we shall try to clarify these questions.

4   Renormalized Effective Action and Free Energy

The complete information concerning the canonical ensemble of black holes with a given inverse temperature β at the boundary is contained in the partition function $Z(\beta)$ given by the Euclidean path integral [11]

$$Z(\beta) = \int D[\phi] \exp(-I[\phi]). \tag{4.1}$$

Here the integration is taken over all fields including the gravitational one that are real on the Euclidean section and are periodic in the imaginary time coordinate τ with period β. The quantity ϕ is understood as the collective variable describing the fields. In particular it contains the gravitational field. Here $D[\phi]$ is the measure of the space of fields ϕ and I_E is the Euclidean action of the field configuration. The action I_E includes the Euclidean Einstein action. The state of the system is determined by the choice of the boundary conditions on the metrics that one integrates over. For the canonical ensemble for the gravitational fields inside a spherical box of radius r_B at temperature T one must integrate over all the metrics inside r_B which are periodically identified in the imaginary time direction with period $\beta = T^{-1}$. Such a partition function must describe in particular the canonical ensemble of black holes. The partition function Z is related with the effective action $\Gamma = -\ln Z$ and with the free energy $F = \beta^{-1}\Gamma = -\beta^{-1}\ln Z$.

By using the stationary-phase approximation one gets

$$\beta F \equiv -\ln Z = I[\phi_0] - \ln Z_1 + \dots. \tag{4.2}$$

Here ϕ_0 is the (generally speaking, complex) solution of classical field equations for action $I[\phi]$ obeying the required periodicity and boundary conditions. Besides the tree-level contribution $I[\phi_0]$, the expression (4.2) includes also one-loop corrections $\ln Z_1$, connected with the contributions of the fields perturbations on the background ϕ_0, as well as higher order terms in loops expansion, denoted by $(\dots)$. The one-loop contribution of a field ϕ can be written as follows $\ln Z_1 = -\frac{1}{2}\mathrm{Tr}\ln(-D)$, where D is the field operator for the field ϕ inside the box r_B. The one-loop contribution contains divergences and required the renormalization. In order to be able to absorb these divergences in the renormalization of the coefficients of the initial classical action we chose the latter in the form

$$I_{cl} = \int d^4x \sqrt{g} L, \tag{4.3}$$

$$L = \left[-\frac{\Lambda_B}{8\pi G_B} - \frac{R}{16\pi G_B} + c_B^1 R^2 + c_B^2 R_{\mu\nu}^2 + c_B^3 R_{\alpha\beta\mu\nu}^2 \right]. \tag{4.4}$$

By using heat-kernel representation for $\ln Z_1$ one can write

$$-\frac{1}{2}\ln\det(-D) = \frac{1}{2}\int_{\delta^2}^{\infty} \frac{ds}{s} \mathrm{Tr}K(s), \tag{4.5}$$

where $K(s)$ is the heat-kernel of the operator D which has the following Schwinger-DeWitt expansion

$$K(s) = e^{-sD} = \frac{1}{16\pi^2 s^2} \sum a_n s^n, \quad s \to 0. \tag{4.6}$$

For the particular case of a scalar massless field

$$a_0 = 1, \qquad a_1 = (1/6 - \xi)R, \tag{4.7}$$

$$a_2 = \frac{1}{180}R_{\alpha\beta\mu\nu}^2 - \frac{1}{180}R_{\mu\nu}^2 - \frac{1}{6}(\frac{1}{5} - \xi)\Box R + \frac{1}{2}(\frac{1}{6} - \xi)^2 R^2 \tag{4.8}$$

By substituting this expansion into (4.5) one can conclude that the one-loop contribution Γ_1 to the effective action can be written in the form

$$\Gamma_1 = \Gamma_1^{div} + \Gamma_1^{fin}. \tag{4.9}$$

where

$$\Gamma_1^{div} = -\frac{1}{32\pi^2} \int d^4x \sqrt{g} \left[\frac{a_0}{2\delta^4} + \frac{a_1}{\delta^2} - 2a_2 \ln(\delta) \right]. \tag{4.10}$$

The divergent part of the one-loop effective action has the same structure as the initial classical action (4.3) and hence one can write

$$\Gamma = \Gamma_{cl}^{ren} + \Gamma_1^{ren}, \tag{4.11}$$

$$\Gamma_1^{ren} = \Gamma_1 - \Gamma_1^{div} = \Gamma_1^{fin}. \tag{4.12}$$

Here Γ_{cl}^{ren} is identical to the initial classical action with the only change that all the bare coefficients Λ_B, G_B, and c_B^i are substituted by their renormalized versions Λ_{ren}, G_{ren}, and c_{ren}^i

$$\frac{\Lambda_{ren}}{G_{ren}} = \frac{\Lambda_B}{G_B} + \frac{1}{8\pi\delta^4}, \tag{4.13}$$

$$\frac{1}{G_{ren}} = \frac{1}{G_B} + \frac{1}{2\pi\delta^2}\left(\frac{1}{6} - \xi\right), \tag{4.14}$$

$$c_{ren}^i = c_B^i + \alpha^i \ln \delta. \tag{4.15}$$

We shall refer to (4.11) as to the loop expansion of the renormalized effective action. After multiplying the the renormalized effective action by β^{-1} we get the expansion for the renormalized free energy.

The effective action Γ contains the complete information about the system under consideration. In particular the variation of Γ with respect to the metric provides one with the equations for the quantum average metric $\bar{g} = \langle g \rangle$:

$$\frac{\delta\Gamma}{\delta\bar{g}} = 0. \tag{4.16}$$

One usually assumes that quantum corrections are small and solves this equation perturbatively:

$$\bar{g} = g_{cl} + \delta g, \tag{4.17}$$

where g_{cl} is a solution of the classical equations. At this point we need to make an important remark. In principle, there exist two possibilities: either to begin with the solution of the classical equations for the action (4.3), or its renormalized version Γ_{cl}^{ren}, which is written in terms of the renormalized constants. One usually assumes that the renormalized values of Λ_{ren} and c_{ren}^i vanish $\Lambda_{ren} = c_{ren}^i = 0$. It means that in general case their initial values were not vanishing unless one is dealing with some special type of theory (e.g. assuming supersimmetry). In other words the global properties of the solutions for I_{cl} and Γ_{cl}^{ren} are generally different. So to provide the condition that δg is small, one is to begin with the metric g_{cl} that is an extremum of Γ_{cl}^{ren}

$$\frac{\delta\Gamma_{cl}^{ren}}{\delta g_{cl}} = 0. \tag{4.18}$$

Figure 4: Embedding diagram for a two-dimensional (τ, r) section of the Gibbons-Hawking instanton. Regularity condition at the Euclidean horizon $r = r_+$ requires $\beta = \beta_H \equiv 8\pi G_{ren} M$.

We assume that the renormalization of the coupling constants in the classical action is made from the very beginning and we shall assume that the 'classical' field g_{cl} is a solution of the equation (4.18). In our case g_{cl} is the Euclidean black hole metric, while the metric $\bar{g}$ describes the Gibbons-Hawking instanton deformed due to the presence of quantum corrections to the metric. The quantity $\Gamma[\bar{g}]$ being expressed as the function of boundary conditions (β and r_B) specifies the thermodynamical properties of a black hole.

5 Thermodynamical Entropy

For the above described canonical ensemble of gravitational fields the leading tree-level contribution to the renormalized effective action is given by the Euclidean gravitational action for the Euclidean black hole solution (the Gibbons-Hawking instanton).

Under the assumption that $\Lambda_{ren} = 0$ and $c^i_{ren} = 0$ the tree-level contribution to the renormalized free energy of the black hole is [12, 13]

$$F_0^{ren} \equiv \beta^{-1} \Gamma_{cl}^{ren}[g_{cl}] = r_B \left(1 - \sqrt{1 - r_+/r_B}\right) - \pi r_+^2 \beta^{-1}. \tag{5.1}$$

Here $r_+ = 2G_{ren} M$ is the gravitational radius of a black hole of mass M, which for a given temperature β^{-1} at the boundary r_B is defined by the relation $\beta =$

$4\pi r_+(1 - r_+/r_B)^{1/2}$. According to definition the thermodynamical entropy of a black hole S_0^{TD} is determined by the response of the free energy of a system including a black hole to the change of the temperature. One can easily verify that

$$S_0^{TD} = -\frac{dF_0^{ren}}{dT} \equiv \beta^2 \frac{dF_0^{ren}}{d\beta} = \frac{A_H}{4l_P^2}, \tag{5.2}$$

and hence it coincides with the Bekenstein-Hawking expression S^{BH}. (It is assumed that r_+ in F_0^{ren} is expressed in terms of β and r_B before differentiation with respect to β.) One-loop contribution in Eq.(4.2) describes quantum correction to the entropy of a black hole as well as the entropy of thermal radiation in its exterior. The latter evidently depends on the radius r_B of the boundary. Since the Euclidean black hole background is regular the corresponding contribution F_1^{ren} is finite. For this reason the quantum corrections to the Bekenstein-Hawking entropy $4\pi M^2$ are also finite. They are small unless the mass of a black hole M is comparable with the Planckian mass [37, 38, 39]. Due to the presence of the conformal anomalies one might expect that the leading one-loop corrections to S^{TD} are of the order $\ln M$ (see, e.g. [40]).

6 Statistical-Mechanical Entropy

The derivation of the thermodynamical entropy of a black hole requires the *on-shell* calculations. It means that one uses only a regular Euclidean metric that is solution of the field equations (4.16). The discussion of the relation of the thermodynamical and statistical-mechanical entropy of a black hole requires *off-shell* calculations. The reason for this is quite simple and can be explained, for example, by using the approach based on the no-boundary wavefunction of BFZ [31]. The matrix elements in the $|\varphi_-\rangle$ basis of the operator $\hat{\rho} \ln \hat{\rho}$ which enters the definition of the statistical-mechanical entropy (1.4) can be obtained by partially differentiating (3.24) with respect to β. On the other hand the Hamiltonian $\hat{H}$ depends on the black-hole geometry, and hence on the mass M of a black hole. That is why to obtain the expression for the statistical-mechanical entropy one needs to be able to use β and M as independent parameters.

Strictly speaking for $\beta \neq \beta_H \equiv 8\pi M$ there are no regular Euclidean solutions with the Euclidean black-hole topology $R^2 \times S^2$. Such solutions can be obtained only if one exclude a horizon sphere S^2. One can also consider the spacetime with included horizon, provided the curvature has there δ-like behavior corresponding to the cone-like singularity of the metric. For $\beta = \beta_H$ the singularity dissapears. In order to be able to discuss the statistical-mechanical entropy and its relation to the thermodynamical entropy one must generalize the calculation of the one-loop contribution to the renormalized free energy to the case of spaces with cone-like singularity. The new feature which arises is that the corresponding renormalized one-loop corrections might contain new type of divergence, which is directly connected with the presence

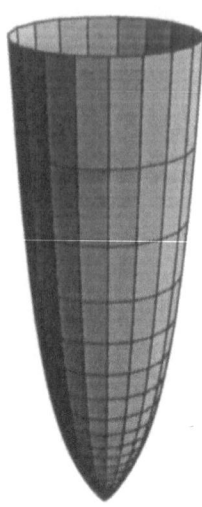

Figure 5: Embedding diagram for a two-dimensional (τ, r) section of the Euclidean black hole with cone-like singularity ($\beta \neq \beta_H \equiv 8\pi G_{ren} M$).

of cone singularity. In order to make the answer finite one might introduce spatial cut-off in the volume integrals near the cone-singularity. It is convenient to restrict the integration by some proper distance l from singularity. This cut-off was present in the 't Hooft's 'brick wall model' [25]. In the model with 'dynamical black-hole-interior' the presence of such a cut-off was connected with the quantum fluctuations of the horizon [30]. Similar cut-off naturally arises in the string theory [33]. The renormalized one-loop contribution F_1^{ren} to the free energy is of the form

$$F_1^{ren} = F_1^{ren}[\beta, \beta_H, \varepsilon], \qquad (6.1)$$

where $\varepsilon = (l/2G_{ren}M)^2$ is the dimensionless cut-off parameter. For $\varepsilon \to 0$ and $\beta \neq \beta_H$ the one-loop free energy F_1^{ren} is divergent $F_1^{ren} \sim \varepsilon^{-1} f(\beta, \beta_H)$. For $\beta = \beta_H$ the divergence dissappears, so that $F_1^{ren}[\beta_H, \beta_H, 0]$ is finite. This quantity is directly related with quantum (one-loop) corrections to the thermodynamical entropy of a black hole

$$S_1^{TD} = \beta_H^2 \frac{d}{d\beta_H} F_1^{ren}[\beta_H, \beta_H, 0] \equiv \left[\beta^2 \frac{\partial F_1^{ren}}{\partial \beta} + \beta_H^2 \frac{\partial F_1^{ren}}{\partial \beta_H} \right]_{\beta=\beta_H}. \qquad (6.2)$$

The expression (6.1) allows one to get the statistical-mechanical entropy S^{SM}.

Dowker and Kennedy [43] and Allen [44] made an important observation that

$$F_1^{ren} = F_{vac}^{ren} + F_{therm}^{ren}, \qquad \frac{\partial F_{vac}^{ren}}{\partial \beta} = 0, \tag{6.3}$$

$$F_{therm}^{ren} = -\beta^{-1} \ln \text{Tr} \left[e^{-\beta \hat{H}} \right] = \ln[\sum \exp(-\beta E_n)], \tag{6.4}$$

where E_n is the energy (eigenvalue of the Hamiltonian $\hat{H}$ of the field φ). By using the expansion in eigenfunctions one can obtain

$$F_{therm}^{ren} = \sum_\lambda f(\beta \omega_\lambda) = \int d\omega N(\omega | \beta_H, \varepsilon) f(\beta \omega). \tag{6.5}$$

$f(\beta \omega) = \beta^{-1} \ln[1 - \exp(-\beta \omega)]$ is free energy of an oscillator of frequency ω at inverse temperature β, and $N(\omega | \beta_H)$ is the density of number of states at the given energy ω in a spacetime of a black hole of mass M. This density of number of states diverges. In order to make it finite we introduced the cut-off ε. We include ε as the argument of N in order to remind about this. The expression (6.5) is usually a starting point for 'brick wall' model.

The statistical-mechanical entropy S^{SM} is

$$S^{SM} = \left[\frac{\partial F_1^{ren}}{\partial \beta} \right]_{\beta_H} = \left[\frac{\partial F_{therm}^{ren}}{\partial \beta} \right]_{\beta_H} \tag{6.6}$$

$$= \int d\omega N(\omega | \beta_H, \varepsilon) s(\beta \omega) \tag{6.7}$$

Here $s(\beta \omega) = \beta \omega / (e^{\beta \omega} - 1) - \ln(1 - e^{-\beta \omega})$ is the entropy of a quantum oscillator of frequency ω with inverse temperature β. S^{SM} is divergent in the limit $\varepsilon \to 0$. The divergence is directly related with the divergency of the density of number of states located in the narrow region in the vicinity of the horizon.

By comparing the expressions (6.2) and (6.6) we can conclude that S^{TD} and S^{SM} differs from one another. It happens for the following two reasons: (1) Vacuum polarization (F_{vac}^{ren}) depends on M and hence on β_H; (2) $d/d\beta$ does not commute with Tr-operation. In general case one gets

$$S_1^{TD} = S^{SM} + \Delta S. \tag{6.8}$$

In the limit $\varepsilon \to 0$ $\Delta S \neq 0$ is also divergent, but S_1^{TD} remains finite and (for $M \gg m_P$) small. The relation (6.8) provides explanation of the entropy renormalization procedure by Thorne and Zurek [20].

Fursaev and Solodukhin [47, 48, 49] recently proposed another approach for off-shell calculations of thermodynamical characteristics of a black hole. Namely, instead of cutting-off the vicinity of a cone singularity, they proposed to calculate the effective action directly on a spacetime with cone-like singularity. In order to make this mathematically well defined one might at first consider a manifold with the topology $R^2 \times S^2$ which is smooth at the fixed-point sphere (horizon) and differs from the cone metric only in the very narrow region of size l near the horizon. The effective action must be considered as the function of l and the parameter l must be finally put to zero. The one-loop correction to the free energy F_1^{cone} can be divergent in this limit, but it is possible to show that the divergence is proportional to $(\beta - \beta_H)^2$. That is why the corresponding contribution $S_1^{cone} = [\beta^2 \partial F_1^{cone} / \partial \beta]_{\beta = \beta_H}$ to the entropy is finite. This off-shell approach gives the same expression for the thermodynamical entropy and might be considered as useful tools for such calculations. The relation of this off-shell entropy S_1^{cone} to the statistical-mechanical entropy S^{SM} is not clear. Among other approaches to the calculation of the entropy we mention also the approach based on the Pauli-Villars regularization [45].

7   Black Hole Thermodynamics and Physics at Planckian Scales

The expression for the statistical-mechanical entropy (2.1) requires cut-off. The value l of the cut-off parameter is of order of the Planckian length.. Does it mean that thermodynamical characteristics of a black hole for their understanding require the knowledge of physics at Planckian scales?

When we are discussing black hole solutions and their properties we use gravitational equations. The coupling constants in these equations are assumed to coincide with 'observable' values. Due to the existence of ultraviolet divergencies the observable coupling constants differ from their initial bare values. Any procedure which gives sense to this renormalization procedure finally must deal with the problems of physics at Planckian scale. But in this sense black hole physics does not differ from the usual Newtonian theory. Calculations in the Newtonian theory also use the renormalized ('observable') gravitational constant, and hence in order to derive the same results in the framework of quantum gravity beginning from some initial background theory one must pass through all the complications connected with the renormalization procedure and redefining the coupling constants. One can do it from very beginning or develop more complicated scheme and made all the renormalizations only at the end. The same is true also for black holes. Besides this in the case of black hole there are situations when quantum gravity becomes really important. It is well known that the final stage of a black hole evaporation as well as the structure near singularity inside a black hole for their consideration require quantum gravity. Quantum gravity might be also useful for study of small quantum corrections to black hole characteristics. But these corrections can essentially change parameters of a black hole when the curvature at the horizon becomes comparable with Planckian curvature, i.e. for black holes of Planckian mass. These remarks are of course trivial. But if we exclude these evident cases do we still need quantum gravity to explain properties of macroscopic black holes?

This question is not new. The standard derivation of Hawking quantum radiation of a black hole formally requires the integration over all (including much higher than Planckian) frequencies of the initial zero-point-fluctuations of a quantum field. There is a belief that one can escape the formal usage of super-Planckian energies in the calculations. This point of view was supported by recent result by Unruh [46]. He considered a model in which due to the presence of dispersion the frequencies of zero-point-fluctuations are restricted. Unruh has shown that nevertheless the Hawking radiation at late time remains constant and with high accuracy thermal.

A similar problem arises in connection of a black-hole entropy. We saw, for example, that the statistical-mechanical entropy S^{SM} is dependent on the cut-off parameter. One might argue that such a cut-off must be provided by quantum gravity. For this reason S^{SM} is (at least potentially) the quantity which for its knowledge requires Planckian scale physics. Does it mean that the study of the thermodynamical properties might give us information about these scales? The above discussion indicates that it is impossible. In the standard gedanken experiments the observable quantity is S^{TD}. S^{SM} (at least in the leading order) does not contribute to S^{TD} and hence one cannot measure it. The statistical-mechanical entropy might be useful for description of excitations in the close vicinity of the horizon (for example of their damping). The main contribution to S^{SM} is given by very high frequency modes inside the gravitational barrier propagating very close to the horizon. It looks like that the only reasonable way to measure S^{SM} is to excite these modes. For example, one can do it by colliding particles of superhigh energy near the horizon. This experiment requires very high (super-Planckian) energies. But having these energies available one can use them for study Planckian physics in usual Minkowski spacetime without any black holes.

To conclude, quantum gravity is required for understanding very fundamental problems of black holes, such as the problem of final state, but it also looks like that the thermodynamics of macroscopical black hole does not provide us with any new powerful tools for verifying the theory of quantum gravity.

8 Acknowledgements

This work was supported by the Natural Sciences and Engineering Research Council of Canada.

References

[1] J. D.Bekenstein, Nuov.Cim.Lett. **4** (1972) 737.

[2] J. D. Bekenstein, Phys.Rev. **D7** (1973) 2333.

[3] J. M. Bardeen, B. Carter, and S. W. Hawking, Comm.Math.Phys. **31**, (1973) 181.

[4] J. D. Bekenstein, Phys.Rev. **D9** (1974) 3292.

[5] K. S. Thorne, W. H. Zurek, and R. H. Price, in *Black Holes: The Membrane Paradigm* , edited by K. S. Thorne, R. H. Price, and D. A. MacDonald (Yale University Press, New Haven, 1986), p.280.

[6] I. Novikov and V. Frolov, *Physics of Black Holes*. (Kluwer Academic Publ., Dordrech-Boston-London), 1989.

[7] R. W. Wald, In:*Black Hole Physics* (Eds.V. DeSabbata and Z. Zhang), (Kluwer Academic Publ., Dordrech-Boston-London), 1992).

[8] V. Frolov and D. N. Page, Phys.Rev.Lett. **71** (1993) 3902.

[9] J.B.Hartle and S.W.Hawking, Phys.Rev. **D13**, 2188 (1976).

[10] G. W. Gibbons and S. W. Hawking, Phys.Rev. **D15** (1976) 2752.

[11] S. W. Hawking, In: *General Relativity: An Einstein Centenary Survey*. (eds. S. W. Hawking and W. Israel), Cambridge Univ.Press, Cambridge, 1979.

[12] J. W. York, Phys.Rev. **D33** (1986) 2092.

[13] H. W. Braden, J. D. Brown, B. F. Whiting, and J. W. Jork, Phys.Rev. **D42** (1990) 3376.

[14] R. M. Wald, Phys.Rev. **D48** (1993) 3427.

[15] T. Jacobson and R. C. Myers, Phys.Rev. Lett. **70** (1993) 3684.

[16] V. Iyer and R. M. Wald, Phys.Rev. **D50** (1994) 846.

[17] T. Jacobson, G. Kang, and R. C. Myers, Phys.Rev. **D49** (1994) 6587.

[18] G. W. Gibbons and R. E. Kallosh, Phys.Rev. **D51** (1995) 2839.

[19] S. W. Hawking, G. T. Horowitz, and S. F. Ross, Phys.Rev. **D51** (1995) 4302.

[20] W. H. Zurek and K. S. Thorne, Phys.Rev.Lett. **54** (1985) 2171.

[21] V. F. Mukhanov, JETP Lett. **44** (1986) 63.

[22] J. D. Bekenstein and V. F. Mukhanov. 'Spectroscopy of quantum black holes' Preprint gr-qc/9505012 (1995).

[23] V. A. Berezin. Phys.Lett., **B241** (1990) 194.

[24] J. D. Bekenstein, preprint gr-qc/9409015 (1994).

[25] G. 't Hooft, Nucl.Phys. **B256** (1985) 727.

[26] L.Bombelli, R.K.Koul, J.Lee, and R.Sorkin, Phys.Rev. **D34** 373 (1986).

[27] M.Srednicki, Phys.Rev.Lett. **71**, 666 (1993).

[28] V. Frolov and I. Novikov, Phys.Rev. **D48** (1993) 1607.

[29] A. G. Doroshkevich and I. D. Novikov, Soviet Phys. JETP. **63** (1993) 1538.

[30] V. Frolov and I. Novikov, Phys.Rev. **D48** (1993) 4545.

[31] A. I. Barvinsky, V. P. Frolov, and A. I. Zelnikov, Phys.Rev.**D51** (1995) 1741.

[32] S. Carlip and C. Teitelboim, Preprint gr-qc/9312002 (1993).

[33] H. J. de Vega and N. Sánchez, Nucl.Phys. **B299** (1988) 818.

[34] L. Susskind and J. Uglum, Phys.Rev. **D50** (1994) 2700.

[35] D. Garfinkle, S. B. Giddings, and A. Strominger, Phys.Rev. **D49** (1994) 958.

[36] J.B.Hartle and S.W.Hawking, Phys.Rev. **D28**, 2960 (1983).

[37] P. R. Anderson, W. A. Hiscock, J. Whitesell, and J. W. York, Phys.Rev. **D50** (1994) 6427.

[38] H. W. Braden, J.D. Brown, B. F. Whiting, J. W. York, Jr. Phys.Rev.**D42** (1990) 3376.

[39] J. L. Louko and B. F. Whiting , Phys.Rev. **D51** (1995) 5583.

[40] D. V. Fursaev, Phys.Rev.**D51** (1995) 5352.

[41] S. W. Hawking, Comm.Math.Phys. **43**, (1975) 199.

[42] V. Frolov, Phys.Rev.Lett. **74** (1995) 3319.

[43] J.S.Dowker and G.Kennedy, J.Phys. **A 11** (1978) 895.

[44] B. Allen, Phys.Rev. **D33** (1986) 3640.

[45] J.-G. Demers, R. Lafrance, and R. C. Myers. Phys.Rev.**D52** (1995) 2245.

[46] W. G. Unruh, Phys.Rev.**D51** (1995) 2827.

[47] D. V. Fursaev and S. N. Solodukhin, On one-loop renormalization of black hole entropy, Preprint E2-94-462, hep-th/9412020 (1994).

[48] S. N. Solodukhin, Phys.Rev.**D51** (1995) 609.

[49] D. V. Fursaev and S. N. Solodukhin, Phys.Rev.**D52** (1995) 2133.

Black Hole Condensation and Duality in String Theory

Andrew Strominger[1]
Department of Physics,
University of California
Santa Barbara, CA 93106-9530

The classical, four-dimensional theories derived by Calabi-Yau compact-ification of string theory bear a striking resemblance to the real world [1]. However, these classical theories are beset by several serious difficulties:

1. There are too many of them. This is aesthetically displeasing because a unified theory should be unique. It also entails a loss of predictive power.

2. The theory breaks down and develops naked singularities at certain "conifold" points in the moduli space of the massless four-dimensional scalar fields [2].

In this talk we shall argue, in the context of type II string theories, that these problems are in part resolved by nonperturbative quantum ef-fects. Thus — unlike e.g. nonabelian gauge theories — string theory needs quantum mechanics for consistency. This suggests that the fundamental for-mulation of quantum string theory may not take the usual form which begins with a classical theory followed by quantization. Rather string theory may be intrinsically quantum in nature and not have a consistent classical limit.

The structure of conifold singularities is an old and beautiful subject in algebraic geometry. The mathematical description will not be repeated here. Relevant aspects and references can be found in [3]. The basic picture is as follows. The space of Calabi-Yau string vacua is the moduli space of Ricci-flat metrics on the Calabi-Yau. For each coordinate Z^i on the moduli space there is a massless 4D scalar $Z^i(x)$ which describes how the size and shape of the Calabi-Yau vary in spacetime. These moduli fields are governed by the 4D effective action

$$\mathcal{L}_{\text{eff}} = \int d^4x \ \sqrt{-g} \ G_{ij}(Z)\nabla_\mu Z^i \nabla_\nu Z^j g^{\mu\nu} . \tag{1}$$

where g is the metric on spacetime and G is the metric on the moduli space.

[1]e-mail: andy@denali.physics.ucsb.edu

N. Sánchez and A. Zichichi (eds.), String Gravity and Physics at the Planck Energy Scale, 209–218.
© *1996 Kluwer Academic Publishers.*

The moduli space metric G is classically determined from Calabi-Yau data [4]. In the (type II) context which we consider, there are no quantum corrections to G due to $N = 2$ supersymmetric nonrenormalization theorems.

DEGENERATING 3-CYCLE

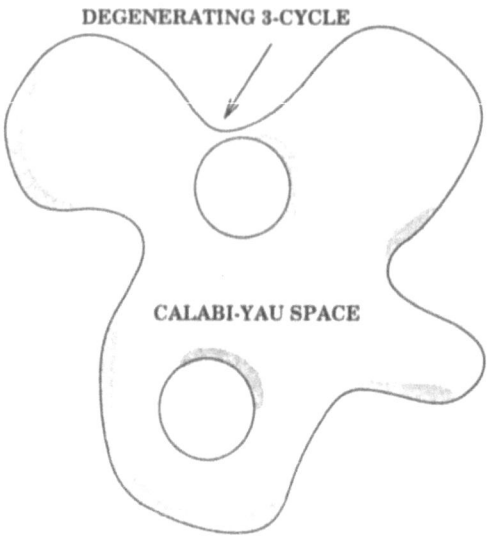

CALABI-YAU SPACE

Figure 1: *Near a conifold, a minimal 3-cycle degenerates to zero volume and the Calabi-Yau space develops a singular node. (The handles are meant to indicate the complex topology involved: real Calabi-Yau spaces have $\pi_1 = 0$.)*

The Z^i's measure the size of topologically non-trivial, minimal-volume cycles (i.e. submanifolds) embedded in the Calabi-Yau [5]. To be definite, let us consider minimal 3-cycles. At "conifold" points in the moduli space, the minimal volume of a topologically non-trivial 3-cycle can actually shrink to zero. We can choose local coordinates so that the conifold singularity is at $Z^1 = 0$. At $Z^1 = 0$ the Calabi-Yau develops a singular node and is no longer a smooth manifold as depicted in figure 1. Conifold singularities generically occur at finite distances in the moduli space.

It is perhaps not surprising that the moduli space metric G in (1) itself turns out to be logarithmically singular at $Z^1 = 0$ [2]

$$G \sim \ln|Z^1| \, . \tag{2}$$

This is a real curvature singularity and cannot be eliminated by a coordinate transformation of the Z^i's. Thus classical string theory breaks down whenever a moduli field happens to run into a conifold singularity. It can be seen that these singularities are real codimension two so it is hard to avoid such collisions.

A curvature singularity is not the only suspicious behavior of type II string theory near a conifold. These theories have extremal, charged black holes whose mass can be exactly determined using $N = 2$ supersymmetry. These masses are proportional to

$$M_{\mathrm{BH}} = |Z^1| \,. \tag{3}$$

Hence the black hole becomes massless at the conifold singularity $Z^1 = 0$.[2]

The black hole degenerates to zero mass for a simple reason. It began life in ten dimensions as a black 3-brane [6]. This is an extended black hole whose horizon is topologically $R^3 \times S^5$, and with a constant mass per unit three volume. In a Calabi-Yau compactification these 3-branes can wrap around a non-trivial 3-cycle. To a low-energy 4D observer, such a configuration will appear to be an ordinary extremal black hole with mass proportional to the volume of the 3-cycle. When this volume degenerates at a conifold the 4D mass will degenerate along with it.

To summarize the picture so far, the conifold is characterized by

$$\begin{aligned}
Z^1 &\rightarrow 0\,, \\
M_{\mathrm{BH}} &\rightarrow 0\,, \\
\mathcal{L}_{\mathrm{eff}} &\rightarrow \infty\,.
\end{aligned} \tag{4}$$

In fact this situation is not as disturbing or unusual as it seems. It is well known that massless particles produce singularities in low-energy effective actions due to infrared divergent loop integrations. A non-singular description of the physics can be found in a Wilsonian effective action $\mathcal{L}_{\mathrm{eff}}^w$. This is obtained (in principle) by starting from the exact microscopic theory and integrating out fluctuations of all fields — massive and massless — down to some Wilsonian cutoff M_c, well below the Planck or string scales, as depicted

---

[2] Far from the conifold the black holes are well-described by semiclassical solutions with horizons. However, in a neighborhood of the conifold, its Compton wavelength exceeds its Schwarzschild radius and the semiclassical description breaks down.

in figure 2. This action differs from the 1PI (one-particle-irreducible) effective action $\mathcal{L}_{\text{eff}}$ usually discussed in string theory in which fluctuations of all wavelengths are integrated out. Divergences in the 1PI effective action arise in integrating out fluctuations of massless fields from M_c down to zero energy. Computations of a scattering process with external momenta of order p using $\mathcal{L}_{\text{eff}}^w$ involves a quantum loop expansion with loop momenta cutoff at M_c. Infrared divergences will then typically be controlled by the external momenta p, and the computation will yield a finite answer.

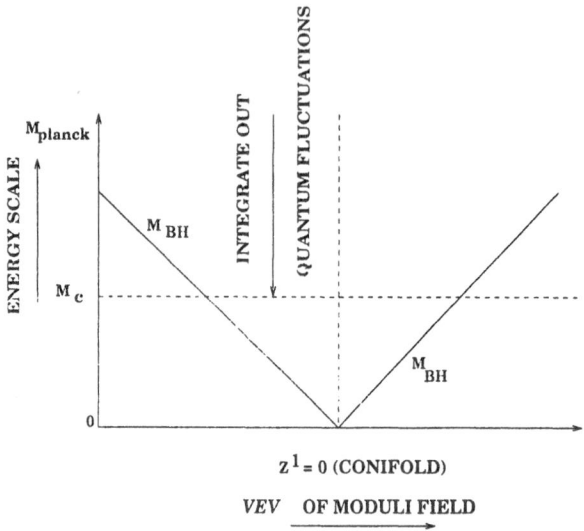

Figure 2: *The smooth Wilsonian effective action is defined by integrating out quantum fluctuations of all fields down to the cutoff M_c. No matter how low M_c is, there is always a region in the moduli space surrounding the conifold in which black holes are lighter than M_c and must be included in the Wilsonian action.*

In the case of conifold singularities, the divergences in $\mathcal{L}_{\text{eff}}$ have precisely the right coefficients to have been produced by integrating out a black hole [3]. This has remarkably been confirmed even for subleading terms in $\mathcal{L}_{\text{eff}}$ [7]. We conclude that the underlying Wilsonian effective action has couplings which are nonsingular as $Z^1 \to 0$. Finite-momentum processes can be computed at the conifold utilizing $\mathcal{L}_{\text{eff}}^w$. Hence we see that classical inconsistencies of string theory are cured by quantum loops of black holes. It is fascinating that the demand for a consistent theory forced us to include these black holes with virtual fluctuations on the same footing as elementary strings.

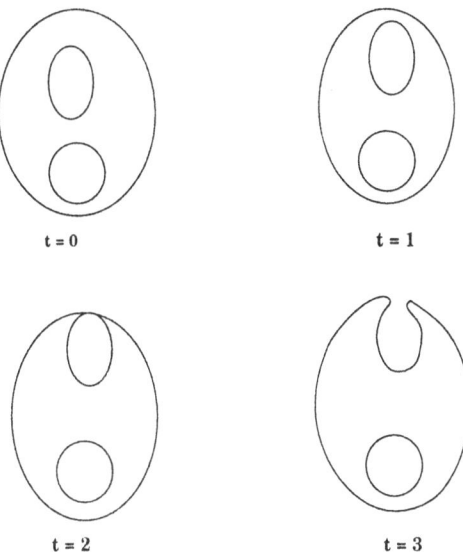

t = 0   t = 1

t = 2   t = 3

Figure 3: *The shape of a Calabi-Yau space slowly changes and develops a node at time $t = 2$. Black holes then condense, implementing a smooth transition to a topologically distinct Calabi-Yau space at time $t = 3$.*

The appearance of a massless particle often signals a phase transition. One may wonder if there is a new phase of string theory characterized by

$$\langle \Phi_{BH} \rangle \neq 0 , \tag{5}$$

where Φ_{BH} is the field whose quanta are the degenerating black holes. This may seem like a difficult question, but in fact the answer is easily determined using $N = 2$ supersymmetry, which fixes the potential for the field Φ_{BH}. In the simple conifold singularities described in [3], the answer is no: black hole condensation is prevented by a quartic potential.

The situation is dramatically different for the more complex conifold singularities analyzed in collaboration with Brian Greene and Dave Morrison [8]. These singularities correspond to multiple degenerations at which P 3-cycles degenerate and P black holes come down to zero mass. The Wilsonian action at the singularity involves P black hole fields, Φ_{BH}^A, $A = 1, \cdots P$. $N = 2$

supersymmetry again determines the potential $V(\Phi_{BH}^A)$. In some cases it is found that V has flat directions along which black holes can condense!

It might appear that a new branch of the string moduli space has been discovered. However, there is overwhelming evidence that $\langle \Phi_{BH}^A \rangle \neq 0$ branches are not new string vacua. Rather they are a new, dual description of old string vacua. The spectrum of massless particles in the $\langle \Phi_{BH}^A \rangle \neq 0$ branches agree in each of the thousands of known examples with the spectrum of a known Calabi-Yau space. Furthermore, pairs of Calabi-Yau's which are connected in this manner by black hole condensation are the same as those pairs previously known from the work of [10] to be connected by a singular conifold transition in which an S^3 is shrunk to zero size and then blown back up as an S^2. Hence black hole condensation in four dimensions corresponds to a change in the topology of the internal Calabi-Yau, as depicted in figure 3. In general relativity the topology of a manifold cannot change in a smooth fashion. String theory is an extension of general relativity in which smooth topology change can occur.

Thousands, and possibly all simply-connected, Calabi-Yau's are connected by such transitions. In this fashion the plethora of disconnected string vacua are unified into a smaller number — possibly one — of moduli spaces as illustrated in figure 4. The long-term aspiration is that, when understood, the dynamics of supersymmetry breaking will select a preferred point(s) in this space.

In the "old", Calabi-Yau, description of the $\langle \Phi_{BH}^A \rangle \neq 0$ phase, Φ_{BH}^A is identified as a field whose quanta are fundamental strings rather than black holes. Thus under the topology-changing phase transitions,

$$\text{Black Holes} \;\rightarrow\; \text{Strings}$$
$$\text{Strings} \;\rightarrow\; \text{Black Holes} \,.$$

Black holes and strings are dual descriptions of the same entity. For decades theorists have pursued the idea that elementary particles are secretly black holes. We have seen that a version of this idea is realized in string theory. Hence string theory succeeds not only in unifying all particles and forces with one another, but in unifying them with black holes as well.

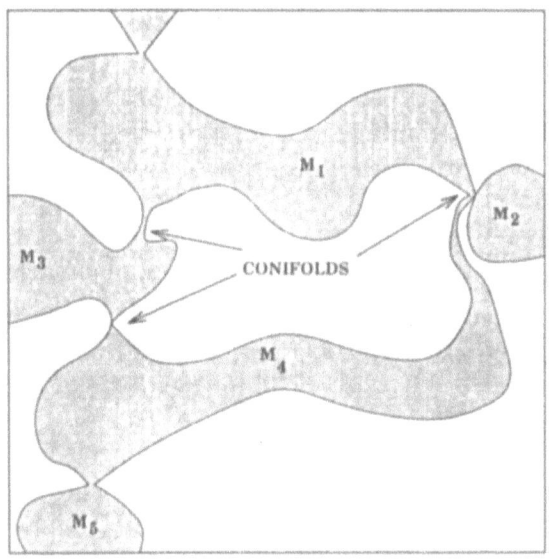

Figure 4: *The vacuum moduli spaces, $M_1, M_2, \cdots$ of topologically distinct Calabi-Yau spaces are branches of a larger moduli space connected via conifold transitions.*

The preceding discussion has close parallels in the beautiful work of Seiberg and Witten on $N = 2$, $d = 4$ gauge theories [11, 12]. In the pure $SU(2)$ gauge theory, [11] there is a conifold singularity which appears at a special point in the moduli space of Higgs *vev*'s. This theory also contains 't Hooft-Polyakov monopoles which degenerate at the conifold. The Wilsonian theory including light monopoles is smooth at the conifold, just like the Wilsonian theory with light black holes described here.

There are also some apparent differences with the work in [11]. The conifold singularity of [11] has an alternate description as a divergence in the *quantum* sum over Yang-Mills instantons, as opposed to the Calabi-Yau conifolds which have an alternate description (utilizing mirror symmetry) as a *classical* sum over worldsheet instantons. This distinction evaporates in the context of a dual description of string theory in which the string is itself a soliton [13]. In such a description the classical worldsheet sum becomes a quantum sum over spacetime instantons [3, 14]. Explicit examples of this have been understood in the context of dualities relating type II-heterotic string compactifications [15]. Duality promotes the analogy to an identity:

the dual transforms of black holes are monopoles, and worldsheet instantons turn into Yang-Mills instantons.

Field theory analogs of the conifold transitions in which the topology of the Calabi-Yau changes also exist [12]. For example, in the $N = 2$ $SU(2)$ gauge theory with two flavors of "quarks" in an $SU(2)$ doublet, the moduli space has several branches. The first is called the Coulomb branch, along which $SU(2)$ is broken to $U(1)$ by an adjoint Higgs *vev* and all quarks are massive. At special conifold points on the Coulomb branch massless charged states appear. These can condense and form a new branch called the Higgs branch along which the $U(1)$ is broken. Condensation of these massless charged states creates a new branch of the gauge theory moduli space in the same fashion that black hole condensation creates a new branch of the string moduli space.

String dualities again promote the analogy to an identity. From the dual, heterotic perspective, the exotic topology-changing conifold transitions of the type II theory are nothing but condensation of various light-charged fields: The moduli space of $N = 2$ heterotic string vacuua contains many special points where charged perturbative string states become light and condense, changing both the massless spectrum and dimension of the moduli space. Hence, a consistent picture of heterotic-type II string duality relies crucially on the existence of black hole condensation in type II theories.

Perhaps the most exciting aspect of recent developments is the deep new puzzles they have raised. I would like to draw attention to one of these puzzles related to the preceding analogy. Consider a moduli field which is slowly rolling in a generic fashion and encounters a conifold singularity. Part of spacetime will spill across the transition, and a bubble of the new phase will form. Inside the bubble a new spectrum of massless particles will appear. Our analysis of the low-energy effective action enables one to obtain the lowest-order approximation to this process.

However, in a complete theory one should, in principle, be able to compute arbitrarily high order corrections to the leading approximation. Clearly, the usual string perturbation rules are useless here because different conformal field theories are relevant to the regions inside and outside the bubble. We do not have a rule for computing these corrections. It is furthermore clear that, whatever those rules are, they are quite different from the usual rules of string theory.

The analogy with the Seiberg-Witten field theory case is again illuminating. In that case the low-energy effective theories on the Higgs and Coulomb branches can be used to give a leading-order description of the formation of a bubble of the Higgs branch inside the Coulomb branch. However, a systematic computation of the corrections can only be made from knowledge of the microscopic $SU(2)$ gauge theory.

In our current understanding of string theory, it is as if we have seen the last equations in the papers of Seiberg and Witten which describe the low-energy effective abelian gauge theories. To fully understand string theory, we must work backwards from these last equations to the first equations in which the theory is fundamentally defined as an $SU(2)$ gauge theory.

Clearly this is an enormous task. At the same time, recent developments have provided us with new tools and concrete questions with which we can address these issues, and progress is being made in leaps and bounds. It is an exciting time for string theory.

ACKNOWLEDGMENTS

This work was supported in part by a grant from the Department of Energy, 91ER40618.

References

[1] P. Candelas, G.T. Horowitz, A. Strominger, and E. Witten, Nucl. Phys. B, 258 (1985) 46.

[2] P. Candelas, X. dela Ossa, P. Green, and L. Parkes, Nucl. Phys. B, 359 (1991) 21.

[3] A. Strominger, Nuc. Phys. B, 451 (1995) 96-108; hep-th/9504090.

[4] A. Strominger, Phys. Rev. Lett., 55 (1985) 2547.

[5] K. Becker, M. Becker and A. Strominger, hep-th/9507158.

[6] G. Horowitz and A. Strominger, Nucl. Phys. B, 360 (1991) 197.

[7] C. Vafa, hep-th/9505023.

[8] B.R. Greene, D.R. Morrison, and A. Strominger, Nucl. Phys. B, 451 (1995) 109-120.

[9] M. Reid, Math. Ann., 278 (1987) 329.

[10] P. Candelas, A. Dale, A. Lutken, and R. Schimmrigk, Nuc. Phys. B, 298 (1988) 493; P. Candelas, P. Green, and T. Hubsch, Nucl. Phys. B, 42 (1990) 246.

[11] N. Seiberg and E. Witten, Nucl. Phys. B, 426 (1994) 19.

[12] N. Seiberg and E. Witten, Nucl. Phys. B, 431 (1991) 484.

[13] A. Dabholkar and J. Harvey, Phys. Rev. Lett., 63 (1989) 478.

[14] A. Klemm, W. Lerche, S. Theisen and S. Yankielowicz, Phys. Lett., B344 (1995) 169, hep-th/9411057.

[15] C.M. Hull and P.K. Townsend, Nucl. Phys. B, 438 (1995) 47, hep-th/9410167; E. Witten, Nucl. Phys. B, 443 (1995), hep-th/9503124; S. Kachru and C. Vafa, hep-th/9505105; S. Ferrara, J.A. Harvey, A. Strominger, and C. Vafa, hep-th/9505152.

Correlation Dynamics of Quantum Fields and Black Hole Information Paradox

B. L. Hu*

Department of Physics, University of Maryland, College Park, MD 20742, USA

(November 15, 1995)

Lectures at the International School of Astrophysics "D. Chalonge", Erice, Sept. 1995

Abstract

In recent years a statistical mechanics description of particles, fields and spacetime based on the concept of quantum open systems and the influence functional formalism has been introduced. It reproduces in full the established theory of quantum fields in curved spacetime and contains also a microscopic description of their statistical properties, such as noise, fluctuations, decoherence, and dissipation. This new framework allows one to explore the quantum statistical properties of spacetime at the interface between the semiclassical and quantum gravity regimes, as well as important non-equilibrium processes in the early universe and black holes, such as particle creation, entropy generation, galaxy formation, Hawking radiation, gravitational collapse, backreaction and the black hole end-state and information lost issues. Here we give a summary of the theory of correlation dynamics of quantum fields and describe how this conceptual scheme coupled with scaling behavior near the infrared limit can shed light on the black hole information paradox.

---

*Email: hu@umdhep.umd.edu

N. Sánchez and A. Zichichi (eds.), String Gravity and Physics at the Planck Energy Scale, 219–232.
© *1996 Kluwer Academic Publishers.*

1 Stochastic Theory of Quantum Fields and Space-time

I gave two lectures at this School. Since they contain material already published, I do not want to repeat them here. Instead, I will mention some on-going projects in our research program and discuss some ideas on how the correlations of quantum fields can be used to address the black hole information paradox issue. The abstracts of these two lectures, and relevant references (where the background material and bibliography can be found) are given below. (For a more comprehensive recent review, see [1])

Lecture 1. Stochastic Analysis of Particles and Fields: The Effective Action and the Influence Functional Methods

We first summarize the in-out (Schwinger-DeWitt) effective action formalism [2] in quantum field theory in curved spacetime [3] and point out the need for generalization to the in-in [4] (Schwinger-Keldysh, or closed-time-path CTP) [5, 6, 7] formalism to treat particle creation and backreaction problems in semiclassical gravity [8]. We then discuss the statistical mechanics of quantum open systems by way of the Brownian motion model, using the influence functional (IF) [9] formalism of Feynman-Vernon [10, 11]. By viewing the subjects of interest in semiclassical gravity as open systems, where the classical geometry is treated as the system and the quantum matter field as the environment, and using the close relation of the CTP and the IF methods, we can examine both the quantum and the statistical mechanical attributes of the processes involved. We derive an expression for the CTP effective action or the IF in terms of the Bogolubov coefficients relating the second quantized operators between two Fock spaces of the field theory [12, 13], thus connecting back with the established theories of quantum fields in curved spacetimes. As example, we show how one can derive the Hawking-Unruh radiations [14, 15] from a non-equilibrium statistical mechanics (NESM) viewpoint [12]. These effects are understood as resulting from the scaling or amplification of quantum noise between observers in different kinematic or dynamical states [16, 17]. Viewing these effects in this light would enable one to consider fully non-equilibrium processes not easily approachable by conventional (e.g., geometric or thermodynamics) means.

Lecture 2. Einstein-Langevin Equation in Semiclassical Gravity: Backreaction as Fluctuation-Dissipation Relation

The influence functional method in NESM (which includes the CTP effective action method in quantum field theory) explicates the meaning and the interrelation [18, 1] of the many quantum processes involved in the backreaction problem, such as particle creation, noise, fluctuations [13], decoherence [19, 20] and dissipation [8]. The backreaction of created particles and their fluctuations is described by an equation of motion derivable from the influence action in the form of an Einstein-Langevin equation [21]. It contains a dissipative term for the dynamics of spacetime and a noise term related to the fluctuations of particle creation in the matter field. As examples, we first study the case of a free quantum scalar field in a spatially flat Friedmann- Robertson-Walker universe and derive the Einstein-Langevin equations for the scale factor for these semiclassical cosmologies [13]. Then using the well-studied model of a quantum scalar field in a Bianchi Type-I universe we illustrate how this Langevin equation and the noise term are derived and show how the creation of particles and the dissipation of anisotropy during the expansion of the very early universe can be understood as the manifestation of a fluctuation-dissipation relation [22]. This theorem, which exists under very general conditions for dissipations in the dynamics of a system, and the noise and fluctuations in its environment, embodies the backreaction effect of matter fields on the spacetime.

Our approach based on statistical field theory extends the conventional theory of semi-classical gravity based on a semiclassical Einstein equation with a source given by the expectation value of the energy-momentum tensor, to that based on a Langevin-type equation, where the dynamics and fluctuations of spacetime are driven by quantum fluctuations of the matter field. This generalized framework is useful for the investigation of quantum processes in the early universe involving fluctuations, vacuum stability and phase transition phenomena and the non-equilibrium statistical mechanics of black holes. It is also essential to an understanding of the cross-over behavior between quantum gravity and general relativity.

Related problems

Let me also mention some related problems currently under investigation:

1) Metric Fluctuations in Semiclassical Gravity [21, 22, 23]
2) Quantum Fluctuations and Structure Formation in the early universe [24, 13, 25]
3) Correlation, Decoherence, Dissipation and Noise of Quantum Fields [26, 27]
4) Stochastic Theory of Accelerated Detectors [28, 29]
5) Backreaction of Unruh Radiation on an Accelerated Detector and a Moving Mirror [30, 31]
6) Fluctuation-Dissipation Relation for a Radiating Black Hole [32]

The first round of attack on Problems 1) to 4) have been completed recently. Problems 5) and 6) are in progress. In the following I will briefly discuss how correlation dynamics in quantum fields might play a role in the black hole information paradox problem. Sec. 2 contains a summary from [27]. Sec. 3 contains entirely new material. The reader should nevertheless be warned that this investigation is still in a preliminary stage, and the ideas are largely speculative. Also since this is more in the nature of a progress report than a review, I have not attempted to include a complete bibliography. Other related work can be found from the reference lists of the quoted papers.

2 Statistical Mechanics of an Interacting Quantum Field

We begin with an analysis of the statistical mechanics of an interacting quantum field. This familiar subject which one learns in the first lessons of quantum field theory is surprisingly rich in its statistical mechanics content, much like the role an ordinary box of gas molecules plays in Boltzmann's sophisticated theory of kinetics and dissipation.

2.1 Correlation, noise, and decoherence in quantum fields

In two recent papers [26, 27] Calzetta and I studied the statistical mechanical properties of interacting quantum fields in terms of the *dynamics of the correlation functions*. We started from the thesis that the full dynamics of an interacting quantum field may be described by means of the Dyson- Schwinger equations governing the infinite hierarchy of Wightman functions which measure the correlations of the field. We showed how this hierarchy of equations can be obtained from the variation of the infinite particle irreducible, or *'master' effective action* (MEA). Truncation of this hierarchy gives rise to a quantum subdynamics governing a finite number of correlation functions (which constitute the 'system'), and expression of the higher order correlation functions (which constitute the 'environment') in terms of the lower-order ones by functional relations ('slaving' or 'factorization') induces *dissipation* in the dynamics of the subsystem driven by the stochastic fluctuations of the environment, which we call the *'correlation noises'*. These two aspects are related by the fluctuation-dissipation relation. This is the quantum field equivalent of the BBGKY hierarchy in Boltzmann's theory. Any subsystem involving a finite number of correlation functions defines an effective theory, which is, by this reasoning, intrinsically dissipative. The relation of loop expansion and correlation order is expounded. We see that ordinary quantum field theory which involves only the mean field and a two-point function, or any finite-loop effective action in a perturbative theory are, by nature, effective theories which possess these properties. Histories defined by lower-order correlation functions can be decohered by the noises from the higher order functions and acquire classical stochastic attributes. We think this scheme invoking the correlation order is a natural way to describe the quantum to classical transition for a closed system as it avoids *ad hoc* stipulation of the system-environment split. It is through decoherence that the subsystem variables become classical and the subdynamics becomes stochastic.

Our viewpoint here is motivated by classical kinetic theory. In the dynamics of a dilute gas [33, 34, 35] the exact Newton's or Hamilton's equations for the evolution of a many body system may be translated into a Liouville equation for the distribution function or the BBGKY hierarchy for the sequence of partial (n- particle) distributions. This reformulation is only formal, which involves no loss of information or predictability. Physical description of the dynamics comes from truncating the BBGKY hierarchy and introducing a factorization condition like the molecular chaos assumption, where the higher order distributions are substituted by functionals of the lower order. Constructed perturbatively, this effective theory follows only approximately the actual dynamics. Moreover, these functionals embody some relevant boundary conditions (such as the 'weakening of correlations' hypothesis [33]), which make them noninvariant upon time reversal. This is how dissipation in the explicitly irreversible Boltzmann's equation appears. [1].

Correlation dynamics described by the Dyson-Schwinger hierarchy derived from the master effective action: Truncation and Factorization

We want to describe a quantum field in terms of the mean field and the (infinite number of) correlation functions. Here, different from the conventional treatment, we view the 'mean' field not as the actual expectation value of the field, but rather as representing the local value of the field within one particular history. Quantum evolution encompasses the coherent superposition of all possible histories [19] and these quantities are subject to fluctuations. This naturally introduces into the theory stochastic elements, which has hitherto been ignored in the usual description of quantum field theory. The theory may be enlarged by including some correlation functions as independent variables along with the 'mean' field, which themselves are subject to fluctuations.

Our starting point is the well-known fact that the set of all Wightman functions (time ordered products of field operators) determines completely the quantum state of a field [38]. Instead of following the evolution of the field in any of the conventional representations (Schrödinger, Heisenberg or Dirac's), we focus on the dynamics of the full hierarchy of Wightman functions. To this end it is convenient to adopt Schwinger's "closed time-path" techniques [4], and consider time ordered Green functions as a subset of all Green functions path- ordered along a closed time loop. The dynamics of this larger set is described by the Dyson - Schwinger equations.

We first showed that the Dyson- Schwinger hierarchy may be obtained via the variational principle from a functional which we call the 'Master Effective Action' (MEA). This is a formal action functional where each Wightman function enters as an independent variable. We then showed that any field theory based on a finite number of (mean field plus) correlation functions can be viewed as a subdynamics of the Dyson- Schwinger hierarchy. The specification of a subdynamics involves two steps: First, the hierarchy is *truncated* at a certain order. A finite set of variables, say, the lowest nth order correlation functions, is identified to be the 'relevant' [37] variables, which constitute the subsystem. Second, the remaining 'irrelevant' or 'environment' variables, say, the n+1 to ∞ order correlation functions are *slaved* to the former. Slaving (or 'factorization' in the Boltzmann theory) means that irrelevant variables are substituted by set functionals of the relevant variables. The process of extraction of a subdynamics from the Dyson- Schwinger hierarchy has a correlate at the level of the effective

---

[1]On closer examination, it is seen that the one- particle distribution function itself describes only the mean number of particles within a certain location in phase space; the actual number is also subject to fluctuations. From the average size of the equilibrium fluctuations, which can be determined from Einstein's formula, and the dissipative element of the dynamics, which is contained in the collision integral, it is possible to compute the stochastic driving force consistent with the fluctuation- dissipation relation near equilibrium [36]

action, where the MEA is truncated to a functional of a finite number of variables. The finite effective actions so obtained (the influence action [9]) are generally nonlocal and complex, which is what gives rise to the noise and dissipation in the subdynamics. Moreover, since the slaving process generally involves the choice of an arrow of time, it leads to irreversibility in the cloak of dissipation in the subdynamics [37].

Decoherence of correlation history and correlation noise; fluctuations and dissipation

Under realistic conditions, one may not be as much concerned with the full quantum evolution of the field as with the development of 'classical' theories where fields are described as c-numbers, plus perhaps a small number of correlation functions to keep track of fluctuations. These classical theories represent the physically observable dynamics after the process of decoherence [19, 20] has destroyed or diminished the coherence of the field. In any case, no actual observation could disclose the infinite number of degrees of freedom of the quantum field, and therefore any conceivable observational situation may be described in the language of a suitable complex 'classical' theory in this sense.

Decoherence is brought by the effect of a coarse-grained environment (or 'irrelevant' sector) on the system (or the 'relevant' sector). In simple models, this split is imposed by hand, as when some of the fields, or the field values within a certain region of spacetime, are chosen as relevant. Here, we will follow the approach of our earlier work on the decoherence of correlation histories [26]. There is no need to select *a priori* a relevant sector within the theory. Instead, we shall seek a natural criterium for successive truncations in the hierarchy of correlations, the degree of truncation depending on the stipulated accuracy of measurement which can be carried out on the system. In this framework, decoherence occurs as a consequence of the fluctuations in the higher order correlations and results in a classical dissipative dynamics of the lower order correlations.

Here, we adopt the consistent histories formulation of quantum mechanics [19] for the study of the quantum to classical transition problem. We consider the full evolution of the field described by the Dyson- Schwinger hierarchy as a fine- grained history while histories where only a finite number of Wightman functions are freely specified (with all others slaved to them) are therefore coarse- grained. We have shown that the finite effective actions obtained for the subsystems of lower-order correlations are related to the decoherence functional between two such histories of correlations [26], its acquiring an imaginary part signifies the existence of noise which facilitates decoherence. Thus decoherence of correlation histories is a necessary condition for the relevance of the c-number theory as a description of observable phenomena. It can be seen that if the c-number theory which emerges from the quantum subdynamics is dissipative, then it must also be stochastic. [2] From our correlation history

---

[2] Because the fundamental variables are quantum in nature, and therefore subject to fluctuations, a classical, dissipative dynamics would demand the accompaniment of stochastic sources (in agreement with the 'fluctuation - dissipation theorems', for, otherwise, the theory would permit unphysical phenomena as the damping away of zero - point fluctuations.) Of course, these uncontrollable fluctuations may be seen at the origin of many phenomena where structure seems to spring 'out of nothing', such as the nucleation of inhomogeneous true vacuum bubbles in a supercooled false vacuum, or the development of inhomogeneities

viewpoint, the stochasticity is in fact not confined to the field distributions– the correlation functions would become stochastic as well [36].

Following Feynman and Vernon [9] in their illustration of how noise can be defined from the influence functional, we can relate the imaginary part of the finite effective actions describing the truncated correlations to the auto-correlation of the stochastic sources, i.e., correlation noises, which drive the c-number fields and their correlation functions via the Langevin- type equations. From the properties of the complete (unitary) field theory which constitutes the closed (untruncated) system, one can show that the imaginary part of the effective action is related to the nonlocal part of the real part of the effective action which depicts dissipation. This is where the fluctuation- dissipation theorem for non-equilibrium systems originates [22].

We thus see once again the intimate connection amongst the three aspects of the theory, decoherence, dissipation, and fluctuations [18, 39], now manifesting in the hierarchy of correlations which defines the subsystems.

3 Correlations in a Quantum Field and Black Hole Information Paradox

We view the black hole and the quantum field with Hawking radiation as a closed system. Even though the quantum field might be assumed to be free in the beginning, interaction still exists in its coupling with the black hole, especially when strong backreaction is included. We can model this complete system by an interacting quantum field. A particle- field system is a particular case of it. Of course a black hole is different from a particle. In this modeling, we will first explain how information is registered or 'lost' in an interacting quantum field, then the distinct features of a black hole and finally, the information loss paradox of black hole systems.

To approach the black hole information loss paradox we need to understand three conceptual points:

1) How does one characterize the information content of a quantum field (interacting, as a model for the black hole - quantum field closed system, with backreactions)

2) How does the information flow from one part of this closed system (hole) to another (field) and vice versa in the lifetime of the black hole? Does information really get 'lost'? If yes, where has it gone? Can it be retrieved? If no, where does it reside?

3) What is special about black hole radiation system as distinct from ordinary particle / field system?

The following is a brief sketch of the picture we have developed based on our studies of these issues in varying depths in the past ten years. The development of the correlation dynamics in quantum fields formalism was done with Esteban Calzetta [27], that of viewing

---

out of a homogeneous early universe.

black hole radiance and inflation as exponential scaling was explored with Yuhong Zhang [16, 40, 41] partly based on work on critical phenomena done earlier with Denjoe O'Connor [42]. Studies of simple models to illuminate various points in this conceptual scheme are being pursued now with Alpan Raval. For general background on this issue, see the review of Page [43] which contains a comprehensive list of references till 1993, and Bekenstein [44]. For string theory related ideas, see, e.g., the recent reviews of [45]. Our approach is closer in spirit to [46, 47].

3.1 Correlation functions as registrar and correlation dynamics as flow-meter of information in quantum fields

The set of correlation functions provides us with the means to register the information content of a quantum field. As mentioned above, the complete set of $n = \infty$ correlation (Wightman) functions carries the complete information about the quantum state of the field. A subset of it which defines the subsystem, such as the mean field and the 2-point function, as is used in the ordinary description of (effective) field theory, carries only partial information. The missing information resides in the correlation noise, and manifests as dissipation in the subsystem dynamics. In this framework, the entropy of an incompletely determined quantum system is simply given by $S = -Tr\rho_{red}ln\rho_{red}$, where the reduced density matrix of the subsystem (say, consisting of the lower correlation orders) is formed by integrating out the environmental variables (the higher correlation orders) after the hierarchy is truncated with factorization conditions.

While the set of correlation functions act as a registrar of information of the quantum system, keeping track of how much information resides in what order, the dynamics of correlations as depicted by the hierarchy of equations of motion derived from the master effective action depicts the flow of information from one order to another, up or down or criss-crossing the hierarchy. Correlation dynamics has been proposed for the description of many body systems before [34], and applied to molecular and plasma kinetics. In the light of the above theoretical description we see this scheme as a potentially powerful way to do quantum information systematics, i.e., keeping track of the content and flow of information in a coherent or partially coherent quantum system.

3.2 Information appears lost to subsystems of lower order correlations – 'missing' information stored in higher order correlations

Most measurements of a quantum field system are of a local or quasilocal nature. If one counts the information content of a system based on the mean field and the lowest order correlation functions, as in the conventional way (of defining quantum field theory in terms of, e.g., 2PI effective action), one would miss out a good portion of the information in the complete system, as much of that now resides in the higher order correlation functions in the hierarchy. These invoke nonlocal properties of the field, which are not easily accessible in the ordinary range of accuracy in measurements. Such an observer would then report on a loss of information in his way of accounting (which is taken to be in agreement with other observers with the same level of accuracy of measurement). Only observers which has access to all orders (the 'master' in the master effective action) would be able to see the complete development of the system and be able to tell when the nth order observer begins to lose track of the information count and reports an information loss. This is more easily seen in molecular dynamics: For observers confined to measuring one particle distribution functions (truncation of the BBGKY hierarchy) and with the molecular chaos assumptions implicitly invoked (factorization condition), he would report on information loss. This is how Boltzmann reasons out the appearance of dissipation in ordinary macroscopic physical phenomena. The same can be said about measurement of quantum systems.

3.3 Exponential scaling in Hawking effect facilitates information transfer to the higher correlations. Black hole with its radiation contains full information, but retrieval requires probing the higher order nonlocal properties of the field

How is this scheme useful in addressing the black hole information problem? How is the black hole / quantum field system different from the ordinary cases? The above scheme can explain the apparent loss of information in a quantum system, but there is an aspect distinct to black holes or systems emitting thermal radiance. Some years ago, I made the observation that all mechanisms of emission of (coherent) thermal radiance such as the Hawking effect in black holes, or the Unruh effect in accelerated detectors, involve an exponential redshifting process of or by the system. This can be compared to the scaling transformation in treating critical phenomena. After sufficient exponential redshifting (at late times of collapse) and the black hole is emitting thermal radiance, the system has reached a state equivalent to the approach to a critical point in phase transition. There, the physical properties of the system are dominated by the infrared behavior, and as such, the lowest order correlation functions are no longer sufficient to characterize the critical phenomena. The contribution of higher order correlation functions would become important. Note that in ordinary situations, only the mean field and the 2 or 3 point correlation functions are needed to give an adequate description of the dynamics of the system. But for black holes or similar systems where exponential red-shifting is at work, higher order correlations are readily activated. The information content profile for a quantum field in the presence of a black hole would be very different from ordinary systems, in that it is more heavily populated in the higher end of the spectrum (of correlation orders). If one carries out measurement at the lower end of the spectrum, one would erroneously conclude that there is information loss.

So, following the correlation dynamics of the black hole / field system, while the state of the combined system remains the same as it had begun, there is a continuous shifting of information content from the black hole to the higher correlations in the field as it evolves. Correlation dynamics of fields can be used to keep track of this information flow. We speculate that the information content of the field will be seen to shift from low correlation orders to the higher ones as Hawking radiation begins and continues. The end state of the system would have a black hole evaporated, and its information content transferred to the quantum field, with a significant portion of it residing in the higher order nonlocal correlations.

This is, however, not the end of the story for the correlation dynamics and information flow in the field. The information contained in the field will continue, as it does in general situations, to shift across the hierarchy. As we know from the BBGKY description of molecular dynamics, after the higher correlation orders in the hierarchy have been populated – and for systems subjected to exponential red-shifting this condition could be reached relatively quickly – the information will begin to trickle downwards in the hierarchy, though far slower than the other direction initially. The time it takes (with many criss-crossing) for the information to return to the original condition is the Poincare recurrence time. This time we suspect is the upper bound for the recoherence time, the time for a coherent quantum system interacting with some environment to regain its coherence [47]. It would be interesting to work out the information flow using the correlation dynamics scheme for a few sample systems, both classical and quantum, so as to distinguish the competing effects of different characteristic processes in these systems, some quantum, some statistical (e.g., decoherence time, relaxation time, recoherence time and recurrence time).

Our depiction above uses the interacting field model. Simpler cases might show somewhat degenerate behavior. [3] Details of these investigations will be reported in journal articles.

Acknowledgement

I thank the director of this School, Prof. Norma Sanchez, for expertly organizing a stimulating and enjoyable meeting. My lectures were based on work I did in the last two years with Esteban Calzetta, Andrew Matacz, Alpan Raval and Sukanya Sinha. Research is supported in part by the National Science Foundation under grant PHY94-21849.

---

[3] An example is the interesting result of recoherence reported by Anglin et al [47]. We think their reported result of a recoherence time of the order of the relaxation time is special to the simple model of particle free-field interaction. As the field modes couple only through their interaction with the particle, and not amongst themselves, there is no structure or dynamics of the information content of the field itself, and the only time scale for it to return is via interaction with the particle, which is why the recoherence time is related to the relaxation time of the particle. We expect in more general and complex systems (thus excluding many spin systems) the recoherence time is much longer than the relaxation time, more in the order of the recurrence time.

References

[1] B. L. Hu, "Quantum Statistical Fields in Gravitation and Cosmology" in *Proc. Third International Workshop on Thermal Field Theory and Applications*, eds. R. Kobes and G. Kunstatter (World Scientific, Singapore, 1994) gr-qc/9403061

[2] J. Schwinger, Phys. Rev. 82, 664 (1951); B. S. DeWitt, Phys. Rep. 19C, 297 (1975)

[3] N. Birrell and P. W. C. Davies, *Quantum Fields in Curved Spaces* (Cambridge University Press, Cambridge, 1982)

[4] J. Schwinger, J. Math. Phys. 2 (1961) 407; P. M. Bakshi and K. T. Mahanthappa, J. Math. Phys. 4, 1 (1963), 4, 12 (1963). L. V. Keldysh, Zh. Eksp. Teor. Fiz. 47 , 1515 (1964) [Engl. trans. Sov. Phys. JEPT 20, 1018 (1965)]; G. Zhou, Z. Su, B. Hao and L. Yu, Phys. Rep. 118, 1 (1985); Z. Su, L. Y. Chen, X. Yu and K. Chou, Phys. Rev. B37, 9810 (1988); B. S. DeWitt, in *Quantum Concepts in Space and Time* ed. R. Penrose and C. J. Isham (Claredon Press, Oxford, 1986); R. D. Jordan, Phys. Rev. D33 , 44 (1986). E. Calzetta and B. L. Hu, *Phys. Rev.* D35, 495 (1987).

[5] E. Calzetta and B. L. Hu, Phys. Rev. D35, 495 (1987).

[6] E. Calzetta and B. L. Hu, Phys. Rev. D37, 2878 (1988)

[7] E. Calzetta and B. L. Hu, Phys. Rev. D40, 656 (1989).

[8] B. L. Hu, Physica A158, 399 (1989).

[9] R. Feynman and F. Vernon, Ann. Phys. (NY) 24, 118 (1963). R. Feynman and A. Hibbs, *Quantum Mechanics and Path Integrals*, (McGraw - Hill, New York, 1965). A. O. Caldeira and A. J. Leggett, Physica 121A, 587 (1983); Ann. Phys. (NY) 149, 374 (1983). H. Grabert, P. Schramm and G. L. Ingold, Phys. Rep. 168, 115 (1988). B. L. Hu, J. P. Paz and Y. Zhang, Phys. Rev. D45, 2843 (1992); D47, 1576 (1993)

[10] B. L. Hu, J. P. Paz and Y. Zhang, Phys. Rev. D45, 2843 (1992)

[11] B. L. Hu, J. P. Paz and Y. Zhang, Phys. Rev. D47, 1576 (1993)

[12] B. L. Hu and A. Matacz, Phys. Rev. D49, 6612 (1994).

[13] E. Calzetta and B. L. Hu, Phys. Rev D49, 6636 (1994).

[14] S.W. Hawking, Commun. Math. Phys. 43, 199 (1975).

[15] W.G. Unruh, Phys. Rev. D 14, 870 (1976). P. C. W. Davies, J. Phys. A: Gen. Phys. 8, 609 (1975). S. A. Fulling, Phys. Rev. D 7, 2850 (1973) .

[16] B. L. Hu, in *Proceedings of the CAP-NSERC Summer Institute in Theoretical Physics, Vol 2* Edmonton, Canada, July 1987, eds K. Khanna, G. Kunstatter and H. Umezawa (World Scientific, Singapore, 1988)

[17] B. L. Hu, in *Proceedings of the Fourth International Workshop on Thermal Field Theory* Dalian, China, August 1995, eds Y. X. Gui and K. Khanna (World Scientific, Singapore, 1996)

[18] B. L. Hu, "Statistical Mechanics and Quantum Cosmology", in *Proc. Second International Workshop on Thermal Fields and Their Applications*, eds. H. Ezawa et al (North-Holland, Amsterdam, 1991).

[19] R. B. Griffiths, J. Stat. Phys. **36**, 219 (1984); R. Omnés, J. Stat Phys. **53**, 893, 933, 957 (1988); Ann. Phys. (N. Y.) **201**, 354 (1990); Rev. Mod. Phys. **64**, 339 (1992); *The Interpretation of Quantum Mechanics*, (Princeton University Press, Princeton (1994)). J. B. Hartle, "Quantum Mechanics of Closed Systems" in *Directions in General Relativity* Vol. 1, eds B. L. Hu, M. P. Ryan and C. V. Vishveswara (Cambridge Univ., Cambridge, 1993); M. Gell-Mann and J. B. Hartle, in *Complexity, Entropy and the Physics of Information*, ed. by W. H. Zurek (Addison-Wesley, Reading, 1990); J. B. Hartle and M. Gell- Mann, Phys. Rev. **D47**, 3345 (1993). J. P. Paz and S. Sinha, Phys. Rev. **D44**, 1038 (1991). H. F. Dowker and J. J. Halliwell, Phys. Rev. **D46**, 1580 (1992). T. Brun, Phys. Rev. **D47**, 3383 (1993). J. Twamley, Phys. Rev. **48**, 5730 (1993). J. P. Paz and W. H. Zurek, Phys. Rev. **48**, 2728 (1993).

[20] W. H. Zurek, Phys. Rev. D24, 1516 (1981); D26, 1862 (1982); in *Frontiers of Nonequilibrium Statistical Physics*, ed. G. T. Moore and M. O. Scully (Plenum, N. Y., 1986); Physics Today 44, 36 (1991); E. Joos and H. D. Zeh, Z. Phys. B59, 223 (1985); A. O. Caldeira and A. J. Leggett, Phys. Rev. A 31, 1059 (1985); W. G. Unruh and W. H. Zurek, Phys. Rev. D40, 1071 (1989). B. L. Hu, J. P. Paz and Y. Zhang, Phys. Rev. **D45**, 2843 (1992); **D47**, 1576 (1993); J. P. Paz, S. Habib and W. H. Zurek, Phys. Rev. **D47**, 488 (1993). W. H. Zurek, J. P. Paz and S. Habib, Phys. Rev. Lett. **70**, 1187 (1993); W. H. Zurek, Prog. Theor. Phys. 89, 281 (1993).

[21] B. L. Hu and A. Matacz, Phys. Rev. **D51**, 1577 (1995).

[22] B. L. Hu and S. Sinha, Phys. Rev. **D51**, 1587 (1995).

[23] A. Campos and E. Verdaguer, Phys. Rev. D53 (1996).

[24] B. L. Hu, J. P. Paz and Y. Zhang, "Quantum Origin of Noise and Fluctuation in Cosmology" in *The Origin of Structure in the Universe* Conference at Chateau du Pont d'Oye, Belgium, April, 1992, ed. E. Gunzig and P. Nardone (NATO ASI Series) (Plenum Press, New York, 1993) p. 227

[25] E. Calzetta and B. L. Hu, Phys. Rev. **D52**, (1995)

[26] E. Calzetta and B. L. Hu, "Decoherence of Correlation Histories" in *Directions in General Relativity, Vol II: Brill Festschrift*, eds B. L. Hu and T. A. Jacobson (Cambridge University Press, Cambridge, 1993)

[27] E. Calzetta and B. L. Hu, "Correlations, Decoherence, Disspation and Noise in Quantum Field Theory", in *Heat Kernel Techniques and Quantum Gravity*, ed. S. A. Fulling (Texas A& M Press, College Station 1995).

[28] Alpan Raval, B. L. Hu and J. R. Anglin, "Stochastic Theory of Accelerated Detectors in a Quantum Field" (1995) gr-qc/9501002

[29] B. L. Hu, Alpan Raval, Don Koks and A. Matacz, "Stochastic Analysis of Nonthermal Radiation from Detectors, Mirrors, Black Holes and Expanding Universe" (1995)

[30] A. Raval and B. L. Hu, "Backreaction of Hawking Radiation on a Moving Mirror" (1996)

[31] P. Johnson, B. L. Hu and A. Raval, "Backreaction of Unruh Radiation on an Accelerating Detector" (1996)

[32] B. L. Hu, A. Raval and S. Sinha, "Fluctuation-Dissipation Relation for a Radiating Black Hole" (1996)

[33] A. I. Akhiezer and S. V. Peletminsky, *Methods of Statistical Physics* (Pergamon, London, 1981).

[34] R. Balescu, *Equilibrium and Nonequilibrium Statistical Mechanics* (John Wiley, New York, 1975)

[35] H. Spohn, *Large Scale Dynamics of Interacting Particles* (Springer-Verlag, Berlin 1991)

[36] M. Kac and J. Logan, "Fluctuations", in *Fluctuation Phenomena*, edited by E. W. Montroll and J. L. Lebowitz (Elsevier, New York, 1979), p.1; *Phys. Rev.* **A13**, 458 (1976).

[37] S. Nakajima, Progr. Theor. Phys. **20**, 948 (1958); R. Zwanzig, J. Chem. Phys. **33**, 1338 (1960); and in *Lectures in Theoretical Physics* III, (ed.) W. E. Britten, B. W. Downes and J. Downs (Interscience, N.Y. 1961) pp. 106-141; H. Mori, Prog. Theor. Phys. **33**, 1338 (1965); C. R. Willis and R. H. Picard, Phys. Rev. **A9**, 1343 (1974); H. Grabert, *Projection Operator Techniques in Non Equilibrium Statistical Mechanics* (Springer-Verlag, Berlin, 1982).

[38] R. Haag, *Local Quantum Physics* (Springer, Berlin, 1992).

[39] M. Gell- Mann and J. B. Hartle, Phys. Rev. **D47**, 3345 (1993).

[40] B. L. Hu and Y. Zhang, "Coarse-Graining, Scaling, and Inflation" Univ. Maryland Preprint 90-186 (1990); B. L. Hu, in *Relativity and Gravitation: Classical and Quantum* Proc. SILARG VII, Cocoyoc, Mexico 1990. eds. J. C. D' Olivo et a (World Scientific, Singapore 1991).

[41] B. L. Hu, Class. Quan. Grav. 10, S93 (1993)

[42] B. L. Hu and D. J. O'Connor, Phys. Rev. **D36**, 1701 (1987).

[43] D. N. Page, in *Proceedings of the 5th Canadian Conference on General Relativity and Relativistic Astrophysics* University of Waterloo, May 1993, eds. R. B. Mann and R. G. McLenaghan (World Scientific, Singapore, 1994) hep-th/9305040

[44] J. D. Bekenstein, in *Proceedings of the 7th Marcel Grossmann Meeting on Recent Developments of General Relativity* Stanford University, July 1994 eds. R. Ruffini (World Scientific, Singapore 1995) gr-qc/9409015

[45] S. B. Giddings, Lectures at the *1994 Trieste Summer School in High Energy Physics and Cosmology*, hep-th/9412138; L. Thorlacius, Lectures at the *1994 Trieste Spring School on String Theory, Gauge Theory and Quantum Gravity*, hep-th/9411020

[46] C. Holzhey and F. Wilczek, Nucl. Phys. B380, 447 (1992); F. Wilczek, in *Black Holes, Membranes, Wormholes and Superstrings*, eds. K. Kalara and D. V. Nanopoulos (World Scientific, Singapore, 1993)

[47] J. R. Anglin, R. Laflamme, W. H. Zurek and J. P. Paz, Phys. Rev. D52, 2221 (1995)

TOPOLOGY AND TIME REVERSAL

A. CHAMBLIN & G.W. GIBBONS
D.A.M.T.P.
University of Cambridge
Silver Street
Cambridge CB3 9EW
U.K.

ABSTRACT

In this lecture we address some topological questions connected with the existence on a general spacetime manifold of diffeomorphisms connected to the identity which reverse the time-orientation.

1. Introduction

If one regards Quantum Gravity as an attempt to unify two distinct but equally fundamental physical theories; quantum mechanics on the one hand and general relativity on the other, one can ask what elements of either theory is it most likely that one will have to sacrifice in the eventual unification. Perhaps the most fundamental innovations of general relativity relate to its treatment of the notion of time. One of most striking features of quantum mechanics is its use of complex amplitudes. One may argue that the introduction of complex numbers into the basic structure of quantum mechanics is closely connected to the treatment in that theory of the notion of change and of time evolution. It therefore seems reasonable to regard the use of complex numbers in conventional quantum mechanics as a potential casuality. More precisely, one may argue that if, as is commonly supposed

N. Sánchez and A. Zichichi (eds.), String Gravity and Physics at the Planck Energy Scale, 233–253.

in quantum cosmology, the classical idea of time is an emergent concept, valid only at late times, low energies and large distances, then so too is our usual idea of a quantum mechanical Hilbert space with its attendant complex structure. *In other words, the complex numbers in quantum mechanics should be thought of as having an essentially historical origin.* Some ideas along these lines were discussed within the context of the semi-classical approach to quantum cosmology in [16–17].

A related question is to ask: how in a theory in which one assumes that spacetime has an everywhere well-defined Lorentzian metric are the properties of quantum fields in those spacetimes affected by such global properties of the spacetime as the existence of closed timelike curves ('CTCs '), a lack of time-orientability or some other pathology which would normally be excluded in a globally hyperbolic spacetime? Are there restrictions on the possible spacetimes for example? One possible restriction comes about by demanding that spacetime admit a spin or pin structure [6]. Another possible restriction arises by demanding that the spacetime has a time-orientation. If it does not, one may argue that one may not be able to construct a quantum mechanical Hilbert space endowed with a complex structure. This suggestion was made some time ago [25] and it has received further support from the work of Bernard Kay [7].

One motivation for asking this question is to try to extend the range of applicability of quantum field theory in a fixed background. Another motivation might be to answer questions about what possibilities the laws of physics in principle allow. This has provided much of the impetus behind recent work on CTCs . Another, and possibly more cogent, reason for considering non-globally-hyperbolic spacetimes is that in the path integral approach to quantum gravity in which one sums over all possible Lorentzian metrics there is *a priori* no good reason for excluding them. One might attempt to perform the functional integral by first freezing the metric and integrating over all matter fields on that spacetime, and then summing over all spacetimes. The first part of the integral is then tantamount to quantizing matter fields on a fixed background. It is customary in the Euclidean formulation to replace the sum over Lorentzian metrics by a sum over Riemannian metrics but one may ask what happens if one tries to avoid this step.

In the Euclidean version one is often concerned with anomalies that may arise when functional determinants fail to be well-defined, for example they may not be invariant under spacetime diffeomorphisms. The diffeomorphisms in question may either be continuously connected to the identity or not. The latter type of global anomalies are closely related to discrete symmetries, or lack of them, such as parity or orientation. They may also be investigated from an Hamiltonian point of view . However, this does not address the possibility of anomalies of a purely Lorentzian kind which manifest themselves only in non-globally-hyperbolic spacetimes. An example is a breakdown of spin structure. If one assumes that spacetime is both time and space-orientable, this can *only* occur in a spacetime which is not globally hyperbolic . If one drops the requirement of space-orientability, however, there may exist pin structure even though the spacetime *is* globally hyperbolic. An example is provided by $\mathbb{RP}^2 \times \mathbb{R}^2$, endowed with the product metric formed from the standard 'round' metric on $\mathbb{RP}^2$ and the Minkowski metric on $\mathbb{R}^2$ (with either signature) [6].

One possible viewpoint on the difficulties experienced with non-time-orientable spacetimes is precisely that there is some sort of anomaly. Roughly speaking, for each complex amplitude in the functional sum one must, if there is no global time-orientation, include its complex conjugate which is associated to the time-reversed amplitude. The result must then necessarily be real and so no true quantum interference is possible. It is interesting to note that this sort of problem would also arise in some attempts to generalize the usual quantum formalism being made by Gell-Mann and Hartle [22] since they also make use of *complex* amplitudes and they incorporate a rule relating complex conjugation to time reversal of a sequence of observables.

The purpose of this lecture is to explore some of these issues in more depth. In particular, we will discuss the relation between the topology of a time-orientable spacetime $\{\mathcal{M}, g\}$ and the existence and properties of various kinds of time-reversing diffeomorphisms. We shall, for the sake of mathematical precision, mainly concentrate on spacetime manifolds $\mathcal{M}$ which are compact and without boundary, but we will comment on the case of non-compact spacetimes and spacetimes with boundaries.

As well as the motivations given above, our results are also relevant to suggestions like that of Sakharov [18] that the early universe may simply be a time reflection of the late universe. Such a viewpoint is essentially a Lorentzian version of the (historically later) no-boundary proposal or the idea of a universe born from nothing [23].

2. Compact Spacetimes

An assumption of compactness in spatial directions is quite natural when discussing topological questions because one has in mind a situation where the non-trivial topology can be localized, at least to the extent that it is not allowed to escape from the spacetime altogether. Compactness in the time direction is less easy to justify (unless there are spacelike boundaries) because it necessarily implies the existence of closed timelike curves . Formerly this was thought to rule out consideration of such spacetimes but more recently, with the advent of studies of the properties of time machines, this view has been abandoned and so we shall not be put off by this feature.

In fact the Euler number $\chi(\mathcal{M})$ of a compact spacetime of arbitrary dimension must vanish:

$$\chi(\mathcal{M}) = 0, \tag{1}$$

and in four dimensions:

$$\chi = 2 - 2b_1 + b_2, \tag{2}$$

where b_i are the Betti numbers. Thus a compact spacetime must have an even second Betti number and infinite fundamental group, and so its universal covering space is non-compact. In this sense it may be thought of as a non-compact spacetime which has been periodically identified and this is indeed typically how examples of time machines are constructed in the literature. However, the reader is cautioned that there is, as we shall see later, no logical connection between whether or not a curve is closed and timelike and whether or not it is homotopically trivial. In general, one expects the fundamantal group $\pi_1(\mathcal{M})$ to be non-Abelian. This is what one expects in the case of two or more time machines, for example if the spacetime has in a connected sum decomposition two summands of the form $S^1 \times S^3$ with time running around the S^1 factors.

In the exceptional case that the fundamental group *is* Abelian, it may be shown [8–11] that the possible Betti numbers (b_1, b_2) must belong to the set: $\{(1,0), (2,2), (3,4), (4,6)\}$. This is because for any closed orientable manifold of any dimension which has an Abelian fundamental group one has the inequality:

$$\frac{1}{2} b_1(b_1 - 1) \le b_2 \tag{3}$$

The result follows from (1) which holds for any spacetime dimension and (2) which holds in four dimensions.

The significance of this non-Abelian-ness in the case that homotopically non-trivial time machines are present is presumably that some physical effects may depend upon the order in which one enters the time machines. It would be interesting to explore this point further. In that connection, it is perhaps worth recalling why it is that non-simply-connected four-manifolds are not classifiable [26]. The point is that by taking the connected sum $\#_k S^1 \times S^3$ of k copies of $S^1 \times S^3$ one obtains a four-manifold whose fundamental group is the free group on k generators (which of course is maximally non-Abelian). One may now perform surgery on this manifold to obtain a new manifold whose fundamemtal group has k generators and r arbitrarily chosen relations. Since there is no algorithm for deciding whether two different presentations give an isomorphic group there can be no algorithm for deciding whether two four-manifolds are homeomorphic .

The process of surgery can be described as follows. Given an element $g \in \pi_1(\mathcal{M}')$ of a four-manifold $\mathcal{M}'$ one can represent it by a closed curve $\gamma \in \mathcal{M}'$. Now surround this closed curve γ by a tube or collared neighbourhood $\mathcal{N}$ of the form $\mathcal{N} = D^3 \times \gamma \equiv D^3 \times S^1$ where D^3 is a closed 3-dimensional disc. The boundary $\partial \mathcal{N}$ of this tube has topology $\partial \mathcal{N} \equiv S^1 \times S^2$. One now removes the tube $\mathcal{N}$ from $\mathcal{M}'$ and replaces it with the simply connected manifold $D^2 \times S^2$ which has the same boundary. The result is a new manifold $\mathcal{M}''$ whose fundamental group differs from that of $\mathcal{M}'$ only by the imposition of the relation $g = 0$. This process is called 'killing an element of the fundamental group'. It may be shown that by a succesion of such killings one may obtain from $\#_k S^1 \times S^3$ a manifold with any desired finitely generated fundamental group.

From a physical point of view it is interesting to note two things. Firstly that the undecidability problem reviewed above may give rise to limitations on what is 'in principle' allowed by the laws of physics when it comes to the sort of wormhole and time machine engineering envisaged by Thorne and others. The possibility arises of having two sets of instructions for building a multiple time machine but having no algorithm for deciding whether the two spacetimes have the same topology. Whether or not this is true is not obvious from the general result quoted above because a compact spacetime must have vanishing Euler number. We do not know whether such manifolds are classifiable or not.

The second point is that the process of surgery gives rise to a manifold which physically looks rather like one containing the creation and annihilation of an extra Einstein-Rosen throat. If the 2-disc D^2 has coordinates $X + iT = r \exp\left(it - \frac{i\pi}{2}\right)$ where the cyclic 'time' coordinate t which parameterizes the original curve γ runs between 0 and 2π then 'half-way round', i.e. on the real axis $T = 0$, the interior of the tube $\mathcal{N}$ has been replaced by a manifold which has the same topology as the Kruskal manifold of a black hole and therefore it has embedded in it a three-manifold which has the topology of a bridge, i.e. of $\mathbb{R} \times S^2$. If these sorts of manifolds do arise in a Lorentzian form of quantum gravity it seems reasonable to think of them as containing 'virtual black holes'.

This interpretation receives some support from the observation that the Riemannian manifolds used as instantons or real tunnelling geometries in the Euclidean approach to vacuum instability and black hole pair creation may be obtained by surgery on a circle , which we would like to associate with the world line of a virtual black hole, from the corresponding false vacuum spacetime. Thus the Euclidean Schwarzschild manifold ($\mathbb{R}^2 \times S^2$) may be obtained from the hot flat space manifold $S^1 \times \mathbb{R}^3$, the Ernst instanton manifold ($S^2 \times S^2 - \{pt\}$) for the creation of pairs of oppositely charged non-extreme black holes from a constant electromagnetic field (topology $\mathbb{R}^4$), and the Nariai and Mellor-Moss Instantons (both with topology $S^2 \times S^2$) are obtained from the De Sitter manifold (S^4). In Kaluza-Klein theory, Witten [27] has argued that the five-dimensional Schwarzschild solution (topology $\mathbb{R}^2 \times S^3$) is the bounce solution which mediates the decay of the Kaluza-Klein vacuum (topology $S^1 \times \mathbb{R}^4$). The five-dimensional manifold corresponding to a magnetic field also has topology $S^1 \times \mathbb{R}^4$. This may decay via Witten's instability but it may also decay into a monopole-anti-monopole pair. The instanton for this process has topology $S^5 - S^1 \equiv \mathbb{R}^2 \times S^3$ and so may also be obtained by surgery on a circle from the false vacuum space-time manifold.

3. Time Reversal in a General Spacetime

Let $\{\mathcal{M}, g^L\}$ be a time-orientable spacetime. Thus the bundle of time-oriented frames $SO_\uparrow(n-1,1)(\mathcal{M}, g^L)$ falls into two connected components. One typically thinks of time reversal Θ as a diffeomorphism:

$$\Theta : \mathcal{M} \to \mathcal{M}$$

which reverses time-orientation, whose lift to $SO_\uparrow(n-1,1)(\mathcal{M}, g^L)$ exchanges the two connected components and is an involution of order two:

$$\Theta^2 = \mathrm{id}.$$

It need not necessarily be an isometry (in general the spacetime will not admit any isometries). One could imagine considering a more general finite group action but presumably one could always find a $\mathbb{Z}_2$ subgroup and we shall assume that this can be done.

In a general non-globally-hyperbolic spacetime it is not obvious whether Θ should reverse space-orientation, or total orientation (assuming $\{\mathcal{M}, g^L\}$ to be space or time-orientable respectively) , whether it should act freely on $\mathcal{M}$ or fix a three-surface for example, or whether it should belong to the identity component $\mathrm{Diff}_0(\mathcal{M})$. The existence and uniqueness and other properties of Θ depends both on the topology of the manifold $\mathcal{M}$ and on the Lorentz metric g^L.

To illustrate these subtleties, consider even the simplest globally hyperbolic spacetime $\mathcal{M} \equiv \mathbb{R} \times \Sigma$ with coordinates $t, \mathbf{x}$, t being timelike and Σ being an orientable $(n-1)$-manifold. Naively we might take

$$\Theta^T : (t, \mathbf{x}) \to (-t, \mathbf{x})$$

but nothing prevents us from considering

$$\Theta^J : (t, \mathbf{x}) \to (-t, \mathbf{x}^*)$$

where

$$J : \mathbf{x} \to \mathbf{x}^*$$

is an involution on the $(n - 1)$-manifold Σ. Clearly Θ^T fixes the three-manifold Σ and reverses total orientation. It therefore lies outside the identity component $\mathrm{Diff}_0(\mathcal{M})$. On the other hand, we might arrange for J to act freely on Σ, possibly reversing or not reversing space-orientation.

These seemingly rather artificial examples actually arise in some applications. In quantum field theory in De Sitter spacetime, dS_n, Σ is the $(n - 1)$-sphere and J its antipodal map. This preserves space-orientation if the spacetime dimension n is even. The map Θ^J is an isometry and is the centre of the isometry group $O(n, 1)$. One may identify points under the action of Θ^J to obtain the 'elliptic interpretation'. This then provides a possible non-singular realisation of Sakharov's ideas of a Lorentzian model of a universe born from nothing. The idea immediately generalizes to a Friedman model whose scale factor is an even function of time. Sakharov's idea was in fact to impose some sort of time-reflection symmetry about a singular big bang at which the scale factor vanishes. He did not use the involution J. In spatially closed models the scale factor often starts from a zero value at the big bang, $t = 0$, rises to a maximum at $t = t_{\max}$ say, and then symmetrically decreases to a vanishing value at the big crunch at $t = 2t_{\max}$. This has led Gold [2] to conjecture that the 'arrow of time reverses' in the contracting phase. In effect he proposed that the entire quantum state is invariant under a time-reversing involution whose action on spacetime is given by:

$$\Theta^G : t \to t_{\max} - t.$$

By contrast Davies [1] (see also Albrow [4]) prefers to continue through the Big Bang and Big Crunch to get a model in which the arrow of time reverses in sucessive cycles. In other words, one imposes invariance under the action of semi-direct product $\mathbb{Z} \odot \mathbb{Z}_2$ given by

$$t \to t + 2t_{\max}$$

and

$$t \to -t.$$

It is clear that similar options are available for non-singular periodic models in which there is neither a Big Bang nor a Big Crunch. Thus for example, in the case of Anti-De Sitter spacetime AdeS_n, the scale factor is a sinusoidal function of cosmic time but the vanishing of the scale factor is an artefact of a poor choice of coordinates. In fact $\mathcal{M} \equiv S^1 \times H^{n-1}$ where $H^{n-1} \equiv \mathbb{R}^{n-1}$ is hyperbolic space and time t runs around the circle, $0 \le t < 2\pi$. The center of the isometry group $O(n-1, 2)$ does not reverse time (it sends $(t, \mathbf{x})$ to $(t + \pi, -\mathbf{x})$). Intuitively, it seems clear that time reversal must have fixed points since we must reverse t and compose with an involution J which may be thought to act on Euclidean space.

In the examples so far (at least if we wish to maintain the boundary conditions) there was no natural choice of Θ in the identity component $\text{Diff}_0(\mathcal{M})$. However in more exotic situations, as we shall see in detail shortly, this seemingly paradoxical situation can occur. Now if no possible Θ lies in the identity component $\text{Diff}_0(\mathcal{M})$ it is reasonable to say that the spacetime $\{\mathcal{M}, g^L\}$ has an intrinsic sense of the passage of time (even though time itself may not be defined!). If however there exists a Θ which does lie in the identity component this is not reasonable. The general situation with respect to $\text{Diff}(\mathcal{M})$ appears to be quite difficult to analyse and so we shall restrict attention here to a simpler question. Is there a homotopy rather than a diffeomorphism carrying the metric g^L with one time-orientation to the same metric with the opposite time-orientation? If there does exist a suitable Θ in the identity component $\text{Diff}_0(\mathcal{M})$ then a homotopy will certainly exist (simply pull back g^L by a curve f_s, $0 \le s \le 1$ of diffeomorphisms joining $f_1 = \Theta$ to the identity $f_0 = \text{id}$). However the converse is not necessarily true. Given a homotopy g_t^L of Lorentz metrics there may exist no diffeomorphism producing it. Now from the point of view of homotopy theory, a closed time-oriented Lorentzian spacetime $\{\mathcal{M}, g^L\}$ contains no more information than a Riemannian manfold $\mathcal{M}$ equipped with a unit vector field $\mathbf{V}$. The spacetime with the opposite time-orientation corresponds homotopically to the same manifold equipped with the negative unit vector field $-\mathbf{V}$.

4. Mathematical Interlude

This following mathematical interlude follows some conversations with Graeme Segal.

4.1. LINEAR AND GENERAL HOMOTOPIES

We suppose that M is a closed, n-dimensional time-orientable Lorentzian manifold. We may, in the standard way, endow M with a Riemannian metric and hence deduce that M admits a global section $\mathbf{V}$ of the bundle $S(M)$

of unit vectors over M. At each point x in M the fibre S_x of $S(M)$ is an $n-1$ sphere.

Pulling the Lorenzian metric back under the action of diffeomorphisms induces an action on $\mathbf{V}$ and we would like to know whether there exists a diffeomorphism $f : M \to M$ which takes $\mathbf{V}$ to its negative, i.e. which reverses the direction of time. In particular we would like to know whether there exits such a diffeomorphism f contained in the identity component $\text{Diff}_0(M)$ of the diffeomorphism group $\text{Diff}(M)$. An easier question to ask is whether there exists a homotopy taking $\mathbf{V}$ to $-\mathbf{V}$ since if there exists a diffomorphism in the identity component a homotopy is given by a curve f_t in $\text{Diff}_0(M)$ joining f to the identity. The converse is however not sufficient because, as we shall see, if one considers $M = S^1 \times S^{2n-1}$ with the vector field running around the S^1 factor one finds that this cannot be reversed by a diffeomorphism but it may be reversed by a homotopy

A homotopy $\mathbf{V}_t$ between $\mathbf{V}$ and $-\mathbf{V}$ thus gives at each point x in M a continuous path $\gamma_x(t)$ from the north pole to the south pole of S^{n-1}. In other words a *general homotopy* $\mathbf{V}_t$ provides a global section s_Z of a bundle $Z(M)$ whose fibres Z_x are the space of paths from the north to the south pole of S^{n-1}. Since any path from the north pole to the south pole of S^{n-1} is homotopic to a closed path on S^{n-1} one sees that from the point of view of homotopy the fibre Z_x is equivalent to the loop space $\Omega(S^{n-1})$ of based loops on S^{n-1} .

Consider now a *special* or *linear* homotopy from $\mathbf{V}$ to $-\mathbf{V}$. By definition this is one for which, at each point x in M, $\mathbf{V}(\mathbf{x})_t$ lies in a an oriented two plane $\pi(x)$ spanned say by the vectors $\mathbf{V}_0$ and $\mathbf{V}_{t_1}$ where $0 < t_1 < 1$. A linear homotopy gives a particular kind of path $\gamma_x(t)$ from the north to the south pole of S^{n-1}, one which is along a great circle in the 2-plane defined by by the vectors $\mathbf{V}_0$ and $\mathbf{V}_{t_1}$. The set of such great circles is parameterized by where the great circle intersects the equatorial S^{n-1}.

The existence of a linear homotopy is thus equivalent to the existence of a global section s_Y of the S^{n-2} bundle $Y(M)$ of unit vectors orthogonal to the vector field $V(x)$. One may think of this S^{n-2} fibre Y_x as the equatorial S^{n-2} in the S^{n-1} fibre S_x of the bundle $S(M)$. It follows that the bundle $Y(M)$ is a sub-bundle of the bundle $Z(M)$. The question of whether every homotopy can be deformed into a linear homotopy then reduces to the question whether every section s_Z may be deformed to a section s_Y.

It should also be clear that the existence of the vector field and a linear homotopy is equivalent to a non-vanishing section of the bundle $V_{n,2}(M)$ of dyads, i.e of ordered pairs of linearly independent vectors $\mathbf{e}_1$ and $\mathbf{e}_2$ say. The fibre of the dyad bundle $V_{n.2}(M)$ is the Stiefel manifold $V_{n,2}$ of dyads. In addition a linear homotopy provides a global section s_G of the bundle $G_{n,2}(M)$ of oriented 2-planes whose fibre is the Grassman manifold $G_{n,2}$.

The existence of a section s_G is, in fact, the necessary and suffient condition that a manifold admit a metric of signature $(n-2, 2)$.

We note *en passant* the following

Lemma *If M is even dimensional a sufficient condition for M to admit a linear homotopy is that it admit an almost complex structure J. In four dimensions this condition is also neccessary.*

The point is that one may then take

$$\mathbf{V}_t = e^{t\pi J}\mathbf{V}_0$$

In four dimensions the existence of an almost complex structure is also necessary since given the dyad field one obtains an almost complex structure by extending the rotation through $\frac{\pi}{2}$ in the two-plane spanned by the two vectors to the unique orthogonal two-plane. The sign ambiguity may be fixed by the convention that the associated two-form is anti-self-dual.

A simple example is provided by the manifold mentioned earlier: $S^1 \times S^{2n-1}$. As is well known this is a complex manifold and hence it certainly admits a complex structure. Thus the vector field which just winds around the S^1 factor can certainly be reversed by a homotopy but it is clear, by using a metric to convert the vector to a one-form and considering the line integral of the one-from around the circle, that it cannnot be reversed by a diffeomorphism. To see that $S^1 \times S^{2n-1}$ is a complex manifold one notes that $S^1 \times S^{2n-1} \equiv \mathbb{C}^{2n}/\mathbb{Z}$ where the integers $\mathbb{Z}$ act on $\mathbb{C}^{2n} \equiv \mathbb{R}^{4n}$ by $(z^1, z^2, \ldots, z^n) \to (\lambda^m z^1, \lambda^m z^2, \ldots \lambda^m z^n,)$ where $m \in \mathbb{Z}$ and λ is a real number not equal to zero or unity.

Now it is known, eg. from Morse theory, that the homotopy type of the fibre Y_x of loops on S^{n-1} is that of a cell-complex corresponding to the geodesic paths. Thus there is a cell corresponding to going once around the sphere, the descending directions parmetrizerized by the equatorial S^{n-1}, and next comes a cell $S^{2(n-2)}$ and so-forth. If $n > 4$ this second cell is higher in dimension than the dimension of the base M of the bundle $Y(M)$. It follows from obstruction theory that there is no obstruction to pushing points of any section s_Y in the fibre Y_x down onto the S^{n-2} of the fibre Y_x of the dyad bundle $Y(M)$. In other words we have the following

Proposition: *In dimensions greater than 4 a general homotopy is deformable to a linear homotopy and thus the necessary and sufficient condition for a general homotopy is the existence a global section of the S^{n-2} bundle $Y(M)$, or equivalently a global section of the dyad bundle $V_{n,2}(M)$.*

Atiyah [15] (see also Thomas [14]) has obtained some necessary conditions for the existence of a non-singular dyad field. From their work one has one has the following

Proposition *A necessary condition that a 4k dimensional manifold, $k > 1$ admit a time revesing homotopy is that the signature $\tau(M)$ be divisible*

by 4. A necessary condition that a $4k + 1$ dimensional manifold admit a time-reversing homotopy is that the real Kervaire semi-characteristic $k(M)$ vanish.

The real Kervaire semi-characteristic is defined by

$$k(M) = \sum b_{2p} \bmod 2$$

where b_{2p} are the Betti numbers, $b_{2p} = \dim H^{2p}(M; \mathbb{R})$.

4.2. FOUR-DIMENSIONAL CASE

In four dimensions the situation is more delicate because the cell-decomposition of the fibre Y_x contains a 4-sphere which has the same dimension as the base. We now turn to a more detailed discussion of the four dimensional case. We begin with some general facts about S^2 and S^3 bundles over oriented four manifolds.

Firstly note that oriented 3-plane, or equivalently S^2 bundles $Q \to M$ have characteristic classes $w_2 \in H^2(M; \mathbb{Z}/2)$ and $p_1 \in H^4(M; \mathbb{Z})$ which satisfy:

$$p_1 = w_2^2 \bmod 4$$

The characteristic classes w_2 and p_1 subject to this condition determine and are determined by the bundle Q. Moreover, the bundle Q admits a cross section if and only if there exists an element $\xi \in H^2(M; \mathbb{Z})$ such that

$$\xi = w_2 \bmod 2$$

and

$$\xi^2 = p_1.$$

The class ξ may be thought of as follows. A non-zero section s of an oriented three-plane bundle gives rise to an oriented 2-plane bundle whose fibres consist of vectors orthogonal to the section s. This oriented two-plane bundle may be thought of as a complex line bundle and ξ is its first Chern class c_1.

Similarly, in four dimensions, real four-dimensional oriented vector bundles $E \to M$ determine and are determined by classes $w_2 \in H^2(M; \mathbb{Z}/2)$ and $p_1, e \in H^4(M; \mathbb{Z})$ such that

$$w_2^2 = p_1 + 2e \bmod 4.$$

Given E one may pass to the bundle of two forms $\Lambda(E)$. Giving the fibres a positive definite metric we obtain two 3-plane bundles, $\Lambda^{\pm}$ of self-dual or anti-self-dual two forms. We have

$$w_2(E) = w_2(\Lambda^+) = w_2(\Lambda^-)$$

and

$$p_1(\Lambda^{\pm}) = p_1(E) \pm 2e(E)$$

We are of course interested in the case when E is the tangent bundle of the manifold M. Then w_2 is the second Steifel-Whitney class $w_2(M)$, e its Euler class $e(M)$, and p_1 its Pontryagin class $p_1(M)$. The Pontryagin class is related to the signature τ by

$$p_1(M) = 3\tau(M),$$

moreover

$$\tau = e \bmod 2$$

Now if E admits a global section s_E (which can happen if and only if the euler class e vanishes) then the bundles $\Lambda^+(E)$ and $\Lambda^-(E)$ are isomorphic. This is because given any vector u orthogonal to the section s_E we get a self or anti-self dual two form, i.e.

$$u \wedge s_E \pm \star u \wedge s_E.$$

Thus both Λ^+ and Λ^- are isomorphic to the bundle of vectors orthogonal to s_E, $E^\perp$. The set of such unit vectors corresponds to the equatorial two-sphere in the three-sphere in our general discussion above.

An almost complex structure or equivalently a linear homotopy therefore exists if and only if there exists a section $u_{E^\perp}$. Such a section exits if and only if there exists an an element $\xi \in H^2(M; \mathbb{Z})$ such that

$$\xi = w_2 \bmod 2$$

and

$$\xi^2 = p_1 = 3\tau.$$

Now in four dimensions $F^2(M) = H^2(M; \mathbb{Z})/\mathrm{Tor}(M)$ is an integral lattice since it is equipped, via the cup product, with an integral valued bilinear product: the intersection form $Q(\ ,\ )$. By Wu's formula the second Stiefel Whitney class satisfies

$$Q(w_2, x) = Q(x, x)$$

for all $x \in F^2(M)$ and thus

$$Q(\xi, x) = Q(x, x) \bmod 2$$

for all $x \in F^2(M)$, in other words the element ξ is a so-called *characteristic* element of the integral lattice $F^2(M)$. It follows on purely arithmetic grounds, by a lemma of Van der Blij [24], that for such an element

$$Q(\xi, \xi) = \tau \bmod 8.$$

The other condition on ξ becomes

$$\xi^2 = p_1 = 3\tau$$

that is, eliminating ξ^2

$$2\tau = 0 \bmod 8$$

or

$$\tau = 0 \bmod 4.$$

Lemma *A neccessary condition that a closed Lorentz 4-manifold admit a linear homotopy is that the signature is divisible by 4*

Moreover we have also shown that

Proposition *A Lorentz 4-manifold admits a linear homotopy if and only if it admits an almost complex structure*

These conditions are non-trivial because, while for a spin manifold $\tau = 0 \bmod 16$ [14], in general one only knows that if the Euler characteristic nanishes then $\tau = 0 \bmod 2$. In fact to obtain an example of a Lorentz 4-manifold which does not admit a linear homotopy consider the connected sum of $2n$ copies of $\mathbb{CP}^2$ with $n + 1$ copies of $S^1 \times S^3$. This has $\tau = 2n$, and does not admit a spin structure, even if n is a multiple of 8. Unless n is divisible by 4 it cannot admit a linear homotopy.

We may relate this discussion to the question of the existence of global sections of the bundle of dyads $V_{4,2}(M)$, a subject studied by Hirzebruch and Hopf [28]. Generically a section will have singularities isolated at points in the manifold M. Surounding each point by small 3-sphere we get a map from $S^3 \to V_{4,2}$. Since $\pi_3(V_4, 2) \equiv \mathbb{Z} \oplus \mathbb{Z}$ one has an index consisting of two integers (a, b) associated with each singularity. If $\alpha = Q(\xi, \xi)$ where ξ is a characteristic element of $H^2(M; \mathbb{Z})$ then the allowed values are given by

$$(a, b) = \frac{1}{4}(\alpha - 3\tau - 2e, \alpha - 3\tau + 2e).$$

In terms of Betti numbers one has

$$\frac{1}{4}(3\tau + 2e) = \frac{1}{4}(b^+ - b^-) + 1 - b_1$$

and

$$\frac{1}{4}(3\tau - 2e) = \frac{5}{4}(b^+ - b^-) - 1 + b_1$$

So the integrality of (a, b) is automatic. In the present case $e = 0$ and if we have a global section then there exists a characteristic element $\xi \in H^2(M; \mathbb{Z})$ such that

$$\xi^2 = 3\tau$$

This is of course the same condition that we used above.

Our necessary condition for the existence of an almost complex structure may also be obtained by considering the index of the associated Dolbeault complex. This is called the arithmetic genus, $ag(M)$. In dimension 4

$$ag(M) = \frac{1}{4}(\chi + \tau) = \frac{1}{2}(b^+ + 1 - b_1)$$

and thus under our assumption that $\chi = 0$ this again leads to the necessary condition for the existence of an almost complex structure is that the signature τ be divisible by 4.

Consider the more general problem of whether a general homotopy exists. This requires the existence of a section of the bundle $Z(M)$. As always, the potential obstructions lie in $H^i(M; \pi_{i-1}(Z_x))$. Since M is four dimensional $H^i(M; \pi_{i-1}(Z_x))$ for $i > 4$, so we need only consider $\pi_i(Z_x)$ for $i \leq 3$. Now Z_x is the space of based loops on S^3 and so

$$\pi_i(Z_x) = \pi_{i+1}(S^3).$$

The possible obstructions are therefore in $H^i(M; \pi_i(S^3))$. Thus there are two potential obstructions: the primary one, which is an element of $H^3(M; \mathbb{Z})$ and a secondary one which is an element of $H^4(M; \mathbb{Z}/2)$.

The primary obstruction coincides with the obstruction for the bundle $Y(M)$ and is the third integral Stiefel Whitney class $W_3(M)$ which is the obstruction to lifting the second $\mathbb{Z}/2$-Stiefel Whitney class $w_2 \in H^2(M; \mathbb{Z}/2)$ to an integral class ξ. It is the obstruction to the introduction of a Spin_c structure, ξ being the Chern class of the circle bundle.. This is well known to vanish for an orientable four-manifold. There remains the secondary obstruction. In the case of the bundle $Y(M)$ this vanishes if $\xi^2 = p_1$. In the case of the bundle $Z(M)$ it vanishes under the weaker condition that

$$\xi^2 = p_1 = 3\tau \bmod 8$$

But as before ξ^2 is congruent to $\tau \bmod 8$ and thus we have the following **Proposition** *The necessary and sufficient condition for general homotopy is*

$$\tau = 0 \bmod 4.$$

The necessary and sufficent condition for a general homotopy is the same as the necessary condition for a linear homotopy obtained above. It is a non-trivial requirement as the examples constructed above illustrate. The remaining question, whose answer is not known at present, is whether the necessary condition is sufficient. This boils down to a purely arithmetic question about the possible intersection forms Q.

5. Some Examples

Every odd dimensional sphere S^{2r+1} admits a time-orientable Lorentz metric g^L. One takes:

$$g^L = g^R - 2\mathbf{V}^\flat \otimes \mathbf{V}^\flat,$$

where g^R is the standard round metric, $\mathbf{V}^\flat$ is the one-form dual to the vector field $\mathbf{V}$ obtained by using the musical isomorphism (i.e. lowering the index with the metric g^R) and the unit vector field $\mathbf{V}$ is tangent to the Hopf fibration. If Z^a, $a = 1, \ldots 2r + 2$ are complex coordinates for $\mathbb{R}^{2r+2} \equiv \mathbb{C}^{r+1}$ then the Hopf fibration corresponds to the $SO(2) \subset SO(2r + 2)$ action :

$$Z^a \to \exp(it) Z^a$$

and

$$\mathbf{V} = \frac{\partial}{\partial t}.$$

The case $r = 1$ should be familiar because it is encountered in the Taub-NUT solutions of Einstein's equations. The general case also arises in higher dimensions as we shall describe later.

Atiyah's result tells us that if r is even, $r = 2k$, then the Lorentz structure described above cannot be obtained by a diffeomorphism which is connected to the identity to the Lorentz structure whose light cones differ merely by being upside down. On the other hand, we may trivially reverse the light cones by using the diffeomorphism Γ consisting of r reflections $\in O(2r + 2)$, i.e. by complex conjugation:

$$\Gamma : \quad Z^a \to \bar{Z}^{\dot{a}}.$$

Now if r is odd then Γ lies in the identity component $SO(2r + 2)$ of $O(2r+2)$ and hence in the identity component $\text{Diff}_0(S^{2r+1})$ of $\text{Diff}(S^{2r+1})$. If however r is even, $r = 2k$, then Γ is not in the identity component of $O(4k + 2)$ and, by Atiyah's result, not in $\text{Diff}_0(S^{4k+2})$ either.

Thus Lorentz metrics on S^{4k+1} of the type we have been considering fall into two classes with opposite time-orientation. This is similar to the situation with respect to orientation (i.e. combined space and time-orientation). An oriented manifold may or may not be diffeomorphic to the same manifold with the opposite orientation. Manifolds which are, are called *reversible*. Manifolds which are not, are called *irreversible* or sometimes *chiral*. Of course in this latter case the diffeomorphism must lie outside the identity component $\text{Diff}_0(\mathcal{M})$.

Chiral manifolds are analogous to enantiomorphic crystal forms, such as seen in quartz for example. In that case, they arise because the point group of the crystal is contained in $SO(3)$ and thus includes no orientation reversing isometries of Euclidean space. In our case, however, we are *not* requiring our diffeomorphism to be an isometry of any metric.

The spheres S^n are obviously reversible because they admit reflections. By contrast, some of the three-dimensional lens spaces, $L_{p,q}$ (with p and q co-prime) are chiral, as first noticed by Kneser. They are obtained from S^3 by identifying points under the action of the cyclic group C_p given by [12–13]:

$$Z^1 \rightarrow \exp\left(\frac{2\pi i}{p}\right) Z^1$$

$$Z^2 \rightarrow \exp\left(\frac{2\pi q i}{p}\right) Z^2$$

Because the action of the cyclic group commutes with the Hopf fibration the time-orientable Lorentz metric described above descends to all of the lens spaces.

The topological classification of the lens spaces depends on the bi-linear map:

$$\lambda : H_1(\mathcal{M}; \mathbb{Z}) \times H_1(\mathcal{M}; \mathbb{Z}) \rightarrow \mathbb{Q}/\mathbb{Z}$$

called the linking form defined on the first homology group $H_1(\mathcal{M}; \mathbb{Z}) \equiv \mathbb{Z}/p\mathbb{Z}$. The linking form λ changes sign under reversal of orientation.

Now the integral curve of the timelike vector field $\mathbf{V}$ gives a generator γ of $H_1(\mathcal{M}, \mathbb{Z})$ with linking invariant:

$$\lambda(\gamma, \gamma) = \frac{q}{p}.$$

The remaining elements α of $H_1(\mathcal{M}, \mathbb{Z})$ are of the form $\alpha = x\gamma$, $x = 0, 1, \ldots p - 1$. The bilinearity of $\lambda(,)$ implies that

$$\lambda(\alpha, \alpha) = x^2 \frac{q}{p}.$$

To exhibit a chiral lens space it suffices to find a pair of co-prime natural numbers (p, q) such that for no $x = 0, 1, \ldots p - 1$ is it true that:

$$x^2 \frac{q}{p} = -\frac{q}{p} \qquad (\text{mod } p).$$

Thus $L_{3,1}$ is an example of a chiral three-dimensional spacetime. On the other hand, the action of the cyclic group C_p may or may not commute with the reflection Γ. If it does, then the action of Γ will descend to the quotient and then we can still reverse time.

We remark here that, consistent with our general idea, the partition function $Z(\mathcal{M}_3)$ for Witten's topological field theory is invariant under all diffeomorphisms whether or not they are in the identity component $\text{Diff}_0(\mathcal{M})$, and obeys:

$$Z(\overline{\mathcal{M}_3}) = \overline{Z(\mathcal{M}_3)}$$

where $\overline{\mathcal{M}}$ is the same manifold as $\mathcal{M}$ but with the opposite orientation. Thus for reversible manifolds it is real, and conversely if it is complex, then the manifold must be chiral.

Turning to four-dimensional manifolds: a standard example of an irreversible four-manifold is $\mathbb{CP}^2$. Notationally one distinguishes between $\mathbb{CP}^2$ and $\overline{\mathbb{CP}}^2$. The Euler characters χ are the same, but the Hirzebruch signatures $\tau = b_2^+ - b_2^-$ are opposite in sign:

$$\tau(\mathbb{CP}^2) = 1 = -\tau(\overline{\mathbb{CP}}^2).$$

Quite generally, a four-manifold with non-vanishing Hirzebruch signature cannot admit an orientation-reversing diffeomorphism. Now consider, for example the connected sum of $K3$ with 12 copies of $S^1 \times S^3$. This has vanishing Euler characteristic and signature 16. It therefore admits no total orientation-reversing diffeomorphism but the Lorentz structure g^L is homotopic to the time-reversed Lorentz structure. We do not know, however, whether there exist diffeomorphisms (connected to the identity or not) which will produce this time-reversal.

6. Generalized Taub-NUT Spacetimes

The four-dimensional Taub-NUT solution of Einstein's vacuum equations has provided many examples of the possible exotic behaviour of Lorentzian metrics. In this section we provide a family of higher-dimensional examples, based on some work by Bais and Batenberg [3] on the associated Riemannian metrics, which serve to illustrate our general results.

Suppose $\{\mathcal{B}, g^{\mathcal{B}}, \omega^{\mathcal{B}}\}$ is a $2p$-dimensional Einstein-Kähler manifold with Kähler form $\omega^{\mathcal{B}}$ which obeys the Dirac quantization condition, i.e., it represents an integral class

$$\left[\frac{1}{2\pi}\omega^{\mathcal{B}}\right] \in H_2(\mathcal{B}; \mathbb{Z})$$

Then $\omega^{\mathcal{B}}$ may be thought of as the curvature of an S^1 bundle over $\mathcal{B}$. Let

$$e^0 = dt + A$$

where $0 \leq t < 2\pi$ be a coordinate on the S^1 fibre and A the connection such that:

$$dA = \omega^{\mathcal{B}}$$

Then the $(2p + 2)$-dimensional time-orientable Lorentzian metric

$$F^{-1}(r)dr^2 + (r^2 + N^2)g^{\mathcal{B}} - 4N^2 F(r)e^0 \otimes e^0$$

is Ricci flat, provided

$$F(r) = \frac{r}{(r^2 + N^2)^p} \int^r (s^2 + N^2)^p \frac{ds}{s^2}$$

The function $F(r)$ contains two arbitrary constants, the generalized 'NUT' charge N and an arbitary constant of integration. If $p = 1$ then $\{\mathcal{B}, g^{\mathcal{B}}, \omega^{\mathcal{B}}\}$ is $\mathbb{CP}^1 \equiv S^2$, the S^1 bundle is S^3 and we recover the usual Taub-NUT solution. Indeed, when $p = 1$ we have

$$F(r) = \frac{r}{(r^2 + N^2)}(r - \frac{N^2}{r} - 2m)$$

where m is a constant of integration. One now recovers the usual Taub-NUT metric [5] with A = $\cos\theta d\phi$, t = ψ and r = t.

For higher values of p one finds that

$$
\begin{aligned}
F(r) &= \frac{1}{(r^2 + N^2)}(\frac{r^2 p}{(2p - 1)} + \frac{pN^2 r^2 p - 2}{(2p - 3)} + ... - N^2 p - 2mr) \\
&= \frac{1}{(r^2 + N^2)}(\frac{r^2}{(2p - 1)}P(r) - N^2 p - 2mr)
\end{aligned}
$$

where $P(r)$ is a polynomial of degree $2(p - 1)$ containing only even powers of r, all of whose coefficients are positive. It follows that the numerator of $F(r)$ has just two real roots.

In these higher p generalisations we can choose $\{\mathcal{B}, g^{\mathcal{B}}, \omega^{\mathcal{B}}\}$ to be $\mathbb{CP}^p$ and then the S^1 bundle becomes S^{2p+1} with its standard Hopf fibration. In this case, the isometry group of the spacetime is $U(p + 1)$ which acts transitively on S^{2p+1} and contains a $U(1)$ factor acting as time translations. The group acting on the base $\mathcal{B}$ is $SU(2p)/\mathbb{Z}_{2p}$.

The resulting $(2p + 2)$-dimensional spacetime may be thought of as a time-orientable two-plane bundle over $\mathcal{B}$ carrying an $SO(2)$-invariant Lorentzian metric on the fibres with local coordinates t, r. Its structure is independent of the particular metric on the base $\mathcal{B}$. Because the numerator of $F(r)$ has only two real roots the structure is qualitatively the same as that of the usual four-dimensional Taub-NUT case. In particular, the Penrose diagram is the same as that shown on page 177 of [5]. Note that if the constant of integration m is chosen to vanish, the metric has an additional discrete isometry r $\rightarrow$ $-$r, interchanging different asymptotic regions.

From the point of view of this paper, we are interested in whether one can find a diffeomorphism which reverses the time-orientation. If we consider the case when $\mathcal{B} \equiv \mathbb{CP}^p$, and we confine ourselves to diffeomorphisms keeping the coordinate r fixed, then we are in the same position as above in our discussion of the odd-dimensional spheres S^{2p+1}. Thus if p is odd, we can and if p is even, we cannot reverse the sense of time by means of a diffeomorphism in the identity component $\mathrm{Diff}_0(\mathcal{M})$. Presumably this means that there is no invariant significance in the sign of the NUT charge N if p is odd but there may be if p is even. *Presumably therefore if p is odd, as it is in the usual four-dimensional case, then a Taub-NUT solution should be considered as its own anti-particle.*

7. Time-Reversal for Dynamical Systems

The topological ideas about time reversal discussed in this lecture may be applied in a different but related context. Suppose we have a finite-dimensional autonomous dynamical system with a compact phase space. That is, we have a symplectic manifold $\{\mathcal{M}_{2r}, \omega\}$ with symplectic form ω and Hamiltonian vector field

$$\mathbf{H} = \omega^{-1} dH.$$

Now time reversal is an *anti-symplectic involution*

$$f : \mathcal{M}_{2r} \rightarrow \mathcal{M}_{2r}; \ f^2 = \mathrm{id}$$

such that

$$f^* \omega = -\omega.$$

and

$$f^* H = H$$

and therefore

$$f_* \mathbf{H} = -\mathbf{H}$$

The standard (non-compact) example is of course $\mathbb{R}^{2r} \equiv T^*(\mathbb{R}^r)$ for which $f : (\mathbf{q}, \mathbf{p}) \rightarrow (\mathbf{q}, -\mathbf{p})$. To get a compact example, one may replace $\mathbb{R}^r$ by any configuration space manifold $\mathcal{Q}$. If $\mathcal{Q}$ were compact and we had a Hamiltonian action of some symmetry group G, we might pass to the symplectic quotient which might be compact.

Since the r-th power:

$$\omega \wedge \ldots \wedge \omega$$

252

defines a volume form on $\mathcal{M}_{2r}$, time reversal is orientation-reversing if r is odd and orientation- preserving if r is even.

Thus if r is odd, it cannot live in the identity component $\mathrm{Diff}_0(\mathcal{M}_{2r})$. Therefore if r is odd and $\mathcal{M}_{2r}$ is irrevesible, then no such f can exist. *Thus if such an $\{\mathcal{M}_{2r}, \omega\}$ exists, it would mean that no dynamical sytem on this phase space, whatever its Hamiltonian function, could be invariant under time reversal!*

8. Some References

[1] P.C.W. Davies, *Nat. Phys. Sci.* **240**, 3–5 (1972)

[2] T. Gold, *Amer. Journ. Phys.*, **30**, 403 (1962)

[3] S. Bais and P. Batenberg, *Nucl. Phys.* **B253**, 162 (1985)

[4] M.C. Albrow, *Nat. Phys. Sci.* **241**, 56–57 (1973)

[5] S.W. Hawking and G.F.R. Ellis, *The large scale structure of spacetime*, CUP Cambridge (1973)

[6] A. Chamblin, *Comm. Math. Phys.* **164**, No. 1, pgs. 65–87 (1994).

[7] B.S. Kay, *Rev. Math. Phys.*, *Special Issue*, 167–195 (1992)

[8] K.K. Lee, *Gen. Rel. Grav.* **4**, 421–433 (1973)

[9] K.K. Lee, *Gen. Rel. Grav.* **5**, 239–242 (1973)

[10] K.T. Chen, *Proc. Amer. Math. Soc.* **26**, 196 (1970)

[11] P.A. Smith, *Ann. Math.* **37**, 526 (1936)

[12] J. Hempel, *Ann. Math. Studies* **86**, Princeton Univ. Press, Princeton (1976)

[13] H. Seifert and W. Threlfall, *A Textbook of Topology*, Academic Press (1980)

[14] E. Thomas, *Bull. Amer. Math. Soc.* **75**, 643–668 (1969)

[15] M.F. Atiyah, *Arbeitsgemeinschaft für Forschung des Landes Nordrhein-Westfalen*, Heft 200

[16] G.W. Gibbons and H.J. Pohle, *Nucl. Phys. B* **410**, 117–142 (1993)

[17] G.W. Gibbons, *Int. Journ. of Mod. Physics D* **3**, No. 1 (1994)

[18] A.D. Sakharov, *Sov. Phys. JETP* **60**, 214 (1984)

[19] A. Chamblin and G.W. Gibbons, *Class. Quant. Grav.* **12**, No. 9, 2243 (1995)

[20] J. Friedman, *Class. Quant. Grav.* **12**, No. 9 2231 (1995)

[21] J. Friedman and A. Higuchi, gr-qc # 9505035

[22] M. Gell-Mann and J. Hartle, *Complexity, Entropy, and the Physics of Information, SFI Studies in the Science of Complexity, Vol. VIII*, pages 425-458. Addison-Wesley, Reading (1990)

[23] A. Vilenkin, *Phys. Rev. D* **27**, 2848 (1983)

[24] J. Milnor and D. Husemoller, *Symmetric Bilinear Forms*, Springer-Verlag (1973)

[25] G.W. Gibbons, *Nucl. Phys. B* **291**, 497 (1986)

[26] W. Massey, *Algebraic Topology, An Introduction*, Harcourt, Brace and World (1967)

[27] E. Witten, *Nucl. Phys. B* **195**, 481 (1982)

[28] F. Hirzebruch and H. Hopf, *Math. Ann.* **136**, 156–172 (1958)

POLYMER GEOMETRY AT PLANCK SCALE
AND QUANTUM EINSTEIN EQUATIONS

ABHAY ASHTEKAR

Center for Gravitational Physics and Geometry
Physics Department, Penn State, University Park, PA 16802

1. Introduction

It is well known that quantum general relativity is perturbatively non-renorma-lizable. Particle theorists often take this to be a sufficient reason to abandon general relativity and seek an alternative which has a better ultraviolet behavior in pertur-bation theory. However, one is by no means forced to this route. For, there do exist a number of field theories which are perturbatively non-renormalizable but are *exactly soluble*. An outstanding example is the Gross-Neveau model in 3 dimensions, $(GN)_3$, which was recently shown to be exactly soluble rigorously [1]. Furthermore, the model does not exhibit any mathematical pathologies. For example, it was at first conjectured that the Wightman functions of a non-renormalizable theory would have a worse mathematical behavior. The solution to $(GN)_3$ showed that this is not the case; as in familiar renormalizable theories, they are tempered distributions. Thus, one can argue that, from a structural viewpoint, perturbative renormalizabil-ity is a luxury even in Minkowskian quantum field theories. Of course, it serves as a powerful guiding principle for selecting physically interesting theories since it ensures that the predictions of the theory at a certain length scale are independent of the potential complications at much smaller scales. But it is *not* a consistency check on the mathematical viability of a theory. Furthermore, in quantum gravity, one is interested precisely in the physics of the Planck scale; the short-distance com-plications are now the issues of primary interest. Therefore, it seems inappropriate to elevate perturbative renormalizability to a viability criterion.

Even if one accepts this premise, however, one is led to ask: Are there specific reasons that suggest that quantum general relativity may exist at a non-perturbative level? The answer, I believe, is in the affirmative. There are growing indications from a number of different directions –computer simulations, canonical quantization and string theory– that the quantum geometry of space-time would be quite different from classical geometry [2,4]. Specifically, in this review we will see that, in a rather wide class of *non-perturbative* quantum gravity theories, the fundamental excitations of the gravitational field are one dimensional rather than three; they resemble one-dimensional networks rather than three dimensional waves. At the Planck scale, geometry has a close similarity with polymers and the three dimensional continuum arises only as a "coarse-grained" approximation. Perturbative treatments, on the other hand, assume the validity of a continuum picture at *all* scales. The ultraviolet

255

N. Sánchez and A. Zichichi (eds.), String Gravity and Physics at the Planck Energy Scale, 255–276.
© *1996 Kluwer Academic Publishers.*

problems one encounters may simply be a consequence of the fact that the true microscopic structure of space-time is captured so poorly in these treatments. Put differently, if the continuum picture is replaced by a more faithful one, the "effective dimension" of space-time could be smaller than four, whence the theory could have a much better behavior non-perturbatively.

In this article, I will accept this premise and consider a non-perturbative quantization of general relativity. Our approach will differ from the traditional methods of constructing quantum field theories in a number of ways. First, we will not use any background metric or connection or indeed any background field. In *this* sense all our constructions will respect the underlying diffeomorphism invariance of the theory. Second, we will always work in the continuum. For actual computations e.g., of the spectra of interesting operators, we will often introduce "floating lattices" or graphs. But these will only serve as computational devices in an *already defined* continuum theory. Third, in the main constructions, we will refer only to the Planck scale; a "macroscopic" scale will not feature in the regularization of operators or construction of the theory itself[1]. In particular, there will be no cut-offs at a fundamental level.

In the light of the text-book treatments of (flat space) quantum field theories, these features may seem surprising. Indeed, if we do not follow one of these standard methods, what then is our strategy? In general terms, the idea is to first re-examine the quantization problem from a somewhat broader perspective. One constructs a suitable algebra of functions on the classical phase space, promotes it to an operator algebra and then seeks representations of this algebra by operators on a Hilbert space. That is, one goes to the "root" of the quantization problem. Due to the presence of an infinite number of degrees of freedom, this problem is difficult in any field theory. For flat space field theories, the standard treatments provide strategies to attack this problem using the powerful machinery associated with the renormalization group. In the present context, where is there is no background metric (or any other field), at least at first sight, these strategies seem not to be as well-suited. What we need is a new functional calculus that respects the diffeomorphism invariance of the underlying theory. Our strategy is to first develop such a calculus and then use it to construct the Hilbert spaces of states and to regularize physically interesting operators thereon, directly in the continuum. The resulting Hilbert spaces and operators are somewhat unconventional. In particular, the fundamental excitations can not be identified with gravitons. To particle physicists, this may seem surprising at first. However, on further thought, one realizes that this is precisely what one should expect: after all gravitons are spin-2 excitations on a Minkowskian background and should therefore arise only as approximate notions in any fully non-perturbative treatment. There already exist detailed re-

---

[1] Such scales may be needed to probe the physical meaning of the theory and will then arise from specific physical problems. For instance, one can seek semi-classical quantum states that, when coarse-grained on, say, the weak interaction scale, reproduce classical geometries. The problem then naturally provides the scale of $\approx 10^{-17}$cm. In the basic theory itself, these "macroscopic" scales do not feature.

sults which indicate how this can come about [5]. However the relation between our approach and the standard constructions based on the renormalization group techniques remains unclear; it would be extremely interesting to investigate this issue in detail.

The general approach followed in this review is being pursued by a large number of researchers in about a dozen different groups. I will not attempt to present a comprehensive or even a systematic survey. Rather, I will focus only on some of the recent mathematical developments. These were motivated in part by earlier exploratory work by a number of colleagues, especially Jacobson, Rovelli and Smolin [6,7]. For brevity, however, I will forego motivational remarks and attempt to present the final picture in a concise fashion focusing on the recent mathematical developments. (For reviews of the earlier work, see [8-11].)

2. Difficulties and Strategies

The non-perturbative approach I wish to discuss is based on canonical quantization. The canonical formulation of general relativity was first obtained in the late fifties and early sixties through a series of papers by Bergmann, Dirac and Arnowitt, Deser and Misner. In this formulation, general relativity arises as a dynamical theory of 3-metrics. The framework was therefore named *geometrodynamics* by Wheeler and used as a basis for canonical quantization both by him and his associates and by Bergmann and his collaborators. The framework of geometrodynamics has the advantage that classical relativists have a great deal of geometrical intuition and physical insight into the nature of the basic variables –3-metrics g_{ab} and extrinsic curvatures K_{ab}. For these reasons, the framework has played a dominant role, e.g., in numerical relativity. However, it also has two important drawbacks. First, it sets the mathematical treatment of general relativity quite far from that of theories of other interactions where the basic dynamical variables are connections rather than metrics. Second, the equations of the theory are rather complicated in terms of metrics and extrinsic curvatures; being non-polynomial, they are difficult to carry over to quantum theory with a reasonable degree of mathematical precision.

For example, consider the standard Wheeler-DeWitt equation:

$$[\sqrt{\tfrac{G\hbar^2}{g}}(g^{ab}g^{cd} - \tfrac{1}{2}g^{ac}g^{bd}) \, \frac{\delta}{\delta g_{ac}} \frac{\delta}{\delta g_{bd}} - \sqrt{\tfrac{g}{G\hbar^2}} \, R(g)] \circ \Psi(g) = 0, \tag{1}$$

where g is the determinant of the 3-metric g_{ab} and R its scalar curvature. As is often emphasized, since the kinetic term involves products of functional derivatives evaluated at the same point, it is ill-defined. However, there are also other, deeper problems. These arise because, in field theory, the quantum configuration space –the domain space of wave functions Ψ– is larger than the classical configuration space. While we can restrict ourselves to suitably smooth fields in the classical theory, in quantum field theory, we are forced to allow distributional field configurations. Indeed, even in the free field theories in Minkowski space, the Gaussian measure that provides the inner product is concentrated on genuine distributions. This is the reason why in quantum theory fields arise as operator-valued distributions. One would expect that the situation would be at least as bad in quantum gravity. If so, even the products of the 3-metrics that appear in front of the momenta and the meaning of the scalar curvature in the potential term are obscure. The left hand side of the Wheeler-DeWitt equation is seriously ill-defined and must be regularized appropriately.

However, as I just said, the problem of distributional configurations arises already in the free field theory in Minkowski space-time. There, we do know how to regularize physically interesting operators. So, why can we not just apply those techniques in the present context? The problem is that those techniques are tied to the presence of a background Minkowski metric. The covariance of the Gaussian measure, for example, is constructed from the Laplacian operator on a space-like plane defined by the induced metric and normal ordering and point-splitting regularizations also make use of the background geometry. In the present case, we do *not* have background fields at our disposal. We therefore need to find another avenue. What is needed is a suitable functional calculus –integral and differential– that respects the diffeomorphism invariance of the theory.

What space are we to develop this functional calculus on? Recall first that, in the canonical approach to diffeomorphism invariant theories such as general relativity or supergravity, the key mathematical problem is that of formulating and solving the quantum constraints. (In Minkowskian quantum field theories, the analogous problem is that of defining the regularized quantum Hamiltonian operator.) It is therefore natural to work with variables which, in the classical theory, simplify the form of the constraints. It turns out that, from this perspective, connections are better suited than metrics [12].

To see this, recall first that, in geometrodynamics, we can choose as our canonical pair, the fields (E_i^a, K_a^i) where E_i^a is a triad (with density weight one) and K_a^i, the extrinsic curvature. Here a refers to the tangent space of the 3-manifold and i is the internal $SO(3)$ –or, $SU(2)$, if we wish to consider spinorial matter– index. The triad is the square-root of the metric in the sense that $E_i^a E^{bi} =: g g^{ab}$, where g is the determinant of the covariant 3-metric g_{ab}, and K_a^i is related to the extrinsic curvature K_{ab} via: $K_a^i = (1/\sqrt{g})K_{ab}E^{bi}$. Let us make a (*real*) transformation:

$$(E_i^a, K_a^i) \mapsto (A_a^i := \Gamma_a^i - K_a^i, E_i^a), \tag{2}$$

where Γ_a^i is the spin connection determined by the triad. It is not difficult to check that this is a canonical transformation on the real phase space [12,13]. It will be convenient to regard A_a^i as the configuration variable and E_i^a as the conjugate momentum so that the phase space has the same structure as in the $SU(2)$ Yang-Mills theory. Let us begin by simply writing down the simplest equations we can, *without any reference to a background field*, using these variables. They are:

$$\mathcal{G}_i := D_a E_i^a = 0$$
$$\mathcal{V}_b := E_i^a F_{ab}^i \equiv \text{Tr } E \times B = 0 \tag{3}$$
$$\mathcal{S} := \epsilon^{ijk} E_i^a E_j^b F_{abk} \equiv \text{Tr } E \cdot E \times B = 0$$

These are the "simplest" equations in the following sense: Among non-trivial gauge covariant expressions, the left side of the first equation is the only one which is at most linear in E and A; that of the second is the only one that is at most linear in E and quadratic in A; and, that of the third is the only one that is at most quadratic in each of E and A. (There is no gauge covariant expression which is linear in A and at most quadratic in E.) Somewhat surprisingly, the first two equations provide us precisely with the 6 of 7 constraints of general relativity: the Gauss constraint $\mathcal{G}_i$ which generates internal frame rotations and the vector constraint $\mathcal{V}_b$ that generates spatial diffeomorphisms [12]. The third equation is almost –but not quite– the last (scalar or Hamiltonian) constraint we seek. To see this, let us translate it in to geometrodynamical variables. It then reduces to the familiar *Euclidean* Hamiltonian constraint [13]

$$R + K^{ab}K_{ab} - K^2 = 0, \tag{3a}$$

rather than the Lorentzian Hamiltonian constraint

$$R - K^{ab}K_{ab} + K^2 = 0, \tag{3b}$$

which, as one might expect, carries opposite signs in front of terms which are quadratic in momenta. (Intuitively, the "Wick" rotation should map the momentum to i times the momentum.) The Gauss and the vector constraints are insensitive to the signature.

To get the Lorentzian theory, the most straightforward avenue is to consider *complex* connections ${}^c A_a{}^i := \Gamma_a^i - iK_a^i$. This avenue has been pursued vigorously in the literature [8-11]. In this article, however, I will present another possibility which has arisen from a recent idea of Thomas Thiemann's and which may allow one to pass from the solutions to the Euclidean quantum constraints to the Lorentzian ones via a generalized Wick transformation. This strategy will be discussed in Section 5. For the moment, let us note only that $SU(2)$ connections can be regarded as the configuration variables also in general relativity. Thus, the classical configuration space is $\mathcal{A}/\mathcal{G}$, the space of suitably smooth connections modulo gauge transformations. Our task is to find the corresponding quantum configuration space and to develop functional calculus on it.

I will conclude this discussion with a curiosity: The Euclidean signature seems to be easier to handle in other approaches as well, in particular, Connes' non-commutative geometry and string theory. In the Connes' "geometrization" of the standard model, one is naturally led to work with the Euclidean signature and the extension of the framework to the Lorentzian regime is still to be worked out. In string theory, while one can allow the target space geometry to be Lorentzian, in all the discussions that I am aware of, the world sheet is kept Euclidean. In perturbation theory off Minkowskian backgrounds, as far as the S-matrix theory is concerned, this seems like an obvious extension of the strategy one adopts in field theory. However, in non-perturbative contexts, such as the discussion of mirror symmetry, restriction to Riemannian world-sheets appears artificial from a physical perspective. As in the above discussion of the Hamiltonian constraint, in both these examples, one needs an appropriate, generalized Wick transform.

3. Tools

The classical configuration space $\mathcal{A}/\mathcal{G}$ is the quotient of the space $\mathcal{A}$ of suitably smooth connections by the group $\mathcal{G}$ of local gauge transformations. $\mathcal{A}/\mathcal{G}$ can be endowed with topology in a standard way, using a Sobolev norm on the space $\mathcal{A}$. The key question now is: What is the *quantum* configuration space? More precisely, in the connection representation, what is the domain space of quantum wave functions? In quantum mechanics –i.e., in the quantum theory of systems with a finite number of degrees of freedom– the classical and the quantum configuration spaces agree. In field theory, as noted in Section 1, the quantum configuration space is a substantial enlargement of the classical one; in scalar field theories, for example, although the classical configurations are smooth (say C^2) functions on a $t = \text{const}$ slice, the quantum configuration space consists of all tempered distributions. Furthermore, this enlargement is not a mere technicality: the set of smooth configurations is of zero measure with respect to the Gaussian measure that determines the inner product! The regularization problems of quantum field theory can be traced back to this fact.

The text-book treatments for the construction of quantum configuration spaces are not directly applicable in the present case because they are geared to the case in which classical configuration spaces are linear. Nonetheless, a natural strategy *is* available [14]. Recall first that the Wilson-loop functions, $T_\alpha(A) := \frac{1}{2}\text{Tr}\mathcal{P} \exp \oint A.dl$, form a natural (over)complete set of functions on $\mathcal{A}/\mathcal{G}$ in the sense that they suffice to separate the points of $\mathcal{A}/\mathcal{G}$. (Trace is taken in the fundamental representation.) They are thus natural candidates for our configuration variables[2]. It is straightforward to construct a C^*-algebra generated by these T_α. It is called the *holonomy algebra* and denoted by $\overline{\mathcal{HA}}$. The first step in the quantization procedure is to develop the representation theory of $\overline{\mathcal{HA}}$. A natural candidate for the quantum configuration space will arise from this theory.

Since elements of this algebra are all configuration variables, the algebra is Abelian and it is equipped with an identity, the Wilson loop function associated with the trivial (point) loop. Now, the representation theory of Abelian C^*-algebras with identity has been developed in detail by Gel'fand and Naimark. One of their basic results is that any such C^*-algebra is naturally isomorphic with the C^*-algebra of *all* continuous functions on a compact, Hausdorff space, called the spectrum of the algebra. Thus, every element a of the algebra is canonically represented by a concrete function $\mathring{a}$ on the spectrum. In our case, the algebra $\overline{\mathcal{HA}}$ is smaller than the algebra of all bounded continuous functions on $\mathcal{A}/\mathcal{G}$ and the spectrum turns out

---

[2] All loops will be based at some arbitrarily chosen but fixed point on Σ. For technical reasons, we will restrict ourselves to piecewise analytic loops. Some of the key results have been extended to the smooth category by Baez and Sawin [15].

to be larger than $\mathcal{A}/\mathcal{G}$. However, since the elements of $\overline{\mathcal{HA}}$ suffice to separate points of $\mathcal{A}/\mathcal{G}$, it follows that $\mathcal{A}/\mathcal{G}$ is densely embedded in the spectrum [16]. Thus, we can regard the spectrum as a completion of $\mathcal{A}/\mathcal{G}$. To emphasize this point, we will denote it by $\overline{\mathcal{A}/\mathcal{G}}$.

Since $\overline{\mathcal{A}/\mathcal{G}}$ is a compact Hausdorff space, using the Riesz representation theorem, one can conclude that every cyclic representation of $\overline{\mathcal{HA}}$ by bounded operators on a Hilbert space is of the following type: The Hilbert space is $L^2(\overline{\mathcal{A}/\mathcal{G}}, d\mu)$ for some measure μ and the operators $\hat{T}_\alpha$ act via:

$$(\hat{T}_\alpha \circ \Psi)(\bar{A}) = \check{T}_\alpha(\bar{A})\Psi(\bar{A}), \tag{4}$$

where the function $\check{W}$ on $\overline{\mathcal{A}/\mathcal{G}}$ is the Gel'fand transform of T_α. Thus, $\overline{\mathcal{A}/\mathcal{G}}$ serves as an universal domain space for quantum states; it is therefore the *quantum configuration space*.

The details of this construction *will not be needed* in what follows. One just needs to note that a natural completion $\overline{\mathcal{A}/\mathcal{G}}$ of $\mathcal{A}/\mathcal{G}$ will serve as the quantum configuration space and that, even though $\mathcal{A}/\mathcal{G}$ is a rather complicated, infinite dimensional space, $\overline{\mathcal{A}/\mathcal{G}}$ is compact.

Our task then is to develop integral and differential calculus on $\overline{\mathcal{A}/\mathcal{G}}$: measures and integration theory are needed to specify the inner-product and differential operators are needed to define observables. The task has been carried out in a series of papers [17-24]. In essence, this development was possible because there are three distinct ways of characterizing $\overline{\mathcal{A}/\mathcal{G}}$, each illuminating a specific aspect of its structure: first, as the Gel'fand spectrum of the holonomy algebra [14]; second, as the space of homomorphisms from the so-called hoop group[3] to the gauge group $SU(2)$ [17]; and third, as a projective limit of configuration spaces associated with graphs in Σ [19,20]. As a result, $\overline{\mathcal{A}/\mathcal{G}}$ has structure which is much richer than what one might expect at first.

That integral calculus can be developed [17-20] is perhaps not surprising: after all, $\overline{\mathcal{A}/\mathcal{G}}$ is a compact, Hausdorff space and therefore admits regular measures. The development of a differential calculus, on the other hand, seems hopeless at first since Gel'fand spectra are only topological spaces; the Gel'fand theory does *not* endow them with a differential structure. In particular, it seems difficult to convert $\overline{\mathcal{A}/\mathcal{G}}$ into an infinite dimensional manifold. However, *because our C^*-algebra is special* –the holonomies have a natural geometrical meaning– $\overline{\mathcal{A}/\mathcal{G}}$ can be regarded as the projective limit of finite dimensional manifolds (the third characterization above). This enables one to introduce the notion of smooth functions on $\overline{\mathcal{A}/\mathcal{G}}$ and use it to

---

[3] The hoop group $\mathcal{HG}$ of the 3-manifold Σ is defined as follows. Consider the space of piecewise analytic, closed loops in Σ which begin and end at the fixed point x. Regard two loops as equivalent if the holonomy of any (smooth) $SU(2)$ connection around one equals that around the other. The space of equivalence classes then has a natural group structure, groups corresponding to different base points being isomorphic. This group is $\mathcal{HG}$. While the definition refers to a specific gauge group, the resulting $\mathcal{HG}$ turns out to be largely independent of this choice: there are only two hoop groups; one Abelian (associated with Abelian gauge groups) and the other non-Abelian [17].

introduce forms, vector fields, and a number of differential and integral operators *without any reference to a background metric* on the underlying 3-manifold Σ [21]. We will see some illustrations of how this is achieved in the next section.

To proceed further in a concrete fashion, one has to choose a specific measure on $\overline{A/G}$ and work in the representation of the holonomy algebra it provides. The algebra itself imposes no restriction on this choice; for *any* regular measure μ, the Wilson loop operators $\hat{T}_\alpha$ are bounded *self-adjoint* operators on the resulting Hilbert space $H \equiv L^2(\overline{A/G}, d\mu)$. It is the requirement that the *momentum* operators be self-adjoint that restricts the measure. (A similar situation occurs already in non-relativistic quantum mechanics: while the position operator $\hat{X}$ is self-adjoint on $L^2(R, f dx)$ for any (regular) function f, the momentum operator $\hat{P} \equiv -i\hbar d/dx$ is self adjoint only if f is a constant. Thus, it is the self-adjointness of the explicitly defined momentum operators that singles out the Lebesgue measure dx.) In our case, the momenta are essentially the electric fields. However, electric fields are not gauge invariant and to obtain an operator which descends to A/G –i.e., has a meaningful action on the space of gauge invariant wave functions– further work is needed[4]. When this is done, one obtains momentum operators that can act on suitably regular functions $\Psi(\bar{A})$ on $\overline{A/G}$. The requirement that these momentum operators be self-adjoint then picks out a measure μ_o on $\overline{A/G}$ [23].

This measure has four important properties [17-19]: i) it is normalized; the total μ_o measure of $\overline{A/G}$ is 1; ii) it is faithful; the integral of every non-negative, continuous function on $\overline{A/G}$ is non-negative and vanishes if and only if the function is identically zero; iii) μ_o is invariant under the (induced action on $\overline{A/G}$ of the) diffeomorphism group on the underlying 3-manifold Σ; and, iv) μ_o is concentrated on genuinely generalized connections; the μ_o-measure of the classical configuration space A/G is zero! The last result is the analog of the standard situation in flat space quantum field theory referred to above; the classical configuration space is topologically dense in the quantum configuration space but measure theoretically sparse.

We can use this measure to construct a Hilbert space $H_o := L^2(\overline{A/G}, d\mu_o)$. This is the space of kinematic states of quantum gravity. It is "kinematic" because we are yet to impose quantum constraints to select physical states which have dynamical information; it is the quantum analog of the *full* phase space of general relativity. To solve quantum dynamics, we have to first regularize quantum constraints and express them as well-defined operators on H_o and then isolate the kernel of these operators. We will turn to this task in Section 5.

We will conclude this section by pointing out two properties of H_o which will

---

[4] The simplest strategy is to use 2-dimensional strips S, parametrized by (σ, τ), which are foliated by closed loops $\tau = $ const., and consider the associated "momentum variables" $P_S(A, E) := \int dS^{ab} \epsilon_{abc} \cdot (\text{Tr} E^c(\sigma, \tau) U(\sigma, \tau))$, where U is the holonomy of A around the loop $\tau = $ const, evaluated at the point (σ, τ) of S. This is a gauge invariant function on the phase space, linear in the momentum E, and depends only on the foliation of S (i.e., not on the details of the parameterization). Since there are no background fields, the Poisson algebra of configuration and momentum variables, T_α and P_S, can be expressed in an elegant fashion, using only the topological properties of loops and strips in 3-dimensions [25].

play an important role in the subsequent discussion.

First, H_o admits [24] an interesting orthonormal basis, obtained by a generalization of Penrose's spin networks [27] (which, although introduced in quite a different context, were also motivated by quantum gravity considerations. See also [26].). A penrose network is a closed graph, each of whose vertices is trivalent (i.e., has three edges), with an assignment of representation of SU(2) (i.e., of a half-integral number j) to each of its edges, such that, if j_1, j_2, j_3 label edges incident at any one vertex, we must have $|j_1 - j_2| \le j_3 \le j_1 + j_2$ (and permutations thereof). Now, each such network N defines a function Ψ_N on $\overline{A/G}$ as follows. First, one can show [18,21] that every (generalized) connection $\bar{A}$ assigns to each edge an element of $SU(2)$, the holonomy. Using this assignment, we can associate a matrix with each edge in the Penrose network. The constraint on the choice of representations ensures that all the resulting matrices can be consistently contracted to produce a number, the value of Ψ_N on the chosen point $\bar{A}$ of $\overline{A/G}$. Remarkably, these states are orthonormal in H_o. However, they are not complete. To obtain a complete set, we have to allow higher valent networks –i.e., the ones which have an arbitrary (but finite) number of edges at any one vertex– assign to each edge a representation of $SU(2)$ *and* to each vertex a suitable contractor (or, an inter-twiner). The resulting space of states can then be used to obtain an orthonormal decomposition of H_o into *finite* dimensional subspaces, each labelled by the network (embedded in Σ) and the assignment of representations [24,23]. (The dimensionality of each subspace is given by the number of independent contractors and one can introduce an orthonormal basis in each subspace using the Schmidt procedure.)

The second property of H_o is that it admits an interesting dense subspace. Fix a closed graph γ (i.e., a collection of vertices joined by analytic edges such that every vertex has at least two edges) in Σ. Every (generalized) connection $\bar{A}$ associates to each edge e_i of γ an element g_i of $SU(2)$. Thus, there is a natural projection from $\overline{A/G}$ to $[SU(2)]^n$, where n is the total number of edges in the graph. Using this projection, we can pull-back to $\overline{A/G}$ functions on $[SU(2)]^n$. These pull-backs,

$$\Psi_\gamma(\bar{A}) := \psi(g_1(\bar{A}), ..., g_n(\bar{A})),\qquad(5)$$

for some smooth function ψ on $[SU(2)]^n$ are called *cylindrical functions* on $\overline{A/G}$. Note that the functions Ψ_γ only know about what the connection $\bar{A}$ is doing at points of Σ which lie in the graph γ. Yet, as we vary the graph and consider more and more vertices and edges, we obtain a bigger and bigger collection of functions on $\overline{A/G}$. They are all square-integrable with respect to μ_o and, furthermore, span a *dense* subspace of H_o. This space $\text{Cyl}(\overline{A/G})$ of all cylindrical functions plays an important role in calculations, rather analogous to that played by the space $C_o^\infty(R)$ of smooth functions of compact support in quantum mechanics on a line. Just as we often first define operators on $C_o^\infty(R)$ and then extend them to (other, better suited dense subspaces of) $L^2(R, dx)$, in the present context, we will often define operators first on $\text{Cyl}(\overline{A/G})$ and then extend them.

These two properties of H_o provide considerable intuition about the nature of quantum states that we have been led to consider. Since an orthonormal basis is provided by networks, it follows that the "elementary excitations" are *1-dimensional* rather than three. They are "loopy" rather than "wavy". A typical excited state looks like a polymer rather than a smooth undulation on flat space. Just as a polymer in a sufficiently complex configuration behaves as if it were a 3-dimensional entity, we will see that the 1-dimensional excitations, if packed densely and superposed coherently, can approximate a 3-dimensional continuum geometry very well. The quantum states we have encountered here are qualitatively different from the ones we come across in Minkowskian quantum field theories. The main reason is the underlying diffeomorphism invariance: In absence of a background geometry it is not possible to introduce the familiar Gaussian measures and associated Fock spaces.

4. Quantum Geometry

The framework constructed in Section 3 provides us with sufficient tools to explore the nature of quantum geometry and, as we will see, the results are quite surprising.

Let us first recall that, in classical general relativity, geometrical observables can be regarded as functions on the phase space. Let us, for example, fix a smooth 2-surface S_o or a 3-dimensional region R_o in the 3-manifold Σ. Then, given any triad field E_i^a, we can assign to S_o an area, $A(S_o)$, and to R_o a volume, $V(R_o)$:

$$A(S_o) := \int_{S_o} |E_i^3 \, E^{3i}|^{\frac{1}{2}} d^2x; \quad \text{and} \quad V(R_o) := \int_{R_o} |\det E|^{\frac{1}{2}} d^3x, \tag{6}$$

where, to simplify the presentation, we have chosen coordinates so that S_o is given by $X_3 = $ const. Thus, $A(S_o)$ and $V(R_o)$ can be regarded as functions on the *full* classical phase space (which happen to depend only on the triad, i.e. happen to be independent of the connection). Our problem now is to promote these functions to operators on the Hilbert space H_o –the quantum analog of the full phase space– and study their properties.

At first sight, the challenge seems unsurmountable. The first obstacle is that even the classical expressions are *non-polynomial* in the triads. Second, our wave functions $\Psi(\bar{A})$ have support on generalized connections which have a "distributional character." Hence, the task of regularizing formal expressions like

$$| - (\frac{\delta}{\delta A_3^i(x)})(\frac{\delta}{\delta A_3^i(x)})|^{\frac{1}{2}}$$

seems formidable. In addition, we do not have a background metric to simplify this task; we thus have the constraint that the final, regularized operators should not depend on any background fields. It turns out, however, that, using the functional calculus on $\overline{A/G}$, all these difficulties can be overcome [28, 23]. (For an approach based on the loop representation, see [29] and, on lattice regularization, see [30]. I should add however that I do not yet know the precise relation to these approaches.)

One begins with the classical expressions, introduces a chart and point splits the triad fields using regulators, i.e., 2-point functions $f_\epsilon(x,y)$ which tend to $\delta^3(x,y)$ in the limit as ϵ tends to zero. One then notices that the resulting expressions can be promoted to operators which have a well-defined action on cylindrical functions on $\overline{A/\mathcal{G}}$. One evaluates this action and, in the final expression, takes the limit $\epsilon \mapsto 0$. The limits are well-defined and yield operators which carry no memory of the background chart or regulators used in the procedure. They map cylindrical functions based on a graph γ to cylindrical functions based on the *same* graph. Let us fix a graph with n edges. Then, given a cylindrical function $\Psi_\gamma(\bar{A})$ which is the pull-back of a function $\psi(g_1,...,g_n)$ of $[SU(2)]^n$, the area operator yields:

$$\hat{A}(S_o) \circ \Psi(\bar{A}) := \ell_P^2 \sum_\alpha | \sum_{I_\alpha J_\alpha} K(I_\alpha, J_\alpha) X_{I_\alpha}^i X_{J_\alpha}^i |^{\frac{1}{2}} \circ \psi(g_1,...,g_n) . \tag{7}$$

Here ℓ_P is the Planck length; α ranges over the vertices of the graph γ which intersect S_o; I_α and J_α denote edges passing through the vertex α; $K(I_\alpha, J_\alpha)$ takes values $0, \pm 1$ depending on the orientation of the two edges relative to S_o; and, X_I^i is the i-th right (left) invariant vector field on the I_α-th copy of the gauge group if the edge I_α is outgoing (incoming) at the vertex α. (Note that since there are three right (left) invariant vector fields on $SU(2)$, the index i runs over $1, 2, 3$.) Each vector field X acts on only one of the arguments in the wave function, i.e., on the copy of the gauge group picked by the associated edge. Similarly, the volume operator is given by:

$$\hat{V}(R_o) \circ \Psi(\bar{A}) := \ell_P^3 \sum_\alpha | \sum_{I_\alpha, J_\alpha, K_\alpha} \epsilon^{ijk} \epsilon(I_\alpha, J_\alpha, K_\alpha) X_{I_\alpha}^i X_{J_\alpha}^j X_{K_\alpha}^k |^{\frac{1}{2}} \circ \psi(g_1,...,g_n) , \tag{8}$$

where the first sum now is over vertices which lie in the region R and $\epsilon(I_\alpha, J_\alpha, K_\alpha)$ is 0 if the three edges are linearly dependent at the vertex α and otherwise ± 1 depending on the orientation they define. Thus, the two operators are well-defined on the space $\mathrm{Cyl}(\overline{A/\mathcal{G}})$ of cylindrical functions on $\overline{A/\mathcal{G}}$. Recall, however, that this space is dense in H_o. Using this fact and the properties of right and left invariant vector fields on $SU(2)$, one can show that they admit unique self-adjoint extensions on H_o.

For our purposes here, *the details of these operators are not important*. It would be sufficient to note just that closed form expressions of the fully regulated operators are available and they involve the actions of right and left invariant vector fields on the appropriate copies of $SU(2)$. Because these actions are completely understood, in practice, it is rather straightforward to compute how these operators act on cylindrical states associated with specific graphs.

A number of remarks are in order.

1. Given that the expressions of the classical observables are already rather involved, why are the expressions of the quantum operators relatively simple? The main reason is that the requirement of diffeomorphism invariance –i.e., absence of background fields– severely restricts the possible operators. Consider, for example, the area operator. Given a graph and a surface, the obvious diffeomorphism invariant notion is that of an intersection of the graph with the surface. Not surprisingly,

the expression of the operator is a sum over intersections. At each intersection, the graph has a number of edges. So, all that the operator can do is to act on the copies of groups associated with these edges. That is, thanks to diffeomorphism invariance, the action of operators is reduced to simple algebraic operations in the representation theory of $SU(2)$!

2. Since the two operators are self-adjoint, one can compute their spectra. *They are purely discrete!* Each of the Penrose spin network state is an eigenstate of *both* the operators. The area operators reduce to a sum of $SU(2)$-Laplacians, each of which has the familiar eigenvalues $\sqrt{j(j+1)}$. The eigenvalue of the volume operator on these states is, however, zero [30, 28]. Thus, higher valent spin networks are essential. Both operators again leave the (finite dimensional) Hilbert space associated with any of these networks invariant. They can therefore be diagonalized separately on each network. (Already in the 4-valent case, there exist states with non-zero volume.) This overall picture shows that *the microscopic geometry at Planck scale is very different from what the continuum picture suggests.* As we argued in the Introduction, this basic fact may well be the essential reason why the perturbation theory fails.

3. The geometrical observables can be defined on the full phase space of the classical theory; one need not restrict oneself to the constraint surface. Similarly, in the quantum theory, the analogous operators have been defined on the kinematic Hilbert space H_o, before imposing quantum constraints. The results are thus robust; they are not sensitive to the details of quantum dynamics. For example, they will continue to hold also in supergravity, and more generally, in any theory in which the triads and connections feature as a basic canonical pair.

4. There is, however, a subtlety. To see this, recall a basic difference between quantum mechanics and quantum field theory. In quantum mechanics, kinematics and dynamics are largely decoupled in the sense the same representation of the canonical commutation relations supports all interesting Hamiltonians. In quantum field theory, on the other hand, kinematics and dynamics are always weakly coupled: there are infinitely many *inequivalent* representations of the canonical commutation relations and, in general, different Hamiltonians are meaningful on different representations. In the present case, then, the key question is whether the quantum constraints of a given theory can be meaningfully imposed on H_o. Results to date indicate (but do not yet prove) that this is likely to be the case for constraints of general relativity. My general expectation is that this will also be the case for a class of theories such as supergravity which are "near general relativity" as far as geometry is concerned. What we have here is a glimpse into the nature of quantum geometry underlying *this* class of quantum gravity theories.

5. Note that $\hat{A}(S)$ and $\hat{V}(R)$ have been defined for *any* surface S and *any* region R in Σ and will therefore not be Dirac observables; they are not expected to commute with the quantum constraints. To obtain Dirac observables, one would have to specify S and R *intrinsically*. (For example, if we had only the Gauss and the diffeomorphism constraints, the total volume of Σ would be a Dirac observable.) A natural strategy is to specify S and R using matter fields (see, e.g.,[31]). In view of the Hamiltonian constraint, the problem of providing an explicit specification of

this type is extremely difficult. However, *if* a surface S or a region R were specified in this manner, Eqs (6) and (7) will provide the expressions of the associated area and volume operators.

6. Note that the expressions of the operators do not involve any free renormalization constants; they are completely unambiguous. For the two operators considered here, this is just a fact of calculations. However, there exist heuristic arguments that indicate that, in a more general context, if the end result of such a regularization procedure is an operator which has no background dependence, then these final operators should be free of arbitrary constants [10]. This reasoning is borne out in the construction of another geometric operator, associated with the classical observable $E(\omega) := \int d^3x (E_i^a E^{bi} \omega_a \omega_b)^{\frac{1}{2}}$, where ω is a smooth 1-form of compact support on Σ. (As in the case of area and volume, a square-root is essential to obtain a density of weight one which one can then integrate without reference to any background metric or volume element. Note incidentally that, since ω_a is arbitrary, classically, the functional $E(\omega)$ has the full information about the metric.) It is not known if there are other well-defined geometrical operators, e.g. one which carries information about the length functional directly.

5. Quantum Einstein Equations

We now turn to quantum dynamics. In the canonical approach, this is captured in the quantum constraints. Our task then is the following: i) Write the regulated constraints as operators on the kinematic Hilbert space H_o; ii) Check for anomalies; iii) If there are none, "solve" the quantum constraints; and, iv) On the space of solutions, introduce an appropriate Hilbert space structure. If all this can be achieved, one would have a coherent mathematical framework; one would say that quantum general relativity does exist non-perturbatively. One would, of course, still have to devise suitable approximation schemes to extract physical predictions.

There is however a key difficulty, of quite a general sort, associated with quantum constraints that needs to be resolved before we proceed further. To illustrate it, let us consider the example of a free relativistic particle in Minkowski space. Then, the classical configuration space is the 4-dimensional Minkowski space-time, the phase space is the cotangent bundle over it and there is a single *dynamical* constraint $C(x,p) := \eta^{ab} p_a p_b + m^2 = 0$. In quantum theory then, we can take $H_o := L^2(R^4, d^4x)$ to be the fiducial, *kinematical* Hilbert space. The constraint can be promoted to a self-adjoint operator $\hat{C} := -(\eta^{ab}\partial_a\partial_b - \mu^2)$ on H_o (where $\mu = m/\hbar$). Thus solutions $\Psi(x)$ to the quantum constraint are easily obtained: they solve the Klein-Gordon equation. The problem is that none of the non-zero solutions lies in the kinematical Hilbert space H_o. (This is obvious in the momentum representation where the solutions $\psi(p)$ are distributions $\psi(p) = g(p)\delta(p.p + m^2)$ for some well-behaved function $g(p)$ and therefore not square-integrable in the L^2-norm.) This is a rather general problem; it is in the exceptional case when the group generated by the constraint functions is compact that the solutions Ψ —i.e., physical states— belong to H_o. Thus, in the generic situation, we face two problems: What is the "home" of the physical quantum states? Once the home has been found and space of suitable solutions isolated, how is one to introduce an inner product on this space?

Fortunately, there exists [32] a general approach to both these problem, based on the idea of "averaging over" the orbit of the group generated by the constraints. In the present case, one first considers the 1-parameter group of unitary transformations, $U(\lambda) := \exp i\lambda\hat{C}$, generated by the constraint $\hat{C}$. The idea is to construct solutions by averaging suitable elements of H_o over the 1-parameter group generated by $U(\lambda)$. For the result to be manageable, however, we have to restrict the initial element to a "nice" (dense) subspace S of H_o. Let us choose this S to be the space of smooth test functions of compact support. Then, given any test field f, one can show that $\Psi_f := \int d\lambda U(\lambda) \circ f$ exists as a well-defined distribution over S and solves the quantum constraint. Thus, the natural home for these solutions is S^*, the topological dual to S we began with. Furthermore, we can now naturally define an inner product: $(\Psi_f, \Psi_h) := \Psi_f(h) \equiv \int d^4x\bar{\Psi}_f(x)h(x)$. It turns out that this is the correct inner product. Thus, in this example, the group averaging procedure provides an answer to both our questions. (As far as I know, however, there are no general theorems to ensure a priori that the method is so directly applicable in more general situations. The practical strategy is to apply it to any given case and see if the final inner-product is well-defined, i.e., positive definite.)

The procedure *is* applicable to the vector (or the diffeomorphism) constraint of general relativity [23]. One first promotes the formal "exponentiated forms" of diffeomorphism constraints to well-defined, unitary operators on the kinematical Hilbert space H_o. (These operators are unitary because the measure μ_o is diffeomorphism invariant.) In the second step, one checks for anomalies. In various lattice regularizations, these were encountered essentially because the lattice regularization manifestly breaks the diffeomorphism invariance. Here, we are working directly in the continuum, have a well-defined Hilbert space and unitary operators $U_{\vec{N}}(\lambda)$ thereon, for each analytic vector field $\vec{N}$ on Σ. We can directly check their algebra. *One finds that there are no anomalies.* One can now proceed to the last two steps: solving the constraint and introducing the inner product on the space of physical states. Here we apply the group averaging technique. As one might expect, the role of the dense sub-space S is played by the space of cylindrical functions $\mathrm{Cyl}(\overline{\mathcal{A}/\mathcal{G}})$ and the physical states lie in its topological dual. They are not normalizable with respect to H_o. Nonetheless, using the procedure outlined above, one can endow them with an inner product and obtain a Hilbert space H_d of solutions to the diffeomorphism constraints[5]. Roughly, states in H_d can be labelled by equivalence classes of spin networks, where two are regarded as equivalent if they are related by the action of the diffeomorphism group. This provides a precise formulation of the interplay between knot theory and general relativity that was anticipated by

---

[5] In the analytic category used in this paper, we have been able to obtain an infinite dimensional family of solutions but it is not clear if this class exhausts all physically interesting situations. This problem would disappear if we could extend the entire discussion to the physically better suited smooth category along the lines initiated by Baez and Sawin [15]. For details, see [23].

Rovelli and Smolin already in [7] and also brings out a number of subtleties. Finally, the inner-product on H_d automatically incorporates the "reality conditions" [8,33] correctly. More precisely, the projection to H_d of operators on H_o which are self-adjoint and commute with the diffeomorphism constraints are guaranteed to be self-adjoint on H_d.

The Gauss constraint is already taken care of since we are working on the space $\overline{\mathcal{A}/\mathcal{G}}$ of gauge equivalent (generalized) connections. Alternatively, as indicated in [23], one could first begin with the space $\overline{\mathcal{A}}$ of (generalized) connections, show that there are no anomalies, impose the Gauss constraint using group averaging, and show that the resulting Hilbert space of states is precisely the space H_o that we began with in this paper.

Finally, we come to the difficult Hamiltonian constraint, the analog of the Wheeler-DeWitt equation. There has been considerable work on solutions to this constraint (see, e.g., [34-37]) and several fascinating results have been obtained. For example, some of the solutions have been identified with well-known knot invariants. However, most of this work is of a rather formal nature; for example, it is generally unclear if the underlying theory is Euclidean or Lorentzian. In keeping with the general spirit of this review, here, I will set these results aside and consider instead a more recent development which focuses more on the structural issues rather than on specific solutions.

Let us begin with the third constraint in Eq (3), even though it refers to the Euclidean –rather than the Lorentzian– theory. The task is to define the corresponding quantum constraint operators on the Hilbert space H_d of solutions to the diffeomorphism constraints. Using a key idea due to Rovelli and Smolin [38], a regulated form of this operator has been constructed and it has been shown that, in the appropriate sense, the constraint algebra closes without anomalies [39]. While this progress is notable, I should point out that, in contrast to the regularization of area and volume operators or the imposition of the diffeomorphism constraint, these calculations should be regarded as preliminary explorations. Specifically, this regularization is "state-dependent" and hence not fully satisfactory, and there is considerable freedom at an intermediate step and no compelling reason to adopt the specific prescription used there. Furthermore, properties of the resulting constraint operator have not been investigated in detail. Even heuristically, it does not appear to be self-adjoint. On the one hand, as Kuchař [40] has pointed out, this would consistent with the closure of the constraint algebra since in finite dimensional models with an analogous algebra, the constraints would not close if the analog of the Hamiltonian constraint were self-adjoint (although this possibility is not ruled out in infinite dimensions due to subtleties associated with regularization.) However, non self-adjointness would also make it impossible to use the group averaging method at least directly. In spite of these important limitations, the calculation does represent a first serious attempt to regularize the Hamiltonian constraint rigorously and therefore holds considerable promise. In geometrodynamics, for example, a comparable stage is yet to be reached with respect to the Wheeler-DeWitt equation.

There is, however, a much more serious difficulty: The classical constraint we began with refers to the *Euclidean* signature while, physically, we need to solve the Lorentzian constraint. Thus, we need a generalized Wick transformation which will map the solutions to the quantum Euclidean constraint to those of the Lorentzian. At first, the problem looks hopelessly difficult. However, recently, Thiemann [41] has suggested a strategy in the context of a "coherent state transform" that will map complex-valued functions of real $SU(2)$ connections to holomorphic functions of an $SL(2,C)$ connection. The same strategy can be adopted in a purely real formulation used in this report (and also extended to allow for the presence of matter sources) [42]. To see how this works, let us return to the classical theory, where the phase space, canonical variables and constraints are all real in both signatures. The idea is to define an automorphism W on the algebra of complex-valued functions on the phase space –i.e., a map which preserves linear combinations, products and Poisson brackets of functions– which *maps the Euclidean Hamiltonian constraint functional (3a) to the Lorentzian one (3b)*. W is generated by a phase-space function T as follows:

$$W \circ f := 1 + \{T, f\} + \tfrac{1}{2!}\{T, \{T, f\}\} + \cdots \equiv \sum_{i=0}^{\infty} \tfrac{1}{n!}\{T, f\}_n\,, \qquad (9)$$

where $\{T, f\}$ is the Poisson bracket between T and f and $\{T, f\}_n$, the repeated Poisson bracket of n T-factors with f and the generating function T is given by:

$$T := \tfrac{i\pi}{2} \int_{\Sigma} d^3x\, K_a^i E_i^a \equiv \tfrac{i\pi}{2}\{V, H_E\}\,. \qquad (10)$$

Here, V denotes the total volume functional and $H_E := \int d^3x g^{-\frac{1}{2}} S$, where, as before, S is the Euclidean scalar or Hamiltonian constraint (see Eq (1.3). Note that T is just the integral of the trace of the extrinsic curvature over Σ.) It is straightforward to verify that $W \circ S = S_L$. Hence, if we could promote T to a quantum operator, we would have the generalized Wick transform $\hat{W} := \exp i\hat{T}$ in the quantum theory which would map the solutions to the Euclidean Hamiltonian constraint to those of the Lorentzian: We would have

$$\hat{S} \circ \Psi = 0 \quad \Longrightarrow \quad \hat{S}_L \circ (\hat{W} \circ \Psi) = 0\,, \qquad (11)$$

$\hat{S}_L$ being the Lorentzian Hamiltonian constraint. That is, while we did not have, a priori, a way of defining the Lorentzian Hamiltonian constraint operator, since classically W maps the Euclidean Hamiltonian constraint functional to the Lorentzian one, we can simply *define* the Lorentzian quantum constraint $\hat{S}_L$ by $\hat{S}_L = \hat{W} \circ \hat{S} \circ \hat{W}^{-1}$. This is an attractive strategy especially since, as Eq (10) shows, $\hat{T}$ could be constructed from the total volume operator (on which we have full control) and the integrated Euclidean Hamiltonian constraint.

To summarize, *if* the Euclidean quantum constraint $\hat{S}$ can be regularized in a more satisfactory fashion (so as to be free of the drawbacks of the present scheme discussed above) and the operator $\hat{T}$ defined and shown to have certain properties on H_d, we would conclude that *all* quantum constraints of general relativity can be imposed consistently, i.e., that quantum general relativity is consistent at a non-perturbative level. Note that it is not essential to exhibit the general solution to the

Hamiltonian constraint; indeed, even in the classical case, we do not have a general solution to the Einstein equations. As in the classical theory, what we need is a few simple solutions which can be interpreted and a degree of control on the structure of the *space* of solutions.

We will conclude with a few remarks on the generalized Wick transformation. First, note that W does *not* arise from a canonical transformation on the real phase space: the generating function T is imaginary, rather than real. (Thus, in quantum theory, $\hat{W}$ is not expected to preserve norms.) As far as I can tell, even on the complex phase space, the canonical transformation generated by T does not have a simple geometric property which can readily enable one to interpret it directly as a "Wick rotation" from the Euclidean phase space to the Lorentzian. In particular, given a specific solution to the *classical* Euclidean constraint, W does not provide us with a solution to the Lorentzian constraint. It is a well-defined mapping on the space of *functions* on the phase space, rather than on the (real) phase space itself. Furthermore, in general, it maps real functions to complex-valued functions; the Hamiltonian constraint is more of an exception than a rule where a real function is mapped to another real function. (For example, while $f(x) \mapsto f(ix)$ will in general map real functions to complex, $\cos x$ is again mapped to a real function, $\cosh x$.) Next, the action of W does preserve the vector and the Gauss constraints (modulo *overall* constants) so that if, as in the construction sketched above, Ψ were to satisfy all Euclidean constraints, $\hat{W} \circ \Psi$ would satisfy *all* Lorentzian constraints. Finally, one can ask whether, on the phase space, W maps the Euclidean action to the Lorentzian. This is indeed the case if one simultaneously transforms lapses and shifts via $(\underline{N}, N^a) \mapsto (-\underline{N}, -N^a)$, where $\underline{N}$ is the lapse field with density weight -1. These transformation properties of lapses and shifts are, however, different from those one encounters in quantum cosmology.

6. Discussion

The new, background independent functional calculus on the quantum configuration space $\overline{\mathcal{A}/\mathcal{G}}$ has enabled us to develop the quantization program systematically. The level of mathematical precision is such that all underlying assumptions are explicit and we can be confident that there are no hidden infinities. I would like to emphasize that, for the problem under consideration, this degree of precision is not a luxury; it is essential. Let me elaborate on this point since it is often overlooked. In theoretical physics in general, and in quantum field theory in particular, we often do formal manipulations, subtract infinities and extract physical answers. We do not worry about defining measures rigorously and are content even if the perturbation series we arrive at diverges uncontrollably so long as individual terms in the series are finite. In quantum gravity, however, I believe that we can not afford to be so cavalier. For, the central problem is somewhat different now. In the case of other three interactions, we have piles of experimental data and the central task is that of organizing it in a coherent fashion and making further predictions that can be tested by experiments. In quantum gravity, we are not blessed with this richness. At the present stage, the key problem is that of consistency: Can the principles of general relativity and quantum theory be unified in a mathematically consistent fashion? To be confident of the answer, we are forced to elevate our mathematical standards. Indeed, perturbative treatments have taught us that it would be extremely unwise to be satisfied with formal manipulations.

The techniques underlying our background-independent functional calculus may seem somewhat unfamiliar to most physicists. The overall situation is rather analogous to the introduction, in the sixties, of global techniques in general relativity. Until then, most relativists were content with local calculations and worked exclusively with coordinates. When the various notions of causality and even the definitions of a singularity were first introduced, they seemed exotic to the practitioners in the field as they were outside of what was then the mainstream. The new techniques did not invalidate the use of coordinate methods for local problems. However, they turned out to be indispensable to the analysis of global issues associated, e.g., with horizons and singularities which were then emerging as the frontier problems. These were the types of problems that the older methods could not handle. Indeed, using local methods, one could not even *define* the notion of the event horizon of a black-hole much less prove theorems about their properties. I believe that the situation with the new functional calculus is quite analogous. These techniques are essential to handle the question of mathematical consistency, the frontier issue at the present stage of our understanding of quantum gravity. Traditional tools, powerful as they are in the analysis of other interactions, appear to be insufficient to face this issue non-perturbatively, without any recourse to a background geometry. With the new calculus, we have tools to meet this challenge.

As we saw, these techniques have already been applied to two problems with considerable success: probing quantum geometry and solving the quantum constraints of the general relativity.

The picture of the geometry at the Planck scale that emerged turned out to be very different from the continuum image. In particular, operators representing areas of 2-surfaces and volumes of 3-dimensional regions have discrete spectra. This departure from the continuum –and particularly the discreteness of the spectrum of the area operators–may well hold the key to some of the current puzzles such as the relation between statistical entropy and the area of black hole horizons. The fundamental, Planck scale excitations of the gravitational field are 1-dimensional and the corresponding geometry is distributional. As was pointed out in Section 3, in this respect, there is a close similarity with polymer physics. Although polymers and the basic phonon excitations in them are 1-dimensional, in suitably complex configurations, they exhibit 3-dimensional properties. The same is true of geometry [3,43]: continuum 3-dimensional geometries *do* arise, but as approximations. More precisely, the Hilbert space H_o admits states which can be interpreted as "semi-classical" in the sense that, when coarse-grained using a macroscopic length scale, they become indistinguishable from smooth continuum geometries in three dimensions. For example, given a flat 3-metric g^o_{ab} on R^3, and a macroscopic length scale L, we can ask for states Ψ in H_o which have the property that for all regions with

g^o-volume of the order of L^3 or bigger and for all surfaces S which are slowly varying on the scale L, the eigenvalues of the volume and area operator approximate the values one would obtain *classically* using g_{ab}^o, up to corrections of the order $O(\ell_P/L)$. Such states exist and can be constructed using techniques from "statistical geometry" that underlie random lattices [43]. These states have been generically called "weaves" because they tell us how to weave a classical geometry from "quantum threads" –the elementary excitations of geometry.

There is an indirect test of these ideas. Although one expects the laboratory physics to be insensitive to the detailed predictions of quantum gravity, it may well be that to do this physics in a coherent fashion, one has to supplement the standard description with some *qualitative* ideas from quantum gravity[6]. For instance, in Minkowskian quantum field theories, the main difficulty comes from the ultra-violet divergences which arise because we integrate over arbitrarily large momenta, i.e., arbitrarily short distances. A more accurate representation of the underlying geometry is given by a weave state. So, rather than doing quantum field theory on a continuum, we should really do it on a weave state. Such a theory would be free of ultra-violet divergences. The key question is whether it reproduces the results of the standard perturbation theory at "low energy." Heuristically, one would expect this to be the case: Since the standard theories are renormalizable, their predictions for phenomena at the 10^{-17}cm scale should be insensitive to the details of the micro-structure of weave states at the Planck scale. (Recall that the weave geometry differs from the continuum only up to terms $O(\ell_P/L) \approx (10^{-33}/10^{-17}) \approx 10^{-16}$.) If one could show this result in detail –i.e., establish that the predictions of a *finite* quantum field theory on a weave state agree with those of the standard perturbation theory for laboratory energies– one could take the rich phenomenological data from particle physics as an indirect evidence for the discrete structure of the Planck scale geometry.

The second main application of the framework is to quantum Einstein equations. The diffeomorphism constraints could be regulated on the kinematical Hilbert space H_o in a manner that is anomaly-free and the space of solutions could be given a Hilbert space structure using the group averaging technique. There is also progress on the Hamiltonian constraint and the idea of using a generalized Wick transformation is tantalizing. However, more work is needed to make these results definitive. It is quite possible that to obtain a satisfactory regularization, one would have to bring in the notion of "framed networks" and replace the gauge group $SU(2)$ by its "quantum version". However, if these last problems could be resolved, one would conclude that quantum general relativity does exist non-perturbatively. It

---

[6] There are numerous examples of such situations in other branches of physics. For example, in astrophysics, one can generally work with Newtonian gravity. However, to discuss the density distributions of stars near the centers of galaxies such as M87 which have large black holes in their centers, one simply changes the boundary conditions on Newtonian equations, allowing stars to disappear once they cross the event horizons. Thus, although the use full general relativity is unnecessary, some "qualitatively new" features of this more accurate theory have to be incorporated.

is already clear that its structure would be very different from that envisaged in perturbative treatments which are all rooted in a continuum picture. My own view is that there may well exist several inequivalent ways of regulating the Hamiltonian constraint, leading to inequivalent theories. Indeed, this is likely to be physically the most important source of non-uniqueness of the quantum theory. If the theory does admit such inequivalent sectors, it would be all the more important to develop approximation schemes to extract their physical content. They will be motivated by the structure of the exact theory and therefore likely to be quite different from the ones used in the standard perturbative treatments. Work is in progress along several lines in this area.

An attractive possibility is to combine the strengths of string theory and the non-perturbative approach discussed here. String theory provides a "tight" strategy to couple matter for which no analogous principle is known within general relativity. On the other hand, at least in practice, string theory is essentially perturbative and needs a background continuum geometry. Quantum general relativity needs no background fields and provides a specific picture of quantum geometry. A natural strategy then would be to investigate strings on these quantum geometries. Indeed, since the excitations of geometry are along 1-dimensional graphs, it it natural to incorporate matter through strings winding around loops in these graphs. This might provide for us a viable perturbation theory. Results of Klebanov and Susskind [44] suggest a concrete direction for this work.

To conclude, let me address an obvious question about the new functional calculus: Since it refers to theories of connections, can it not be used for a non-perturbative treatment of Yang-Mills theories? Unfortunately, in the general case, there is an obstruction: While our emphasis has been on diffeomorphism invariance and absence of background fields, Yang-Mills theories depend on the background Minkowskian metric rather heavily. Thus, to make the framework directly amenable to *general* Yang-Mills theories, we would have to develop it further in a substantial way by adding new techniques which are tailored to the presence of a flat background metric. There is, however, an exception: two space-time dimensions where one only needs an area element, rather than a metric, to specify the Yang-Mills action. Thus, the theories are now invariant under area preserving diffeomorphisms. In this case, our techniques *are* directly applicable and have led to a number of *new* results [45]. These include: explicit expressions of the Schwinger functions for Wilson loops and a direct proof of the equivalence of the Euclidean path integral formulation and the Hamiltonian quantum theory. Furthermore, unlike in other approaches, the invariance of the quantum theory under area preserving diffeomorphisms is manifest in this treatment.

Acknowledgements

Most of the recent results reported here were obtained in collaboration with C. Isham, D. Marolf, J. Mourão, T. Thiemann and, especially, J. Lewandowski. I am most grateful to them for constant intellectual stimulation over the past two years and for their patience with my slow pace in writing up the results. I am grateful also to John Baez, Rodolfo Gambini, Karel Kuchař, Jorge Pullin, Carlo Rovelli, and Lee Smolin for their numerous suggestions, comments and criticisms which were often vital. This work was supported in part by the NSF grant PHY 93-96246 and by the Eberly Research Funds of the Penn State University.

References

1. P. Faria de Veiga, Ecole Polytechnique thesis (1990); C. de Calan, P. Faria de Veiga, J. Magnen and R. Sénéor, *Phy. Rev. Lett.* **66** (1991) 3233; A. S. Wightman, in: *Mathematical Physics Towards XXIst Century*, eds R. N. Sen and A. Gersten (Ben Gurion University Press, 1994).
2. D. Amati, M. Ciafolini and G. Veneziano, *Nucl. Phys.* **B347** (1990) 550.
3. A. Ashtekar, C. Rovelli and L. Smolin, *Phys. Rev. Lett.* **69** (1992) 237;
4. A. Agishtein and A. Migdal, *Mod. Phys. Lett.* **7** (1992) 85.
5. J. Iwasaki and C. Rovelli, *Int. J. Mod. Phys.* **D1** (1993) 533; *Class. & Quantum Grav.***11** (1994) 1653.
6. T. Jacobson and L. Smolin, *Nucl. Phys.* **B299** (1988) 295.
7. C. Rovelli and L. Smolin, *Phys. Rev. Lett.* **72** (1994) 446.
8. A. Ashtekar, *Non-Perturbative Canonical Gravity* (World Scientific, Singapore, 1991); in *Gravitation and Quantization* eds B. Julia and j. Zinn-Justin (Elsevier, Amsterdam 1995).
9. R. Rovelli, *Class. & Quantum Grav.* **8** (1991) 1613.
10. L. Smolin, in *Quantum Gravity and Cosmology* eds J. P. Mercader, H. Solà and E. Verdaguer (World Scientific, Singapore, 1992).
11. R. Gambini and J. Pullin *Loops, Knots, Gauge Theories and Quantum Gravity* (Cambridge University Press, Cambridge, 1995).
12. A. Ashtekar, *Phys. Rev. Lett.* **57** (1986) 2244; *Phys. Rev.***D36** (1987) 1587.
13. A. Ashtekar, in *Mathematics and General Relativity* (AMS, Providence, 1987), J.F. Barbero G. *Phys. Rev* **D51** (1995) 5507.
14. A. Ashtekar and C. J. Isham, *Class. & Quantum Grav.* **9** (1992) 1433.
15. J. C. Baez and S. Sawin, *Functional integration of spaces of connections*, q-alg/9507023.
16. A. Rendall, *Class. & Quantum Grav.* **10** (1993) 605.
17. A. Ashtekar and J. Lewandowski, in *Knots and Quantum Gravity*, ed J. Baez (Oxford University Press, Oxford, 1994).
18. J. Baez, *Lett. Math. Phys.***31** (1994) 213; in *The Proceedings of the Conference on Quantum Topology*, ed D. N. Yetter (World Scientific, Singapore, in press).
19. D. Marolf and J. Morão, *Commun. Math. Phys.* **170** (1995) 583.
20. A. Ashtekar and J. Lewandowski, *J. Math. Phys.* **36** (1995) 2170.
21. A. Ashtekar and J. Lewandowski, *J. Geo. & Phys.* (in press).
22. A. Ashtekar, J. Lewandowski, D. Marolf, J. Mourão and T. Thiemann, *Coherent state transform on the space of connections, J. Funct. Analysis* (in

press).

23. A. Ashtekar, J. Lewandowski, D. Marolf, J. Mourão and T. Thiemann, *J. Math. Phys.* (in press).

24. J. C. Baez, *Spin networks in gauge theory*, *Adv. Math.* (in press); *Spin networks in non-perturbative quantum gravity*, gr-qc/9504036.

25. L. Smolin (private communication).

26. C. Rovelli and L. Smolin, *Spin networks and quantum gravity*, preprint CGPG-95/4-1.

27. R. Penrose, in *Quantum Theory and Beyond*, ed T. Bastin, Cambridge University Press, Cambridge 1971).

28. A. Ashtekar and J. Lewandowski, *Quantum geometry*, (preprint).

29. C. Rovelli and L. Smolin *Nucl. Phys.***B442**, 593 (1995).

30. R. Loll, *The volume operator in discretized gravity*, (pre-print).

31. C. Rovelli, *Nucl. Phys.***B405** (1993) 797; L. Smolin, *Phys. Rev.***D49** (1994) 4028.

32. A. Higuchi, *Class. & Quantum Grav.***8** (1991) 1983; 2023; N. P. Landsman *J. Geo. & Phys.***15** (1995) 285.

33. A. Ashtekar and R. S. Tate, *J. Math. Phys.* **34** (1994) 6434.

34. B. Brügmann and J. Pullin, *Nucl. Phys.* **B363** (1991) 221; **B390** (1993) 399.

35. B. Brügmann, R. Gambini and J. Pullin, *Phys. Rev. Lett.* **68** (1992) 431.

36. H. Nicolai and H. J. Matschull, *J. Geo. and Phys.* **11** (1993) 15; H. J. Matschull, pre-print gr/qc 9305025;

37. H. A. Morales-Técotl and C. Rovelli, *Phys. Rev. Lett.***72** (1994) 3642.

38. C. Rovelli and L. Smolin (private communication).

39. A. Ashtekar and J. Lewandowski (in preparation).

40. K. Kuchař (private communication).

41. T. Thiemann (in preparation).

42. A. Ashtekar (in preparation).

43. A. Ashtekar and L. Bombelli (in preparation).

44. I. Klebanov and L. Susskind, *Nucl. Phys.* **B309** (1988) 175.

45. A. Ashtekar, J. Lewandowski, D. Marolf, J. Morão and T. Thiemann, in *Geometry of Constrained Dynamical Systems*, ed. J. Charap (Cambridge University Press, Cambridge, 1994); *Quantum Yang-Mills theory in two dimensions: A complete solution* (pre-print).

MODULAR COSMOLOGY

Thomas Banks
Rutgers University

This talk is a summary of work[1] done in collaboration with Micha Berkooz, Greg Moore, Steve Shenker and Paul Steinhardt on a cosmology whose early history is described in terms of the moduli fields of string theory. There have been a number of other approaches to string cosmology[2], which differ from ours in a number of important respects. I will not have time to compare and contrast these approaches, and will mention only a very clever idea of Lyth and Stewart[3] that may solve one of the most serious problems of modular cosmology.

Theories of slow roll inflation depend on the existence of scalar fields with very flat potentials. Essentially, the potential should not vary appreciably when the field changes by amounts of order the Planck mass. This means that slow roll inflation *can never be discussed in terms of effective field theories whose validity is limited to scales much lower than the Planck scale.* All effective field theories of inflation contain fine tuned dimensionless constants, and violate the principle of *naturalness* which is one of the cornerstones of modern quantum field theory.

It is fortunate then that string theory, our only extant theory of Planck scale physics, contains fields, the moduli, which have the requisite properties for driving inflation. The homogeneous modes of these fields are the various sizes of internal Kaluza-Klein manifolds, the dilaton, and their superpartners. They parametrize the continuous classical vacuum degeneracies of string theory. The natural range of variation of these fields is the Planck scale[1]. To all orders in perturbation theory the potential for the moduli fields vanishes. Therefore, if nonperturbative physics gives rise to a modular potential, it will have the form:

$$M^4 v\left(\frac{\Phi}{M_P}\right)$$

where

$$M = e^{-\frac{ReS}{k}} M_P$$

---

[1]... actually the string scale, but the difference between these will not be significant in our discussion.

N. Sánchez and A. Zichichi (eds.), String Gravity and Physics at the Planck Energy Scale, 277–284.

is a nonperturbative scale generated by the strong dynamics of some low energy gauge theory (k is related to the β function of the gauge theory). $S \equiv \frac{8\pi^2}{g_S^2} + ia$ is the scalar component of the dilaton superfield.

For a potential of this form, the number of e-foldings of inflation, and the amplitude of the, approximately scale invariant, spectrum of density fluctuations are given by formulae of the form

$$N_e \sim \int dx \frac{v'(x)}{v(x)}$$

$$\frac{\delta\rho}{\rho} \sim \frac{M^2}{M_P^2}$$

Matching the first of these to the minimum of 60 e-foldings that we need for an acceptable inflationary model, we realize that the 60 will have to come from dimensionless pure numbers in the theory. These may have to do with the relative normalizations of the Einstein lagrangian and the modular kinetic energy near the point in moduli space where inflation takes place, or with an unusual degree of flatness of the potential near this point. At the moment this looks like a mild fine tuning of parameters.

The condition of acceptable density fluctuations leads us to expect that $M \sim 10^{-2.5} M_P$. There are several things to note about this scale. It is near the putative scale of SUSY unification, and it is 5 or 6 orders of magnitude higher than the highest scale of SUSY breaking allowed by the SUSY solution of the hierarchy problem. Early discussions of string phenomenology usually assumed that all nonperturbative physics was connected to SUSY breaking, but this is not a sacred principle. There is however one possible way for the same potential which breaks SUSY to give rise to inflation. In string theory, the coupling is a dynamical variable. The scale of SUSY breaking is determined by the value of M at the minimum of the potential for the dilaton and the rest of the moduli. Inflation occurs when some of the scalars are sitting far from their vacuum values. If the dilaton is one of these inflationary scalars, there is no reason for M to be of order the SUSY breaking scale during inflation.

The main problem with this scenario was pointed out by Brustein and Steinhardt[4] before our work on modular cosmology. It is a cosmological version of the Dine-Seiberg[5] problem. The current vacuum state of the world has zero cosmological constant and is degenerate with respect to the state at infinitely weak coupling. The barrier between these states is of order the SUSY breaking scale. In the scenario of the preceding paragraph, postinflationary history begins at an energy density 22 orders of magnitude higher than this barrier, and it is hard to see how we can prevent the dilaton from flying over the barrier into the weak coupling region. Actually, as pointed out in the appendix of [1] things are not as bad as they seem. If, as we expect, the potential varies exponentially (or more rapidly) as a function of the dilaton, then the distance in dilaton space travelled when the energy density drops from E_0 to E is $\Delta S = C Ln(\frac{E_0}{E})$. To compute the constant C we have to know more about the details of the lagrangian for the moduli. If C is not too large, the dilaton will not penetrate the barrier into the weak coupling region. Given our current inability to compute the modular potential in string theory, it is hard to tell what will really happen.

We are thus led to the suggestion that inflationary dynamics is primarily due to SUSY preserving nonperturbative effects with a higher scale than the nonperturbative dynamics that violates SUSY. Within the context of inflationary cosmology, this scenario is very strongly constrained. The formula for the scalar field potential of gauge singlet fields in supergravity is, schematically,

$$V = e^K (|F|^2 - 3|W|^2)$$

where F is the SUSY order parameter, W is the superpotential, and K is the Kahler potential (F is a one form on field space and its square is taken with the Kahler metric). Supersymmetry requires that $F_i = 0$. These are N complex equations for N complex unknowns, and generically have a solution. However, it is not generally true that the additional equation $W = 0$ is satisfied at the supersymmetric point. In tree level string theory this is often the case, but that is probably because the tree level superpotential vanishes on a high dimensional subspace of field space. If a nonperturbative superpotential is generated for all of the moduli, we would expect that most supersymmetric points have nonvanishing superpotential and negative vacuum energy.

If the nonperturbative superpotential comes from low energy gauge theory, we can make an even stronger statement. The dependence of the superpotential on the moduli is then completely determined by the one loop renormalization of the gauge coupling[2]. The superpotential has the form

---

[2]Here we make the assumption that there can be no cancellations of contributions to the superpotential coming from gauge groups which get strong at different scales. These

$e^{-cS}f(\Phi)$, where Φ represents the nondilatonic moduli. $f(\Phi_0)$ can vanish only if massive chiral superfields transforming under the gauge group become massless at Φ_0. This does not often happen, and supersymmetric vacua with vanishing superpotential are correspondingly rare.

Generically then, we can expect that SUSY preserving nonperturbative dynamics will lead to a negative cosmological constant. In a postinflationary universe, such a minimum of the effective potential does not even lead to a stationary solution of the equations of cosmology. Instead of coming to rest at the minimum, the moduli fields cause a recollapse of the universe on microscopic time scales, and shoot off to infinite energy density. This may be viewed as a cosmological explanation of why SUSY *must* be broken. Inflation of the universe to large size requires that the cosmological constant in the nonperturbative vacuum be nonnegative.

All of this has interesting implications for our search for SUSY preserving nonperturbative dynamics as a driving force for inflation. It implies that this idea is only consistent in a very special class of string vacua: namely those points of moduli space where chiral superfields charged under some strong low energy gauge group become massless. This is a powerful cosmological selection principal for string vacua. Mike Dine and I have recently found examples of such points[8] . Unfortunately, they do not provide us with interesting models of the real world. The low energy dynamics is exactly supersymmetric and has a moduli space of vacua. In part, this failure is connected to the fact that our gauge group just below the Planck scale, is simple. The extra massless multiplets destroy asymptotic freedom and it is not possible to break SUSY at a lower scale.

A more realistic model would have a direct product gauge group. The first factor would become strong at the inflation scale and generate a super-potential for some subset of the moduli without breaking SUSY. Inflation would take place when these moduli were displaced from their vacuum con-figuration. It is important to realize that the dilaton cannot be included in this first group of moduli. The superpotential depends exponentially on the dilaton, and the equation $W = 0$, if it is satisfied at all, is satisfied for all values of the dilaton. When the "inflationary" moduli are in their vacuum state, the dilaton potential (from the inflationary gauge group) vanishes iden-tically. During inflation however, the dilaton will feel a potential which will

---

are effects of different exponential order in the weak coupling expansion.

rapidly attract it to a point $S_0(\Phi)$. As the inflationary moduli approach their minimum (call it $\Phi = 0$), the superpotential generated by the second factor of the gauge group will come into play. This is the potential which will freeze the dilaton and break SUSY. It provides a barrier of height M_{SUSY}^4, between its minimum S_m and the weak coupling region. If $S_0(0)$ lies on the strong coupling side of the S_m then the dilaton will begin its motion in this potential with energy of order the height of the barrier. It will therefore remain in the strong coupling region, avoiding the Brustein-Steinhardt problem.

Modular cosmology thus leads naturally to inflationary universes with a small amplitude of density fluctuations, and can easily avoid the Brustein-Steinhardt problem. However, reheating, the nemesis of many supersymmetric inflationary models, is also a problem here. Any moduli (and this certainly includes the dilaton) which do not get mass from SUSY preserving dynamics, will have masses of order $\frac{F_{SUSY}}{M_P} < 1$ TeV. They have only gravitational strength couplings to ordinary matter. Postinflationary history will start with these moduli shifted by an amount of order M_P from their minima (e.g in the model above, the dilaton is at $S_0(0) \neq S_m$ when inflation ends.). The energy stored in these fields will remain constant until the Hubble parameter falls to the scale of the modular masses. At this time, the modular energy density is a finite fraction of the total energy density of the universe. Subsequently it will red shift like the density of nonrelativistic matter, until the moduli decay. This interferes with the standard calculation of nucleosynthesis for any allowed value of the moduli masses[6] . If the moduli are light enough (as they would be if SUSY breaking occurs only at the weak scale) they would be predicted to dominate the current energy density of the universe by many orders of magnitude. Clearly these predictions are at variance with observations.

There have been many proposals for resolving this problem but the only one that does not suffer from fatal flaws is due to Lyth and Stewart[3] . The proposal is somewhat intricate so I will not review it here. It remains to be seen whether their model can be derived from string theory. I will mention one other possible resolution of the problem here since I have discussed it in many seminars but it has not appeared in the literature. It is not implausible to assume that the dilaton is the only field whose expectation value is not determined by supersymmetric dynamics. It is somewhat less plausible to assume that the order of magnitude estimate $\frac{F}{M_P}$ for the dilaton mass is off by two orders of magnitude. If however we make these assumptions, then the dilaton will decay in time to provide a radiation dominated universe at the time of nucleosynthesis. Baryogenesis may result from the decays of the dilaton itself, since there is no reason for its interactions with ordinary matter to preserve either CP or baryon number. Scott Thomas has pointed out to me that it takes two orders of baryon violating interactions to produce an asymmetry. Since the dilaton couplings are all Planck suppressed, this means that we need a renormalizable B violating interaction in order for this mechanism to generate a large enough asymmetry. Of course, such couplings are possible in the SUSY standard model, and are not ruled out by observation. Their existence would imply that R parity is strongly violated, and there is unlikely to be a stable lightest supersymmetric particle. The detailed phenomenology of this proposal has not yet been worked out. A final word on this subject: both of these proposals require a high scale of SUSY breaking which is gravitationally transmitted to the standard model. This is one of the strongest arguments against low energy SUSY breaking in string theory.

To conclude, let me discuss the issue of initial conditions. This is of course a highly controversial subject. Our assumption was that at energy densities just below the Planck scale, quantum string theory produces a probability distribution for classical initial conditions of scalar fields and the metric. There is a hidden assumption in this sentence, namely that one can find spacelike surfaces on which the energy density is constant, and near the Planck scale. We also assumed that there was no restriction on the size of the classical universe at this time. If the theory turns out to have an *a priori* restriction to closed universes, it might be natural to find that the quantum probability for Planck scale density was correlated with Planck size. We have no justification for the assumptions that we have made about the quantum probability distribution.

We then asked whether generic initial conditions, homogeneous over a horizon size, lead to inflation. An important point in this discussion is that although moduli space in noncompact, it is of finite volume[7] . Thus, we might expect a finite probability for the homogeneous modes of the moduli fields to lie in finite regions of the space. We are not forced to be in the weak coupling or large internal dimension region where inflation is impossible. However, the potential is a bounded function on moduli space, and the amplitude of the density fluctuations constrain it to be much smaller than the Planck scale. Thus at very early times we are dealing with scalar fields with no potential. Such a system becomes more and more inhomogeneous as time goes on. However, it is still possible to have inflation driven by domain walls. String theory has several different types of domain walls: 1)Walls associated with the degeneracy between the true vacuum and the extreme weak coupling region. (these are of interest primarily in the scenario in which inflation is driven at strong coupling by the SUSY breaking potential itself.) 2)Metastable domain walls associated with orbifold points in moduli space (as long as the minimum of the potential does not lie at the orbifold points). 3)Possible topologically stable domain walls that wind around conifold points in moduli space. If the potential is sufficiently flat [3] near the top of the domain wall, then the wall will be homogeneous over a horizon volume and its core will begin to inflate. The inhomogeneity that evolves from generic initial conditions, and the finite volume of moduli space, guarantee that many of these walls will be formed as the energy density drops below the scale of the potential. Thus, if we can believe in one mild fine tuning of parameters, generic initial conditions will lead to inflation in modular cosmology.

Our exploratory investigation of modular cosmology is far from a definitive treatment of the subject. We believe that we have demonstrated that cosmology puts strong and interesting constraints on string theory. Consequently, it is likely to be an area in which observational tests of string theory are possible.

References

[1] T.Banks, M.Berkooz, P.J.Steinhardt, *Phys. Rev.*D52, (1995),705; T.Banks, M.Berkooz, G.Moore, S.H.Shenker, P.J.Steinhardt, *Modular Cosmology*, RU-94-93, hep-th/9503114, to be published in *Phys. Rev.*D.

[2] A.Tseytlin,C.Vafa,*Nucl.Phys.*B372,(1992),443;M.Gasperini,N.Sanchez, G.Veneziano,*Int. J.Mod. Phys.* A6,(1991),3853; *Nucl. Phys.* B364, (1991),365; M.Gasperini, G.Veneziano, *Phys.Lett.* B296, (1992), 51; *Astropart. Phys.* 1, (1993), 317; *Mod. Phys. Lett.* A8, (1993), 3701; E.J.Copeland, A.R.Liddle, D.H.Lyth, E.D.Stewart, D.Wands, *Phys. Rev.*D49, (1994),6410; E.D.Stewart, *Inflation Supergravity and Super-*

---

[3] Sufficiently flat means dimensionless curvature $\sim \frac{1}{50}$. This is the same fine tuning discussed above.

strings, hep-ph/9405389,(to appear in Phys. Rev. D), *Mutated Hybrid Inflation*, astro-ph/9407040, (to appear in Phys. Lett.), D.H.Lyth, E.D.Stewart, *Cosmology with a TeV Mass GUT Higgs* hep-ph/9502417.

[3] D.H.Lyth, E.D.Stewart, *ibid.*

[4] R.Brustein, P.J.Steinhardt, *Phys. Lett.***B302**,(1993),196.

[5] M.Dine, N.Seiberg, *Phys. Lett.***162B**,(1985),299.

[6] T.Banks, D.Kaplan, A.Nelson, *Phys. Rev.***D49**,(1994),779; B.DeCarlos, J.A.Casas, F.Quevedo, E.Roulet, *Phys. Lett.***B318**,(1993),447.

[7] J.Horne, G.Moore, *Nucl. Phys.***B432**,(1994),109.

[8] T.Banks, M.Dine, *Quantum Moduli Spaces of N = 1 String Theories*, hep-th/9508071.

STATUS OF STRING COSMOLOGY:
BASIC CONCEPTS AND MAIN CONSEQUENCES

G. VENEZIANO

Theoretical Physics Division, CERN
CH - 1211 Geneva 23

Abstract. After recalling a few basic concepts from cosmology and string theory, I will discuss the main ideas/assumptions underlying string cosmology and show how these lead to a two-parameter family of "minimal" models. I will then explain how to compute, in terms of those parameters, the spectrum of scalar, tensor and electromagnetic perturbations, point at their (T and S-type) duality symmetries, and mention their most relevant physical consequences.

1. Basic Facts about Cosmology and Inflation

It is well known[1] that the Standard Cosmological Model (SCM) works well at "late" times, its most striking successes being perhaps the red shift, the cosmic microwave background (CMB), and primordial nucleosynthesis.

However, the SCM suffers from various problems. At the theoretical level the most serious of these is the initial singularity problem, which basically tells us that we cannot have theoretical control over the initial conditions of the SCM. At a phenomenological level, the SCM cannot explain naturally:

i) the homogeneity and isotropy of our Universe as manifested, in particular, through the small value of $\Delta T/T = O(10^{-5})$ observed with COBE[2];

ii) the flatness problem, i.e. the fact that, within an order of magnitude, $\Omega \equiv \rho / \rho_{crit} \sim 1$;

N. Sánchez and A. Zichichi (eds.), String Gravity and Physics at the Planck Energy Scale, 285–304.
© *1996 Kluwer Academic Publishers.*

iii) the origin of large-scale structure.

Inflation, i.e. a long phase of accelerated expansion of the Universe ($\dot{a}, \ddot{a} > 0$, where a is the scale factor), is the only way known at present of solving the above-mentioned phenomenological problems. Various types of inflationary models have been proposed [for a review, see 3), 4)] each one supposedly mending the problems of the previous version. Particularly severe are the constraints coming from demanding:

a) a graceful exit with the right amount of reheating;

b) the right amount of large-scale inhomogeneities.

In order to satisfy such constraints, fine-tuned initial conditions and/or inflaton potentials are necessary. And this without mentioning the fact that inflation is not addressing at all the initial singularity problem.

Actually, Kolb and Turner, after reviewing the prescriptions for a successful inflation, add[4]:

"Perhaps the most important – and most difficult – task in building a successful inflationary model is to ensure that the inflaton is an integral part of a sensible model of particle physics. The inflaton should spring forth from some grander theory and not vice versa".

I will argue below that superstring theory could be the sought-after grander theory (what could be better than a theory of everything?) naturally providing an inflation-driving scalar field in the general sense defined again in ref. 4):

"It is now apparent that inflation, which was originally so closely related to Spontaneous Symmetry Breaking, is a much more general phenomenon.... Stated in its full generality, inflation involves the dynamical evolution of a very weakly-coupled scalar field that was originally displaced from the minimum of its potential."

I hope to convince you that this will be precisely the picture that we claim takes place in string cosmology. In order to substantiate this claim, I will have to digress and recall a few basic facts in Quantum String Theory.

2. Basic facts in quantum string theory (QST)

I am listing below a few basic properties of strings, emphasizing those that are most relevant for our subsequent discussion. These are:

1. Unlike its classical counterpart, quantum string theory contains a fundamental length scale λ_s representing[5] the ultraviolet, short-distance cut-off (equivalently, a high-momentum cut-off at $E = M_s c^2 \equiv \hbar c / \lambda_s$).

2. Tree-level masses are either zero or $O(M_s)$. Quantum mechanics allows massless strings with non-zero angular momentum[6] while, classically, $M^2 > \text{const.} \times J$. The existence of such states is obviously a crucial property of QST, without which it could not pretend to be a candidate theory of all known interactions.

3. The effective interaction of the massless fields at $E \ll M_s$ takes the form of a classical, gauge-plus-gravity field theory with specified parameters. It is described by an effective action[7],[8] of the (schematic) type:

$$\Gamma_{eff} = \frac{1}{2} \int d^4 x \sqrt{-g} \, e^{-\phi} \left[\lambda_s^{-2} (\mathcal{R} + \partial_\mu \phi \partial^\mu \phi) + F_{\mu\nu}^2 + \bar{\psi} \not{D} \psi + R^2 + \ldots \right]$$
$$+ \left[\text{higher orders in } e^\phi \right] . \tag{1}$$

Equation (1) contains two dimensionless expansion parameters. One of them, $g^2 \equiv e^\phi$, controls the analogue of QFT's loop corrections, while the other, $\lambda^2 \equiv \lambda_s^2 \cdot \partial^2$, controls string-size effects, which are of course absent in QFT.

4. As indicated in (1), QST has (actually needs!) a new particle/field, the so-called dilaton ϕ, a scalar massless particle (at the perturbative level). It appears in Γ_{eff} as a Jordan–Brans–Dicke[9] scalar with a "small" negative ω_{BD} parameter, $\omega_{BD} = -1$.

5. The dilaton's VEV provides[8],[10] a unified value for:

a) The gauge coupling(s) at $E = O(M_s)$.
b) The gravitational coupling in string units.
c) Yukawa couplings, etc., at the string scale.

In formulae:

$$\ell_p^2 \equiv 8\pi G_N \hbar = e^\phi \lambda_s^2 ,$$
$$\alpha_{GUT}(\lambda_s^{-1}) \simeq \frac{e^\phi}{4\pi} , \tag{2}$$

implying (from $\alpha_{GUT} \approx 1/20$) that the string-length parameter λ_s is about 10^{-32} cm. Note, however, that, in a cosmological context in which ϕ evolves in time, the above formulae can only be taken to give the *present* values of α and ℓ_p/λ_s. In the scenario we will advocate, both quantities were much smaller in the very early Universe!

288

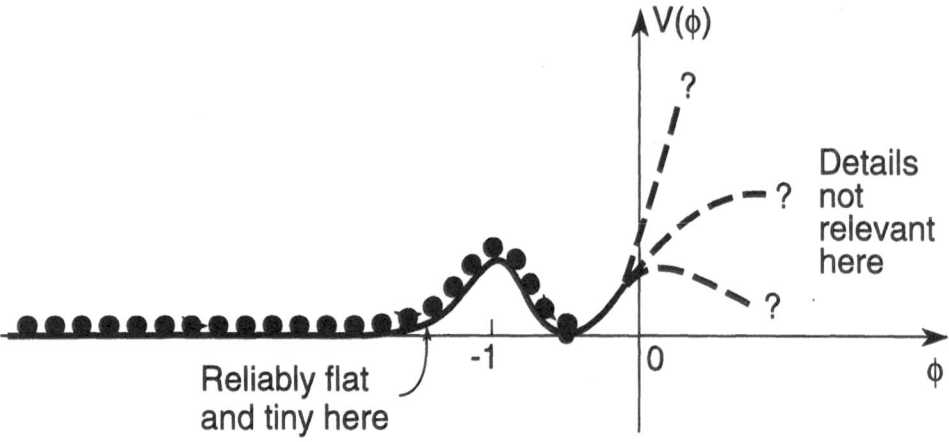

Figure 1. A possible dilaton potential with illustration of an inflation-driving rolling dilaton (large dots).

6. Dilaton couplings at large distance are such[11] that a massless dilaton is most likely ruled out[11,12] by precision tests[13] of the equivalence principle [for a possible way out see, however, ref. 14)], i.e.

$$M_\phi > 10^{-4} \text{ eV} . \tag{3}$$

7. Details about the dilaton potential are unknown, yet:

a) On theoretical grounds, in critical superstring theory, the dilaton potential has to go to zero as a double exponential as $\phi \to -\infty$ (weak coupling):

$$V(\phi) \sim \exp\left(-c^2 \exp(-\phi)\right) = \exp\left(-\frac{c^2}{4\pi\alpha_{GUT}}\right) , \tag{4}$$

with c^2 a positive (but model-dependent) constant.

b) On physical grounds it should have a non-trivial minimum at its present value ($\langle\phi\rangle = \phi_0 \sim 0$) with a vanishing cosmological constant, $V(\phi_0) = 0$.

A typical potential satisfying a) and b) is shown in Fig. 1. The dotted lines at $\phi > 0$ represent our ignorance about strongly coupled string theory. Fortunately, the details of what happens in that region will not be very relevant for our subsequent discussion.

8. There is an exact (all-order) vacuum solution for (critical) superstring theory. Unfortunately, it corresponds to a free theory ($g = 0$ or $\phi = -\infty$)

in flat, ten-dimensional, Minkowski space-time, nothing like the world we seem to be living in!

Before closing this section I would like to comment briefly on a point which appears to be the source of much confusion even among experts: it is the debate between working in the (so-called) String and Einstein "frames" (not to be confused with different coordinate systems). Since the two frames are related by a local field redefinition (a conformal, dilaton-dependent rescaling of the metric to be precise) all physical quantities are independent of the frame one is using. The question is: what should we call the metric? Although, to a large extent, this is a question of taste, one's intuition may work better with one definition than with another. Note also that, since the dilaton is time-independent today, the two frames coincide now.

Let us compare the virtues and problems with each frame.

A) STRING FRAME This is the metric appearing in the fundamental (Polyakov) action for the string. Classical, weakly coupled strings sweep geodesic surfaces with respect to this metric. Also, the dilaton dependence of the low energy effective action takes the simple form indicated in (1) only in the string frame. The advantage of this frame is that the string cut-off is fixed and the same is true for the value of the curvature at which higher orders in the σ-model coupling λ become relevant. The main disadvantage is that the gravitational action is not so easy to work with.

B) EINSTEIN FRAME In this frame the pure gravitational action takes the standard Einstein-Hilbert form. Consequently, this is the most convenient frame for studying the cosmological evolution of metric perturbations. The Planck length is fixed in this frame while the string length is dilaton (hence generally time) dependent. In the Einstein frame Γ_{eff} takes the form:

$$\Gamma_{eff} = \frac{1}{16\pi G_N} \int d^4 x \sqrt{-g} \left[R + \partial_\mu \phi \partial^\mu \phi + e^{-\phi} F_{\mu\nu}^2 + \partial_\mu A \partial^\mu A + e^\phi m^2 A^2 \right] + \left[G_N e^{-\phi} R^2 + \ldots \right], \qquad (5)$$

showing that, in this frame, masses are dilaton dependent (even at tree level) and so is the value of R at which higher order stringy corrections become important. It is for the above reasons that I will choose to base my discussion (although not always the calculations) in the String frame.

3. Main ideas/assumptions of string cosmology

The very basic postulate of (our own version of) String Cosmology[15),16)] is that the Universe did indeed start near its trivial vacuum mentioned at the end of the previous section.

Fortunately, if one looks at the space of homogeneous (and for simplicity spacially-flat) perturbative vacuum solutions, one finds that the trivial vacuum is a very special, *unstable* solution. This is depicted in Fig. 2a for the simplest case of a ten-dimensional cosmology in which three spatial dimensions evolve isotropically while six "internal" dimensions are static (it is easy to generalize the discussion to the case of dynamical internal dimensions, but then the picture becomes multidimensional).

The straight lines in the $H, \bar{\phi}$ plane (where $\bar{\phi} \equiv \dot{\phi} - 3H$) represent the evolution of the scale factor and of the coupling constant as a function of the cosmic time parameter (arrows along the lines show the direction of the time evolution). As a consequence of a stringy symmetry, known[15),17)] as "Scale Factor Duality (SFD)", there are two branches (two straight lines). Furthermore, each branch is split by the origin in two time-reversal-related parts (time reversal changes the sign of both H and $\dot{\bar{\phi}}$).

The origin (the trivial vacuum) is an "unstable" fixed point: a small perturbation in the direction of positive $\dot{\bar{\phi}}$ makes the system evolve further and further from the origin, meaning larger and larger coupling and absolute value of the Hubble parameter. This means an accelerated expansion or an accelerated contraction, i.e. in the latter case, inflation. It is tempting to assume that those patches of the original Universe that had the right kind of fluctuation have grown up to become (by far) the largest fraction of the Universe today.

In order to arrive at a physically interesting scenario, however, we have to connect somehow the top-right inflationary branch to the bottom-right branch, since the latter is nothing but the standard FRW cosmology, which has presumably prevailed for the last few billion years or so. Here the so-called "exit problem" arises. At lowest order in λ^2 (small curvatures in string units) the two branches do not talk to each other. The inflationary (also called $+$) branch has a singularity in the future (it takes a finite cosmic time to reach ∞ in our gragh if one starts from anywhere but the origin) while the FRW ($-$) branch has a singularity in the past (the usual big-bang singularity).

It is widely believed that QST has a way to avoid the usual singularities

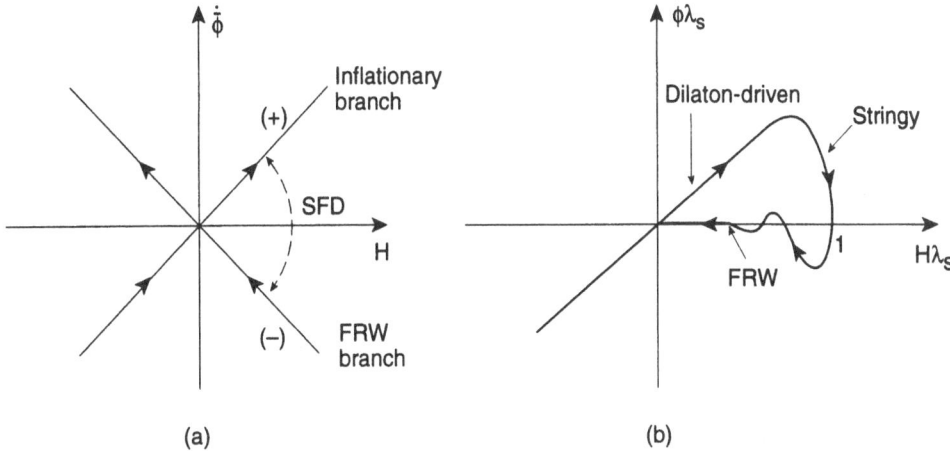

Figure 2.   Phase diagrams for the perturbative (a) regime and a conjectured non-perturbative solution (b) to the branch-change problem.

of Classical General Relativity or at least a way to reinterpret them[18),19)]. It thus looks reasonable to assume that the inflationary branch, instead of leading to a non-sensical singularity, will evolve into the FRW branch at values of λ^2 of order unity. This is schematically shown in Fig. 2b, where we have gone back from $\bar{\phi}$ to $\dot{\phi}$ and we have implicitly taken into account the effects of a non-vanishing dilaton potential at small ϕ in order to freeze the dilaton at its present value. The need for the branch change to occur at large λ^2, first argued for in[20)], has been recently proved in ref. 21).

There is a rather simple way to parametrize a class of scenarios of the kind defined above. They contain (roughly) three phases and two parameters. Indeed:

In phase I the Universe evolves at $g^2, \lambda^2 \ll 1$ and thus is close to the trivial vacuum. This phase can be studied using the tree-level low-energy effective action (1) and is characterized by a long period of dilaton-driven inflation. The accelerated expansion of the Universe, instead of originating from the potential energy of an inflaton field, is driven by the growth of the coupling constant (i.e. by the dilaton's kinetic energy, see ref. 22) for a similar kind of inflationary scenario) with $\dot{\phi} = 2\dot{g}/g \sim H$ during the whole phase.

Phase I supposedly ends when the coupling λ^2 reaches values of $O(1)$, so that higher-derivative terms in the effective action become relevant. Assuming that this happens while g^2 is still small (and thus the potential is still negligible), the value g_s of g at the end of phase I (the beginning of phase II) is an arbitrary parameter (a modulus of the solution).

During phase II, the stringy version of the big bang, the curvature, as well as $\dot{\phi}$, are assumed to remain fixed at their maximal value given by the string scale (i.e. we expect $\lambda \sim 1$). The coupling g will instead continue to grow from the value g_s until it is its own turn to reach values $O(1)$. At that point, assuming a branch change to have occurred at large curvatures, the dilaton will be attracted to the true non-perturbative minimum of its potential; the standard FRW cosmology can then start, provided the Universe was heated-up and filled with radiation (this is not a problem, see below). The second important parameter of this scenario is the duration of phase II or better the total red-shift, $z_s \equiv a_{end}/a_{beg}$, which has occurred from the beginning to the end of the stringy phase.

Our present ignorance about this most crucial phase (and in particular about the way the exit can be implemented) prevents us from having a better description of this phase which, in principle, should not introduce new arbitrary parameters (z_s should be eventually determined in terms of g_s).

During Phase III, the Universe evolves towards smaller and smaller curvatures but stays at moderate-to-strong coupling. This is the regime in which usual QFT methods are applicable. The details of the particular gauge theory emerging from the string's non-perturbative vacuum will be very important in determining the subsequent evolution and in particular the problem of structure formation, dark matter and the like.

Our scenario contains implicitly an arrow of time, which points in the direction of increasing entropy, inhomogeneity and structure. As a result of the amplification of primordial vacuum fluctuations, the Universe is not coming back to its initial simple (and unique) state (the origin in Fig. 2), but to the much more structured (and interesting) state in which we are living today. Actually, the arrow of time itself should be determined by the direction in which entropy (and complexity) are growing. This will force us to identify (*by definition*) the perturbative vacuum with the initial state of the Universe!

4. Observable consequences

All the observable consequences I will discuss below have something to do with the well-known phenomenon[23] of amplification of vacuum quantum fluctuations in cosmological backgrounds. Any conformally flat cosmological background is known:

a) to amplify tensor perturbations, i.e. to produce a stochastic background of gravitational waves;
b) to induce scalar-metric perturbations from the coupling of the metric either to a fluid or to scalar particles (in our context to the dilaton).

By contrast, because of the scale-invariant coupling of gauge fields in four dimensions, electromagnetic (EM) perturbations are *not* amplified in a conformally flat cosmological background (even if inflationary). In string cosmology, the presence of a time-dependent dilaton in front of the gauge-field kinetic term yields, on top of the two previously mentioned effects,

c) an amplification of EM perturbations corresponding to the creation of macroscopic magnetic (and electric) fields.

Various physically interesting questions arise in connection with the three effects I have just mentioned. These include the following:

1. Does the Universe remain quasi-homogeneous during the whole string-cosmology history?
2. Does one generate a phenomenologically interesting (i.e. measurable) background of GW?
3. Can one produce large enough seeds for generating the observed galactic (and extragalactic) magnetic fields?
4. Can scalar, tensor (and possibly EM) perturbations explain the large-scale anisotropy of the CMB observed by COBE?
5. Do these perturbations have anything to do with the CMB itself?

In the rest of this talk I will first explain, on the toy example of the harmonic oscillator, the common mechanism by which quantum fluctuations are amplified in cosmological backgrounds. I will then give our present answers to the questions listed above, leaving details and derivations to the talk by M. Gasperini[24].

Consider a one-dimensional (non-relativistic) harmonic oscillator moving in a cosmological background of the simplest kind, characterized by a scale factor $a(t)$. In units in which the mass of the oscillator is 1, the Lagrangian reads:

$$L = \frac{1}{2}a^2(\dot{x}^2 - \omega^2 x^2) \tag{6}$$

while the canonical momentum and Hamiltonian are given by

$$p = a^2\dot{x} , \quad H = \frac{1}{2}(a^{-2}p^2 + a^2\omega^2 x^2) . \tag{7}$$

Let us first discuss the solutions of the classical equations of motion:

$$\ddot{x} + 2\,\dot{a}/a\,\dot{x} + \omega^2 x = 0 \,, \qquad \ddot{y} + (\omega^2 - \ddot{a}/a)y = 0 \,, \tag{8}$$

where $y \equiv ax$ is the proper (physical) amplitude as opposed to the comoving amplitude x.

Solutions to Eqs. (8) simplify in two opposite regimes:

a) For $\omega^2 \gg \ddot{a}/a$ there is "adiabatic damping" of the comoving amplitude (the name is clearly unappropriate in the case of contraction):

$$x \sim a^{-1}\,e^{\pm i\omega t} \,, \qquad p \sim a\,\omega\,e^{\pm i\omega t} \tag{9}$$

which means that, in this regime, the proper amplitude y and the proper momentum p/a stay constant (and so does the Hamiltonian).

b) For $\omega^2 \ll \ddot{a}/a$ one finds the so-called "freeze-out" regime in which:

$$x \sim B + C \int_0^t dt'a^{-2}(t')$$
$$p \sim C + \ldots \tag{10}$$

where the comoving amplitude and momentum are fixed. In this regime the Hamiltonian (the energy) of the system tends to grow at late times whenever a increases or decreases by a large factor during the freeze-out regime. In the former case the energy is dominated asymptotically by the term proportional to x^2 and is due to the "stretching" of the oscillator caused by the fast expansion, while in the latter case the term proportional to p^2 dominates because of the large blue-shift suffered by the momentum in a contracting background.

Consider now a cosmology such that

$$\omega^2 > \ddot{a}/a \,, \qquad t < t_{ex}, t > t_{re}$$
$$\omega^2 < \ddot{a}/a \,, \qquad t_{ex} < t < t_{re} \tag{11}$$

where, anticipating our subsequent discussion, we have defined the moments of exit and re-entry by the condition $\omega \sim H$. Such an example will be typical of our scenario, since a given scale will be well inside the horizon at the beginning (small Hubble parameter), outside during the high-curvature regime, and then inside again after re-entry. By joining smoothly the two asymptotic solutions, we easily find that the energy of the harmonic oscillator (which is constant during the initial and final phases) has been amplified during the intermediate phase by a factor:

$$|c|^2 = Max\left(\frac{a^2(t_{re})}{a^2(t_{ex})} \,, \quad \frac{a^2(t_{ex})}{a^2(t_{re})}\right) \tag{12}$$

corresponding to the two above-mentioned cases.

The excercise can be repeated at the quantum level starting, for instance, from a harmonic oscillator in its ground state. Quantum mechanics fixes the size of the initial amplitude, momentum and energy:

$$|x| \sim a^{-1}\sqrt{\frac{\hbar}{\omega}}, \quad E \sim \hbar\omega . \tag{13}$$

The quantum mechanical interpretation of eq. (12) is that c is the Bogoliubov coefficient transforming the initial ground state into the final excited quantum state ($|c|^2$ being the average occupation number for the latter). Note that the final state ends up being highly "squeezed", i.e. having a large Δx or Δp depending on the sign of H. If, because of coarse-graining, the squeezed coordinate is not measured, the final state will look like a high-entropy, statistical ensemble of quasi-classical oscillators.

Note, finally, the (Scale-Factor) duality invariance of the resulting amplification. Under $a \to a^{-1}$, position and momentum operators swap their role as the variable in which sqeezing or amplification occurs. Thus the final amplification remains the same.

Up to technical complications, things work out pretty much in the same way for strings[25] and for the three kinds of perturbations mentioned at the beginning of this section. In particular, for each one of the latter, one can define[26] a canonical variable ψ^i (similar to the harmonic oscillator's y) satisfying an equation of the type

$$\psi_k'' + [k^2 - V_i(\eta)]\psi_k = 0 , \tag{14}$$

where the label i on ψ has been suppressed, k is the comoving wave number, and derivatives with respect to conformal time η are denoted by a prime.

Since, for each i, the "potential" V_i is very small at very early times, grows to a maximum during the stringy era and, finally, drops rapidly to zero at the beginning of the radiation era, a given scale (k) begins and ends inside the horizon with an intermediate phase outside. Larger scales exit earlier and re-enter later. Also, in our scenario, larger scales exit and re-enter at smaller values of H. Very short scales exit during the stringy era and, for those, our predictions will not be as solid as for the scales that leave the horizon during the perturbative dilatonic phase I. The fact that the amplification of perturbation depends just on some ratios of fields evaluated at exit and re-entry (and not on the details of the evolution in between) makes us believe that our detailed results are trustworthy for those larger scales. This being said, I present below some results on the five issues mentioned above (see, again, ref. 24) for derivations and/or details).

1. Does the Universe remain quasi-homogeneous during the whole string-cosmology history?

The answer to this question turns out to be yes! This is not a priori evident since, in commonly used gauges[26] for scalar perturbations of the metric (e.g. the so-called longitudinal gauge in which the metric remains diagonal), such perturbations appear to grow very large during the inflationary phase and to destroy homogeneity or, at least, to prevent the use of linear perturbation theory. Similar problems had been encountered earlier in the context of Kaluza-Klein cosmology[27]. In ref. 28) it was shown that, by a suitable choice of gauge (an "off-diagonal" gauge), the growing mode of the perturbation can be tamed. This can be double-checked by using the so-called gauge-invariant variables of Bruni and Ellis[29]. The bottom line is that scalar perturbations in string cosmology behave no worse than tensor perturbations, to which we now turn our attention.

2. Does one generate a phenomenologically interesting (i.e. measurable) background of GW?

The canonical variable ψ for tensor perturbations (i.e. for GW) is defined by:

$$g_{\mu\nu} = a^2[\eta_{\mu\nu} + h_{\mu\nu}]$$
$$\psi = (a/g)\, h = a\, e^{-\phi/2} h \,, \tag{15}$$

where h stands for either of the two transverse-traceless polarizations of the gravitational wave. As long as the perturbation is inside the horizon, ψ remains constant while h is adiabatically damped. By contrast, outside the horizon, ψ is amplified according to

$$\psi_k \sim (a/g)[C_k + D_k \int_{\eta_{ex}}^{\eta} d\eta'\, g^2(\eta')\, a^{-2}(\eta')] \tag{16}$$

where, for each Fourier mode of (comoving) wave number k, $\eta_{ex} = k^{-1}$. The first term in (16) clearly corresponds to the freezing of h itself, while the second term represents the freezing of its associated canonical momentum. In standard (non-dilatonic) inflationary models, the first term dominates since a grows very fast. In our case, the second term dominates since the growth of a is over-compensated by the growth of g (i.e. of ϕ). This is equivalent to saying that, in the Einstein frame, our background describes a contracting Universe.

After matching the result (16) with the usual oscillatory, damped behaviour of the radiation-dominated epoch, one arrives at the final
result[28),30)] for the magnitude of the stochastic background of GW
today:

$$|\delta h_\omega| \equiv k^{3/2}|h_k| \quad \sim \quad \sqrt{\frac{H_0}{M_s}} z_{eq}^{-1/4} z_s \left(\frac{g_s}{g_1}\right) \left(\frac{\omega}{\omega_s}\right)^{1/2}$$

$$\left[\ln\left(\frac{\omega_s}{\omega}\right) + (z_s)^{-3}\left(\frac{g_s}{g_1}\right)^{-2}\right], \quad \omega < \omega_s \quad (17)$$

where $\omega = k/a$ is the proper frequency, $z_{eq} \sim 10^4$, $\omega_s \sim z_s^{-1}(g_1)^{1/2} \times 10^{11}$ Hz $\equiv z_s^{-1}\omega_1$.

The above result can be converted into a spectrum of energy density
per logarithmic interval of frequency. In critical density units:

$$\frac{d\Omega_{GW}}{d\ln\omega} = z_{eq}^{-1}(g_s)^2 \left(\frac{\omega}{\omega_s}\right)^3 \left[\ln\left(\frac{\omega_s}{\omega}\right) + (z_s)^{-3}\left(\frac{g_s}{g_1}\right)^{-2}\right]^2, \quad \omega < \omega_s .$$

$$(18)$$

The above spectrum looks quasi-thermal al large scales (i.e. at $\omega < \omega_s$),
but is amplified by a large factor relative to a Planckian spectrum of
temperature ω_s. In analogy with the harmonic oscillator case, there is a
duality symmetry of the spectrum, this time under the transformation
$(z_s, g_s) \rightarrow (z_s, z_s^{-3}g_s^{-1})$. The transformation corresponds to changing
ϕ into $-\phi$ i.e. to what we may call $\bar{S}$-duality. As with the harmonic
oscillator, the metric perturbation and its canonically conjugate momentum variable swap their role under such transformation.

In Fig. 3 we show the spectrum of stochastic gravitational waves expected from our two-parameter model. For a given pair g_s, z_s one identifies a point in the $\omega, \delta h_\omega$ plane as illustrated explicitly in the case
of $g_s = 10^{-3}, z_s = 10^6$. The resulting point (indicated by a large dot)
represents the end-point $\omega_s, \delta h_{\omega_s}$ of the $\omega^{1/2}$ spectrum corresponding
to scales crossing the horizon during the dilatonic era.

Although the rest of the spectrum is more uncertain, one can argue
that it has to join smoothly the point $\omega_s, \delta h_{\omega_s}$ to the true end-point
$\delta h \sim 10^{-30}, \omega \sim 10^{11} Hz$. The latter corresponds to a few gravitons
produced at the maximal amplified frequency ω_1, the last scale to go
outside the horizon during the stringy phase. The full spectrum is also
shown in the figure for the case $g_s = 10^{-3}, z_s = 10^6$, with the wiggly
line representing the less well known high frequency part.

Curves of constant Ω_{GW} are also shown. If $g_s < 1$, as we have assumed,
spectra will always lie below the $\Omega_{GW} = 10^{-4}$ line corresponding to
as many photons as gravitons been produced. On the other hand, by

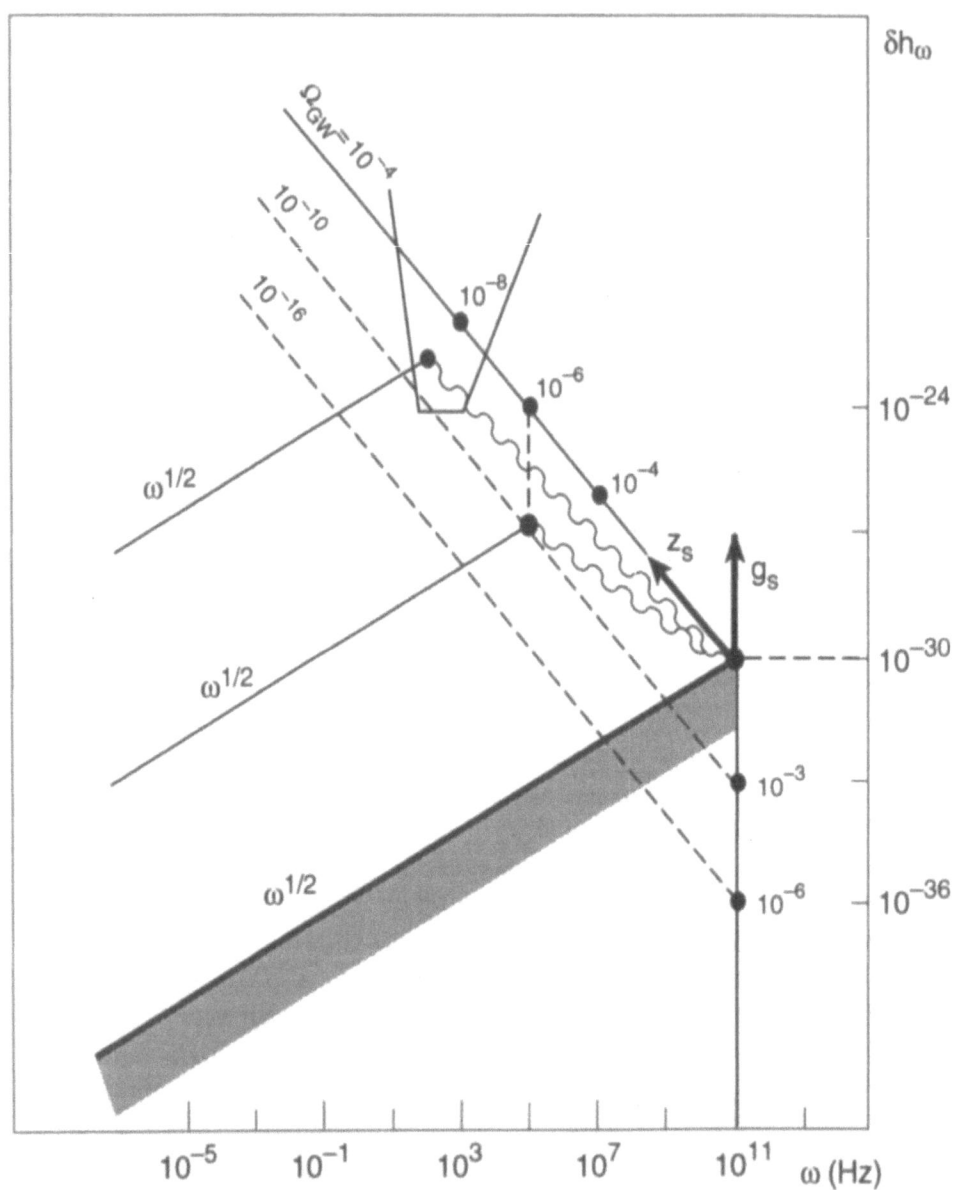

Figure 3. GW spectra from string cosmology against interferometric sensitivity.

invoking $\bar{S}$-duality, one can argue that the actual spectrum, by containing two duality-related contributions, will never lie below the self-dual spectrum ending at $\delta h \sim 10^{-30}, \omega \sim 10^{11} Hz$ (the thick line bordering the shaded region). In conclusion all possible spectra sweep the angular wedge inside the two abovementioned lines.

The odd-shaped region in Fig. 3 shows the expected sensitivity of the so-called "Advanced LIGO" project[31]. While there is no hope to detect our spectrum at LIGO if $g_s = 10^{-3}, z_s = 10^6$, perspectives would be better for, say, $g_s = 10^{-1}, z_s = 10^8$ (the corresponding spectrum is also sketched).

Resonant bars might also be able to reach comparable sensitivity in the kHz region, while microwave cavities, if conveniently developed, could be used in the region 10^6-10^9 Hz [33]. Another interesting possibility consists of coincidence experiments between an interferometer and a bar. The quoted sensitivity[34] to a stochastic background, as a function of the frequency f, of the individual sensitivities of the bar and of the interferometer δh, and of the observation time T_{obs}, is:

$$\delta\Omega_{GW} = 1.5 \ 10^{-5}(f/10^3\text{Hz})^3(10^{21}\delta h_{int})(10^{21}\delta h_{bar})(T_{obs}/10^7 \text{ s})^{-1/2}. \tag{19}$$

Obviously, detecting a stochastic backgound like ours is a formidable challenge. Also, the physical range of our parameters g_s, z_s could be such that no observable signal will be produced. What is interesting, however, is the mere existence of cosmological models predicting a non-negligible yield of GW in a range of frequencies where other sources predict just a "desert". More complete studies of the sensitivity of various detectors to a stochastic, coloured spectrum of our kind are presently under way.

3. **Can large enough seeds be produced for the generation of observed galactic (and extragalactic) magnetic fields?**

As already mentioned, seeds for generating the galactic magnetic fields through the so-called cosmic dynamo mechanism[35] can be generated in our scenario by the amplification of the quantum fluctuations of the EM field. In this case the canonical variable is just the (Fourier transform of the) usual A_μ potential. In analogy with (16) its amplification, while outside the horizon, is described by the asymptotic solution:

$$A_k \sim g^{-1}\left[C_k + D_k \int_{\eta_{ex}}^{\eta} d\eta' g^2(\eta')\right], \tag{20}$$

which leads[36],[37] to an overall amplification of the electromagnetic field by a factor $|c_k|^2 \sim (g_{re}/g_{ex})^2 + (g_{ex}/g_{re})^2$. This time the spectrum is invariant under $g \to g^{-1}$ i.e. under ordinary S or electric-magnetic duality. In our cosmological scenario we have excluded the possibility of a decreasing coupling constant and, therefore, the main contribution to the amplification comes from the second term on the r.h.s. of eq.(20) which gives $|c_k|^2 \sim (g_{re}/g_{ex})^2$.

One can express this result in terms of the fraction of electromagnetic energy stored in a unit of logarithmic interval of ω normalized to the one in the CMB, ρ_γ. One finds:

$$r(\omega) = \frac{\omega}{\rho_\gamma} \frac{d\rho_B}{d\omega} \simeq \frac{\omega^4}{\rho_\gamma} |c_-(\omega)|^2 \equiv \frac{\omega^4}{\rho_\gamma} (g_{re}/g_{ex})^2 . \tag{21}$$

The ratio $r(\omega)$ stays constant during the phase of matter-dominated as well as radiation-dominated evolution, in which the Universe behaves like a good electromagnetic conductor[38]. In terms of $r(\omega)$ the condition for seeding the galactig magnetic field through ordinary mechanisms of plasma physics is[38]

$$r(\omega_G) \geq 10^{-34} \tag{22}$$

where $\omega_G \simeq (1 \text{ Mpc})^{-1} \simeq 10^{-14}$ Hz is the galactic scale. Using the known value of ρ_γ, we thus find, from (21, 22):

$$g_{ex} < 10^{-33} , \tag{23}$$

i.e. a very tiny coupling at the time of exit of the galactic scale.

The conclusion is that string cosmology stands a unique chance in explaining the origin of the galactic magnetic fields. Indeed, if the seeds of the magnetic fields are to be attibuted to the amplification of vacuum fluctuations, their present magnitude can be interpreted as prime evidence that the fine structure constant has evolved to its present value from a tiny one during inflation. The fact that the needed variation of the coupling constant ($\sim 10^{30}$) is of the same order as the variation of the scale factor needed to solve the standard cosmological problems, can be seen as further evidence for scenarios in which coupling and scale factor grow roughly at the same rate during inflation.

4. Can scalar, tensor (and possibly EM) perturbations explain the large-scale anisotropy of the CMB observed by COBE?

The answer here is certainly negative as far as scalar and tensor perturbations are concerned. The reason is simple: for spectra that are normalized to $O(1)$ (at most) at the maximal amplified frequency $\omega_1 \sim 10^{11}$ Hz, and that grow like $\omega^{1/2}$, one cannot have any substantial power at the scales $O(10^{-18}\text{Hz})$ to which COBE is sensitive. The origin of $\Delta T/T$ at large scale would have to be attributed to other effects (e.g. topological defects).

Fortunately, there is a possibility [39] that the EM perturbations themselves might explain the anisotropies of the CMB since their spectrum turns out to be flatter (and also more model-dependent) than that of metric perturbations. Assuming this to be the case, an interesting relation is obtained[39] between the magnitude of large scale anysotropies and the slope of the power spectrum. Such a relation turns out to be fully consistent, with present bounds on the spectral index.

5. **Do all these perturbations have anything to do with the CMB itself?**

Stated differently, this is the question of how to arrive at the hot big bang of the SCM starting from our "cold" initial conditions. The reason why a hot universe can emerge at the end of our inflationary epochs (phases I and II) goes back to an idea of L. Parker[40], according to which amplified quantum flluctuations can give origin to the CMB itself if Planckian scales are reached.

Rephrasing Parker's idea in our context amounts to solving the following bootstrap-like condition: at which moment, if any, will the energy stored in the perturbations reach the critical density?

The total energy density ρ_{qf} stored in the amplified vacuum quantum fluctuations is given by:

$$\rho_{qf} \sim N_{eff} \frac{M_s^4}{4\pi^2} (a_1/a)^4 \ , \tag{24}$$

where N_{eff} is the number of effective (relativistic) species, which get produced (whose energy density decreases like a^{-4}) and a_1 is the scale factor at the (supposed) moment of branch-change. The critical density (in the same units) is given by:

$$\rho_{cr} = e^{-\phi} M_s^2 H^2 \ . \tag{25}$$

At the beginning, with $e^\phi \ll 1$, $\rho_{qf} \ll \rho_{cr}$ but, in the $(-)$ branch solution, ρ_{cr} decreases faster than ρ_{qf} so that, at some moment, ρ_{qf} will become the dominant sort of energy while the dilaton kinetic term will become negligible. It would be interesting to find out what sort of initial temperatures for the radiation era will come out of this assumption.

5. Conclusions

I want to conclude by listing which are, in my opinion, the pluses and minuses of the scenario I have advocated:

The Goodies

- Inflation comes naturally, without ad-hoc fields and fine-tuning: there is even an underlying symmetry yielding inflationary solutions.
- Initial conditions are natural, yet a simple universe would evolve into a rich and complex one.

- The kinematical problems of the SCM are solved.
- Perturbations do not grow too fast to spoil homogeneity.
- An interesting characteristic spectrum of GW is generated.
- Larger-than-usual electromagnetic perturbations are easily generated and could explain the galactic magnetic fields.
- A hot big bang could be a natural outcome of our inflationary scenario.

The Baddies

- A scale-invariant spectrum is all but automatic (unlike what happens in normal vacuum-energy-driven inflation).
- Our understanding of the high curvature (stringy) phase and of the crucially needed change of branch is still poor in spite of recent progress in Conformal Field Theory.

Acknowledgements

I would like to acknowledge the help and encouragement of my collaborators in the work reported here: Ramy Brustein (CERN–Beer Sheva), Maurizio Gasperini (Turin), Massimo Giovannini (CERN–Turin), and Slava Mukhanov (Zurich–Moscow). This work has also benefited from earlier collaborations with Jnan Maharana (Bhubaneswar), Kris Meissner (Trieste–Varsaw), Roberto Ricci (CERN–Rome), Norma Sanchez (Paris) and Nguyen Suan Han (Hanoi).

References

1. S. Weinberg, *Gravitation and Cosmology*, John Wiley & Sons, Inc., New York (1972).
2. G. Smoot et al., *Astrophys. J.* **396** (1992) L1.
3. L.F. Abbott and So-Young Pi (eds.), *Inflationary Cosmology*, World Scientific, Singapore (1986).
4. E. Kolb and M. Turner, *The Early Universe*, Addison-Wesley, New York (1990).
5. G. Veneziano, *Europhys. Lett.* **2** (1986) 133.
6. G. Veneziano, "Quantum strings and the constants of Nature", in *The Challenging Questions* (Erice, 1989), ed. A. Zichichi, Plenum Press, New York (1990).
7. C. Lovelace, *Phys. Lett.* **B135** (1984) 75;
   C.G. Callan, D. Friedan, E.J. Martinec and M.J. Perry, *Nucl. Phys.* **B262** (1985) 593.
8. E.S. Fradkin and A.A. Tseytlin, *Nucl. Phys.* **B261** (1985) 1.
9. P. Jordan, *Z. Phys.* **157** (1959) 112;
   C. Brans and R.H. Dicke, *Phys. Rev.* **124** (1961) 925.
10. E. Witten, *Phys. Lett.* **B149** (1984) 351.
11. T.R. Taylor and G. Veneziano, *Phys. Lett.* **B213** (1988) 459.
12. J. Ellis et al., *Phys. Lett.* **B228** (1989) 264.

13. See, for instance, E. Fischbach and C. Talmadge, *Nature* **356** (1992) 207.
14. T. Damour and A. M. Polyakov, *Nucl. Phys.* **B423** (1994) 532.
15. G. Veneziano, *Phys. Lett.* **B265** (1991) 287; Proceeding 4th PASCOS Conference (Syracuse, May 1994), K.C. Wali ed., World Scientific, Singapore, p. 453.
16. M. Gasperini and G. Veneziano, *Astropart. Phys.* **1** (1993) 317; *Mod. Phys. Lett.* **A8** (1993) 3701; *Phys. Rev.* **D50** (1994) 2519.
17. A.A. Tseytlin, *Mod. Phys. Lett.* **A6** (1991) 1721;
    A.A. Tseytlin and C. Vafa, *Nucl. Phys.* **B372** (1992) 443.
18. E. Kiritsis and C. Kounnas, *Phys. Lett.* **B331** (1994) 51;
    A.A. Tseytlin, *Phys. Lett.* **B334** (1994) 315.
19. P. Aspinwall, B. Greene and D. Morrison, *Phys. Lett.* **B303** (1993) 249;
    E. Witten, *Nucl. Phys.* **B403** (1993) 159.
20. R. Brustein and G. Veneziano, *Phys. Lett.* **B329** (1994) 429.
21. N. Kaloper, R. Madden and K. A. Olive, *Towards a singularity-free inflationary universe?*, Univ. Minnesota preprint UMN-TH-1333/95 (June 1995).
22. J. Levin and K. Freese, *Nucl. Phys.* **B421** (1994) 635.
23. L.P. Grishchuk, *Sov. Phys. JEPT* **40** (1975) 409;
    A.A. Starobinski, *JEPT Lett.* **30** (1979) 682;
    V.A. Rubakov, M. Sazhin and A. Veryaskin, *Phys. Lett.* **B115** (1982) 189;
    R. Fabbri and M. Pollock, *Phys. Lett.* **B125** (1983) 445.
24. M. Gasperini, *Status of string cosmology: phenomenological aspects*, these proceedings.
25. M. Gasperini, N. Sanchez and G. Veneziano, *Int. J. Theor. Phys.* **A6** (1991) 3853;
    Nucl. Phys. **B364** (1991) 365;
    M. Gasperini, *Phys. Lett.* **B258** (1991) 70;
    G. Veneziano, *Helv. Phys. Acta* **64** (1991) 877.
26. See, e.g. V. Mukhanov, H.A. Feldman and R. Brandenberger, *Phys. Rep.* **215** (1992) 203.
27. R.B. Abbot, B. Bednarz and S.D. Ellis, *Phys. Rev.* **D33** (1986) 2147.
28. R. Brustein, M. Gasperini, M. Giovannini, V. Mukhanov and G. Veneziano, *Phys. Rev.* **D51** (1995) 6744.
29. G. F. R. Ellis and M. Bruni, *Phys. Rev.* **D40** (1989) 1804;
    M. Bruni, G. F. R. Ellis and P. K. S. Dunsby, *Class. Quant. Grav.* **9** (1992) 921.
30. R. Brustein, M. Gasperini, M. Giovannini and G. Veneziano, *Relic gravitational waves from string cosmology*, CERN-TH/95-144 (1995), *Phys. Lett.* in press;
    see also M. Gasperini and M. Giovannini, *Phys. Rev.* **D47** (1992) 1529.
31. R.E. Vogt et al., *Laser Interferometer Gravitational-Wave Observatory*, proposal to the National Science Foundation (Caltech, 1989);
    C. Bradascia et al., in *Gravitational Astronomy*, eds. D.E. McClelland and H. Bachor, World Scientific, Singapore, (1991).
32. G. V. Pallottino and V. Pizzella, *Nuovo Cim.* **C4** (1981) 237;
    M. Cerdonio et. al, *Phys. Rev. Lett.* **71** (1993) 4107.
33. F. Pegoraro, E. Picasso and L. Radicati, *J. Phys.* **A11** (1978) 1949;
    C. M. Caves, *Phys. Lett.* **B80** (1979) 323;
    C. E. Reece et al., *Phys. Lett.* **A104** (1984) 341.
34. P. Astone, J. A. Lobo and B. F. Schutz, *Class. Quant. Grav.* **11** (1994) 2093.
35. E. N. Parker, *Cosmical Magnetic Fields*, Clarendon, Oxford (1979);
    Y. B. Zeldovich, A. A. Ruzmaikin and D. D. Sokoloff, *Magnetic fields in astrophysics*, Gordon and Breach, New York (1983).
36. M. Gasperini, M. Giovannini and G. Veneziano, *Primordial magnetic fields from string cosmology*, CERN-TH/95-85 (April 1995) *Phys. Rev. Lett.* in press.
37. D. Lemoine and M. Lemoine, *Primordial magnetic fields in string cosmology*, Inst. d'Astrophysique de Paris preprint (April 1995).
38. M. S. Turner and L. M. Widrow, *Phys. Rev.* **D37** (1988) 2743.
39. M. Gasperini, M. Giovannini and G. Veneziano, *Electromagnetic origin of the CMB*

anisotropy in string cosmology, CERN-TH/95-102 (April 1995) *Phys. Rev.* in press.

40. L. Parker, *Nature* **261** (1976) 20.

STATUS OF STRING COSMOLOGY: PHENOMENOLOGICAL ASPECTS

M. GASPERINI

Dipartimento di Fisica Teorica, Università di Torino,
Via P.Giuria 1, 10125 Turin, Italy

1. Introduction

Inspired by the basic ideas of string theory, we have recently started the investigation of a cosmological scenario in which the standard big-bang singularity is smoothed out and is replaced by a phase of maximal (finite) curvature. Such a phase is preceeded in time by a "pre-big-bang" epoch [1-3], which has (approximately) specular properties with respect to the present phase of standard decelerated evolution.

This scenario was originally motivated by the solutions of the string equations of motion in curved backgrounds [4], and by the duality symmetries of the string effective action [5-7] (see for instance [8] for a more detailed introduction). Its basic ingredients are an initial weak coupling, perturbative regime characterized by a very small background curvature (in string units), and a final transition to the radiation era controlled by the dilaton dynamics and by the two basic parameters (string length, string coupling) of string theory. The most revolutionary aspect of this scenario (with respect to the conventional, even inflationary, picture) is probably the fact that the early, pre-Planckian universe can be consistently described in terms of a semiclassical low energy effective action, with the vacuum as the most natural initial conditions for the quantum fluctuations of all the background fields (more details on this scenario can be found in Gabriele Veneziano's contribution to these proceedings [9]).

According to the above scenario, the pre-big-bang epoch is characterized by an accelerated (i.e. inflationary) evolution. In all inflationary

N. Sánchez and A. Zichichi (eds.), String Gravity and Physics at the Planck Energy Scale, 305–343.
© 1996 *Kluwer Academic Publishers.*

models, the most direct (and probably most spectacular) phenomeno-logical predictions is the parametric amplification of perturbations [10], and the corresponding generation of primordial spectra, directly from the quantum fluctuations of the background fields. In a string cosmol-ogy context such an effect is even more spectacular, as the growth of the curvature scale during the pre-big-bang phase is associated with a perturbation spectrum which grows with frequency [11,12]; moreover, the growth of the curvature may also force the comoving perturbation amplitude to grow (instead of being frozen) outside the horizon, as first noted in [2]. This leads to a more efficient amplification of perturba-tions, but the amplitude could grow too much, during the pre-big-bang phase, so as to prevent us from applying the standard linearized for-malism.

In view of this aspect, the aim of this lecture is twofold. On one hand I will discuss how, in some case, this anomalous growth can be gauged away, so that perturbations can consistently linearized in an ap-propriate frame. On the other hand I will show that, even if the growth is physical, and we have to restrict ourself to a reduced portion of pa-rameter space in order to apply a linearized approach, such enhanced amplification is nevertheless rich of interesting phenomenological con-sequences.

I will discuss in particular the following points: *i)* the growing mode of scalar metric perturbations in a dilaton-driven background, which appears in the standard longitudinal gauge and which seems to complicate the computation of the spectrum [13]; *ii)* the production of a relic gravity wave background with a spectrum strongly enhanced in the high frequency sector, and its possible observation by large inter-ferometric detectors [14] (such as LIGO and VIRGO); *iii)* the amplifi-cation of electromagnetic perturbations due to their direct coupling to the dilaton background, and the generation of primordial "seeds" for the galactic and extragalactic magnetic field [15]; *iv)* the generation of the large scale CMB anisotropy directly from the vacuum fluctuations of the electromagnetic field [16]. Finally, I will present some specu-lations about a possible geometric origin of the CMB radiation itself [17]. This lecture is an extended version of previous lectures given at the Observatory of Paris [18] and at the Gaeta Workshop on the "Very Early Universe" (Gaeta, August 1995, unpublished). The main results reported here are based on recent work done in collaboration with R. Brustein, M. Giovannini, V. Mukhanov and G. Veneziano.

Throughout this lecture the evolution of perturbations will be dis-cussed in a type of background which I will call, for short, "string cosmology background". At low energy such background represents a solution [3,5] of the gravi-dilaton effective action S, at tree-level in the string coupling $e^{\phi/2}$, and to zeroth order in the inverse string tension α',

$$S = - \int d^{d+1}x \sqrt{|g|} e^{-\phi} (R + \partial_\mu \phi \partial^\mu \phi) \qquad (1.1)$$

(with the possible addition of string matter sources). The background describes the accelerated evolution from the string perturbative vacuum, with flat metric and vanishing dilaton coupling ($\phi = -\infty$), towards a phase driven by the kinetic energy of the dilaton field ($H^2 \sim \dot{\phi}^2$), with negligible contribution from the dilaton potential (and other matter sources). In this initial phase the curvature scale H^2 and the dilaton coupling e^ϕ are both growing at a rate uniquely determined by the action (1.1), and the possible presence of matter, in the form of a perfect gas of non-mutually interacting classical strings, is eventually diluted [3,19].

The background evolution can be consistently described in terms of the action (1.1), however, only up to the time $t = t_s$ when the curvature reaches the string scale, namely when $H \simeq H_s = (\alpha')^{-1/2} \equiv \lambda_s^{-1}$. At that time all higher orders in α' (i.e. all higher-derivative corrections to the string effective action) become important, and the background enters a truly "stringy" phase, whose kinematic details cannot be predicted on the ground of the previous simple action. The presence of this high-curvature phase cannot be avoided, as it is required [20] to stop the growth of the curvature, to freeze out the dilaton, and to arrange a smooth transition (at $t = t_1$) to the standard radiation-dominated evolution (where $\phi =$const).

In previous works (see for instance [3,8]) we assumed that the time scales t_s and t_1 (marking respectively the beginning of the string and of the radiation era) were of the same order, and we computed the perturbation spectrum in the sudden approximation, by matching directly the radiation era to the dilaton-driven phase. Here I will consider a more general situation in which the duration of the string era (t_1/t_s) is left completely arbitrary, and I will discuss its effects on the perturbation spectrum.

During the pre-big-bang epoch, from the flat and cold initial state to the highly curved (and strongly coupled) final regime, the background evolution is accelerated, and can be invariantly characterized, from a kinematic point of view, as a phase of shrinking event horizons [1-3]. If we parameterize the pre-big-bang scale factor (in cosmic time) as

$$a(t) \sim (-t)^\beta, \qquad -\infty < t < 0 \qquad (1.2)$$

then the condition for the existence of shrinking event horizons

$$\int_t^0 \frac{dt'}{a(t')} < \infty \qquad (1.3)$$

is simply $\beta < 1$. As a consequence, there are two physically distinct classes of backgrounds in which the event horizon is shrinking.

If $\beta < 0$ we have a metric describing a phase of accelerated expansion and growing curvature,

$$\dot{a} > 0, \quad \ddot{a} > 0, \quad \dot{H} > 0 \qquad (1.4)$$

of the type of pole-inflation [21], also called super-inflation ($H = \dot{a}/a$, and a dot denotes differentiation with respect to the cosmic time t). If $0 < \beta < 1$ we have instead a metric describing accelerated contraction and growing curvature scale,

$$\dot{a} < 0, \quad \ddot{a} < 0, \quad \dot{H} < 0 \qquad (1.5)$$

The first type of metric provides a representation of the pre-big-bang scenario in the String (or Brans-Dicke) frame, in which test strings move along geodesic surfaces. The second in the Einstein frame, in which the gravi-dilaton action is diagonalized in the standard canonical form (see for instance [2,3]).

In both types of backgrounds the computation of the metric perturbation spectrum may become problematic, but the best frame to illustrate the difficulties is probably the Einstein frame, where the metric is contracting. It should be recalled, in this context, that the tensor perturbation spectrum for contracting backgrounds was first given in [22], but the possible occurrence of problems, due to a growing solution of the perturbation equations, was pointed out only much later [23], in the context of dynamical dimensional reduction. The problem, however, was left unsolved.

2. The "growing mode" problem

Consider the evolution of tensor metric perturbations, $\delta g_{\mu\nu} = a^2 h_{\mu\nu}$, in a $(3+1)$-dimensional conformally flat background, parameterized in conformal time ($\eta = \int dt/a$) by the scale factor

$$a(\eta) \sim (-\eta)^\alpha, \qquad -\infty < \eta < 0 \qquad (2.1)$$

Define the correctly normalized variables $u_{\mu\nu} = aM_p h_{\mu\nu}$ (M_p is the Planck mass), whose Fourier components obey canonical commutation relations, $[u_k, \dot{u}_{k'}] = i\delta_{k,k'}$. The modes u_k satisfy, for each of the two physical (transverse traceless) polarizations, the well known perturbation equation [10]

$$u_k'' + (k^2 - \frac{a''}{a})u_k = 0 \tag{2.2}$$

(a prime denotes differentiation with respect to η). In a string cosmology background the horizon is shrinking, so that all comoving length scales k^{-1} are "pushed out" of the horizon. For a mode k whose wavelength is larger than the horizon size (i.e. for $|k\eta| << 1$), we have then the general (to leading order) asymptotic solution

$$h_k = \frac{u_k}{aM_p} = A_k + B_k|\eta|^{1-2\alpha}, \qquad \eta \to 0_- \tag{2.3}$$

where A_k and B_k are integration constants.

The asymptotic behavior of the perturbation is thus determined by α. If $\alpha < 1/2$ the perturbation tends to stay constant outside the horizon, and the typical amplitude $|\delta_h|$ at the scale k^{-1}, for modes normalized to an initial vacuum fluctuation spectrum,

$$\lim_{\eta \to -\infty} u_k \sim \frac{1}{\sqrt{k}}e^{-ik\eta} \tag{2.4}$$

can be given as usual [24] in terms of the Hubble factor at horizon crossing $(k\eta \sim 1)$

$$|\delta_h| = k^{3/2}|h_k| \simeq \left(\frac{H}{M_p}\right)_{HC} \tag{2.5}$$

In this case the amplitude is always smaller than one provided the curvature is smaller than Planckian. This case includes in particular $\alpha < 0$, namely all backgrounds describing, according to eq. (2.1), accelerated inflationary expansion.

If, on the contrary, $\alpha > 1/2$, then the second term is the dominant one in the solution (2.3), the perturbation amplitude tends to grow outside the horizon,

$$|\delta_h| = k^{3/2}|h_k| \simeq \left(\frac{H}{M_p}\right)_{HC}|k\eta|^{1-2\alpha}, \qquad \eta \to 0_- \tag{2.6}$$

and may become larger than one, thus breaking the validity of the perturbative approach. Otherwise stated: the energy density (in critical units) stored in the mode k, i.e. $\Omega(k) = d(\rho/\rho_c)/d\ln k$, may become larger than one, in contrast with the hypothesis of negligible back-reaction of the perturbations on the initial metric.

One might think that this problem - due to the dominance of the second term in eq. (2.3) - appears in the Einstein frame because of the contraction (which corresponds to $\alpha > 0$), but disappears in the String frame where the metric is expanding. Unfortunately this is not true because, in the String frame, the different metric background is compensated by a different perturbation equation, in such a way that the perturbation spectrum remains exactly the same [2,3].

This important property of perturbations can be easily illustrated by taking, as a simple example, an isotropic solution of the $(d+1)$-dimensional gravi-dilaton equations [2,3], obtained from the action (1.1) supplemented by a perfect gas of long, stretched strings as sources (with equation of state $p = -\rho/d$).

In the Einstein frame the solution describes a metric background which is contracting for $\eta \to 0_-$,

$$a = (-\eta)^{2(d+1)/(d-1)(3+d)}, \quad \phi = -\frac{4d}{3+d}\sqrt{\frac{2}{d-1}}\ln(-\eta) \qquad (2.7)$$

and the tensor perturbation equation

$$h_k'' + (d-1)\frac{a'}{a}h_k' + k^2 h_k = 0 \qquad (2.8)$$

has an asymptotic solution (for $|k\eta| << 1$) which grows, according to eq. (2.3), as

$$\lim_{\eta \to 0_-} h_k \sim |\eta|^{(1-d)/(d+3)} \qquad (2.9)$$

In the String frame the metric is expanding,

$$\tilde{a} = (-\eta)^{-2/(3+d)}, \quad \tilde{\phi} = -\frac{4d}{3+d}\ln(-\eta) \qquad (2.10)$$

but the perturbation is also coupled to the time-variation of the dilaton background [12],

$$h_k'' + \left[(d-1)\frac{\tilde{a}'}{\tilde{a}} - \tilde{\phi}'\right]h_k' + k^2 h_k = 0 \qquad (2.11)$$

As a consequence, the explicit form of the perturbation equation is exactly the same as before,

$$h_k'' + \frac{2(d+1)}{d+3}\frac{h_k'}{\eta} + k^2 h_k = 0 \qquad (2.12)$$

so that the solution is still growing, asymptotically, with the same power as in eq. (2.9).

It may be noted that in the String frame the growth of perturbations outside the horizon is due to the joint contribution of the metric and of the dilaton background to the "pump" field responsible for the parametric amplification process [25]. Such an effect is thus to be expected in generic scalar-tensor backgrounds, as noted also in [26]. The particular example chosen above is not much relevant, however, for a realistic scenario in which the phase of pre-big-bang inflation is long enough to solve the standard cosmological problems. In that case, in fact, all scales which are inside our present horizon crossed the horizon (for the first time) during the dilaton-driven phase or during the final string phase, in any case when the contribution of matter sources was negligible [3,8].

We shall thus consider, as a more significant (from a phenomenological point of view) background, the vacuum, dilaton-driven solution of the action (1.1), which in the Einstein frame (and in $d = 3$) can be explicitly written as

$$a = (-\eta)^{1/2}, \qquad \phi = -\sqrt{3}\ln(-\eta), \qquad -\infty < \eta < 0 \qquad (2.13)$$

In such a background one finds that the growth of tensor perturbations is simply logarithmic [13],

$$|\delta_h(\eta)| \simeq \left|\frac{H}{M_p}\right|_{HC} \ln|k\eta| \simeq \frac{H_s}{M_p}|k\eta_s|^{3/2}\ln|k\eta|, \qquad |k\eta_s| < 1, \ |\eta| > |\eta_s| \qquad (2.14)$$

so that it can be easily kept under control, provided the curvature scale $H_s \sim (a_s\eta_s)^{-1}$ at the end of the dilaton phase is bounded.

The problem, however, is with scalar perturbations, described in the longitudinal gauge by the variable ψ such that [24]

$$(g_{\mu\nu} + \delta g_{\mu\nu})dx^\mu dx^\nu = a^2(1 + 2\psi)d\eta^2 - a^2(1 - 2\psi)(dx_i)^2 \qquad (2.15)$$

The canonical variable v associated to ψ is defined (for each mode k) by [24]

$$\psi_k = -\frac{\phi'}{4k^2 M_p}\left(\frac{v_k}{a}\right)' \qquad (2.16)$$

and satisfies a perturbation equation

$$v_k'' + (k^2 - \frac{a''}{a})v_k = 0 \qquad (2.17)$$

which is identical to eq. (2.2) for the tensor canonical variable, with asymptotic solution

$$\frac{v_k}{a} \simeq \frac{1}{\sqrt{k}} \frac{\ln|k\eta|}{a_{HC}}, \qquad |k\eta| \ll 1 \qquad (2.18)$$

Because of the different relation between canonical variable and metric perturbation, however, it turns out that the amplitude of longitudinal perturbations, normalized to an initial vacuum fluctuation spectrum,

$$\lim_{\eta \to -\infty} v_k \sim \frac{1}{\sqrt{k}} e^{-ik\eta} \qquad (2.19)$$

grows, asymptotically, like η^{-2}. We have in fact, from (2.16),

$$|\delta_\psi(\eta)| = k^{3/2}|\psi_k| \simeq \left|\frac{H}{M_p}\right| |k\eta|^{-1/2} \simeq \left.\left|\frac{H}{M_p}\right|\right|_{HC} |k\eta|^{-2} \simeq$$

$$\simeq \left(\frac{H_s}{M_p}\right) \frac{|k\eta_s|^{3/2}}{|k\eta|^2} \sim \frac{1}{\eta^2}, \qquad \eta \to 0_- \qquad (2.20)$$

This growth, as we have seen, cannot be eliminated by passing to the String frame. Neither can be eliminated in a background with a higher number of dimensions. In fact, in $d > 3$, the isotropic solution (2.13) is generalized as [3]

$$a = (-\eta)^{1/(d-1)}, \qquad \phi = -\sqrt{2d(d-1)} \ln a, \qquad -\infty < \eta < 0 \quad (2.21)$$

and the scalar perturbation equation in the longitudinal gauge [3]

$$\psi_k'' + 3(d-1)\frac{a'}{a}\psi_k' + k^2\psi_k = 0 \qquad (2.22)$$

has the generalized asymptotic solution

$$\psi_k = A_k + B_k \eta a^{-3(d-1)} \qquad (2.23)$$

By inserting the new metric (2.21) one thus finds the same growing time-behavior, $\psi_k \sim \eta^{-2}$, exactly as before. The same growth of ψ_k is also found in anisotropic, higher-dimensional, dilaton-dominated backgrounds [13].

Because of the growing mode there is always (at any given time η) a low frequency band for which $|\delta_\psi(\eta)| > 1$. In $d = 3$, in particular, such band is defined [from eq.(2.20)] as $k < \eta^{-1}(H/M_p)^2$. For such modes the linear approximation breaks down in the longitudinal gauge, and a full non-linear treatment would seem to be required in order to compute the spectrum. In spite of this conclusion, a linear description of scalar perturbations may remain possible provided we choose a different gauge, more appropriate to linearization than the longitudinal one.

A first signal that a perturbative expansion around a homogeneous background can be consistently truncated at the linear level, comes from an application of the "fluid flow" approach [27,28] to the perturbations of a scalar-tensor background. In this approach, the evolution of density and curvature inhomogeneities is described in terms of two covariant scalar variables, Δ and C, which are gauge invariant to all orders [29]. They are defined in terms of the momentum density of the scalar field, $\nabla\phi$, of the spatial curvature, $^{(3)}R$, and of their derivatives. By expanding around our homogeneous dilaton-driven background (2.13) one finds [13] for such variables, in the linear approximation, the asymptotic solution ($|k\eta| << 1$),

$$\Delta_k = const, \qquad\qquad c_k = const + A_k \ln|k\eta| \qquad (2.24)$$

This solution shows that such variables tend to stay constant outside the horizon, with at most a logarithmic variation (like in the tensor case), which is not dangerous.

As a consequence, the amplitude of density and curvature fluctuations can be consistently computed in the linear approximation (for all modes) in terms of Δ and C, and their spectral distribution (normalized to an initial vacuum spectrum) turns out to be exactly the same as the tensor distribution (2.14), which is bounded.

What is important, moreover, is the fact that such a spectral distribution could also be obtained directly from the asymptotic solution of the scalar perturbation equations in the longitudinal gauge [13],

$$\psi_k = c_1 \ln|k\eta| + \frac{c_2}{\eta^2} \qquad (2.25)$$

simply by neglecting the growing mode contribution (i.e. setting $c_2 = 0$). This may suggests that such growing mode has no direct physical meaning, and that it should be possible to get rid of it through an appropriate coordinate choice.

A good candidate to do the job is what we have called [13] "off-diagonal" gauge,

$$(g_{\mu\nu} + \delta g_{\mu\nu})dx^\mu dx^\nu = a^2 \left[(1 + 2\varphi)d\eta^2 - (dx_i)^2 - 2\partial_i B dx^i d\eta\right] \quad (2.26)$$

which represents a complete choice of coordinates, with no residual degrees of freedom, just like the longitudinal gauge. In this gauge there are two variables for scalar perturbations, φ and B, and their asymptotic solution is, in the linear approximation,

$$\varphi_k = c_1 \ln|k\eta| \sim \psi_k, \qquad B_k = \frac{c_2}{\eta} \sim \eta\psi_k, \qquad (\partial B)_k \sim |k\eta|\psi_k \quad (2.27)$$

(c_1 and c_2 are integration constants). The growing mode is thus completely gauged away for homogeneous perturbations (for which $\partial_i B = 0$). It is still present for non-homogeneous perturbations in the off-diagonal part of the metric, but it is "gauged down" by the factor $k\eta$ which is very small, asymptotically.

Fortunately this is enough for the validity of the linear approximation, as the amplitude of the off-diagonal perturbation, in this gauge, outside the horizon,

$$|\delta_B| \simeq |k\eta||\delta_\psi| \simeq \left(\frac{H_s}{M_p}\right)|k\eta_s|^{1/2}\left|\frac{\eta_s}{\eta}\right| \quad (2.28)$$

stays smaller than one for all modes $k < |\eta_s|^{-1}$, and for the whole duration of the dilaton-driven phase, $|\eta| > |\eta_s|$. We have explicitly checked that quadratic corrections are smaller than the linear terms in the perturbation equations, but a full second order computation requires a further coordinate transformation [13]. The higher order problem is very interesting in itself, but a complete discussion of the problem is outside the scope of this lecture. Having established that the vacuum fluctuations of the metric background, amplified by a phase of dilaton-driven evolution, can be consistently described (even in the scalar case) as small corrections of the homogeneous background solution, let me discuss instead some phenomenological consequence of such amplification. Scalar perturbations and dilaton production were discussed in [3,8,30]. Here I will concentrate, first of all, on graviton production.

3. The graviton spectrum from dilaton-driven inflation

Consider the amplification of tensor metric perturbation in a generic string cosmology background, of the type of that described in Sect. 1. Their present spectral energy distribution, $\Omega(\omega, t_0)$, can be computed in terms of the Bogoliubov coefficient determining their amplification (see Sect. 5 below), or simply by following the evolution of the typical amplitude $|\delta_h|$ from the time of horizon crossing down to the present time t_0. For modes crossing the horizon in the inflationary dilaton-driven phase $(t < t_s)$, and re-entering the horizon in the decelerated radiation era $(t > t_1)$, one easily finds, from eq. (2.14),

$$\Omega(\omega, t) \equiv \frac{\omega}{\rho_c} \frac{d\rho}{d\omega} \simeq A\Omega_\gamma \left(\frac{H_s}{M_p}\right)^2 \left(\frac{\omega}{\omega_s}\right)^3 \ln^2\left(\frac{\omega}{\omega_s}\right), \quad t > t_1, \quad \omega < \omega_s$$

(3.1)

Here $\omega = k/a$ is the red-shifted proper frequency for the mode k at time t, $\rho_c = M_p^2 H^2$ is the critical energy density, $\Omega_\gamma = (H_1/H)^2 (a_1/a)^4 = \rho_\gamma/\rho_c$ is the radiation energy density in critical units, and $\omega_s = H_s a_s/a$ is the maximum frequency amplified during the dilaton-driven phase. Finally, A is a possible amplification factor due to the subsequent string phase $(t_s < t < t_1)$, in case that the perturbation amplitude grows outside the horizon (instead of being constant) during such phase. This additional amplification does not modify however the slope of the spectrum, as we are considering modes that crossed the horizon before the beginning of the string phase (see Sect. 5).

An important property of the spectrum (3.1) is the universality of the slope ω^3 with respect to the total number d of spatial dimensions, and with respect to their possible anisotropy. Actually, the spectrum is also duality-invariant [14], in the sense that it is the same for all backgrounds, including those with torsion, obtained via $O(d, d)$ transformations [6] from the vacuum dilaton-driven background.

The spectrum (3.1) has also the same slope (modulo logarithmic corrections) as the low frequency part of a thermal black body spectrum, which can be written (in critical units) as

$$\Omega_T(\omega, t) = \frac{\omega^4}{\rho_c} \frac{1}{e^{\omega/T} - 1} \simeq B\Omega_\gamma \left(\frac{H_s}{M_p}\right)^2 \left(\frac{\omega}{\omega_s}\right)^3 \frac{T}{\omega_s}, \quad \omega < T \quad (3.2)$$

Here $B = (H_s/H_1)^2 (a_s/a_1)^4$ is a constant factor which depends on the time-gap between the beginning of the string phase and the beginning of the string era. We can thus parameterize the graviton spectrum (3.1) in terms of an effective temperature

$$T_s = (A/B)\omega_s$$

(3.3)

which depends on the initial curvature scale H_s, and on the subsequent kinematic of the high-curvature string phase.

For a negligible duration of the string phase, $t_s \sim t_1$, we have in particular $H_s \sim H_1 \sim M_p$, and the spectrum (3.1) is peaked around a maximal amplified frequency $\omega_s \sim H_1 a_1/a \sim 10^{11}$Hz, while it is exponentially decreasing at higher frequencies (where the parametric amplification is not effective). Moreover, $T_s \sim \omega_s \sim 1^\circ K$, so that this spectrum, produced by a geometry transition, is remarkably similar to that of the observed cosmic black body radiation [17] (see also Sect. 9 below).

The problem, however, is that we don't know the duration and the kinematics of the high curvature string phase. As a consequence, we know the slope (ω^3) of this "dilatonic" branch of the spectrum, but we don't know the position, in the (Ω, ω) plane, of the peak frequency ω_s. This uncertainty is, however, interesting, because the effects of the string phase could shift the spectrum (3.1) to a low enough frequency band, so as to overlap with the possible future sensitivity of large interferometric detectors such as LIGO [31] and VIRGO [32]. I will discuss this possibility in terms of a two-parameter model of background evolution, presented in the following Section.

4. Two-parameter model of background evolution

Consider the scenario described in Sect. 1 (see also [3,8,9]), in which the initial (flat and cold) vacuum state, possibly perturbed by the injection of an arbitrarily small (but finite) density of bulk string matter, starts an accelerated evolution towards a phase of growing curvature and dilaton coupling, where the matter contribution becomes eventually negligible with respect to the gravi-dilaton kinetic energy. Such a phase is initially described by the low energy dilaton-dominated solution,

$$a = |\eta|^{1/2}, \qquad \phi = -\sqrt{3}\ln|\eta|, \qquad -\infty < \eta < \eta_s \qquad (4.1)$$

up to the time η_s, when the curvature reaches the string scale $H_s = \lambda_s^{-1}$, at a value of the string coupling $g_s = \exp(\phi_s/2)$. Provided the value of ϕ_s is sufficiently negative (i.e. provided the coupling g_s is sufficiently small to be still in the perturbative regime), such a value is also completely arbitrary, since there is no perturbative potential to break invariance under shifts of ϕ.

For $\eta > \eta_s$ the background enters a high curvature string phase of arbitrary (but unknown) duration, in which all higher-derivative (higher-order in $\alpha' = \lambda_s^2$) contributions to the effective action become important. During such phase the dilaton keeps growing towards the strong coupling regime, up to the time $\eta = \eta_1$ (at a curvature scale H_1), when a non-trivial dilaton potential freezes the coupling to its present constant value $g_1 = \exp(\phi_1/2)$. We shall assume, throughout this paper, that the time scale η_1 marks the end of the string era as well as the (nearly simultaneous) beginning of the standard, radiation-dominated evolution, where $a \sim \eta$ and $\phi = \text{const}$ (see however Sect. 9 for a possible alternative).

During the string phase the curvature is expected to stay controlled by the string scale, so that

$$|H| \simeq gM_p = \frac{e^{\phi/2}}{\lambda_p} = \frac{1}{\lambda_s}, \qquad \eta_s < \eta < \eta_1 \qquad (4.2)$$

where λ_p is the Planck length. As a consequence, the curvature is increasing in the Einstein frame (where λ_p is constant), while it keeps constant in the string frame, where λ_s is constant and the Planck length grows like g from zero (at the initial vacuum) to its present value $\lambda_p \simeq 10^{-19}(GeV)^{-1}$. In both cases the final scale $H_1 \simeq g_1 M_p$ is fixed, and has to be of Planckian order to match the present value of the ratio λ_p/λ_s. Using standard estimates [33]

$$g_1 \simeq \frac{H_1}{M_p} \simeq \left(\frac{\lambda_p}{\lambda_s}\right)_{t_0} \simeq 0.3 - 0.03 \qquad (4.3)$$

In analogy with the dilaton-driven solution (4.1), let us now parameterize, in the Einstein frame, the background kinematic during the string phase with a monotonic metric and dilaton evolution,

$$a = |\eta|^\alpha, \qquad \phi = -2\beta \ln|\eta|, \qquad \eta_s < \eta < \eta_1 \qquad (4.4)$$

representing a sort of "average" time-behavior. Note that the two parameters α and β cannot be independent since, according to eq. (4.2),

$$\left|\frac{H_s}{H_1}\right| \simeq \frac{g_s}{g_1} \simeq \left|\frac{\eta_1}{\eta_s}\right|^{1+\alpha} \simeq \left|\frac{\eta_1}{\eta_s}\right|^\beta \qquad (4.5)$$

from which

$$1 + \alpha \simeq \beta \simeq -\frac{\log(g_s/g_1)}{\log|\eta_s/\eta_1|} \qquad (4.6)$$

(note also that the condition $1 + \alpha = \beta$ cannot be satisfied by the vacuum solutions of the lowest order string effective action [3], in agreement with the fact that all orders in α' are full operative in the high curvature string phase [20]).

The background evolution, for this class of models, is thus completely determined in terms of two parameters only, the duration (in conformal time) of the string phase, $|\eta_s/\eta_1|$, and the shift of the dilaton coupling (or of the curvature scale in Planck units) during the string phase, $g_s/g_1 = (H_s/M_p)/(H_1/M_p)$. I will use, for convenience, the decimal logarithm of these parameters,

$$
x = \log_{10} |\eta_s/\eta_1| = \log_{10} z_s
$$
$$
y = \log_{10}(g_s/g_1) = \log_{10} \frac{(H_s/M_p)}{(H_1/M_p)} \tag{4.7}
$$

Here $z_s = |\eta_s/\eta_1| \simeq a_1/a_s$ defines the total red-shift associated to the string phase in the String frame, where the curvature is approximately constant and the metric undergoes a phase of nearly de Sitter expansion. It should be noted, finally, that the parameters (4.7) are completely frame-independent, as conformal time and dilaton field are exactly the same both in the String and Einstein frame.

5. Parameterized graviton spectrum

Consider the background discussed in the previous Section, characterized by the dilaton-driven evolution (4.1) for $\eta < \eta_s$, by the string evolution (4.4) for $\eta_s < \eta < \eta_1$, and by the standard radiation-dominated evolution for $\eta > \eta_1$. In these three regions, eq. (2.2) for the canonical variable u_k has the general exact solution

$$
\begin{array}{ll}
u_k = |\eta|^{1/2} H_0^{(2)}(|k\eta|), & \eta < \eta_s \\[2mm]
u_k = |\eta|^{1/2} \left[A_+(k) H_\nu^{(2)}(|k\eta|) + A_-(k) H_\nu^{(1)}(|k\eta|) \right], & \eta_s < \eta < \eta_1 \\[2mm]
u_k = \frac{1}{\sqrt{k}} \left[c_+(k) e^{-ik\eta} + c_-(k) e^{ik\eta} \right], & \eta > \eta_1
\end{array}
$$
$$\tag{5.1}$$

where $\nu = |\alpha - 1/2|$, and $H^{(1,2)}$ are the first and second kind Hankel functions. We have normalized the solution to an initial vacuum fluctuation spectrum, containing only positive frequency modes at $\eta = -\infty$

$$\lim_{\eta \to -\infty} u_k = \frac{e^{-ik\eta}}{\sqrt{k}} \qquad (5.2)$$

The asymptotic solution for $\eta \to +\infty$ is however a linear superposition of positive and negative frequency modes, determined by the so-called Bogoliubov coefficients $c_\pm(k)$ which parameterize, in a second quantization approach, the unitary transformation connecting $|in\rangle$ and $|out\rangle$ states. So, even starting from an initial vacuum state, it is possible to find a non-vanishing expectation number of produced particles (in this case gravitons) in the final state, given (for each mode k) by $\langle n_k \rangle = |c_-(k)|^2$.

We shall compute $c_\pm$ by matching the solutions (5.1) and their first derivatives at η_s and η_1. We observe, first of all, that the required growth of the curvature and of the coupling during the string phase (in the Einstein frame) can only be implemented for $|\eta_1| < |\eta_s|$, i.e. $\beta = 1 + \alpha > 0$ [see eq.(4.5)]. This leads to an inflationary string phase, characterized in the Einstein frame by accelerated expansion ($\dot{a} > 0$, $\ddot{a} > 0$, $\dot{H} > 0$) for $-1 < \alpha < 0$, and accelerated contraction ($\dot{a} < 0$, $\ddot{a} < 0$, $\dot{H} < 0$) for $\alpha > 0$. As a consequence, modes which "hit" the effective potential barrier $V(\eta) = a''/a$ of eq.(2.2) (otherwise stated: which cross the horizon) during the dilaton-driven phase, i. e. modes with $|k\eta_s| < 1$, stay under the barrier also during the string phase, since $|k\eta_1| < |k\eta_s| < 1$. In such case the maximal amplified proper frequency

$$\omega_1 = \frac{k_1}{a} \simeq \frac{1}{a\eta_1} \simeq \frac{H_1 a_1}{a} \simeq \left(\frac{H_1}{M_p}\right)^{1/2} 10^{11} Hz = \sqrt{g_1} 10^{11} Hz \qquad (5.3)$$

is related to the highest frequency crossing the horizon in the dilaton phase, $\omega_s = H_s a_s/a$, by

$$\omega_s = \omega_1 \left|\frac{\eta_1}{\eta_s}\right| < \omega_1 \qquad (5.4)$$

For an approximate estimate of c_- we may thus consider two cases.

If $\omega_s < \omega < \omega_1$, i.e. if we consider modes crossing the horizon in the string phase, we can estimate $c_-(\omega)$ by using the large argument limit of the Hankel functions when matching the solutions at $\eta = \eta_s$, using however the small argument limit when matching at $\eta = \eta_1$. In this case the parametric amplification is induced by the second background transition only, as $A_+ \simeq 1$ and $A_- \simeq 0$, and we get

$$|c_-(\omega)| \simeq \left(\frac{\omega}{\omega_1}\right)^{-\nu-1/2}, \qquad \omega_s < \omega < \omega_1 \qquad (5.5)$$

(modulo numerical coefficients of order of unity). If, on the contrary, $\omega < \omega_s$, i.e. we consider modes crossing the horizon in the dilaton phase, we can use the small argument limit of the Hankel functions at both the matching epochs η_s and η_1. This gives $A_\pm = b_\pm |k\eta_s|^{-\nu} \ln |k\eta_s|$ ($b_\pm$ are numbers of order one), and

$$|c_-(\omega)| \simeq \left|\frac{\eta_s}{\eta_1}\right|^\nu \left(\frac{\omega}{\omega_1}\right)^{-1/2} \ln\left(\frac{\omega_s}{\omega}\right), \qquad \omega < \omega_s \qquad (5.6)$$

We can now compute, in terms of $\langle n \rangle = |c_-|^2$, the spectral energy distribution $\Omega(\omega, t)$ (in critical units) of the produced gravitons, defined in such a way that the total graviton energy density ρ_g is obtained as $\rho_g = \rho_c \int \Omega(\omega) d\omega / \omega$. We have then

$$\Omega(\omega, t) \simeq \frac{\omega^4}{M_p^2 H^2} |c_-(\omega)|^2 \simeq$$

$$\simeq \left(\frac{H_1}{M_p}\right)^2 \left(\frac{H_1}{H}\right)^2 \left(\frac{a_1}{a}\right)^4 \left(\frac{\omega}{\omega_1}\right)^{3-2\nu}, \qquad \omega_s < \omega < \omega_1$$

$$\simeq \left(\frac{H_1}{M_p}\right)^2 \left(\frac{H_1}{H}\right)^2 \left(\frac{a_1}{a}\right)^4 \left(\frac{\omega}{\omega_1}\right)^3 \left|\frac{\eta_s}{\eta_1}\right|^{2\nu} \ln^2\left(\frac{\omega_s}{\omega}\right), \qquad \omega < \omega_s$$

$$(5.7)$$

According to eqs. (4.6) and (4.7), moreover, $2\nu = |2\alpha - 1| = |3 + 2y/x|$ and $|\eta_s/\eta_1| = 10^x$. The tensor perturbation spectrum (5.7) is thus completely fixed in terms of our two free parameters x, y, of the (known) fraction of critical energy density stored in radiation at time t, $\Omega_\gamma(t) = (H_1/H)^2 (a_1/a)^4$, and of the (in principle known) present value of the ratio $g_1 = \lambda_p / \lambda_s$, as

$$\Omega(\omega, t) = g_1^2 \Omega_\gamma(t) \left(\frac{\omega}{\omega_1}\right)^{3 - \left|\frac{2y}{x} + 3\right|}, \qquad 10^{-x} < \frac{\omega}{\omega_1} < 1 \qquad (5.8)$$

$$\Omega(\omega, t) = g_1^2 \Omega_\gamma(t) \left(\frac{\omega}{\omega_1}\right)^3 10^{|2y + 3x|}, \qquad \frac{\omega}{\omega_1} < 10^{-x} \qquad (5.9)$$

The same spectrum can also be obtained, with a different approach, working directly in the String frame (see [14]).

The first branch of this spectrum, with unknown slope, is due to modes crossing the horizon in the string phase, the second to modes crossing the horizon in the dilaton phase (I have omitted, for simplicity, the logarithmic term in eq. (5.9), because it is not much relevant for the order of magnitude estimate that I want to discuss here). Note that, as previously stressed, the cubic slope of the dilatonic branch of the spectrum is completely insensitive to the kinematic details of the string phase. Such details can only affect the overall intensity of the perturbation spectrum, and their effects can thus be absorbed by rescaling the total duration of the string phase. For the case $\alpha < 1/2$, in which the amplitude of perturbations is constant outside the horizon, we recover in particular eq.(3.1) with $A = 1$.

Let us now impose on such spectrum the condition of falling within the possible future sensitivity range of large interferometric detectors (such as that of the "Advanced LIGO" project [34]), namely

$$\Omega(\omega_I) \gtrsim 10^{-10}, \qquad \omega_I = 10^2 Hz \qquad (5.10)$$

It implies

$$\left| y + \frac{3}{2}x \right| > \frac{3}{2}x - \frac{(3 + \log_{10} g_1)x}{9 + \log_{10} g_1}, \qquad x > 9 + \frac{1}{2}\log_{10} g_1$$

$$|2y + 3x| > 21 - \frac{1}{2}\log_{10} g_1, \qquad x < 9 + \frac{1}{2}\log_{10} g_1 \qquad (5.11)$$

These conditions define an allowed region in our parameter space (x, y) which has to be further restricted, however, by the upper bound obtained from pulsar-timing measurements [35], namely

$$\Omega(\omega_P) \lesssim 10^{-6}, \qquad \omega_P = 10^{-8} Hz \qquad (5.12)$$

which implies

$$\left| y + \frac{3}{2}x \right| < \frac{3}{2}x - \frac{(1 + \log_{10} g_1)x}{19 + \frac{1}{2}\log_{10} g_1}, \qquad x > 19 + \frac{1}{2}\log_{10} g_1$$

$$|2y + 3x| < 55 - \frac{1}{2}\log_{10} g_1, \qquad x < 19 + \frac{1}{2}\log_{10} g_1 \qquad (5.13)$$

We have to take into account, in addition, the asymptotic behavior of tensor perturbations outside the horizon. During the dilaton phase the growth is only logarithmic, but during the string phase the growth is faster (power-like) for backgrounds with $\alpha > 1/2$. Since the above spectrum has been obtained in the linear approximation, expanding around a homogeneous background, we must impose for consistency that the perturbation amplitude stays always smaller than one, so that perturbations have a negligible back-reaction on the metric. This implies $\Omega < 1$ at all ω and t. This bound, together with the slightly more stringent bound $\Omega < 0.1$ required by standard nucleosynthesis [36], can be automatically satisfied - in view of the g_1^2 factor in eqs. (5.8), (5.9) - by requiring a growing perturbation spectrum, namely

$$y < 0, \qquad\qquad y > -3x \qquad\qquad (5.14)$$

The conditions (5.11), (5.13) and (5.14) determine the allowed region of our parameter space compatible with the production of cosmic gravitons in the interferometric sensitivity range (5.10) (denoted by LIGO, for short). Such a region is plotted in **Fig.1**, by taking $g_1 = 1$ as a reference value. It is bounded below by the condition of nearly homogeneous background (5.14), and above by the same condition plus the pulsar bound (5.13). The upper part of the allowed region corresponds to a class of backgrounds in which the tensor perturbation amplitude stays constant, outside the horizon, during the string phase ($\alpha < 1/2$). The lower part corresponds instead to backgrounds in the the amplitude grows, outside the horizon, during the string phase ($\alpha > 1/2$). The area within the full bold lines refers to modes crossing the horizon in the dilaton phase; the area within the thin lines refers to modes crossing the horizon in the string phase, where the reliability of our predictions is weaker, as we used field-theoretic methods in a string-theoretic regime. Even neglecting all spectra referring to the string phase, however, we obtain a final allowed region which is non-vanishing, though certainly not too large.

The main message of this figure and of the spectrum (5.7) (irrespective of the particular value of the spectral index) is that graviton production, in string cosmology, is in general strongly enhanced in the high frequency sector (kHz-GHz). Such a frequency band, in our context, could be in fact all but the "desert" of relic gravitational radiation that one may expect on the ground of the standard inflationary scenario. This conclusion is independent of the kinematical details of the string phase, which can only affect the shape of the high frequency tail of the spectrum, but not the peak intensity of the spectrum, determined by the fundamental string parameter λ_s. As a consequence, a sensitivity of $\Omega \sim 10^{-4} - 10^{-5}$ in the KHz region (which does not seem out of reach in coincidence experiments between bars and interferometers [37]) could be already enough to detect a signal, so that a null result (in that band, at that level of sensitivity) would already provide a significant constraint on the parameters of the string cosmology background. This should encourage the study and the development of gravitational detectors (such as, for instance, microwave cavities [38]) with large sensitivity in the high frequency sector.

Fig. 1

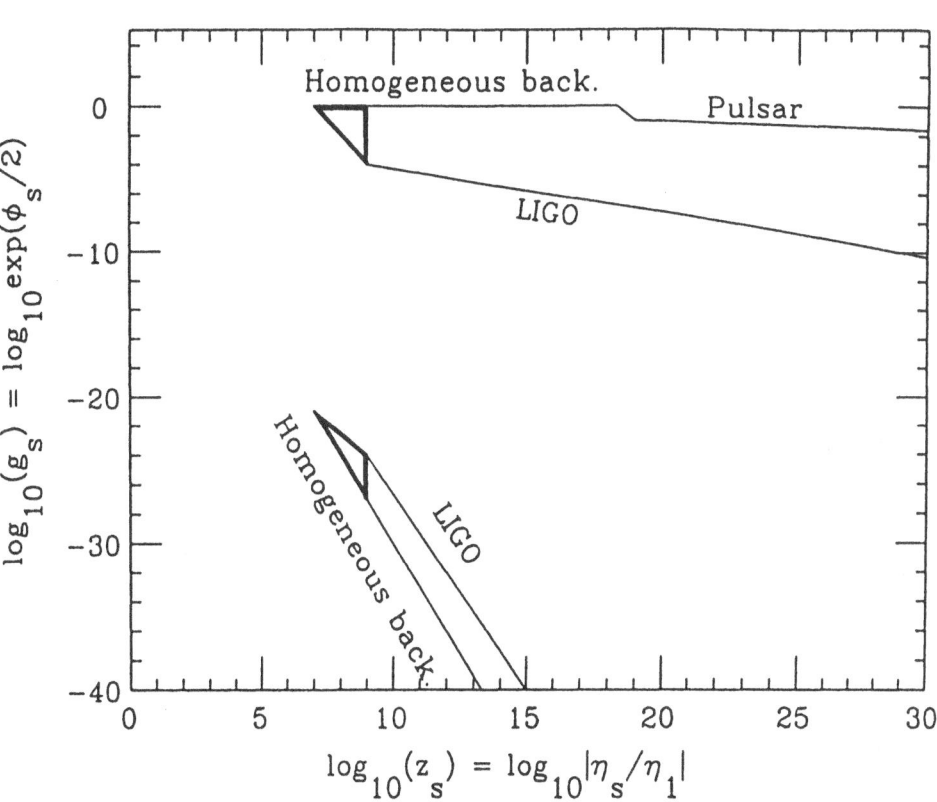

In the following Section I will compare the allowed region of **Fig. 1**, relative to graviton production (and their possible detection), to the allowed region relative to the amplification of electromagnetic perturbations (and to the production of primordial magnetic fields).

6. Parameterized electromagnetic spectrum

In string cosmology, the electromagnetic field $F_{\mu\nu}$ is directly coupled to the dilaton background. To lowest order, such coupling is represented by the string effective action as

$$\int d^{d+1}x\sqrt{|g|}e^{-\phi}F_{\mu\nu}F^{\mu\nu} \qquad (6.1)$$

The electromagnetic field is also coupled to the metric background $g_{\mu\nu}$, of course, but in $d = 3$ the metric coupling is conformally invariant, so that no parametric amplification of the electromagnetic fluctuations is possible in a conformally flat background, like that of a typical inflationary model. One can try to break conformal invariance at the classical or quantum level - there are indeed various attempts in this sense [39,40] - but it turns out that it is very difficult, in general, to obtain a significant electromagnetic amplification from the metric coupling in a natural way, and in a realistic inflationary scenario.

In our context, on the contrary, the vacuum fluctuations of the electromagnetic field can be directly amplified by the time evolution of the dilaton background [15,41]. Consider in fact the correct canonical variable ψ^μ representing electromagnetic perturbations [according to eq. (6.1)] in a $d = 3$, conformally flat background, i.e. $\psi^\mu = A^\mu e^{-\phi/2}$, where $F_{\mu\nu} = \partial_\mu A_\nu - \partial_\nu A_\mu$. The Fourier modes ψ_k^μ satisfy, for each polarization component, the equation

$$\psi_k'' + \left[k^2 - V(\eta)\right]\psi_k = 0, \qquad V(\eta) = e^{\phi/2}(e^{-\phi/2})'' \qquad (6.2)$$

obtained from the action (6.1) by imposing the standard radiation gauge for electromagnetic waves in vacuum, $A^0 = 0$, $\nabla \cdot \mathbf{A} = 0$. Such equation is very similar to the tensor perturbation equation (2.2), with the only difference that the Einstein scale factor, in the effective potential $V(\eta)$, is replaced by the inverse of the string coupling $g^{-1} = e^{-\phi/2}$.

Consider now the string cosmology background of Sect. 4, in which the dilaton-driven phase (4.1) and the string phase (4.4) are followed by the radiation-driven expansion. For such background, the effective potential (6.2) is given explicitly by

$$V = \frac{1}{4\eta^2}(3 - \sqrt{12}), \qquad \eta < \eta_s$$

$$V = \frac{\beta(\beta - 1)}{\eta^2}, \qquad \eta_s < \eta < \eta_1 \qquad (6.3)$$

$$V = 0, \qquad \eta > \eta_1$$

The exact solution of eq. (6.2), normalized to an initial vacuum fluctuation spectrum ($\psi_k \to e^{-ik\eta}/\sqrt{k}$ for $\eta \to -\infty$), is thus

$$\psi_k = |\eta|^{1/2} H_\sigma^{(2)}(|k\eta|), \qquad \eta < \eta_s$$

$$\psi_k = |\eta|^{1/2} \left[B_+(k) H_\mu^{(2)}(|k\eta|) + B_-(k) H_\mu^{(1)}(|k\eta|) \right], \qquad \eta_s < \eta < \eta_1$$

$$\psi_k = \frac{1}{\sqrt{k}} \left[c_+(k) e^{-ik\eta} + c_-(k) e^{ik\eta} \right], \qquad \eta > \eta_1$$

$$(6.4)$$

where $\sigma = (\sqrt{3} - 1)/2$, and $\mu = |\beta - 1/2|$.

For this model of background the effective potential grows in the dilaton phase, keeps growing in the string phase where it reaches a maximum $\sim \eta_1^{-2}$ around the transition scale η_1, and then goes rapidly to zero in the subsequent radiation phase, where $\phi = \phi_1 =$const. The maximum amplified frequency is of the same order as before, $\omega_1 = H_1 a_1/a = |\eta_s/\eta_1|\omega_s > \omega_s$, where $\omega_s = H_s a_s/a$ is the last mode hitting the barrier (or crossing the horizon) in the dilaton phase. For modes with $\omega > \omega_s$ the amplification is thus due to the second background transition only: we can evaluate $|c_-|$ by using the large argument limit of the Hankel functions when matching the solutions at η_s (which gives $B_+ \simeq 1$, $B_- \simeq 0$), using however the small argument limit when matching at η_1, which gives

$$|c_-(\omega)| \simeq \left(\frac{\omega}{\omega_1} \right)^{-\mu - 1/2}, \qquad \omega_s < \omega < \omega_1 \qquad (6.5)$$

Modes with $\omega < \omega_s$, which exit the horizon in the dilaton phase, stay outside the horizon also in the string phase, so that we can use the small argument limit at both the matching epochs: this gives $B_\pm = b_\pm |k\eta_s|^{-\sigma - \mu}$ ($b_\pm$ are numbers of order of unity) and

$$|c_-(\omega)| \simeq \left(\frac{\omega}{\omega_s} \right)^{-\sigma} \left(\frac{\omega}{\omega_1} \right)^{-1/2} \left| \frac{\eta_s}{\eta_1} \right|^\mu, \qquad \omega < \omega_s \qquad (6.6)$$

We are interested, in particular, in the ratio

$$r(\omega) = \frac{\omega}{\rho_\gamma} \frac{d\rho}{d\omega} \simeq \frac{\omega^4}{\rho_\gamma} |c_-(\omega)|^2 \qquad (6.7)$$

measuring the fraction of electromagnetic energy density stored in the mode ω, relative to the total radiation energy ρ_γ. By using the parameterization of Sect. 4 we have $2\mu = |2\beta - 1| = |1 + 2y/x|$ and $|\eta_s/\eta_1| = \omega_1/\omega_s = 10^x$, so that the electromagnetic perturbation spectrum is again determined by two parameters only, the duration of the string phase $|\eta_s/\eta_1|$, and the initial value of the string coupling, $g_s = g_1 10^y$. We find

$$r(\omega) = g_1^2 \left(\frac{\omega}{\omega_1}\right)^{3 - \left|\frac{2y}{x} + 1\right|}, \qquad 10^{-x} < \frac{\omega}{\omega_1} < 1 \qquad (6.8)$$

for modes crossing the horizon in the string phase, and

$$r(\omega) = g_1^2 \left(\frac{\omega}{\omega_1}\right)^{4 - \sqrt{3}} 10^{x(1-\sqrt{3}) + |2y+x|}, \qquad \frac{\omega}{\omega_1} < 10^{-x} \qquad (6.9)$$

for modes crossing the horizon in the dilaton phase. The same spectrum has been obtained, with a different approach, also in the String frame [15,16] (in this paper, the definition of g_1 has been rescaled with respect to [16], by absorbing into g_1 the 4π factor). Finally, this spectrum can be easily generalized to include the effects of an arbitrary number of shrinking internal dimensions during the pre-big-bang phase [15]. The basic qualitative result presented in the following Sections hold however independently of such a generalization.

7. Seed magnetic fields

The above spectrum of amplified electromagnetic vacuum fluctuations has been obtained in the linear approximation, expanding around a homogeneous background. We have thus to impose on the spectrum the consistency condition of negligible back-reaction, $r(\omega) < 1$ at all ω. For $g_1 < 1$ this condition requires a growing perturbation spectrum, and imposes a rather stringent bound on parameter space,

$$y < x, \qquad y > -2x \qquad (7.1)$$

[note that a growing spectrum also automatically satisfies the nucleosynthesis bound $r < 0.1$, in view of the g_1^2 factor which normalizes the strength of the spectrum, and of eq. (4.3)].

It becomes now an interesting question to ask whether, in spite of the above condition, the amplified fluctuations can be large enough to seed the dynamo mechanism which is widely believed to be responsible for the observed galactic (and extragalactic) magnetic fields [42]. Such a mechanism would require a primordial magnetic field coherent over the intergalactic Mpc scale, and with a minimal strength such that [39]

$$r(\omega_G) \gtrsim 10^{-34}, \qquad \omega_G = 10^{-14} Hz \qquad (7.2)$$

This means, in terms of our parameters,

$$|y + \frac{x}{2}| > \frac{3}{2}x - \frac{(17 + \log_{10} g_1)x}{25 + \frac{1}{2}\log_{10} g_1}, \quad x > 25 + \frac{1}{2}\log_{10} g_1$$

$$x(1 - \sqrt{3}) + |2y + x| > 23 - 0.87\log_{10} g_1, \quad x < 25 + \frac{1}{2}\log_{10} g_1 \quad (7.3)$$

Surprisingly enough the answer to the previous question is positive, and this marks an important point in favor of the string cosmology scenario considered here, as it is in general quite difficult - if not impossible - to satisfy the condition (7.2) in other, more conventional, inflationary scenarios.

The allowed region of parameter space, compatible with the production of seed fields [eq. (7.3)] in a nearly homogeneous background [eq. (7.1)], is shown in **Fig. 2** (again for the reference value $g_1 = 1$). In the region within the full bold lines the seed fields are due to modes crossing the horizon in the dilaton phase, in the region within the thin lines to modes crossing the horizon in the string phase. In both cases the background satisfies $y < -x/2$, i.e. $\beta > 1/2$, so that the whole allowed region refers to perturbations which are always growing outside the horizon, even in the string phase.

We may see from **Fig. 2** that the production of seed fields require a very small value of the dilaton coupling at the beginning of the string phase,

$$g_s = e^{\phi_s/2} \lesssim 10^{-20} \qquad (7.4)$$

This initial condition is particularly interesting, as it could have an important impact on the problem of freezing out the classical oscillations of the dilaton background (work is in progress). It also requires a long enough duration of the string phase,

$$z_s = |\eta_s/\eta_1| \gtrsim 10^{10} \qquad (7.5)$$

Fig. 2

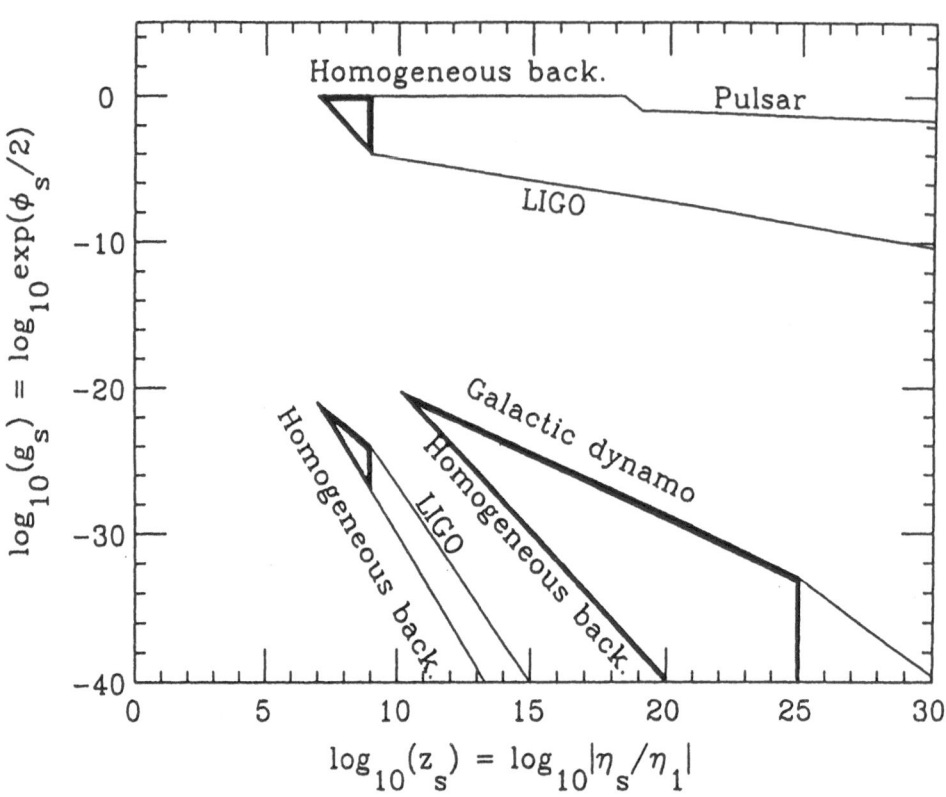

which is not unreasonable, however, when z_s is translated in cosmic time and string units, $z_s = \exp(\Delta t/\lambda_s)$, namely $\Delta t \gtrsim 23\lambda_s$.

Also plotted in **Fig. 2**, for comparison, are the allowed regions for the production of gravitons falling within the interferometric sensitivity range, taken from the previous picture. Since there is no overlapping, a signal detected (for instance) by LIGO would seem to exclude the possibility of producing seed fields, and vice-versa. Such a conclusion should not be taken too seriously, however, because the allowed regions of **Fig. 2** actually define a "minimal" allowed area, obtained within the restricted range of parameters compatible with a linearized description of perturbations. If we drop the linear approximation, then the allowed area extends to the "south-western" part of the plane (x, y), and an overlap between electromagnetic and gravitational regions becomes possible. In that case, however, the perturbative approach around a homogeneous background could not be valid any longer, and we would not be able to provide a correct computation of the spectrum.

The above discussion is based on the low-energy string effective action (6.1). Of course there are corrections to this action coming from the loop expansion in the string coupling $g = e^{\phi/2}$, and higher curvature corrections coming from the α' expansion. Loop corrections are not important, however, until we work in a region of parameter space in which the dilaton is deeply in its perturbative regime ($g \ll 1$), as is indeed the case for the production of seed fields (moreover, the non-perturbative dilaton potential is known to be extremely small, as $V(\phi) \sim \exp(-1/g^2)$ in the weak coupling regime [9]). The α' corrections may be important, but only for modes which crossed the horizon during the string phase (for modes crossing the horizon before the slope of the spectrum is unaffected by the subsequent background kinematics, as already stressed in Sect. 5 for the case of tensor perturbations). It should be clearly stressed, therefore, that the main result of this Section - namely the existence of a wide region of parameter space in which seed fields are efficiently produced - is completely independent of the unknown details of the string phase, which can influence in a direct way only the high frequency tail of the spectrum, and only control the possible extension of the allowed region in the limit of very large z_s.

8. The anisotropy of the CMB radiation

In the inflationary models based on the low energy string effective action, the spectrum of scalar and tensor metric perturbations grows in general too fast with frequency [3,12,13] to be able to explain the large scale anisotropy detected by COBE [43,44]. If we insist, however, in looking for an explanation of the anisotropy in terms of the quantum fluctuations of some primordial field (amplified by the background evolution), a possible - even if unconventional - explanation in a string cosmology context is provided by the vacuum fluctuations of the electromagnetic field [16].

Consider in fact the electromagnetic perturbations re-entering the horizon ($|k\eta| \sim 1 \sim \omega/H$) after amplification. At the time of reentry H^{-1} they provide a field coherent over the horizon scale, which can seed the cosmic magnetic fields, as discussed in the previous Section. If reentry occurs before the decoupling era, the perturbations may be expected to thermalize and to become homogeneous very rapidly soon after reentry, because of their interactions with matter sources in thermal equilibrium. Modes crossing the horizon after decoupling, on the contrary, contribute to the formation of a stochastic perturbation background with a spectrum which remains frozen until the present time t_0, and which may significantly affect the isotropy and homogeneity of the CMB radiation. For a complete electromagnetic origin of the observed anisotropy, $\Delta T/T$, at the present horizon scale, $\omega_0 \sim 10^{-18}$Hz, the perturbation amplitude should satisfy in particular the condition

$$r(\omega_0)\Omega_\gamma(t_0) \sim (\Delta T/T)_0^2 \tag{8.1}$$

namely

$$r(\omega_0) \simeq 10^{-6}, \qquad \omega_0 = 10^{-18} Hz \tag{8.2}$$

According to our spectrum [eqs.(6.8) and (6.9)] this condition can be satisfied consistently with the homogeneity bound (7.1), and without fine-tuning of parameters, provided the string phase is so long that all scales inside our present horizon crossed the horizon (for the first time) during the string phase, i. e. for $\omega_0 > \omega_s$ (or $z_s > 10^{29}$). If we accept this electromagnetic explanation of the anisotropy, we have then two important consequences.

The first follows from the fact that the peak value of the spectrum (6.8) is fixed, so that the spectral index n, defined by

$$r(\omega) = g_1^2 \left(\frac{\omega}{\omega_1}\right)^{n-1} \tag{8.3}$$

can be completely determined as a function of the amplitude at a given scale. For the horizon scale, in particular, we have from eqs. (8.1) and (5.3)

$$n \simeq \frac{25 + \frac{5}{2}\log_{10} g_1 - 2\log_{10}(\Delta T/T)_{\omega_0}}{29 + \frac{1}{2}\log_{10} g_1} \tag{8.4}$$

I have taken explicitly into account here the dependence of the spectrum on the the present value of the string coupling g_1 (which is illustrated in **Fig. 3**), to stress that such dependence is very weak, and that our estimate of n from $\Delta T/T$ is quite stable, in spite of the rather large theoretical uncertainty about g_1^2 (nearly two order of magnitude, recall eq. (4.3)).

In order to match the observed anisotropy, $\Delta T/T \sim 10^{-5}$, we obtain from eq. (8.4) (see also **Fig. 3**, where the relation (8.4) is plotted for three different values of g_1)

$$n \simeq 1.11 - 1.17 \tag{8.5}$$

This slightly growing (also called "blue" spectrum) is flat enough to be well compatible with the present analyses of the COBE data [43,44].

The second consequence follows from the fact that fixing a value of n in eq. (8.3) amounts to fix a relation between the parameters x and y of our background, according to eq. (6.8). If we accept, in particular, a value of n in the range of eq. (8.5), then we are in a region of parameter space which is also compatible with the production of seed fields, according to eq. (7.2). This means that we are allowed to formulate cosmological models in which cosmic magnetic fields and CMB anisotropy have the same common origin, thus explaining (for instance) why the energy density ρ_B of the observed cosmic magnetic fields is of the same order as that of the CMB radiation:

$$\rho_B \sim \rho_\gamma \int^{\omega_1} r(\omega)d(\ln \omega) \sim \rho_{CMB} \tag{8.6}$$

A coincidence which is otherwise mysterious, to the best of my knowledge.

It is important to mention that the values of the parameters leading to eq.(8.5) are also automatically consistent with the bound following from the presence of strong magnetic fields at nucleosynthesis time [45], which imposes $r(\omega_N) \lesssim 0.05$ at the scale corresponding to the end of nucleosynthesis, $\omega_N \simeq 10^{-12}$Hz. By comparing photon and graviton production [eqs. (6.8) and (5.8)] we find, moreover, that for a background in which n lies in the range (8.5) the graviton spectrum grows fast enough with frequency ($\Omega \sim \omega^m$, $m = n+1 = 2.11-2.17$) to be well compatible with the pulsar bound (5.12). Note that, with such a value of m, the metric perturbation contribution to the COBE anisotropy is completely negligible.

It should be stressed, finally, that (in contrast with what discussed in the previous Section) the details of the string phase are of crucial importance for a possible electromagnetic explanation of the CMB

Fig. 3

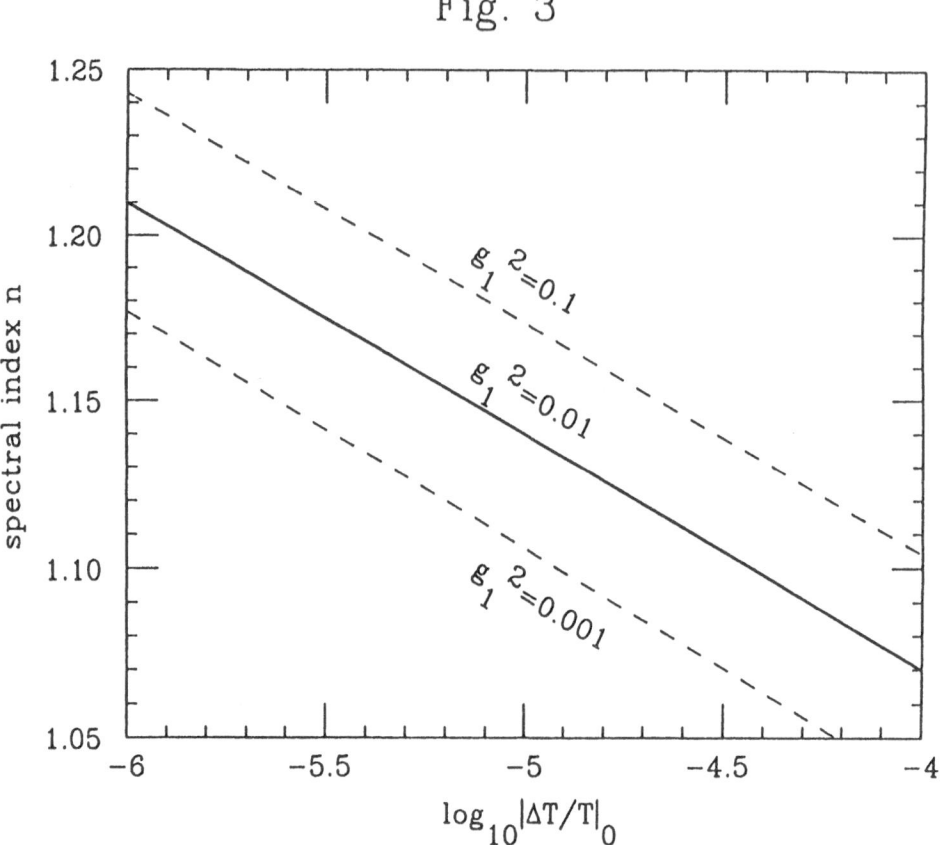

anisotropy. However, once we accept a model in which the background curvature stays nearly constant in the String frame, the anisotropy discussed in this Section is again generated in a range of parameters for which the dilaton is deeply inside the perturbative regime ($g \ll 1$), so that the electromagnetic perturbation equations are certainly stable with respect to loop corrections. Moreover, since we are expanding around the vacuum background ($F_{\mu\nu} = 0$), the perturbation equations are also stable in the linear approximation against α' corrections, provided such corrections appear in the form of powers of the Maxwell field strength with no higher derivative term. This behaviour, on the other hand, is typical of the Eulero forms (like the Gauss-Bonnet invariant) which are conjectured to contribute to the correct (ghost-free) higher curvature expansion of the gravity sector [46]. The possibility discussed in this Section can thus find consistent motivations in a string cosmology context.

9. Possible origin of the CMB radiation

In the standard inflationary picture the density perturbations of the homogeneous and isotropic cosmological model, and the $3°K$ thermal radiation background, have physically distinct origin. The former arise from the vacuum fluctuations of the metric (and possibly other fields), amplified by the external action of the cosmological background, which performs a transition from an inflationary phase to a phase of decelerated evolution. The latter arises instead from the reheating era subsequent to inflation, with a production mechanism which is strongly model-dependent [47] (for instance, collisions of bubbles produced in the phase transition, and/or inflaton decay). As first pointed out by Parker [48], however, also the thermal black-body radiation could have a geometric origin, with a production mechanism closely related to that which amplifies inhomogeneities. Indeed, in a string cosmology context, the class of backgrounds able to provide an electromagnetic explanation of the CMB anisotropy can also account for the production of the CMB radiation itself, directly from the amplification of the vacuum fluctuations of the electromagnetic (and other gauge) fields.

Without introducing "ad hoc" some radiation source, suppose in fact that the background accelerates up to some maximum (nearly Planckian) scale H_1, corresponding to the peak of the effective potential $V(\eta)$ in the perturbation equations, and then decelerates, with corresponding decreasing of the potential barrier. In such a context, the modes of comoving frequency k which "hit" the effective potential barrier $|V(\eta)|$ (namely with $|k\eta_1|\lesssim 1$), are parametrically amplified by the external "pump" field. In a second quantization language this corresponds, as discussed in Sect. 5, to a copious production of particles with a power-like spectral distribution [10],

$$\langle n_k \rangle = |c_-(k)|^2 \simeq |k\eta_1|^{-\alpha} \tag{9.1}$$

(the power α depends on the slope of the potential barrier, and then on the kinematical behavior of the background). The production of particles is instead exponentially suppressed for those modes which never hit the barrier, $|k\eta_1|\gtrsim 1$. In that case one obtains [48,49]

$$\frac{|c_-(k)|^2}{1 + |c_-(k)|^2} \simeq e^{-k\eta_T} \tag{9.2}$$

where η_T is the scale of (conformal) time characterizing the transition from the accelerated to the decelerated regime. For a transition occurring at the time η_1, in particular, it is natural to have $\eta_T \sim |\eta_1|$. For $|k\eta_1|\gtrsim 1$, all the produced particles are thus characterized by a distribution of thermal type,

$$\langle n_k \rangle = |c_-(k)|^2 \simeq \frac{1}{e^{k|\eta_1|} - 1} \tag{9.3}$$

as first noted by Parker [48], at a proper temperature $T_1(t)$ determined by the transition curvature scale, H_1, as

$$T_1(t) \sim \frac{1}{a|\eta_1|} \sim \frac{H_1 a_1}{a(t)} \tag{9.4}$$

The change in the background evolution thus leads to the production of a mixture of all kinds of ultra-relativistic particles, with a spectrum which is thermal (at a temperature T_1) at high frequency ($\omega > T_1$), and possibly distorted by parametric amplification effects at low frequency. For a typical (smaller than Planckian) inflationary scale, the low frequency part of the spectrum remains frozen for those particles (like gravitons and dilatons) which interact only gravitationally, and then decouple immediately after the background transition; on the contrary, the spectrum at low frequency may be expected to thermalize rapidly for all the other produced particles which go on interacting among themselves (and with the background sources) for a long enough period of time after the transition. For such particles the spectrum eventually approaches a thermal distribution in the whole amplified frequency band, with a total their energy density Ω_T (in critical units) fixed by T_1 as

$$\Omega_T(t) \sim \frac{GT_1^4}{H^2} \sim \left(\frac{H_1}{M_p}\right)^2 \left(\frac{H_1}{H}\right)^2 \left(\frac{a_1}{a}\right)^4 \qquad (9.5)$$

Even if, initially, $\Omega_T < 1$ (as $H_1 < M_p$), this thermal component of the produced radiation may then become dominant provided, at the beginning of the decelerated epoch, the scale factor $a(t)$ grows in time more slowly than $H^{-1/2}$ (this is the case, for instance, of the time-reversed dilaton-dominated solution (4.1) which expands like $a \sim t^{1/3}$ for $t \to +\infty$). In that case the relics of such radiation might be identified with the (presently observed) cosmic thermal background, with a red-shifted temperature $T_1(t_0) \simeq 3^0 K$.

It is important to stress that, if the thermal radiation produced in the transition becomes a dominant source of the background, such identification is always possible, in principle, quite independently of the kinematic details of the inflationary phase, of the transition scale H_1, and of the scale $H_r = H_1(H_1/M_p)(a_1/a_r)^2$ at which the radiation becomes dominant. In this context, however, the scale H_1 also determines the amplitude of those fluctuations whose spectrum is not thermalized at low frequencies, but remains frozen after the transition. Therefore, if we identify the observed thermal radiation with that produced in the transition, then the energy density at the scale $\omega_T = T_1$ turns out to be uniquely fixed, also for the decoupled perturbations, in terms of the energy density of the observed CMB spectrum. By calling $\Omega_P(\omega, t) \simeq \omega^4 \langle n(\omega) \rangle / M_p^2 H^2$ the energy density (in critical units) of such perturbations, we have in fact from eqs. (9.1), (9.4) and (9.5)

$$\Omega_P(\omega_T, t) \sim \Omega_T(t) \qquad (9.6)$$

At the present time $t = t_0$ we thus obtain the constraint $\Omega_P(\omega_T) \sim 10^{-4}$ at $\omega_T \sim 10^{11} \text{Hz}$ (modulo factors of order of unity). This condition must be satisfied, in particular, by the energy density stored in gravitational (tensor) perturbations which, as discussed in Sect. 5, is constrained to be much smaller, $\Omega << 10^{-4}$, at lower frequencies (the large scale degree of isotropy [43,44] implies, for instance, $\Omega_P \lesssim 10^{-10}$ at $\omega \simeq 10^{-18} \text{Hz}$). The identification of the observed thermal radiation with that produced in an inflationary background transition is thus compatible with such phenomenological bounds, only if the transition amplifies perturbations with a growing spectrum (which is indeed the case for the string cosmology scenario discussed here).

The discussion of an explicit example in which thermal radiation is produced, and eventually becomes dominant, would require however a model of smooth background evolution. Such a model cannot be constructed to the lowest order in α' from the effective gravi-dilaton action, even including an arbitrary dilaton potential [20]. It may be constructed, however, by including the antisymmetric torsion tensor (equivalent to the axion in four dimensions) among the string background fields, at least if we accept a model of background which is initially contracting at the beginning of the decelerated regime. There is no compelling reason, after all, why the phase of decelerated expansion should start immediately after the change from the negative to the positive time branch of a solution of the string cosmology equations. In particular, a background transition from accelerated expansion to decelerated contraction is also an efficient source of radiation, whose energy density is naturally led to become dominant, as shown in the following example (in an expansion $\rightarrow$ expansion transition it is instead more difficult, for the produced radiation, to dominate over other conventional background sources).

Consider in fact the background field equations of motion [50], at tree-level in the string loop expansion parameter e^ϕ, and to zeroth order in α', written in the String frame

$$R_\mu^\nu + \nabla_\mu \nabla^\nu \phi - \frac{1}{2}\frac{\partial V}{\partial \phi}\delta_\mu^\nu - \frac{1}{4}H_{\mu\alpha\beta}H^{\nu\alpha\beta} = 8\pi G e^\phi T_\mu^\nu$$

$$R - (\nabla_\mu \phi)^2 + 2 \nabla_\mu \nabla^\mu \phi + V - \frac{\partial V}{\partial \phi} - \frac{1}{12}H_{\mu\nu\alpha}H^{\mu\nu\alpha} = 0$$

$$\partial_\nu(\sqrt{|g|}e^{-\phi}H^{\nu\alpha\beta}) = 0 \qquad (9.7)$$

Here $H_{\mu\nu\alpha} = \partial_\mu B_{\nu\alpha}+$ cyclic permutations is the field strength of the antisymmetric (torsion) tensor $B_{\mu\nu} = -B_{\nu\mu}$. I have included a general dilaton potential, $V(\phi)$, and the possible phenomenological contribution of other sources, represented generically by T_μ^ν. By setting $V = 0$ and $T_{\mu\nu} = 0$ we find for the system (9.7) the particular exact (anisotropic) solution [7] (with non-trivial torsion $H_{\mu\nu\alpha} \neq 0$)

$$g_{ij} = \begin{pmatrix} \frac{\alpha+\beta b^2 t^2}{\beta+\alpha b^2 t^2} & \frac{\sqrt{\alpha\beta}(1+b^2 t^2)}{\beta+\alpha b^2 t^2} & 0 \\ \frac{\sqrt{\alpha\beta}(1+b^2 t^2)}{\beta+\alpha b^2 t^2} & 1 & 0 \\ 0 & 0 & 1 \end{pmatrix}$$

$$B_i{}^j = \begin{pmatrix} 0 & \frac{\sqrt{\alpha\beta}(1+b^2 t^2)}{\beta+\alpha b^2 t^2} & 0 \\ -\frac{\sqrt{\alpha\beta}(1+b^2 t^2)}{\beta+\alpha b^2 t^2} & 0 & 0 \\ 0 & 0 & 0 \end{pmatrix}$$

$$e^\phi = e^{\phi_0} \left[1 + b^2 t^2 \coth^2\left(\frac{\gamma}{2}\right)\right]^{-1}, \qquad \alpha = \cosh\gamma + 1, \quad \beta = \cosh\gamma - 1.$$

$$(9.8)$$

Such solution can also be obtained by "inverting" and appropriately "boosting" (through scale factor duality and $O(2,2)$ transformations) the globally flat metric [7]

$$ds^2 = dt^2 - (bt)^2 dx^2 - dy^2 - dz^2 \qquad (9.9)$$

(ϕ_0, β and γ are free parameters).

The background (9.8) is non-trivial only in the (x, y) part of its spatial sections. In order to characterize its kinematic properties, consider the rate-of-change H_x of the relative distance along the x direction, between two observers at rest with a congruence of comoving geodesics u^μ. By projecting the expansion tensor, $\theta_{\mu\nu} = (\nabla_\mu u_\nu + \nabla_\nu u_\mu)/2$, on the unitary vector n_x^μ along x ($n_x^\mu u_\mu = 0$, $n_x^\mu n_{x\mu} = -1$), one easily finds

$$H_x = -\theta_{\mu\nu} n_x^\mu n_x^\nu = -\frac{4t \cosh\gamma}{\alpha\beta(1 + t^4) + (\alpha^2 + \beta^2)t^2} \qquad (9.10)$$

(I have set $b = 1$, for simplicity).

In the $t \to -\infty$ limit $H_x, \dot{H}_x, \dot{\phi}, \ddot{\phi}$ are all positive. In the $t \to +\infty$ limit we have instead $H_x < 0$, $\dot{\phi} < 0$, while $\dot{H}_x, \ddot{\phi}$ are still positive. The time evolution of H_x, (which is the analog of the Hubble parameter of isotropic cosmological backgrounds) shows that the solution (9.8) connects smoothly a phase of accelerated expansion of the superinflationary type, with growing dilaton and curvature scale, to a phase of decelerated contraction, with decreasing dilaton and curvature scale, passing through a phase of maximal (finite) curvature. The solution is defined over the whole time range $-\infty \leq t \leq +\infty$, without any singularity in the curvature and dilaton coupling [7].

In spite of its regular behavior, this solution would seem to be of little phenomenological interest as it is anisotropic, and the phase of contraction (and dilaton rolling) continues for ever down to $t = +\infty$. Suppose, however, to perturb the above background by taking into account the back-reaction of the produced radiation. Consider, for instance, the amplification of the quantum fluctuations of the metric background, whose tensor part satisfies the equation [12,14]

$$\ddot{h}_\omega - \dot{\bar{\phi}} \dot{h}_\omega + \omega^2 h_\omega = 0 \qquad (9.11)$$

for each of the two physical polarization modes of proper frequency ω. In the background (9.8)

$$\bar{\phi} = \phi - \ln \det |g_{\mu\nu}|^{1/2} = -\ln |bt| + const \qquad (9.12)$$

so that, as discussed in Sect. 3, tensor fluctuations are amplified with a nearly thermal spectrum, peaked around a frequency which is of the same order as the maximal curvature scale reached by the background.

Assuming that such scale is determined by λ_s, the energy density ρ_r of the produced radiation provides an initial contribution (for $t \sim \lambda_s$) which is certainly subdominant, as it is suppressed with respect to the other terms of eq. (9.7) by the factor

$$\left(\frac{8\pi G e^\phi \rho_r}{|H_{\mu\nu\alpha} H^{\mu\nu\alpha}|} \right)_{t\sim\lambda_s} \sim \left(\frac{\lambda_p}{\lambda_s} \right)^2 \ll 1 \qquad (9.13)$$

[the expected numerical value of λ_s, in standard Planck units, is given in eq.(4.3)]. However, the radiation contribution decreases in time more slowly than that the torsion-generated shear terms present in eq. (9.7), as

$$\frac{e^\phi \rho_g}{|H_{\mu\nu\alpha} H^{\mu\nu\alpha}|} \sim |\det g_{\mu\nu}|^{-4/6} \sim t^{4/3}, \quad t \to +\infty \qquad (9.14)$$

The two contributions are of the same order for $t = t_r \sim \lambda_s (\lambda_s/\lambda_p)^{3/2}$. For $t \gg t_r$ the produced radiation becomes then dominant and tends to isotropize the initial solution (a well known consequence of the radiation back-reaction [51]). In that limit the the torsion tensor becomes negligible, and putting in eq.(9.7)

$$g_{\mu\nu} = diag(1, -a^2(t)\delta_{ij}), \quad T_\mu^\nu = diag(\rho, -p\delta_i^j), \quad p = \rho/3, \quad H_{\mu\nu\alpha} = 0 \qquad (9.15)$$

we can approximately describe the background evolution through the radiation dominated, isotropic gravi-dilaton equations

$$\dot{\overline{\phi}}^2 - dH^2 = 8\pi G \overline{\rho} e^{\overline{\phi}}$$

$$\dot{H} - H\dot{\overline{\phi}} = \frac{4\pi}{3} G \overline{\rho} e^{\overline{\phi}}$$

$$\dot{\overline{\rho}} + H\overline{\rho} = 0 \qquad (9.16)$$

(we have introduced the "shifted" variables $\overline{\phi} = \phi - \ln \sqrt{|g|}$, $\overline{\rho} = \rho\sqrt{|g|}$ where, in the three-dimensional isotropic case, $\sqrt{|g|} = a^3$).

By selecting the positive time branch of the general solution [2,3] of (9.16), and imposing as initial condition a state of decelerated contraction (to match with the previous regime of background evolution), we are led to a particular solution which can be written, in conformal time,

$$a = \frac{1}{L}\left[\frac{\eta}{\eta + \eta_0(3 + \sqrt{3})}\right]^{\frac{-\sqrt{3}}{2}}\left[\frac{2}{3}\eta^2 + 2\eta\eta_0(1 + \frac{1}{\sqrt{3}})\right]^{1/2}$$

$$e^\phi = e^{\phi_0}\left[\frac{\eta}{\eta + \eta_0(3 + \sqrt{3})}\right]^{-\sqrt{3}}, \qquad \eta > 0 \qquad (9.17)$$

(ϕ_0, η_0, L are integration constants). This solution starts from a singularity (that has been fixed, by time translation, at $\eta = 0$), and evolves from an initial contracting, decreasing dilaton state, towards a final state of standard radiation-dominated expansion, $a \sim t^{1/2}$, with $\phi =$const.

If we consistently take into account the back-reaction of the produced radiation, the background evolution may thus approximately described by the solution (9.8) for $t < t_r$, and by the solution (9.17) for $t > t_r$. The initial contraction is stopped and eventually driven to expansion. The bounce in the scale factor, in the radiation-dominated part of the background, marks the beginning of the standard "post-big-bang" regime. In this simple example the evolution fails to be continuous at $t = t_r$, because of the sudden approximation used to match the torsion-driven to the radiation-driven solution. There are no background singularities at the matching point, however, as we are joining two different solutions within the same time branch, $t \to +\infty$. The dominating radiation, moreover, is entirely produced - via quantum effects - by the classical background evolution.

10. Conclusion

In inflationary string cosmology backgrounds perturbations can be amplified more efficiently than in conventional inflationary backgrounds, as the perturbation amplitude my even grow, instead of being constant, outside the horizon. In some case, like scalar metric perturbations in a dilaton-driven background, the effects of the growing mode can be gauged away. But in other cases the growth is physical, and can prevent a linearized description of perturbations.

In any case, such enhanced amplification is interesting and worth of further study, as it may lead to phenomenological consequences which are unexpected in the context of the standard inflationary scenario. For instance, the production of a relic graviton background strong enough to be detected by the large interferometric detectors, or the production of primordial magnetic fields strong enough to seed the galactic dynamo. Moreover, the possible existence of a relic stochastic electromagnetic background, due to the amplification of the vacuum fluctuations of the electromagnetic field, strong enough to be entirely responsible for the observed large scale CMB anisotropy.

The main problem, in this context, is that a rigorous and truly unambiguous discussion of all these interesting effects would require a complete model of background evolution, including a smooth transition from the accelerated to the decelerated regime, through a quantum string era of Planckian curvature. A solid string-theoretic treatment of such an era at present is still lacking, even if recent progress and suggestions [52] may prove useful. The understanding of singularities in string theory would certainly put on a firmer ground the phenomenological model discussed in this lecture, and might even provide a framework for the calculation of our basic parameters z_s, g_s.

I have shown, nevertheless, that by including torsion in the low energy effective action it seems possible (even to lowest order in α') to formulate very simple models of background implementing a "graceful exit" from the pre-big-bang regime, at least if we accept a contracting metric in the post-big-bang evolution. In that context a thermal radiation background is automatically produced as a consequence of the classical evolution, and the associate quantum back-reaction may eventually become dominant, thus driving the background towards a final expanding, constant dilaton regime. Moreover, thanks to the growth of the perturbation spectrum, the identification of that radiation with the observed CMB one is perfectly consistent with the presently known phenomenological bounds, thus providing a framework for a unified explanation of the $3^\circ K$ background and of its anisotropies.

Acknowledgments.

I would like to thank R. Brustein, M. Giovannini, J. Maharana, K. Meissner, V. Mukhanov, N. Sánchez and G. Veneziano for fruitful and enjoyable collaborations which led to develop the "pre-big-bang" cosmological scenario and to investigate its possible phenomenological consequences.

11. References

1. M. Gasperini and G. Veneziano, Astropart. Phys. 1, 317 (1993)
2. M. Gasperini and G. Veneziano, Mod. Phys. Lett. A8, 3701 (1993)
3. M. Gasperini and G. Veneziano, Phys. Rev. D50, 2519 (1994)
4. M. Gasperini, N. Sanchez and G. Veneziano, Nucl. Phys. B364, 3 65 (1991);
   Int. J. Mod. Phys. A6, 3853 (1991).
5. G. Veneziano, Phys. Lett. B265, 287 (1991)
6. K. A. Meissner and G. Veneziano, Phys. Lett. B267, 33 (1991);
   Mod. Phys. Lett. A6, 3397 (1991);
   M. Gasperini and G.Veneziano, Phys. Lett. B277, 256 (1992).
7. M. Gasperini, J. Maharana and G. Veneziano, Phys. Lett. B272, 277 (1991)
8. M. Gasperini, in "Proc. of the 2nd Journée Cosmologie" (Observatoire de Paris, June 1994), ed. by N. Sanchez and H. de Vega (World Scientific, Singapore), p.429
9. G. Veneziano, *String cosmology: basic ideas and general results*, these proceedings
10. L. P. Grishchuk, Sov. Phys. JEPT 40, 409 (1975);
    A. A. Starobinski, JEPT Lett. 30, 682 (1979).
11. M. Gasperini and M. Giovannini, Phys. Lett. B282, 36 (1992)
12. M. Gasperini and M. Giovannini, Phys. Rev. D47, 1529 (1992)
13. R. Brustein, M. Gasperini, M. Giovannini, V. F. Mukhanov and G. Veneziano, Phys. Rev. D51, 6744 (1995)
14. R. Brustein, M. Gasperini, M. Giovannini and G. Veneziano, Phys. Lett. B (1995), in press (hep-th/9507017)
15. M. Gasperini, M. Giovannini and G. Veneziano, *Primordial magnetic fields from string cosmology*, CERN-TH/95-85 (hep-th/9504083)
16. M. Gasperini, M. Giovannini and G. Veneziano, *Electromagnetic origin of the CMB anisotropy in string cosmology*, CERN-TH/95-102 (astro-ph/9505041)
17. R. Brustein, M. Gasperini and M. Giovannini, *Possible common origin of primordial perturbations and of the cosmic microwave background*, Essay written for the 1995 Awards for Essays on Gravitation (Gravity Research Foundation, Wellesley Hills, Ma), and selected for Honorable Mention (unpublished)
18. M. Gasperini, *Amplification of vacuum fluctuations in string cosmology backgrounds*, in "Proc. of the 3rd Colloque Cosmologie" (Observatoire de Paris, June 1995), ed. by N. Sanchez and H. de Vega (World Scientific, Singapore) (hep-th/9506140)

342

19. M. Gasperini, M. Giovannini, K. A. Meissner and G. Veneziano, *Evolution of a string network in backgrounds with rolling horizons* (CERN-TH/95-40), to appear in "New developments in string gravity and physics at the Planck energy scale", ed. by N. Sanchez (World Scientific, Singapore, 1995)
20. R. Brustein and G. Veneziano, Phys. Lett. B329, 429 (1994); N. Kaloper, R. Madden and K. A. Olive, *Towards a singularity free inflationary universe?*, UMN-TH-1333/95 (June 1995).
21. D. Shadev, Phys. Lett. B317, 155 (1984); R. B. Abbott, S. M. Barr and S. D. Ellis, Phys. Rev. D30, 720 (1984); E. W. Kolb, D. Lindley and D. Seckel, Phys. Rev. D30, 1205 (1984); F. Lucchin and S. Matarrese, Phys. Lett. B164, 282 (1985).
22. A. A. Starobinski, Rel. Astr. Cosm., Byel. SSR Ac. Sci. Minsk (1976), p.55 (in russian)
23 R. B. Abbott, B. Bednarz and S. D. Ellis, Phys. Rev. D33, 2147 (1986)
24. V. Mukhanov, H. A. Feldman and R. Brandenberger, Phys. Rep. 215, 203 (1992)
25. L. P. Grishchuk and Y. V. Sidorov, Phys. Rev. D42, 3413 (1990)
26. J. D. Barrow, J. P. Mimoso and M. R. de Garcia Maia, Phys. Rev. D48, 3630 (1993)
27. S. W. Hawking, Astrophys. J. 145, 544 (1966)
28. A. R. Liddle and D. H. Lyth, Phys. Rep. 231, 1 (1993)
29. M. Bruni, G. F. R. Ellis and P. K. S. Dunsby, Class. Quantum Grav. 9, 921 (1992)
30. M. Gasperini, Phys. Lett. B327, 214 (1994)
31. A. Abramovici et al., Science 256, 325 (1992)
32. B. Caron et. al., *Status of the VIRGO experiment*, Lapp-Exp-94-15
33. V. Kaplunowski, Phys. Rev. Lett. 55, 1036 (1985)
34. K. S. Thorne, in "300 Years of Gravitation", ed. by S. W. Hawking and W. Israel (Cambridge Univ. Press, Cambridge, 1987)
35. D. R. Stinebring et al., Phys. Rev. Lett. 65, 285 (1990)
36. V. F. Schwarztmann, JEPT Letters 9, 184 (1969)
37. P. Astone, J. A. Lobo and B. F. Schutz, Class. Quantum Grav. 1, 2093 (1994)
38. F. Pegoraro, E. Picasso, L. Radicati, J. Phys. A1, 1949 (1978); C. M. Caves, Phys. Lett. B80, 323 (1979); C. E. Reece et al., Phys. Lett A104, 341 (1984).
39. M. S. Turner and L. M. Widrow, Phys. Rev. D37, 2743 (1988)
40. B. Ratra, Astrophys. J. Lett. 391, L1 (1992); A. D. Dolgov, Phys. Rev. D48, 2499, (1993).

41. D. Lemoine and M. Lemoine, *Primordial magnetic fields in string cosmology*, April 1995
42. E. N. Parker, "Cosmical Magnetic fields" (Clarendon, Oxford, England, 1979)
43. G. F. Smoot et al., Astrophys. J. 396, L1 (1992)
44. C. L. Bennett et al., Astrophys. J. 430, 423 (1994)
45. D. Grasso and H. R. Rubinstein, Astropart. Phys. 3, 95 (1995)
46. B. Zwiebach, Phys. Lett. B156, 315 (1985)
47. E. W. Kolb and M. S. Turner, *The Early Universe* (Addison-Wesley, Redwood City, Ca, 1990)
48. L. Parker, Nature 261, 20 (1976)
49. N. D. Birrel and P. C. W. Davies, *Quantum fields in curved spaces* (Cambridge Univ. Press, Cambridge, England, 1982);
    B. Allen, Phys. Rev. D37 (1988) 2078;
    J. Garriga and E. Verdaguer, Phys. Rev. D39 (1989) 1072;
    C. R. Stephens, Phys. Lett. A142 (1989) 68.
50. C. Lovelace, Phys. Lett. B135 (1984) 75;
    E. S. Fradkin and A. A. Tseytlin, Nucl. Phys. B261 (1985) 1;
    C. G. Callan et al., Nucl. Phys. B262 (1985) 593.
51. J. B. Zeldovich and I. D. Novikov, *Relativistic astrophysics*, (Chicago University Press, 1983)
52. E. Kiritsis and K. Kounnas, Phys. Lett. B331, 51 (1994);
    A. A. Tseytlin, Phys. Lett. B334, 315 (1994);
    E. Martinec, Class. Quantum Grav. 12, 941 (1995).

PREDICTIONS FROM QUANTUM COSMOLOGY

ALEXANDER VILENKIN

Institute of Cosmology, Department of Physics and Astronomy, Tufts University, Medford, MA, 02155, USA

1. Introduction

If the cosmological evolution is followed back in time, we come to the initial singularity where the classical equations of general relativity break down. This led many people to believe that in order to understand what actually happened at the origin of the universe, we should treat the universe quantum-mechanically and describe it by a wave function rather than by a classical spacetime. This quantum approach to cosmology was initiated by DeWitt [1] and Misner [2], and after a somewhat slow start has become very popular in the last decade or so. The picture that has emerged from this line of development [3, 4, 6, 5, 7, 8, 9] is that a small closed universe can spontaneously nucleate out of nothing, where by 'nothing' I mean a state with no classical space and time. The cosmological wave function can be used to calculate the probability distribution for the initial configurations of the nucleating universes. Once the universe nucleated, it is expected to go through a period of inflation, which is a rapid (quasi-exponential) expansion driven by the energy of a false vacuum. The vacuum energy is eventually thermalized, inflation ends, and from then on the universe follows the standard hot cosmological scenario. Inflation is a necessary ingredient in this kind of scheme, since it gives the only way to get from the tiny nucleated universe to the large universe we live in today.

Another possible use for quantum cosmology is to determine the probability distribution for the values of the constants of Nature. The constants can vary from one universe to another due to a different choice of the vacuum state, a different compactification scheme in higher-dimensional theories, or to Planck-scale wormhole effects [10]. The cosmological wave function will then be a superposition of terms corresponding to all possible values of the constants.

345

N. Sánchez and A. Zichichi (eds.), String Gravity and Physics at the Planck Energy Scale, 345–367.
© 1996 *Kluwer Academic Publishers.*

In these lectures, I would like to review where we stand in this program. The general ideas of quantum cosmology and predictions for the initial state are discussed in Section 2-4, followed by a discussion of predictions for the constants of Nature in Sections 5,6. Due to the time constraints, some important topics will be left out. These include topology-changing processes, third quantization, consistent histories approach, and decoherence.

2. A Simple Model

2.1. 'THE GREATEST MISTAKE OF MY LIFE'

First I would like to illustrate how the nucleation of a universe can be described in a very simple model. The model is defined by the action

$$S = \int d^4x \sqrt{-g} \left(\frac{R}{16\pi G} - \rho_v \right), \tag{1}$$

where ρ_v is a constant vacuum energy and the universe is assumed to be homogeneous, isotropic, and closed,

$$ds^2 = \sigma^2[-dt^2 + a^2(t)d\Omega_3^2]. \tag{2}$$

Here, $d\Omega_3^2$ is the metric on a unit three-sphere, and $\sigma^2 = 2G/3\pi$ is a normalizing factor chosen for later convenience. The scale factor $a(t)$ satisfies the evolution equation

$$\dot{a}^2 + 1 - H^2 a^2 = 0, \tag{3}$$

where

$$H = 4G\rho_v^{1/2}/3. \tag{4}$$

The solution of Eq.(3) is the de Sitter space,

$$a(t) = H^{-1} \cosh(Ht). \tag{5}$$

The universe contracts at $t < 0$, reaches the minimum radius $a = H^{-1}$ at $t = 0$, and re-expands at $t > 0$.

This is similar to the behavior of a particle bouncing off a potential barrier, with a playing the role of particle coordinate. Now, we know that in quantum mechanics particles can not only bounce off, but can also tunnel through potential barriers. This suggests the possibility that the negative-time part of the evolution in (5) may be absent, and that the universe may instead tunnel from $a = 0$ directly to $a = H^{-1}$.

When I suggested this idea in 1982, I made an attempt to estimate the tunneling probability in the semiclassical approximation. To describe

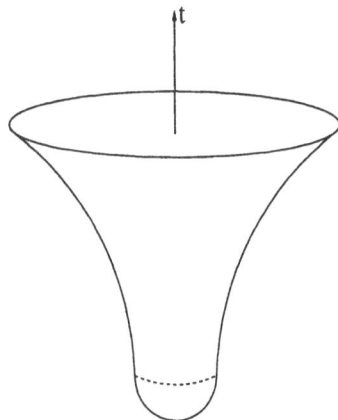

Figure 1. A schematic representation of the birth of inflationary universe.

the tunneling process, I used the bounce solution of the Euclidean field equations, which can be obtained by substituting $t = -i\tau$ in Eqs.(2),(5),

$$ds^2 = \sigma^2[d\tau^2 + \tilde{a}^2(\tau)d\Omega_3^2], \tag{6}$$

$$\tilde{a}(\tau) = H^{-1}\cos(H\tau). \tag{7}$$

This metric describes a four-sphere S^4 of radius H^{-1}. The nucleation of the universe is schematically represented in Fig. 1, where the bounce solution (7) connects to the Lorentzian solution (5) at the turning point $\tau = t = 0$.

For 'normal' quantum tunneling, the tunneling probability $\mathcal{P}$ is proportional to $\exp(-S_E)$, where S_E is the Euclidean action for the corresponding bounce. In our case,

$$S_E = \int d^4x\sqrt{-g}\left(-\frac{R}{16\pi G} + \rho_v\right) = -2\rho_v\Omega_4\sigma^4 H^{-4} = -3/8G^2\rho_v, \tag{8}$$

where

$$R = 12H^2 = 32\pi G\rho_v \tag{9}$$

is the scalar curvature, and $\Omega_4 = 4\pi^2/3$ is the volume of a unit four-sphere. Hence, I concluded in Ref. [3] that

$$\mathcal{P} \propto \exp\left(\frac{3}{8G^2\rho_v}\right). \tag{10}$$

Following fashion, I might declare this 'the greatest mistake of my life' [11].

2.2. THE TUNNELING WAVE FUNCTION

What I now think is the correct answer is given by $\mathcal{P} \propto \exp(-|S_E|)$. In the case of 'normal' quantum tunneling, the Euclidean action is positive-definite, and $|S_E| = S_E$, but for quantum gravity this is no longer so. The reason for using the absolute value of S_E can be understood by considering the tunneling wave function for our problem. To write the corresponding wave equation, we first substitute (2) into (1), and after integrating by parts find the Lagrangian

$$\mathcal{L} = \frac{1}{2}a(1 - \dot{a}^2 - H^2a^2). \tag{11}$$

The momentum conjugate to a is

$$p_a = -a\dot{a}, \tag{12}$$

and the Hamiltonian is

$$\mathcal{H} = -\frac{1}{2a}(p_a^2 + a^2 - H^2a^4). \tag{13}$$

The evolution equation (3) implies that

$$\mathcal{H} = 0. \tag{14}$$

Quantization of this model amounts to replacing $p_a \rightarrow -i\partial/\partial a$ and imposing the Wheeler-DeWitt equation

$$\mathcal{H}\psi = 0. \tag{15}$$

This gives

$$\left[\frac{d^2}{da^2} - U(a)\right]\psi(a) = 0, \tag{16}$$

where

$$U(a) = a^2(1 - H^2a^2), \tag{17}$$

and I have ignored the ambiguity in the ordering of non-commuting operators a and p_a. (This ambiguity is unimportant in the semiclassical domain which we will be mainly concerned with in these lectures).

Eq.(16) has the form of a one-dimensional Schrodinger equation for a 'particle' described by a coordinate $a(t)$, having zero energy, and moving in a potential $U(a)$. The classically allowed region is $a \geq H^{-1}$, and the WKB solutions of (16) in this region are

$$\psi_{\pm}(a) = [p(a)]^{-1/2}\exp\left[\pm i \int_{H^{-1}}^{a} p(a')da' \mp i\pi/4\right], \tag{18}$$

where $p(a) = [-U(a)]^{1/2}$. The under-barrier, $a < H^{-1}$, solutions are

$$\tilde{\psi}_{\pm}(a) = |p(a)|^{-1/2}\exp\left[\pm \int_{a}^{H^{-1}} |p(a')|da'\right]. \tag{19}$$

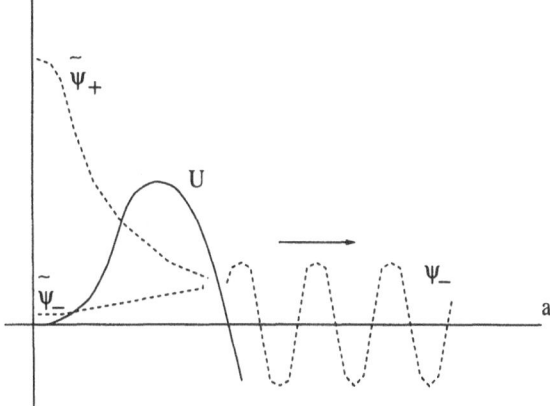

Figure 2. Tunneling wave function for the de Sitter minisuperspace model. The 'potential' $U(a)$ is shown by a solid line and the wave function by a dashed line.

For $a \gg H^{-1}$,

$$\hat{p}_a \psi_\pm(a) \approx \pm p(a)\psi_\pm(a), \qquad (20)$$

and Eq.(12) tells us that $\psi_-(a)$ and $\psi_+(a)$ describe an expanding and a contracting universe, respectively. In the tunneling picture, it is assumed that the universe originated at small size and then expanded to its present, large size. This means that the component of the wave function describing a universe contracting from infinitely large size should be absent:

$$\psi(a > H^{-1}) = \psi_-(a). \qquad (21)$$

The under-barrier wave function is found from the WKB connection formula,

$$\psi(a < H^{-1}) = \tilde{\psi}_+(a) - \frac{i}{2}\tilde{\psi}_-(a). \qquad (22)$$

The growing exponential $\tilde{\psi}_-(a)$ and the decreasing exponential $\tilde{\psi}_+(a)$ have comparable amplitudes at the nucleation point $a = H^{-1}$, but away from that point the decreasing exponential dominates (see Fig. 2). The 'nucleation probability' can be estimated as

$$\mathcal{P} \sim \exp\left(-2\int_0^{H^{-1}} |p(a')| da'\right) = \exp(-|S_E|) = \exp\left(-\frac{3}{8G^2\rho_v}\right). \qquad (23)$$

The use of the semiclassical approximation is justified as long as $|S_E| = 3/8G^2\rho_v \gg 1$, or $\rho_v \ll \rho_p$, where $\rho_p = G^{-2} = m_p^4$ is the Planck density and m_p is the Planck mass.

Eq.(23) was obtained independently by Linde [5], Zel'dovich and Starobinsky [6], Rubakov [7], and myself [8]. But the story does not end here. Not everybody agrees that my first answer (10) was a mistake. The same expression for $\mathcal{P}$ is obtained in the Euclidean approach developed by Hawking and collaborators. We shall return to this on-going debate after discussing the general formalism of quantum cosmology.

3. Wave Function of the Universe

3.1. WHEELER-DE WITT EQUATION

In the general case, the wave function of the universe is defined on super-space, which is the space of all 3-dimensional geometries and matter field configurations,

$$\psi[h_{ij}(\mathbf{x}), \varphi(\mathbf{x})], \tag{24}$$

where h_{ij} is the 3-metric, and matter fields are represented by a single scalar field φ. The wave function ψ satisfies the Wheeler-DeWitt (WDW) equation,

$$\mathcal{H}\psi(h_{ij}, \varphi) = 0, \tag{25}$$

which can be thought of as representing the fact that the energy of a closed universe is equal to zero. The WDW equation can be symbolically written in the form

$$(\nabla^2 - U)\psi = 0, \tag{26}$$

which is similar to the Klein-Gordon equation. Here, ∇^2 is the superspace Laplacian, and the functional $U(h_{ij}, \varphi)$ can be called 'superpotential'. (We shall not need explicit forms of ∇^2 and U).

We have no idea how to solve the WDW equation in the general case. Most of what we know about quantum cosmology has been found using minisuperspace models in which the infinite number of degrees of freedom in Eq.(25) is reduced to a few independent variables. The de Sitter model (16) with a single variable is the simplest example of minisuperspace. The minisuperspace approach is justified when the remaining degrees of freedom can be treated as small perturbations. The corresponding wave function can then be calculated perturbatively [12, 13].

Quantum cosmology is based on quantum gravity and shares all of its problems. In addition, it has some extra problems which arise when one tries to quantize a closed universe. The first problem stems from the fact that ψ is independent of time. This can be understood [1] in the sense that the wave function of the universe should describe everything, including the clocks which show time. In other words, time should be defined intrinsically in terms of the geometric or matter variables. However, no general prescription has yet been found that would give a function $t(h_{ij}, \varphi)$ that would be, in some sense, monotonic. A related problem is the definition of probability.

Given a wave function ψ, how can we calculate probabilities? One can try to use the conserved current [1, 2]

$$J = i(\psi^*\nabla\psi - \psi\nabla\psi^*), \qquad \nabla \cdot J = 0. \tag{27}$$

The conservation is a useful property, since we want probability to be conserved. But one runs into the same problem as with Klein-Gordon equation: the probability defined in this way is not positive-definite. Although we do not know how to solve these problems in general, they can both be solved in the semiclassical domain. In fact, it is possible that this is all we need.

3.2. SEMICLASSICAL UNIVERSES

Let us consider the situation when some of the variables $\{c\}$ describing the universe behave classically, while the rest of the variables $\{q\}$ must be treated quantum-mechanically. Then the wave function of the universe can be written as a superposition

$$\psi = \sum_k A_k(c)e^{iS_k(c)}\chi_k(c,q) \equiv \sum_k \psi_k^{(c)}\chi_k, \tag{28}$$

where the classical variables are described by the WKB wave function $\psi_k^{(c)} = A_k e^{iS_k}$. In the semiclassical regime, ∇S is large, and substitution of (28) into the WDW equation (26) yields the Hamilton-Jacobi equation for $S(c)$,

$$\nabla S \cdot \nabla S + U = 0. \tag{29}$$

The summation in (28) is over different solutions of this equation. Each solution of (29) is a classical action describing a congruence of classical trajectories (which are essentially the gradient curves of S). Hence, a semiclassical wave function $\psi_c = Ae^{iS}$ describes an ensemble of classical universes evolving along the trajectories of $S(c)$. A probability distribution for these trajectories can be obtained using the conserved current (27). Since the variables c behave classically, these probabilities do not change in the course of evolution and can be thought of as probabilities for various initial conditions. The time variable t can be defined as any monotonic parameter along the trajectories, and it can be shown [1, 14] that in this case the corresponding component of the current J is non-negative, $J_t \geq 0$. Moreover, one finds [15, 16, 12] that the 'quantum' wave function χ satisfies the usual Schrodinger equation,

$$i\partial\chi/\partial t = H_\chi\chi \tag{30}$$

with an appropriate Hamiltonian H_χ. Hence, all the familiar physics is recovered in the semiclassical regime.

This semiclassical interpretation of the wave function ψ is valid to the extent that the WKB approximation for ψ_c is justified and the interference between different terms in (28) can be neglected. Otherwise, time and probability cannot be defined, suggesting that the wave function has no meaningful interpretation. In a universe where no object behaves classically (that is, predictably), no clocks can be constructed, no measurements can be made, and there is nothing to interpret.

3.3. BOUNDARY CONDITIONS

As (almost) any differential equation, the WDW equation has an infinite number of solutions. To get a unique solution, one has to specify some boundary conditions in superspace. In ordinary quantum mechanics, the boundary conditions for the wave function are determined by the physical setup external to the system under consideration. In quantum cosmology, there is nothing external to the universe, and it appears that a boundary condition should be added to Eq.(25) as an independent physical law.

Several candidates for this law of boundary conditions have been proposed. Hartle and Hawking [4] suggested that $\psi(h, \varphi)$ should be given by a path integral over compact, Euclidean 4-geometries $g_{\mu\nu}(\mathbf{x}, \tau)$ bounded by the 3-geometry $h_{ij}(\mathbf{x})$ with the field configuration $\varphi(\mathbf{x})$:

$$\psi = \int^{(h,\varphi)} [dg][d\varphi] \exp[-S_E(g, \varphi)]. \tag{31}$$

In this path-integral representation, the boundary condition corresponds to specifying the class of histories integrated over in Eq.(31). Compact 4-geometries can be thought of as histories interpolating between a point ('nothing') and a finite 3-geometry h_{ij}.

Alternatively, I proposed [8, 17] that $\psi(h, \varphi)$ should be obtained by integrating over Lorentzian histories interpolating between a vanishing 3-geometry $\emptyset$ and (h, φ) and lying to the past of (h, φ):

$$\psi(h, \varphi) = \int_{\emptyset}^{(h,\varphi)} [dg][d\varphi] e^{iS}. \tag{32}$$

This wave function is closely related to Teitelboim's causal propagator [18] $K(h_2, \varphi_2 | h_1, \varphi_1)$:

$$\psi(h, \varphi) = K(h, \varphi | \emptyset). \tag{33}$$

Linde [5] suggested that, instead of the standard Euclidean rotation $t \to -i\tau$, the action S_E in (31) should be obtained by rotating in the opposite sense, $t \to +i\tau$.

Halliwell and Hartle [19] discussed a path integral over complex metrics which are not necessarily purely Lorentzian or purely Euclidean. This encompasses all of the above proposals and opens new possibilities. However, the space of complex metrics is very large, and no obvious choice of integration contour suggests itself as the preferred one.

In addition to these path-integral no-boundary proposals, one candidate law of boundary conditions has been formulated directly as a boundary condition in superspace. This is the so-called tunneling boundary condition [20, 21] which requires that ψ should include only outgoing waves at boundaries of superspace. It has been argued [22] that, in a wide class of models, this boundary condition is equivalent to the Lorentzian path integral proposal (32).

For the simple de Sitter model of Sec.2, the tunneling wave function $\psi_T(a)$ is given by Eqs.(21),(22), the Hartle-Hawking wave function is [23]

$$\psi_H(a > H^{-1}) = \psi_+(a) - \psi_-(a), \tag{34}$$

$$\psi_H(a < H^{-1}) = \tilde{\psi}_-(a), \tag{35}$$

and the Linde wave function is [5, 24, 25]

$$\psi_L(a > H^{-1}) = \frac{1}{2}[\psi_+(a) + \psi_-(a)], \tag{36}$$

$$\psi_L(a < H^{-1}) = \tilde{\psi}_+(a). \tag{37}$$

Unlike the tunneling wave function, both Hartle-Hawking and Linde wave functions include expanding and contracting universe components with equal amplitudes (see Fig. 3).

4. Predictions for the Initial State

4.1. INITIAL VACUUM ENERGY

To see what kind of cosmological predictions we can get from different boundary conditions, I would like to consider a somewhat more realistic model. Instead of a constant vacuum energy ρ_v, I introduce a scalar field φ with a potential $V(\varphi)$. Since vacuum energy is very small in our part of the universe, $V(\varphi)$ should have a minimum with $V \approx 0$. The WDW equation for this two-dimensional model can be solved assuming that $V(\varphi)$ is a slowly-varying function and is well below the Planck density,

$$|V'/V| \ll m_p^{-1}, \qquad V \ll \rho_p \equiv m_p^4. \tag{38}$$

A slowly-varying $V(\varphi)$ helps to simplify the equation, but is also necessary for the inflationary scenario. If the condition $V(\varphi) \ll \rho_p$ is violated, then the semiclassical approximation is not valid and higher-order corrections to quantum gravity are important.

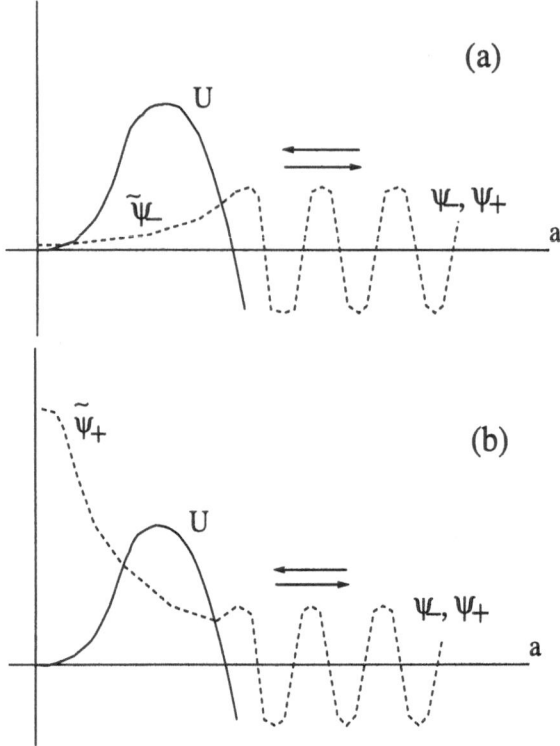

Figure 3. Hartle-Hawking (a) and Linde (b) wave functions for de Sitter minisuperspace model.

After an appropriate rescaling of the scale factor a and the scalar field φ, the WDW equation can be written as

$$\left[\frac{\partial^2}{\partial a^2} - \frac{1}{a^2}\frac{\partial^2}{\partial \varphi^2} - U(a,\varphi)\right]\psi(a,\varphi) = 0, \tag{39}$$

where

$$U(a,\varphi) = a^2[1 - a^2 V(\varphi)]. \tag{40}$$

With the assumptions (38), one finds [21] that Hartle-Hawking, Linde, and tunneling solutions of this equation are given essentially by the same expressions as for the simple model (16), but with ρ_v replaced by $V(\varphi)$. The only difference is that the wave function is multiplied by a factor $C(\varphi)$, such that $\psi(a,\varphi)$ becomes φ-independent in the limit $a \to 0$ (with $|\varphi| < \infty$).

The initial state of the nucleating universe in this model is characterized by the value of the scalar field φ, with the initial value of a given by $a = V^{-1/2}(\varphi)$. The probability distribution for φ can be found using the conserved current (27),

$$\partial_a J^a + \partial_\varphi J^\varphi = 0. \tag{41}$$

With a proper normalization, the quantity $\rho(a, \varphi)d\varphi$, where

$$\rho(a, \varphi) = J^a(a, \varphi) = i(\psi^* \partial_a \psi - \psi \partial_a \psi^*), \tag{42}$$

can be interpreted as the probability for the scalar field to be between φ and $\varphi + d\varphi$ when the scale factor is equal to a.

For the tunneling wave function one finds

$$\rho_T(\varphi) \approx C_T \exp \left(-\frac{3}{8G^2 V(\varphi)} \right), \tag{43}$$

where C_T is a normalization constant. This is the same as Eq.(23) with ρ_v replaced by $V(\varphi)$. ρ is independent of a because φ remains approximately constant along the classical trajectories (with a playing the role of time). The probability distribution (43) is strongly peaked at the value $\varphi = \varphi_{max}$ where $V(\varphi)$ has a maximum. Thus, the tunneling wave function 'predicts' that the universe is most likely to nucleate with the largest possible vacuum energy. This is just the right initial condition for inflation. The high vacuum energy drives the inflationary expansion, while the field φ gradually 'rolls down' the potential hill, and ends up at the minimum with $V(\varphi) \approx 0$, where we are now.

For Linde's wave function, evaluation of the current for the expanding-universe component of ψ_L gives the same probability distribution (43). The Hartle-Hawking wave function gives a similar distribution, but with a crucial difference in sign,

$$\rho_H(\varphi) = C_H \exp \left(+\frac{3}{8G^2 V(\varphi)} \right). \tag{44}$$

Note that ρ_H is the same as the nucleation probability (10) found using the instanton method. This is not surprising: the Hartle-Hawking proposal involves the same Euclidean rotation as the one used to obtain the instanton. The distribution (44) is peaked at $V(\varphi) \approx 0$, and thus the Hartle-Hawking wave function appears to predict an empty universe with $V \approx 0$. Such initial condition does not lead to inflation, and is therefore inconsistent with observations.

Hawking and Page [26] have pointed out that things may be not so bad in models of 'chaotic' inflation, where $V(\varphi) \to \infty$ at $\varphi \to \infty$ (a typical example is $V(\varphi) \propto \varphi^{2k}$ with k an integer). In such models, $\rho(\varphi \to \infty) \to$ *const*, the distribution (44) is not normalizable, and the ensemble described by this distribution is dominated by universes with arbitrarily large initial values of φ. The problem with this argument is that, in order to outweigh the exponentially large values of $\rho_H(\varphi)$ at small φ, one has to go to extremely large values of φ, for which the potential $V(\varphi)$ will far exceed the

Planck energy density [except, perhaps, for a very special shape of $V(\varphi)$]. The semiclassical approximation, on which the derivation of Eq.(44) was based, cannot be trusted in this regime. For a futher discussion of this issue see Refs. [27, 28].

4.2. INITIAL STATE OF THE MODULI

The tunneling *vs.* Hartle-Hawking debate takes an interesting turn in superstring theories, where the relevant part of the low-energy effective action has the form

$$ S = \int d^4 x \sqrt{-g} \left[\frac{R}{16\pi G} - K_{AB}(\varphi) g^{\mu\nu} \partial_\mu \varphi^A \partial_\nu \varphi^B - V(\varphi) \right]. \quad (45) $$

Here, φ^A are the moduli fields, $K_{AB}(\varphi)$ is the metric of moduli space, $A, B = 1, 2, ..., n$, and n is the number of moduli. The potential $V(\varphi)$ vanishes to all orders of perturbation theory, but is expected to be generated non-perturbatively, with a characteristic scale well below the Planck mass. It has been argued that moduli are natural candidates for the role of the inflaton in superstring cosmology [29, 30, 31].

As before, we shall restrict ourselves to a closed Robertson-Walker universe with homogeneous moduli fields. Then, after an appropriate rescaling of t, a, φ^A, and $V(\varphi)$, the Lagrangian for our model can be written as

$$ \mathcal{L} = \frac{1}{2} a(1 - \dot{a}^2) + a^3 \left[\frac{1}{2} K_{AB}(\varphi) \dot{\varphi}^A \dot{\varphi}^B - V(\varphi) \right] \quad (46) $$

The momenta conjugate to a and φ^A are

$$ p_a = -a\dot{a}, \qquad p_A = a^3 K_{AB} \dot{\varphi}^B, \quad (47) $$

and the Hamiltonian is

$$ \mathcal{H} = \frac{1}{2a} [-p_a^2 + a^{-2} K^{AB} p_A p_B - U(a, \varphi)], \quad (48) $$

where $U(a, \varphi)$ is given by (40) and K^{AB} is related to K_{AB} by the standard relation $K^{AB} K_{BC} = \delta_C^A$.

The WDW equation is obtained by replacing $p_a \to -i\partial/\partial a$, $p_A \to -i\partial/\partial\varphi_A$,

$$ \left[-\frac{\partial^2}{\partial a^2} + \frac{1}{a^2} |K|^{-1/2} \frac{\partial}{\partial \varphi^A} \left(|K|^{1/2} K^{AB} \frac{\partial}{\partial \varphi^B} \right) + U(a, \varphi) \right] \psi = 0, \quad (49) $$

where $K = det(K_{AB})$ and the ordering of factors φ^A and $\partial/\partial\varphi^A$ has been chosen so that the equation is invariant with respect to reparametrizations of the moduli space, $\varphi^A \to \tilde{\varphi}^A(\varphi^B)$ [32]. The probability distribution for φ^A is then $d\mathcal{P} = \rho(a, \varphi)d^n\varphi$, where

$$\rho = J^a = i|K|^{1/2}(\psi^*\partial_a\psi - \psi\partial_a\psi^*). \tag{50}$$

For a slowly-varying potential, Eq.(49) is solved in the same way as Eq.(39) for a single scalar field, and one finds

$$\rho(\varphi) = C|K|^{1/2}\exp\left(\pm\frac{3}{8G^2V(\varphi)}\right), \tag{51}$$

where the upper sign corresponds to Hartle-Hawking, and the lower sign to the tunneling wave function.

Now, the moduli space is non-compact, and one could expect that the remedy suggested by Hawking and Page [26] to avoid the empty universe problem should work in this case as well. However, Horne and Moore have argued [33] that, despite the existence of non-compact regions, the volume of the moduli space is finite,

$$\int |K|^{1/2}d^n\varphi < \infty. \tag{52}$$

Moreover, it is expected that the moduli potential $V(\varphi)$ asymptotically vanishes in all non-compact directions. If either of these expectations is correct, then the Hartle-Hawking probability distribution is unavoidably peaked at very low densities, so that the initial states leading to inflation are highly unlikely.

5. Predictions for the Constants of Nature

5.1. VARIABLE CONSTANTS

In theories that allow variation of the constants of Nature, the cosmological wave function is a superposition

$$\psi = \sum_\alpha \psi_\alpha(a, \varphi), \tag{53}$$

where the subscript α is a collective symbol for the constants $\{\alpha_j\}$. In superstring theories, some of the constants parametrize different compactifications of extra dimensions, and different sets of $\{\alpha_j\}$ correspond to different moduli spaces with their own potentials $V(\varphi)$. The number of different compactifications is believed to be $\gtrsim 10^4$. Hence, the spectrum of $\{\alpha_j\}$ can be rather dense.

The wave function $\psi_\alpha(a, \varphi)$ gives the amplitude for a universe to nucleate with a set of constants $\{\alpha_j\}$ and to have the values a, φ for the scale factor and moduli fields. The relative normalization of different components is fixed for both Hartle-Hawking and tunneling wave functions if one uses the path integral formulation of the corresponding boundary conditions, Eqs.(31),(32). The overall normalization is determined by

$$\sum_{\alpha} \int \rho_\alpha(\varphi)d\varphi = 1. \tag{54}$$

We can think of the probability distribution $\rho_\alpha(\varphi)$ as describing an ensemble of universes which, following Gell-Mann [34], I will call 'multiverse'. The probability that a universe arbitrarily picked in this multiverse will have a particular set of $\{\alpha_j\}$ is

$$w_\alpha = \int \rho_\alpha(\varphi)d\varphi. \tag{55}$$

Adopting the tunneling boundary condition for ψ, we expect

$$w_\alpha \propto \exp\left(-\frac{3}{8G^2(\alpha)V_{max}(\alpha)}\right), \tag{56}$$

where $V_{max}(\alpha) = max\{V(\varphi)\}$ for the constants $\{\alpha\}$.

It has been recently suggested [35] that most, if not all, moduli spaces may actually be connected to one another. The potential on this interconnected web of moduli spaces may still have a large number of maxima and minima, with different low-energy physics at each minimum. The constants $\{\alpha_j\}$ then parametrize different minima of $V(\varphi)$, and the probabilities w_α are obtained by integrating $\rho(\varphi)$ over the basin of attraction of the corresponding minimum. We still expect the estimate (56) to apply, with $V_{max}(\alpha)$ being the highest maximum of $V(\varphi)$ in the basin of attraction of the minimum α.

The potential $V(\varphi)$ on the moduli space may have some flat directions. The associated massless fields may not be in conflict with observations if they are very weakly coupled. The values of these fields, which affect the 'constants' of low-energy physics, will then be determined by the initial conditions at nucleation and by the following cosmological evolution. Such fields should be included in $\{\alpha_j\}$ as continuous variables parametrizing the constants of Nature [36].

Finally, Coleman [10] has argued that all constants appearing in sub-Planckian physics may become totally undetermined due to Planck-scale wormholes connecting distant regions of spacetime. Then the spectrum of the constants is also continuous, but unlike massless moduli, they cannot vary from one spacetime point to another, but only from one universe to another (disconnected) universe. To simplify the discussion, I will assume a discrete spectrum of the constants.

5.2. PRINCIPLE OF MEDIOCRITY

It is quite possible that a randomly picked universe will be unsuitable for life, and therefore the distribution (56) is not adequate for predicting the observed values of the constants. Moreover, the number of civilizations in some of the universes may be much greater than in the others, and this difference should also be taken into account when evaluating the probabilities [40]. The probability distribution of constants for a *civilization* randomly picked in the multiverse is

$$\mathcal{P}_\alpha = C^{-1} w_\alpha \mathcal{N}_\alpha, \tag{57}$$

where $\mathcal{N}_\alpha$ is the average number of civilizations in a universe with a set of constants $\{\alpha_j\}$ and $C = \sum_\alpha w_\alpha \mathcal{N}_\alpha$ is a normalization constant. $\mathcal{N}$ is taken to be the total number of civilizations through the entire history of the universe and is assumed to be finite. The case of eternal inflation, where $\mathcal{N} = \infty$, will be discussed in Section 6.

If we assume that our civilization is a 'typical' inhabitant of the multiverse, then we 'predict' that the constants of Nature in our universe are somewhere near the maximum of the distribution (57). The assumption of being typical was called 'the principle of mediocrity' in Ref. [39]. It is a version of the 'anthropic principle' which has been extensively discussed in the literature [42].

The number $\mathcal{N}$ can be expressed as

$$\mathcal{N}_\alpha = \mathcal{V}_\alpha \nu_{civ}(\alpha), \tag{58}$$

where $\mathcal{V}_\alpha$ is the volume of the universe at the end of inflation (that is, the 3-volume of the hypersurface that divides the spacetime into inflating and thermalized parts), and $\nu_{civ}(\alpha)$ is the average number of civilizations originating per unit thermalized volume.

The definition of probability (57) based on the number of civilizations is somewhat arbitrary. One could, for example, assign a weight to each civilization, depending on its lifetime and/or the number of individuals. We shall deal with this uncertainty by concentrating on stable 'predictions' from (57) which are not sensitive to the choice of the definition.

The concept of 'naturalness' that is commonly used to assess the plausibility of elementary particle models is based on the assumption that the probability distribution for the constants is nearly flat, $\mathcal{P}_\alpha \approx const$. The principle of mediocrity gives a very different perspective on what is natural and what is not. It predicts that the constants $\{\alpha_j\}$ are likely to be such that the product

$$\mathcal{P}_\alpha \propto w_\alpha \mathcal{V}_\alpha \nu_{civ}(\alpha) \tag{59}$$

is maximized. The factors in this product have a strong (exponential) dependence on $\{\alpha_j\}$, and the distribution $\mathcal{P}_\alpha$ can be strongly peaked in some region of α-space.

It should be emphasized that predictions of the principle of mediocrity are not guaranteed to be correct. After all, our civilization may be special in some respects. The predictions can be expected to have only statistical accuracy. That is, with a large number of predictions, only few of them are likely to be wrong.

5.3. PREDICTIONS FOR FINITE INFLATION

From Eq.(56), the nucleation probability is maximized when the maximum of the potential approaches the Planck scale, $V_{max}(\alpha) \sim \rho_p$. (I assume that $V(\varphi)$ cannot get much greater than ρ_p).

The volume factor $\mathcal{V}$ is given by $\mathcal{V} = \mathcal{V}_0 Z^3$, where $\mathcal{V}_0 \sim (G V_{max})^{-3/2}$ is the initial volume at nucleation and Z is the expansion factor during inflation. The maximum of Z is achieved by maximizing the highest value of the potential V_{max}, where inflation starts, and minimizing the slope of $V(\varphi)$: the field φ takes longer to roll down for a flatter potential.

The cosmological literature abounds with remarks on the 'unnaturally' flat potentials required by inflationary scenarios. With the principle of mediocrity the situation is reversed: flat is natural. Instead of asking why $V(\varphi)$ is so flat, one should now ask why it is not flatter.

The 'human factor' $\nu_{civ}(\alpha)$ may impose stringent constraints on the constants $\{\alpha_j\}$. We do not know what other forms of intelligent life are possible, but the principle of mediocrity favors the hypothesis that our form is the most common in the multiverse. The conditions required for life of our type to exist [the low-energy physics based on the symmetry group $SU(3) \times SU(2) \times U(1)$, the existence of stars and planets, supernova explosions] may then fix, by order of magnitude, the values of the fine structure constant, and of electron, nucleon, and W-boson masses, as discussed in Ref. [42]. Anthropic considerations also impose a bound on the allowed flatness of the potential $V(\varphi)$. If it is too flat, then the thermalization temperature after inflation is too low for baryogenesis. The lowest temperature at which baryogenesis can still occur is set by the electroweak scale, $T_{min} \sim m_W$. Hence, if other constraints do not interfere, we expect the universe to thermalize at $T \sim m_W$.

Superflat potentials required by the principle of mediocrity typically give rise to density fluctuations which are many orders of magnitude below the strength needed for structure formation. This means that the observed structures must have been seeded by some other mechanism. An alternative mechanism is based on topological defects: strings, global monopoles, and textures, which could be formed at a symmetry breaking phase transition [43]. The required symmetry breaking scale for the defects is $\eta \sim 10^{16} \, GeV$. With 'natural' (in the traditional sense) values of the couplings, the transition temperature is $T_c \sim \eta$, which is much higher than the thermalization temperature $(T_{th} \sim m_W)$, and no defects are formed after inflation. It is possible for the phase transition to occur during inflation, but the resulting defects are inflated away, unless the transition is sufficiently close to the end of inflation. To arrange this requires some fine-tuning of the constants. However, the alternative is to have thermalization at a much higher temperature and to cut down on the amount of inflation. Since the dependence of the volume factor V on the duration of inflation is exponential, we expect that the gain in the volume will more than compensate for the decrease in 'α-space' due to the fine-tuning. We note also that in some supersymmetric models the critical temperature of superheavy string formation can 'naturally' be as low as m_W [44].

Another possibility is to use more complicated models of inflation, such as 'hybrid' inflation [45], which involve several scalar fields and can give reasonably large density fluctuations even when the potentials are very flat in some directions in the field space [46]. The amount of fine-tuning required in these models appears to be comparable to that in the case of topological defects.

The symmetry breaking scale $\eta \sim 10^{16} \, GeV$ for the defects is suggested by observations, but we have not explained why this particular scale has been selected. The value of η determines the amplitude of density fluctuations, which in turn determines the time when galaxies form, the galactic density, and the rate of star formation in the galaxies. Since these parameters certainly affect the chances for civilizations to develop, it is quite possible that η is significantly constrained by the anthropic factor $\nu_{civ}(\alpha)$. It would therefore be interesting to study how structure formation would proceed in a universe with a very different amplitude of density fluctuations (and a very different baryon density). Some steps in this direction have been made in Ref. [47].

If ν_{civ} is indeed sharply peaked at some value of η and thus fixes the amplitude of density fluctuations and the epoch of active galaxy formation, then an upper bound on the cosmological constant can be obtained by requiring that it does not disrupt galaxy formation until the end of that epoch. An anthropic bound on the cosmological constant has been first discussed by Weinberg [48]. He argued that, since there is evidence for the existence of quasars and protogalaxies as early as $z \sim 4$, the anthropic principle cannot rule out vacuum energy domination at $z \lesssim 4$. The matter density at $z = 4$ is greater than the present matter density by a factor $(1 + z)^3 = 125$, and he concluded that the anthropic bound on ρ_v cannot be stronger than $\rho_v/\rho_{m0} \lesssim 100$. This falls short of the observational upper bound [49], $|\rho_v/\rho_{m0}|_{obs} \lesssim 10$, by a factor ~ 10 [50].

On the other hand, the principle of mediocrity suggests that we look not for the value of ρ_v that makes galaxy formation barely possible, but for the value that maximizes the amount of matter in galaxies [51]. This amount grew substantially after $z = 4$, and it is quite possible that it increased, say, by a factor of ~ 2 as late as $z \sim 1$. Requiring that ρ_v does not dominate before $z \sim 1$, we obtain $\rho_v/\rho_{m0} \lesssim 10$. The actual value of ρ_v is likely to be comparable to this upper bound. Negative values of ρ_v are bounded by requiring that our part of the universe does not recollapse while stars are still shining and new civilizations are being formed. This gives a bound comparable to that for positive ρ_v (by absolute value). A more detailed discussion of the bounds on the cosmological constant will be given elsewhere.

Let us now summarize the 'predictions' of the principle of mediocrity for the case of finite inflation [39]. The preferred models have very flat inflaton potentials, thermalization and baryogenesis at the electroweak scale, non-negligible cosmological constant, and density fluctuations seeded either by topological defects, or by quantum fluctuations in models like hybrid inflation (as long as these features are consistent with the spectrum of the constants $\{\alpha_j\}$).

6. Predictions for Eternal Inflation

I have assumed so far that inflation has a finite duration, so that the thermalized volume V and the number of civilizations $\mathcal{N}$ are both finite. This, however, is not generally the case. The evolution of the inflaton field φ is influenced by quantum fluctuations, and as a result thermalization does not occur simultaneously in different parts of the universe. In many models it can be shown that at any time there are parts of the universe that are still inflating [52, 53, 60]. The conclusions of Section 5.3 are directly applicable only if inflation is finite for all the allowed values of the constants $\{\alpha_j\}$. For eternally inflating universes the situation is substantially more complicated. This subject is now under active investigation, and I will have time only for a quick review.

6.1. DISCONNECTED UNIVERSES

Let us first suppose that different sets of constants $\{\alpha_j\}$ correspond to different, disconnected universes. In order to calculate the probabilities (57), we should then be able to compare the thermalization volumes V_α.

In an eternally inflating universe, the thermalization volume V is infinite and has to be regulated. If one simply cuts it off by including only parts of the volume that thermalized prior to some moment of time t_c, with the same value of t_c for all universes, then one finds that the results are extremely sensitive to the choice of the time coordinate t. For example, cutoffs at a fixed proper time and at a fixed scale factor a give drastically different results [54]. An alternative procedure [55] is to introduce the cutoff at the time when all but a small fraction ϵ of the initial (co-moving) volume of the universe has thermalized. The value of ϵ is taken to be the same for all universes, but the corresponding cutoff times t_c are generally different. The limit $\epsilon \rightarrow 0$ is taken after calculating the probability distribution $\mathcal{P}_\alpha$. It was shown in [55] that the resulting distribution is not sensitive to the choice of t.

The regularized volume $\mathcal{V}$ can be calculated in terms of the distribution function $\rho(\varphi, a)$, which is defined so that $\rho(\varphi, a)d\varphi$ gives the fraction of the co-moving volume where the scalar field(s) takes values between φ and $\varphi + d\varphi$, with the scale factor a playing the role of a time coordinate. The function $\rho(\varphi, a)$ satisfies a 'diffusion' equation [52, 56], which I will not reproduce here. The important thing for us to know is that the asymptotic form of ρ at large a is

$$\rho(\varphi, a \rightarrow \infty) \approx f(\varphi)a^{-\gamma}. \tag{60}$$

The positive constant γ can be found by solving an eigenvalue problem [56], and $d = 3 - \gamma$ has the meaning of the fractal dimension of the inflating region [57]. (It can be shown that $\gamma \leq 3$). I will omit the calculations performed in Ref. [55] and even the rather lengthy expression for $\mathcal{V}$ obtained as a result of those calculations. The essence of the result can be expressed as

$$\mathcal{V} \propto \epsilon^{-(3-\gamma)/\gamma} Z^3. \tag{61}$$

Here, Z is the expansion factor during the slow-roll phase of inflation, when quantum fluctuations are small.

In the limit $\epsilon \rightarrow 0$, non-vanishing probabilities are obtained only for $\{\alpha_j\}$ corresponding to the smallest value of γ,

$$\gamma(\alpha) = min. \tag{62}$$

The eigenvalue γ decreases as the potential $V(\varphi)$ becomes flatter [57, 54], and thus the condition (62) tends to select maximally flat potentials.

It is possible that the condition (62) selects a unique set of $\{\alpha_j\}$. Then all constants of Nature can, at least in principle, be predicted with 100% certainty. On the other hand, it is conceivable that the minimum of γ is strongly degenerate, so that Eq.(62) selects a large subset of all $\{\alpha_j\}$. Then all values of α not in this subset have a vanishing probability, and the probability distribution within the subset is proportional to $w_\alpha Z_\alpha^3 \nu_{civ}(\alpha)$ [see Eq.(59)]. The probability maximum is then determined by the same considerations as in the case of finite inflation.

It should be emphasized that the conditions of minimizing $\gamma(\alpha)$ and maximizing $w_\alpha Z_\alpha^3 \nu_{civ}(\alpha)$ are not on an equal footing, with the first of these conditions always taking precedence. Even a tiny decrease in γ leads to an infinite increase of the thermalization volume in the limit $\epsilon \to 0$. Suppose, for example, that we have two sets of constants, $\{\alpha_j^{(1)}\}$ and $\{\alpha_j^{(2)}\}$, such that $\gamma(\alpha^{(1)}) < \gamma(\alpha^{(2)})$, but the thermalization temperature for the constants $\alpha^{(1)}$ is too low for baryogenesis, while for $\alpha^{(2)}$ it is sufficiently high. We would still have to conclude that the constants $\alpha^{(1)}$ are infinitely more probable than $\alpha^{(2)}$. In a universe described by the constants $\alpha^{(1)}$, life can appear only as a result of a huge fluctuation of the baryon density. The probability of such a fluctuation per unit volume is incredibly small, but its smallness is more than compensated for when the volume is increased by an infinite factor.

6.2. MULTIPLE VACUA IN A SINGLE UNIVERSE

Let us now consider eternal inflation in a single universe where the potential $V(\varphi)$ has a large number of minima, parametrized by the constants $\{\alpha_j\}$. Thermalization will then ocur in different minima in different parts of the universe. The asymptotic form of the distribution function $\rho(\varphi, a)$ in this case is still given by Eq.(60), but now γ has the same value everywhere and is independent of $\{\alpha_j\}$ [54]. From Eq.(61), the regularized thermalization volumes are $\mathcal{V}_\alpha \propto Z_\alpha^3$, and the corresponding probabilities are

$$\mathcal{P}_\alpha \propto Z_\alpha^3 \nu_{civ}(\alpha). \tag{63}$$

The probability distribution (63) has the same dependence on the slow-roll expansion factor Z and on the anthropic factor ν_{civ} as we found in the case of finite inflation. The predictions for $\{\alpha_j\}$ are, therefore, also the same (see Section 5.3).

6.3. ETERNAL INFLATION AND QUANTUM COSMOLOGY

The ideas of eternal inflation and quantum cosmology have always had a somewhat uneasy coexistence. The picture of an eternally inflating universe, with new islands of thermalization constantly being formed, makes one wonder about the possibility of extending this picture to the infinite past. The universe would then be in a steady state of eternal inflation without a beginning, the problem of the initial singularity would be avoided, and there would be no need for quantum cosmology. However, it has been shown [58] that, under rather general assumptions, inflation cannot be eternal in both future and past directions. Hence, an eternally inflating universe must still have a beginning, and we probably need quantum cosmology to describe it.

On the other hand, the eternal nature of inflation can make the initial state at the nucleation of the universe completely irrelevant. The universe eventually reaches the steady-state regime described by the asymptotic form (60) and stays in this regime thereafter. If transitions between vacua with different $\{\alpha_j\}$ are in principle possible, then the universe completely forgets its initial conditions. In this case, all one needs from the cosmological wave function is that the probability for eternal inflation to start should be non-zero. Hartle-Hawking and tunneling wave functions are then in equally good agreement with observations, and the probability distribution for $\{\alpha_j\}$ is given by Eq.(63). (This equation was derived without relying on quantum cosmology).

Thus, we see that eternal inflation is quite capable of determining the values of $\{\alpha_j\}$ on its own, without any help from quantum cosmology. The inverse is also true: in models where inflation is finite for all allowed values of $\{\alpha_j\}$, quantum cosmology can determine the probability distribution for the initial states and for the constants of Nature, without any need for eternal inflation [59]. At this time it is hard to tell which of the two approaches is more promising, and both are probably worth pursuing.

Acknowlegements

It is a pleasure to thank the Organizers of the Course for their warm hospitality. I am also grateful to Arvind Borde, Allen Everett, Cumrun Vafa, Serge Winitzki, and particularly Andrei Linde and Don Page for discussions and comments on the manuscript. This work was supported in part by the National Science Foundation.

References

1. De Witt, B.S. (1967) *Phys. Rev.* **160**, 1113.
2. Misner, C.W. (1972) in *Magic Without Magic*, Freeman, San Francisco.
3. Vilenkin, A. (1982) *Phys. Lett.* **117B**, 25.
4. Hartle, J.B. and Hawking, S.W. (1983) *Phys. Rev.* **D28**, 2960.
5. Linde, A.D. (1984) *Lett. Nuovo Cim.* **39**, 401.
6. Zel'dovich, Y.B. and Starobinsky, A.A. (1984) *Sov. Astron. Lett.* **10**, 135.
7. Rubakov, V.A. (1984) *Phys. Lett.* **148B**, 280.
8. Vilenkin, A. (1984) *Phys. Rev.* **D30** 509.
9. The idea that a closed universe could be a vacuum fluctuation was first suggested by E.P. Tryon (1973) *Nature* **246**, 396, and independently by P.I. Fomin (1973) *ITP Preprint*, Kiev; (1975) *Dokl. Akad. Nauk Ukr. SSR* **9A**, 831. However, these authors offered no mathematical description for the nucleation of the universe. Quantum tunneling of the entire universe through a potential barrier was first discussed by D. Atkatz and H. Pagels (1982) *Phys. Rev.* **D25**, 2065.

10. Coleman, S. (1988) *Nucl. Phys.* **B307**, 867. Similar ideas were explored by Giddings, S.B. and Strominger, A. (1988) *Nucl. Phys.* **B307**, 854 and by Banks, T. (1988) *Nucl. Phys.* **B309**, 493.

11. We note that calling something the greatest mistake of one's life may be a mistake. For example, the introduction of the cosmological constant, which Einstein called the greatest mistake of his life, now appears to be not such a bad idea.

12. Halliwell, J.J. and Hawking, S.W. (1985) *Phys. Rev.* **D31**, 1777.

13. Vachaspati, T. and Vilenkin, A. (1988) *Phys. Rev.* **D37**, 898.

14. Vilenkin, A. (1989) *Phys. Rev.* **D39**, 1116.

15. Lapchinsky, V. and Rubakov, V.A. (1979) *Acta Phys. Polon.* **B10**, 1041.

16. Banks, T. (1985) *Nucl. Phys.* **B249**, 332.

17. Vilenkin, A. (1985) *Nucl. Phys.* **B252**, 141.

18. Teitelboim, C. (1982) *Phys. Rev.* **D25**, 3159.

19. Halliwell, J.J. and Hartle, J.B. (1990) *Phys. Rev.* **D41**, 1815.

20. Vilenkin, A. (1986) *Phys. Rev.* **D33**, 3560.

21. Vilenkin, A. (1988) *Phys. Rev.* **D37**, 888.

22. Vilenkin, A. (1994) *Phys. Rev.* **D50**, 2581.

23. Hawking, S.W. (1984) *Nucl. Phys.* **B239**, 257.

24. Linde, A.D. (1990) *Particle Physics and Inflationary Cosmology*, Harwood Academic, Chur.

25. I am grateful to Slava Mukhanov for pointing out to me that Linde's contour rotation and the tunneling boundary condition give different wave functions and to Andrei Linde for a discussion of this point.

26. Hawking, S.W. and Page, D.N. (1986) *Nucl. Phys.* **B264**, 185.

27. Grishchuk, L.P. and Rozhansky, L.V. (1988) *Phys. Lett.* **B208**, 369.

28. Barvinsky, A.O. and Kamenshchik, A.Y. (1994) *Phys. Lett.* **B332**, 270.

29. Binnetruy, P. and Gaillard, M.K. (1986) *Phys. Rev.* **D34**, 3069.

30. Banks, T. *et. al.* (1994) Modular Cosmology, *Rutgers Preprint* RU-94-93.

31. Thomas, S. (1995) Moduli Inflation from Dynamical Supersymmetry Breaking, *SLAC Preprint* SLAC-PUB-95-6762.

32. The most general factor ordering consistent with reparametrization invariance allows an extra term $\xi a^{-2} \mathcal{R}(\varphi)$ in Eq.(49). Here, $\mathcal{R}$ is the scalar curvature of the moduli space and ξ is a numerical coefficient. Addition of such a term modifies the potential $U(a, \varphi)$ but does not change the conclusions of Sec. 4.2.

33. Horne, J. and Moore, G. (1994) *Nucl. Phys.* **B432**, 109.

34. Gell-Mann, M. (1994) *The Quark and the Jaguar*, Freeman, New York.

35. Strominger, A. (1995) Massless Black Holes and Conifolds in String Theory, hep-th/9504047.

36. The probability distribution for a Brans-Dicke field (which is similar to the dilaton of superstring theories) was discussed, using a different approach, by Garcia-Bellido and Linde [37, 38].

37. Garcia-Bellido, J. and Linde, A.D. (1995) *Phys. Rev.* **D51**, 429.

38. Garcia-Bellido, J., Linde, A.D. and Linde, D.A. (1994) *Phys. Rev.* **D50**, 730

39. Vilenkin, A. (1995) *Phys. Rev. Lett.* **74**, 846.

40. This and the following sections are partly based on my papers [39, 55]. Related ideas were discussed by Albrecht [41] and by Garcia-Bellido and Linde [37]

41. Albrecht, A. (1995) in *The Birth of the Universe and Fundamental Forces*, ed. by F. Occhionero, Springer-Verlag.

42. Carter, B. (1974) in *I.A.U. Symposium*, Vol. **63**, ed. by M.S. Longair, Reidel, Dordrecht; (1983) *Philos. Trans. R. Soc. London* **A310**, 347; Carr, B.J. and Rees, M.J. (1979) *Nature (London)* **278**, 605; Barrow, J.D. and Tipler, F.J. (1986) *The Anthropic Cosmological Principle*, Clarendon, Oxford.

43. Vilenkin, A. and Shellard, E.P.S. (1994) *Cosmic Strings and Other Topological Defects*, Cambridge University Press, Cambridge.

44. Lazarides, G., Panagiotakopoulos, C. and Shafi, Q.(1986) *Phys. Rev. Lett.* **56**, 432;

(1987) *Phys. Lett.* **183B**, 289.

45. Linde, A.D. (1994) *Phys. Rev.* **D49**, 748; Copeland, E.J. *et. al.* (1994) *Phys. Rev.* **D49**, 6410.
46. I am grateful to Andrei Linde for pointing out to me that hybrid inflation can give sufficiently large density fluctuations, even with flat potentials.
47. Rees, M.J. (1983) *Philos. Trans. R. Soc. London* **A310**, 311.
48. Weinberg, S. (1987) *Phys. Rev. Lett.* **59**, 2607; (1989) *Rev. Mod. Phys.* **61**, 1.
49. See, e.g., Carroll, S.M., Press, W.H. and Turner, E.L. (1992) *Ann. Rev. Astron. Astrophys.* **30**, 499.
50. This argument assumes that the probability distribution for ρ_v in the range of interest is nearly flat. It is possible, however, that the 'fundamental' variable that has a flat distribution at sub-Planckian scales is the characteristic energy scale $\eta = \rho_v^{1/4}$. Then the discrepancy between the anthropic and observational bounds on η is only by a factor ~ 2.
51. More exactly, we look for the values of ρ_v that achieve a balance between fine-tuning and maximizing the amount of matter in galaxies. To make this quantitative, let $w(\rho_v)d\rho_v$ be the probability distribution for ρ_v for the nucleating universes, and let $f(\rho_v)$ be the fraction of baryonic matter that ends up in galaxies at a given value of ρ_v. (Here I assume that ρ_v has a continuous spectrum). Then the most probable values of ρ_v are found by maximizing the product $f(\rho_v)w(\rho_v)\rho_v$.
52. Vilenkin, A. (1983) *Phys. Rev.* **D27**, 2848.
53. Linde, A.D. (1986) *Phys. Lett.* **B175**, 395.
54. Linde, A.D., Linde, D.A. and Mezhlumian, A. (1994) *Phys. Rev.* **D49**, 1783.
55. Vilenkin, A. (1995) *Making Predictions in Eternally Inflating Universe*, gr-qc/9505031.
56. Starobinsky, A.A. (1986) in *Current Topics in Field Theory, Quantum Gravity and Strings*, ed. by H.J. de Vega and N. Sanchez, Springer, Heidelberg.
57. Aryal, M. and Vilenkin, A. (1987) *Phys. Lett.* **B199**, 351.
58. Borde, A. and Vilenkin, A. (1994) *Phys. Rev. Lett.* **72**, 3305; Borde, A. (1994)*Phys. Rev.* **D50**, 3392.
59. A combined approach using both quantum cosmology and eternal inflation would be necessary only if $\{\alpha_j\}$ split into groups, such that transitions between different groups are disallowed, *and* the absolute minimum of $\gamma(\alpha)$ is attained in more than one group.
60. Linde, A.D. (1994) *Phys. Lett.* **B327**, 208; Vilenkin, A. (1994) *Phys. Rev. Lett.* **72**, 3137.

STATISTICS OF THE MICROWAVE BACKGROUND ANISOTROPIES CAUSED BY COSMOLOGICAL PERTURBATIONS OF QUANTUM-MECHANICAL ORIGIN

L. P. GRISHCHUK
McDonnell Center for the Space Sciences, Physics Department
Washington University, St. Louis, Missouri 63130
and
Sternberg Astronomical Institute, Moscow University
119899 Moscow, V234, Russia

The genuine quantum gravity effects can already be around us. It is likely that the observed large-angular-scale anisotropies in the microwave background radiation are induced by cosmological perturbations of quantum-mechanical origin. Such perturbations are placed in squeezed vacuum quantum states and, hence, are characterized by large variances of their amplitude. The statistical properties of the anisotropies should reflect the underlying statistics of the squeezed vacuum quantum states. In this paper, the theoretical variances for the temperature angular correlation function are described in detail. It is shown that they are indeed large and must be present in the observational data, if the anisotropies are truly caused by the perturbations of quantum-mechanical origin. Unfortunately, these large theoretical statistical uncertainties will make the extraction of cosmological information from the measured anisotropies a much more difficult problem than we wanted it to be. This contribution to the Proceedings is largely based on references [42,8]. The Appendix contains an analysis of the "standard" inflationary formula for density perturbations.

1. Introduction

In the context of cosmology, the quantum gravitational physics is usually understood as the early Universe physics at the Planck scale. It is assumed that the gravitational field is fully quantized and the Universe itself is, in a sense, quantized. (For a comprehensive recent review of the structural issues

369

N. Sánchez and A. Zichichi (eds.), String Gravity and Physics at the Planck Energy Scale, 369–408.
© 1996 *Kluwer Academic Publishers.*

in quantum gravity see [1]). However, there exists also another meaning in which we can speak of quantum gravity effects. To clarify the difference, let us consider an analogy from condensed matter physics. Imagine a crystal and various quantized excitations in it: phonons, rotons, excitons, etc.. The creation and annihilation operators that the condensed matter theorists write do not create and annihilate the crystal, they create and annihilate excitations in the crystal. The excitations should not necessarily be linear, one can take account of the higher-order corrections too. The theory of the crystal excitations is a fully legitimate quantum theory which does not attempt to quantize "everything in sight". Similarly, in our study below, we do not write the creation and annihilation operators that create and destroy universes. Our operators create and destroy perturbations in the Universe. Nevertheless, the effects that we are studying are genuine quantum gravity effects in the sense that they inherently contain all the three fundamental constants. The gravitational constant and velocity of light enter because we deal with a gravitational field (its energy-momentum tensor), the Planck constant enters because we normalize the vacuum energy of the field to have "a half of the quantum" in each mode. All three fundamental constants combine in the Planck length l_{Pl}, and this quantity naturally appears as the coefficient in the most of our formulas. The sending of the l_{Pl} to zero would eliminate the entire expression.

Now, why at all do we think that the anisotropies in the microwave background radiation may have something to do with gravitational quantum physics? Why, in the first place, is there such a considerable interest to the measured large-angular-scale anisotropies [2]? After all, we have always knew that the observed part of the Universe is homogeneous and isotropic only approximately, on average, and we are certainly aware of large deviations on smaller scales.

The point is that the photons of the microwave background have been traveling to us for almost all duration of that characteristic time which we call the age of the Universe and which is determined by the value of the measurable quantity — the Hubble parameter. The anisotropies on largest angular scales are specific in that they are produced by cosmological perturbations on largest spatial scales. In our context, the largest spatial scale means definitely larger than any directly studied distance, and comparable or larger than the characteristic cosmological distance — the Hubble radius — associated with the Hubble parameter. In terms of the crystal analogy, and assuming that the size of the crystal varies at some time scale, we are dealing with excitations whose wavelengths are comparable and longer than the light travel spatial scale associated with the time variation.

The next question is whether we will be attempting to explain the origin and nature of such long-wavelength cosmological perturbations or we will be happy to simply accept their existence. There is no logical contradiction in the second position. One can study the perturbations with whatever accessible observational accuracy one has, and then extrapolate back in time their evolution according to classical dynamical laws. Typically, we will end up with a very anisotropic and inhomogeneous universe at some very early times. As an explanation, we will have nothing more to say except that this was the Universe that was given to us for our future life and study. A more appealing position, at least aesthetically, is to try and find out a universal and reliable mechanism which could be capable of generating the required perurbations in the originally homogeneous and isotropic Universe. The problem is not to increase the wavelengths of the perturbations up to the size of the present Hubble radius (in the expanding Universe, the wavelength is always growing with time) but to generate them. In principle, one can imagine that such a mechanism may have operated in the relatively recent Universe and may have not be related to quantum gravity. In practice, taking into account everything what we know about cosmology and the necessity to produce perturbations with extremely long wavelengths, it is difficult to suggest such a mechanism, especially if we do not want to make many additional hypotheses. It appears that we can only rely on the quantum processes in the very early Universe. The parametric interaction of quantized cosmological perturbations with strong variable gravitational field of the very early Universe provides us with such a possibility. The perturbations are generated as a result of amplification of their zero-point quantum oscillations. Returning to the laboratory physics analogy, one can recall that a cavity filled with a dielectric medium and initially free of electromagnetic radiation will eventually contain the radiation if the parameters of the dielectric medium vary properly in time. The quantum-mechanical (parametric) mechanism of generating cosmological perturbations relies only on the validity of the general relativity and the basic principles of quantum field theory. The observational consequences of this phenomenon we will study below.

The line of reasoning in this study can be summarized as follows.

We see the anisotropies in the microwave background radiation at the largest angular scales [2]. Observers convincingly argue that this is a genuine cosmological effect.

If the large-angular-scale anisotropy in the microwave background is really produced by cosmological perturbations (density perturbations, rotational perturbations, gravitational waves), then their today's wavelengths are of the order and longer than today's Hubble radius l_H. Strictly speaking, all wavelengths give contributions to the anisotropy at every given angular scale. But if the spectrum of the perturbations is not excessively "red" or "blue", the dominant contribution is provided by wavelengths indicated above. For instance, the major contribution to the quadrupole anisotropy is provided by wavelengths somewhat longer than l_H.

In the expanding Universe, the wavelengths of perturbations increase in proportion to the cosmological scale factor. The wavelengths that are longer than some length scale today have always been longer than that scale in the past. Moreover, the wavelengths of the perturbations of our interest are much longer than the Hubble radius defined at the previous times, when one goes back in time up to the era of primordial nucleosynthesis — the earliest era of which we have observational data. It is hard to imagine (although it does not seem to be logically impossible) that cosmological perturbations of our interest, with such long wavelengths, could have been generated by local physical processes during the interval of time between the era of primordial nucleosynthesis and now. We are bound to conclude that these perturbations were generated in the very early Universe, before the era of primordial nucleosynthesis. There is still 80 orders of magnitude, in terms of energy density, to go from the era of primordial nucleosynthesis to the Planck era; a lot of things could have happened in between.

The law of evolution of the very early Universe is not known, but it is likely that it could have been significantly different from the law of expansion of the radiation-dominated Universe. If so, some amount of cosmological perturbations must have been generated quantum-mechanically, as a result of parametric interaction of the quantized perturbations with strong variable gravitational field of the very early Universe. Gravitational waves have been generated inevitably, while density and rotational perturbations — if we were lucky; see [3] and references therein. (If the cosmological scale factor has always been the one of the radiation-dominated Universe, we must stop here, because the parametric coupling vanishes in this case, and cosmological perturbations cannot be amplified classically and cannot be generated quantum-mechanically.) The amount and spectrum of the generated perturbations depend on the law of evolution of the very early Universe (the strength and variability of the gravitational pump field), and this is how we can learn about what was going on there. In particular, the law of evolution of the very early Universe could have been of inflationary type.

If the cosmological perturbations were generated quantum-mechanically, they should be placed in the squeezed vacuum quantum states [4] (for an introduction to squeezed states see, for example, Ref. [5,6] and the pioneering works quoted there). Squeezing of cosmological perturbations might have degraded by now at short wavelengths but should survive at long wavelengths, especially in the case of gravitational waves.

The squeezed vacuum quantum states can only be squeezed in the variances of phase which unavoidably means the increased variances in amplitude. The statistical properties of the squeezed vacuum quantum states are significantly different from the statistical properties of the "most classical" quantum states — coherent states. This is well illustrated by the fact that the variance of the number of quanta in a strongly squeezed vacuum quantum state is much larger than the variance of the number of quanta in the coherent state with the same mean number of quanta $\langle N \rangle$, $\langle N \rangle \gg 1$. For a squeezed vacuum state the variance is $\langle N^2 \rangle - \langle N \rangle^2 = 2\langle N \rangle (\langle N \rangle + 1) \gg \langle N \rangle$, while for a coherent state it is $\langle N^2 \rangle - \langle N \rangle^2 = \langle N \rangle$. In cosmology, the mean number $\langle N \rangle$ is a characteristic of the expected mean square amplitude of cosmological perturbations, while the variance $\langle N^2 \rangle - \langle N \rangle^2$ is a characteristic of theoretical uncertainties in the amplitude. These two characteristics are independent properties of a quantum state or a stochastic process. Theoretical models may agree on $\langle N \rangle$ and disagree on variance or agree on variance and disagree on $\langle N \rangle$.

The statistical properties of squeezed cosmological perturbations will inevitably be reflected in statistical properties of the microwave background anisotropies caused by them. Squeezing is a phase-sensitive phenomenon, and to fully extract its properties the quantum optics experimenters use the phase-sensitive detecting techniques based on a local oscillator. In cosmology, we are very far from being able to build a local oscillator, except of maybe, in the distant future, for short gravitational waves. Besides, in our study of the microwave background anisotropies, we are interested in so long-wavelength perturbations that it would take billions of years to wait for seeing the time dependent oscillations of variances in the quadrature components of the perturbation field. On the other hand, the amount of cosmological squeezing is enormously greater than what is achieved in quantum optics laboratory experiments. In cosmology, we can only rely on the phase-insensitive, direct detection. One can expect that the underlying large variances of the amplitude of cosmological perturbations should result in large statistical deviations from the mean values for the microwave anisotropies.

A detailed study and proof of this statement is the purpose of this work.

At this point it is necessarry to comment on the possible numerical values of $\langle N \rangle$ (and, hence, the amplitude) for cosmological perturbations of different nature which can be generated quantum-mechanically in one and the same cosmological model. A consistent quantum theory provides us, of course, with both, the mean value of N and its variance. According to the calculations of Ref. [3], the contribution of quantum-mechanically generated gravitational waves to the large-angular-scale anisotropy is somewhat greater (even in the limit of the de Sitter expansion) than the contribution of quantum-mechanically generated density perturbations. It is argued in [7,8] that the "standard" inflationary formula for density perturbations, which requires in this limit an arbitrarily large excess of density perturbations over gravitational waves, is based on errors. We additionally discuss this issue in some detail in the Appendix of this paper. However, the major emphasis of this paper is on the statistical properties of cosmological perturbations which are determined by their common origin — quantum mechanics and squeezing. These properies are related to the variance of N rather than to its mean value. Our discussion will be equally well applicable to the perturbations of any nature (density perturbations, rotational perturbations, gravitational waves) if they have the same origin.

2. The General Equations for Quantized Cosmological Perturbations

Here we will briefly summarize some basic information about quantized cosmological perturbations (see [3,7] and references therein). The squeezed field operator derived in this Section is a basic mathematical construction for our further discussion of statistical properties.

The metric of the homogeneous isotropic universe can be written in the form

$$ds^2 = -a^2(\eta)(d\eta^2 - \gamma_{ij}\,dx^i\,dx^j) \quad , \tag{1}$$

where γ_{ij} is the spatial metric. For reasons of simplicity, we will be considering only spatially flat universes, that is $\gamma_{ij} = \delta_{ij}$.

Following Lifshitz, it is convenient to write the perturbed metric in the form

$$ds^2 = -a^2(\eta)[d\eta^2 - (\delta_{ij} + h_{ij})dx^i\,dx^j] \quad , \tag{2}$$

where h_{ij} are functions of η-time and spatial coordinates. By writing the perturbed metric in this form we do not lose anything in the physical content of the problem, but we gain considerably in the mathematical tractability of the perturbed Einstein equations. The one who is interested in solving equations will certainly be interested in their simpler form. Those who prefer "gauge-invariant formalisms" are welcome to take the found solution and compute with its help whichever gauge-invariant quantity they like. These quantities, being gauge-invariant, have the same values in all gauges.

The components h_{ij} of the perturbed gravitational field can be classified in terms of scalar, vector, and tensor eigenfunctions of the Laplace differential operator. The components of the perturbed energy-momentum tensor can also be classified in the same manner. After that, the linearized Einstein equations reduce to a set of ordinary differential equations, separately for scalar (density perturbations), vector (rotational perturbations), and tensor (gravitational waves) parts.

The number of independent unknown functions of time that can potentially be present (on grounds of the classification scheme) in the perturbed

Einstein equations is always greater than the number of independent equations. It is 6 functions and 4 equations for density perturbations, 3 functions and 2 equations for rotational perturbations, and 2 functions and 1 equation for gravitational waves. In order to make the system of equations closed, it is necessary to say something about the perturbed components of the energy-momentum tensor or to specify from the very beginning the form of the energy-momentum tensor. The popular choices are perfect fluids and scalar fields. Even for gravitational waves, it is not a totally trivial question what their definition is (see, for example, Ref. [9]). However, after everything is being set, and as soon as the scale factor $a(\eta)$ (the background solution) is known, the general solution to the perturbed equations can be found. In practice, exact solutions are being found piecewise, at the intervals of evolution where the energy-momentum tensor has simple prescribed forms.

We can now write the quantum-mechanical operator for the perturbations of the gravitational field h_{ij} in the following universal form:

$$ h_{ij} = \frac{C}{a(\eta)} \frac{1}{(2\pi)^{3/2}} \int_{-\infty}^{\infty} d^3n \sum_{s=1}^{2} \overset{s}{p}_{ij}(\mathbf{n}) \frac{1}{\sqrt{2n}} \left[\overset{s}{c}_{\mathbf{n}}(\eta) e^{i\mathbf{n}\mathbf{x}} + \overset{s\dagger}{c}_{\mathbf{n}}(\eta) e^{-i\mathbf{n}\mathbf{x}} \right]. \quad (3) $$

We will start the explanation of Eq. (3) from the polarization tensors $\overset{s}{p}_{ij}$. Let us introduce, in addition to the unit wave-vector $\mathbf{n}/n$, two more unit vectors l_i, m_i, orthogonal to each other and to $\mathbf{n}$:

$$ \frac{n_i}{n} = (\sin\theta\cos\phi,\ \sin\theta\sin\phi,\ \cos\theta), \qquad l_i = (\sin\phi,\ -\cos\phi,\ 0), $$
$$ m_i = \pm(\cos\theta\cos\phi,\ \cos\theta\sin\phi,\ -\sin\theta), \qquad (4) $$

$+$ for $\theta < \frac{\pi}{2}$, $-$ for $\theta > \frac{\pi}{2}$.

The two independent polarization tensors, $s = 1, 2$, for each class of perturbations, can be written as follows. For gravitational waves:

$$ \overset{1}{p}_{ij}(\mathbf{n}) = (l_i l_j - m_i m_j), \qquad \overset{2}{p}_{ij}(\mathbf{n}) = (l_i m_j + l_j m_i). $$

For rotational perturbations:

$$ \overset{1}{p}_{ij}(\mathbf{n}) = \frac{1}{n}(l_i n_j + l_j n_i), \qquad \overset{2}{p}_{ij}(\mathbf{n}) = \frac{1}{n}(m_i n_j + m_j n_i). $$

For density perturbations:

$$ \overset{1}{p}_{ij}(\mathbf{n}) = \sqrt{\frac{2}{3}} \delta_{ij}, \qquad \overset{2}{p}_{ij}(\mathbf{n}) = -\sqrt{3} \frac{n_i n_j}{n^2} + \frac{1}{\sqrt{3}} \delta_{ij}. $$

The polarization tensors of each class satisfy the conditions $\overset{s}{p}_{ij}\overset{s'}{p}{}^{ij} = 2\delta_{ss'}$, $\overset{s}{p}_{ij}(-\mathbf{n}) = \overset{s}{p}_{ij}(\mathbf{n})$. In practical handling of the density perturbations it proves convenient to use sometimes, in addition to the *scalar* polarization component $\overset{1}{p}_{ij}$, the *longitudinal-longitudinal* component (proportional to $n_i n_j$) instead of $\overset{2}{p}_{ij}$. The explicit functional dependence of the polarization tensors is needed for the calculation of various angular correlation functions.

The evolution of the creation and annihilation operators $\overset{s}{c}_\mathbf{n}(\eta)$, $\overset{s\dagger}{c}_\mathbf{n}(\eta)$, for each class of perturbations and for each polarization state, is defined by the Heisenberg equations of motion:

$$\frac{dc_\mathbf{n}(\eta)}{d\eta} = -i[c_\mathbf{n}(\eta), H], \qquad \frac{dc_\mathbf{n}^\dagger(\eta)}{d\eta} = -i[c_\mathbf{n}^\dagger(\eta), H]. \tag{5}$$

The dynamical content of the problem is determined by the Hamiltonian H. Its form depends on the class of perturbations and additional assumptions about the energy-momentum tensor which we have to make, as was discussed above.

Under the simplest assumptions about gravitational waves (waves interact only with the background gravitational field, there is no anisotropic material sources) the Hamiltonian for each polarization component takes on the form

$$H = nc_\mathbf{n}^\dagger c_\mathbf{n} + nc_{-\mathbf{n}}^\dagger c_{-\mathbf{n}} + 2\sigma(\eta)c_\mathbf{n}^\dagger c_{-\mathbf{n}}^\dagger + 2\sigma^*(\eta)c_\mathbf{n}c_{-\mathbf{n}} \tag{6}$$

where the coupling function $\sigma(\eta)$ is $\sigma(\eta) = \frac{i}{2}\frac{a'}{a}$.

For rotational perturbations, assuming that the primeval matter is capable of supporting torque oscillations, assuming that the oscillations are minimally coupled to gravity, and assuming that the torsional velocity of sound is equal to the velocity of light, the Hamiltonian for each polarization component reduces to exactly the same form (6) with the same coupling function $\sigma(\eta)$.

For density perturbations, we consider specifically a minimally coupled scalar field with arbitrary scalar field potential as a model for matter in the very early Universe, and perfect fluids at the later eras. The quantization is based on the *scalar* polarization component (the function of time responsible for another polarization state is not independent). There is only one independent sort of creation and annihilation operators in this case. The operators $\overset{s}{c}_\mathbf{n}(\eta)$, $\overset{s\dagger}{c}_\mathbf{n}(\eta)$ are expressible in terms of the operators $d_\mathbf{n}(\eta)$, $d_\mathbf{n}^\dagger(\eta)$ for which the Hamiltonian has again the same form (6) but with the coupling function $\sigma(\eta) = \frac{i}{2} \frac{(a\sqrt{\gamma})'}{a\sqrt{\gamma}}$, where

$$\gamma(\eta) = 1 + \left(\frac{a}{a'}\right)' .$$

For density perturbations, it is the operators $d_{\mathbf{n}}(\eta)$, $d_{\mathbf{n}}^{\dagger}(\eta)$ that participate in Eqs. (5), (6).

Now, let us turn to the constant C in Eq. (3). Its value is determined by the normalization of the field of each class to the "half of the quantum in each mode". Under the assumptions listed above, one derives $C = \sqrt{16\pi} \, l_{Pl}$ for gravitational waves, $C = \sqrt{32\pi} \, l_{Pl}$ for rotational perturbations, and $C = \sqrt{24\pi} \, l_{Pl}$ for density perturbations, where l_{Pl} is the Planck length, $l_{Pl} = (G\hbar/c^3)^{1/2}$.

The form of the Hamiltonian (6) dictates the form of the solution (Bogoliubov transformation) to Eq. (5):

$$
\begin{aligned}
c_{\mathbf{n}}(\eta) &= u_n(\eta)c_{\mathbf{n}}(0) + v_n(\eta)c_{-\mathbf{n}}^{\dagger}(0) \\
c_{\mathbf{n}}^{\dagger}(\eta) &= u_n^*(\eta)c_{\mathbf{n}}^{\dagger}(0) + v_n^*(\eta)c_{-\mathbf{n}}(0)
\end{aligned}
\tag{7}
$$

where $c_{\mathbf{n}}(0)$, $c_{\mathbf{n}}^{\dagger}(0)$ are the initial values of the operators taken long before the interaction with the pump field became important ($\sigma(\eta)/n \to 0$) and which define the vacuum state $c_{\mathbf{n}}(0)|0\rangle = 0$. The complex functions $u_n(\eta)$, $v_n(\eta)$ obey coupled first-order differential equations following from Eq. (4) and satisfy the condition $|u_n|^2 - |v_n|^2 = 1$ which guarantees that the commutator relationship $[c_{\mathbf{n}}(0), c_{\mathbf{m}}^{\dagger}(0)] = \delta^3(\mathbf{n} - \mathbf{m})$ is satisfied at all times, $[c_{\mathbf{n}}(\eta), c_{\mathbf{m}}^{\dagger}(\eta)] = \delta^3(\mathbf{n} - \mathbf{m})$. If one introduces the function $\mu_n(\eta) = u_n(\eta) + v_n^*(\eta)$, one recovers from the equations for $u_n(\eta)$, $v_n(\eta)$ the classical equations of motion. For gravitational waves:

$$\mu_n'' + \left[n^2 - \frac{a''}{a}\right]\mu_n = 0 .\tag{8}$$

For rotational perturbations:

$$\mu_n'' + \left[n^2 \frac{v_t^2}{c^2} - \frac{a''}{a}\right]\mu_n = 0 .\tag{9}$$

where v_t is the torsional velocity of sound which we assumed above to be c. For the scalar field density perturbations:

$$\mu_n'' + \left[n^2 - \frac{(a\sqrt{\gamma})''}{a\sqrt{\gamma}}\right]\mu_n = 0 .\tag{10}$$

378

If the pump field is such that the γ function is independent of time, Eq. (10) reduces to exactly the same form as Eq. (8) for gravitational waves.

In the Schrödinger picture, the initial vacuum quantum state $|0_\mathbf{n}\rangle\,|0_{-\mathbf{n}}\rangle$ evolves into a two-mode squeezed vacuum quantum state. In our problem, each of the two-mode squeezed vacuum quantum states is a product of two identical one-mode squeezed vacuum quantum states which correspond to the decomposition of the real field h_{ij} over real spatial harmonics $\sin \mathbf{nx}$ and $\cos \mathbf{nx}$. In the Heisenberg picture, the initial vacuum quantum state does not evolve in time and is the same now.

By using Eq. (7) one can present the field (3) in the form

$$
h_{ij}(\eta, \mathbf{x}) = C\frac{1}{(2\pi)^{3/2}} \int_{-\infty}^{\infty} d^3n \sum_{s=1}^{2} \overset{s}{p}_{ij}(\mathbf{n})\frac{1}{\sqrt{2n}}
$$
$$
\times \left[\overset{s}{h}_n(\eta)e^{i\mathbf{nx}} \overset{s}{c}_\mathbf{n}(0) + \overset{s*}{h}_n(\eta)e^{-i\mathbf{nx}} \overset{s\dagger}{c}_\mathbf{n}(0) \right] , \qquad (11)
$$

where the functions $\overset{s}{h}_n(\eta)$ are $\overset{s}{h}_n(\eta) = \frac{1}{a(\eta)}[\overset{s}{u}_n(\eta)+\overset{s*}{v}_n(\eta)]$. For gravitational waves and rotational perturbations, the functions $\overset{s}{h}_n$ are simply $\overset{s}{h}_n = \overset{s}{\mu}_n/a$ where $\overset{s}{\mu}_n$ are solutions to Eqs. (8), (9) with appropriate initial conditions. For density perturbations, the functions $\overset{s}{h}_n$ are derivable from solutions to Eq. (10) in accord with the relationship between c and d operators. Besides, for density perturbations, we should regard $\overset{1}{c}_\mathbf{n}(0) = \overset{2}{c}_\mathbf{n}(0)$, $\overset{1\dagger}{c}_\mathbf{n}(0) = \overset{2\dagger}{c}_\mathbf{n}(0)$ in Eq. (11). In all cases, for a given cosmological model, that is for a model in which the scale factor $a(\eta)$ is known from the very early times and up to now, the functions $\overset{s}{h}_n$ can be found from the classical equations of motion with appropriate initial conditions.

It follows from Eq. (11) that the mean quantum-mechanical value of the field h_{ij} is zero at every spatial point and at every moment of time, $\langle 0|h_{ij}|0\rangle = 0$. One can also calculate variances of the field, that is the expectation values of its quadratic combinations. One useful quantity is $h_{ij}h^{ij}$. By manipulating with the product of two expressions (11), using the summation properties of the polarization tensors, and remembering that the only nonvanishing correlation function is

$$
\langle 0|\overset{s}{c}_\mathbf{n}(0)\overset{s'\dagger}{c}_{\mathbf{n}'}(0)|0\rangle = \delta_{ss'}\delta^3(\mathbf{n} - \mathbf{n}') \quad ,
$$

one can derive the formula

$$
\langle 0|h_{ij}(\eta, \mathbf{x})h^{ij}(\eta, \mathbf{x})|0\rangle = \frac{C^2}{2\pi^2} \int_{0}^{\infty} n \sum_{s=1}^{2} |\overset{s}{h}_n(\eta)|^2 \, dn \quad . \qquad (12)
$$

Equation (12) shows that the variance is independent of the spatial point **x** but does depend on time.

The expression under the integral in formulas such as Eq. (12) is usually called the power spectrum (in this case, it is the power spectrum of the quantity $h_{ij}h^{ij}$):

$$P(n) = \frac{C^2}{2\pi^2} n \sum_{s=1}^{2} |\overset{s}{h}_n(\eta)|^2 \quad . \tag{13}$$

In cosmology, it is common to use the power spectrum defined in terms of the logarithmic frequency interval, that is the function

$$P_Z(n) = \frac{C^2}{2\pi^2} n^2 \sum_{s=1}^{2} |\overset{s}{h}_n(\eta)|^2 \tag{14}$$

(Z from Zeldovich). We are mostly interested in the power spectrum of cosmological perturbations in the present Universe, at the matter-dominated stage. This spectrum is never smooth as a function of frequency (wavenumber) n. Squeezing and associated standing wave pattern of the field make the spectrum an oscillating function of n for each moment of time. In their turn, the oscillations in the power spectrum will produce oscillations in the distribution of the higher-order multipoles of the angular correlation function for the temperature anisotropies. However, the spectrum is smooth for sufficiently long waves. At a given moment of time, it applies to all perturbations whose wavelengths are of the order and longer than the Hubble radius defined at that time. Moreover, the smooth part of the spectrum is power-law dependent on n if the scale factor $a(\eta)$ of the very early Universe (the pump field) was power-law dependent on η-time.

Let us assume that the scale factor at the initial stage of expansion was

$$a(\eta) = l_0 |\eta|^{1+\beta} \tag{15}$$

where l_o and β are constants. If the evolution is governed by a scalar field, the Einstein equations require the constant β to be $\beta \leq -2$. The value $\beta = -2$ corresponds to the de Sitter expansion. At later times, the scale factor changed to the laws of the radiation-dominated and matter-dominated universes. From solutions for $\overset{s}{h}(\eta)$ traced up to the matter-dominated stage, one can find

$$\sum_{s=1}^{2} |\overset{s}{h}_n(\eta)|^2 \sim \frac{1}{l_o^2} n^{2\beta+2} \quad \text{and} \quad P_Z(n) \sim \frac{l_{Pl}^2}{l_o^2} n^{2(\beta+2)} \ .$$

It is convenient to introduce the *characteristic* amplitude $h(n)$ of the metric perturbations defining this amplitude as the standard deviation (square

root of variance) of the perturbed gravitational field per logarithmic frequency interval. In the long-wavelength limit under discussion, this quantity
is universally expressed (both, for gravitational waves and density perturbations) by the formula [21]:

$$h(n) \sim \frac{l_{Pl}}{l_o} n^{\beta+2} \quad . \qquad (16)$$

Note that the functional form of $h(n)$ is the same for gravitational waves and
density perturbations, the difference is in the numerical coefficient (omitted
in this discussion) which is somewhat in favor of gravitational waves [3,7].
The numerical level of $h(n)$ is mainly controlled by the constant l_o.

The spectra of other quantities can be found in the same manner. For
instance, in case of density perturbations, one can derive the spectrum of
perturbations in the matter density $\delta\rho/\rho$. Since the relationship between
$\delta\rho/\rho$ and the metric perturbations involves the factor $(n\eta)^2$ and, hence,
involves two extra powers of n, $\frac{\delta\rho}{\rho}(n) \sim n^2 h(n)$, one finds

$$\langle 0|\frac{\delta\rho}{\rho}\frac{\delta\rho}{\rho}|0\rangle \sim \int_0^\infty P_Z^\rho(n)\frac{dn}{n}$$

where $P_Z^\rho(n) \sim (l_{Pl}^2/l_o^2)n^{2(\beta+4)}$ and

$$\frac{\delta\rho}{\rho}(n) \sim \frac{l_{Pl}}{l_o}n^{\beta+4} \quad . \qquad (17)$$

It follows from Eq. (16) that $h(n)$ is independent of n if $\beta = -2$. This
independence corresponds to the original Zeldovich's definition of the "flat"
spectrum: all waves enter the Hubble radius with the same amplitude. If
the gravitational field perturbations $h(n)$, regardless of their wavelength,
have equal amplitudes upon entering the Hubble radius, the matter density
perturbations $\frac{\delta\rho}{\rho}(n)$ do also have equal amplitudes (the extra factor $(n\eta)^2$
is of the order of 1 at the time when a given wave n enters the Hubble
radius). For models of the very early Universe governed by a scalar field,
the spectral index $\beta + 2$ in Eq. (16) can never be positive.

Formula (16) and the associated formula (17) should be compared with
the "standard" inflationary formula which requires that the amplitudes of
density perturbations taken at the time of entering the Hubble radius should
go to infinity in the limit of the de Sitter inflation, $\beta \to -2$. It should also be
noted that a "disgusting convention" (the term is borrowed from Ref. [10])
is often being used according to which one and the same Harrison-Zeldovich
spectrum is described by the spectral index $n_t = 0$ for gravitational waves
and by the spectral index $n_s = 1$ for density perturbations. Of course, there
is no need in this convention. In both cases, the metric perturbations with

the Harrison-Zeldovich spectrum are described by the same spectral index (zero), see Eq. (16).

We will finish this section with a short discussion of coherent states. There is no natural mechanism for the generation of cosmological perturbations in coherent states, but if there were one it would be reflected in many parts of the theory. The interaction part of the Hamiltonian (6) would be linear (not quadratic) in the creation and annihilation operators. The analogue of Eq. (7) would read

$$
\begin{aligned}
c_{\mathbf{n}}(\eta) &= e^{-in\eta} c_{\mathbf{n}}(0) + \alpha_n(\eta) \\
c_{\mathbf{n}}^{\dagger}(\eta) &= e^{in\eta} c_{\mathbf{n}}^{\dagger}(0) + \alpha_n^*(\eta) \quad ,
\end{aligned}
\tag{18}
$$

where the complex function $\alpha_n(\eta)$ is determined by the coupling function in the Hamiltonian. On the position — momentum diagram, the evolution (18) of the field operators corresponds to *displacing* the vacuum state without *squeezing* whereas the evolution (7) corresponds to *squeezing* the vacuum state without *displacing*. In terminology of mechanics, coherent states are produced by a force acting on the oscillator whereas squeezed vacuum states are produced by a parametric influence. In coherent states, the mean value of the field is not zero. The correlation functions (at least, for some quantities) would also be different from the squeezed state case. In cosmological context, this would eventually be reflected in the differing statistical properties of perturbations and induced microwave background anisotropies.

The calculations of the next Section are based in an essential way on the field operator (11) for the squeezed vacuum perturbations.

3. Quantum-Mechanical Expectation Values for the Microwave Background Anisotropies

The microwave background anisotropies are a subject of intense study [22].

In absence of cosmological perturbations, the temperature of the microwave background radiation seen in all directions on the sky would be the same, T. Let us denote a direction on the sky by a unit vector $\mathbf{e}$. The presence of cosmological perturbations makes the temperature seen in the direction $\mathbf{e}$ differing from T. The temperature perturbation produced by density perturbations or gravitational waves can be described by the formula [11]:

$$
\frac{\delta T}{T}(\mathbf{e}) = \frac{1}{2} \int_0^{w_1} \frac{\partial h_{ij}}{\partial \eta} e^i e^j \, dw
\tag{19}
$$

where $\partial h_{ij}/\partial \eta$ is taken along the integration path $x^i = e^i w$, $\eta = \eta_R - w$, from the event of reception $w = 0$ to the event of emission $w = w_1 =$

$\eta_R - \eta_E$. The formula for rotational perturbations is more complicated than (19) [11], and we will leave rotational perturbations aside.

For quantized cosmological perturbations, the $\frac{\delta T}{T}(\mathbf{e})$ becomes a quantum-mechanical operator. Using Eq. (11) we can write this operator as

$$
\frac{\delta T}{T}(\mathbf{e}) = \quad \frac{C}{2} \frac{1}{(2\pi)^{3/2}} \int_0^{w_1} dw \int_{-\infty}^{\infty} d^3 n \sum_{s=1}^{2} \overset{s}{p}_{ij}(\mathbf{n}) e^i e^j
$$
$$
\times \left[\overset{s}{c}_{\mathbf{n}}(0) \overset{s}{f}_n(w) e^{iw\mathbf{n}\mathbf{e}} + \overset{s\dagger}{c}_{\mathbf{n}}(0) \overset{s*}{f}_n(w) e^{-iw\mathbf{n}\mathbf{e}} \right] \tag{20}
$$

where

$$
\overset{s}{f}_n(w) \equiv \frac{1}{\sqrt{2n}} \left. \frac{d\overset{s}{h}_n}{d\eta} \right|_{\eta = \eta_R - w} \quad .
$$

Having defined the observable $\frac{\delta T}{T}(\mathbf{e})$ and knowing the quantum state $|0\rangle$ we can compute various quantum-mechanical expectation values. In the laboratory quantum mechanics, the verification of theoretical predictions expressed in terms of the expectation values would require experiments on many identical systems. An immediate generalization of this principle to cosmology would require speculations about outcomes of experiments performed in "many identical universes". Without having access to "many universes" we can only rely on the mean (expected) values of the observables and on the probability distribution functions as indicators of what is likely or not to be observed in our own single Universe. We will return to this point in Sec. IV.

The expected value of the temperature perturbation to be observed in every fixed direction on the sky is zero:

$$
\langle 0 | \frac{\delta T}{T}(\mathbf{e}) | 0 \rangle = 0 \quad .
$$

One particular measured temperature map is the result of the measurement performed over one particular realization of the random process describing cosmological perturbations of quantum-mechanical origin. For this realization, the temperature perturbations may, should, and in fact are, present. Many measurements will not help (except of reducing the instrumental noises) in the sense that they all should give identical results, because the timescale of the perturbations under discussion is so enormously larger than an interval of time between the experiments. If the COBE's map is correct, we will have to live with this map practically forever.

Let us now compute the expected angular correlation function for the temperature perturbations seen in two given directions on the sky, $\mathbf{e}_1$ and

e_2. This correlation function is defined as the mean value for the product of $\frac{\delta T}{T}(e_1)$ and $\frac{\delta T}{T}(e_2)$:

$$K(e_1, e_2) = \langle 0| \frac{\delta T}{T}(e_1) \frac{\delta T}{T}(e_2) |0 \rangle \quad . \tag{21}$$

By manipulating with the product of two expressions (20) one can derive the formula

$$K(e_1, e_2) = \frac{1}{4} C^2 \frac{1}{(2\pi)^3} \int_0^{w_1} dw \int_0^{w_1} d\bar{w} \int_{-\infty}^{\infty} d^3 n \, e^{in(e_1 w - e_2 \bar{w})}$$

$$\times \sum_{s=1}^{2} \left(\overset{s}{p}_{ij}(n) e_1^i e_1^j \right) \left(\overset{s}{p}_{ij}(n) e_2^i e_2^j \right) \overset{s}{f}_n(w) \overset{s*}{f}_n(\bar{w}) \quad . \tag{22}$$

The next step is the formidable task of taking the integrals over angular variables in 3-dimensional wave-vector n space. However, it can be done (see Ref. [12] for gravitational waves, Ref. [13] for rotational perturbations, and Ref. [3] for density perturbations). The final expression reduces, without making any additional assumptions whatsoever, to the form

$$K(e_1, e_2) = K(\delta) = l_{Pl}^2 \sum_{l=l_{min}}^{\infty} K_l \, P_l(\cos \delta) \quad . \tag{23}$$

We see that the correlation function depends only on the angle δ between the directions e_1, e_2 not directions themselves. The coefficient l_{Pl}^2 is taken from C^2, other numerical coefficients are included in K_l. The quantities K_l involve the integration of $\overset{s}{f}_n(w)$ over the parameter w and the remaining integration over the wave-numbers n. The numerical values of K_l depend on a chosen sort of cosmological perturbations and a chosen cosmological model; so far, the formula (23) is totally general. $P_l(\cos \delta)$ are the Legendre polynomials. The lowest multipole l_{min} follows automatically from the theory and it turns out to be, not surprisingly, $l_{min} = 0$ for density perturbations, $l_{min} = 2$ for gravitational waves (and $l_{min} = 1$ for rotational perturbations). For the separation angle $\delta = 0$, Eq. (23) reduces to the variance of $\frac{\delta T}{T}(e)$, that is

$$\langle 0| \frac{\delta T}{T}(e) \frac{\delta T}{T}(e) |0 \rangle = K(0) = l_{Pl}^2 \sum_{l=l_{min}}^{\infty} K_l \quad . \tag{24}$$

Formula (23) gives the expected value of the observable $\frac{\delta T}{T}(e_1) \frac{\delta T}{T}(e_2)$. If the experimenter measured this observable in "many universes" and averaged the measured numbers, he/she would get the result (23). Moreover, formula (23) says that if the experimenter made the measurements at any other pair of directions, but with the same separation angle δ, he/she would again get, after the averaging over "many universes", the result (23). Without having access to "many universes", we can ask what is the theoretical standard deviation of the quantity $\frac{\delta T}{T}(\mathbf{e}_1)\frac{\delta T}{T}(\mathbf{e}_2)$. (In practice, for deriving $K(\delta)$ we need a kind of ergodic hypothesis allowing us to replace the averaging over "universes" by the averaging over pixels on a single map.) The variance $V(\mathbf{e}_1, \mathbf{e}_2)$ of this quantity is, by definition,

$$V(\mathbf{e}_1, \mathbf{e}_2) = \langle 0|\frac{\delta T}{T}(\mathbf{e}_1)\frac{\delta T}{T}(\mathbf{e}_2)\frac{\delta T}{T}(\mathbf{e}_1)\frac{\delta T}{T}(\mathbf{e}_2)|0\rangle - \left[\langle 0|\frac{\delta T}{T}(\mathbf{e}_1)\frac{\delta T}{T}(\mathbf{e}_2)|0\rangle\right]^2 .$$
(25)

The standard deviation is the square root of this number.

The calculation of $V(\mathbf{e}_1, \mathbf{e}_2)$ requires us to deal with the product of four expressions (20). However, the mean values of the products of four creation and annihilation operators are easy to handle. One can show that $V(\mathbf{e}_1, \mathbf{e}_2)$ depends only on the separation angle δ and

$$V(\mathbf{e}_1, \mathbf{e}_2) = V(\delta) = \left[\langle 0|\frac{\delta T}{T}(\mathbf{e}_1)\frac{\delta T}{T}(\mathbf{e}_2)|0\rangle\right]^2 + \left[\langle 0|\frac{\delta T}{T}(\mathbf{e})\frac{\delta T}{T}(\mathbf{e})|0\rangle\right]^2 , \quad (26)$$

that is

$$V(\delta) = K^2(\delta) + K^2(0) . \quad (27)$$

The standard deviation for the observable $\frac{\delta T}{T}(\mathbf{e}_1)\frac{\delta T}{T}(\mathbf{e}_2)$ is

$$\sigma(\delta) = [V(\delta)]^{1/2} = \sqrt{K^2(\delta) + K^2(0)} . \quad (28)$$

In a similar fashion one can derive the higher order correlation functions for two directions $\mathbf{e}_1$, $\mathbf{e}_2$ and the correlation functions for larger number of directions, but we will not need this information.

In the limit $\delta = 0$ Eqs. (25), (26) say that

$$\langle 0|\left[\frac{\delta T}{T}(\mathbf{e})\right]^4|0\rangle = 3\left[\langle 0|\left[\frac{\delta T}{T}(\mathbf{e})\right]^2|0\rangle\right]^2 . \quad (29)$$

The familiar factor 3 relating the fourth-order moment with the square of the second-order moment (given that the first-order moment is equal to zero) is the reflection of the underlying Gaussian nature of the squeezed vacuum wavefunctions associated with the Hamiltonian (6).

By examining Eq. (28) one can conclude that for each separation angle δ the standard deviation of the angular correlation function is very big. Even at those separation angles at which $K(\delta)$ vanishes, the standard deviation is as big as the variance for $\frac{\delta T}{T}(\mathbf{e})$ itself. However, the value of the standard deviation for a given variable is not very informative *per se*, as long as the probability density function for this variable is not known. If the probability density function (p.d.f.) were normal, we could say that the probability to find a result outside of 1σ interval is 32%. Without knowing the p.d.f. we could resort to the Chebyshev inequality, but it would only tell us that this probability is less than 1. To get more information about possible deviation of the angular correlation function from its mean values we will consider in Sec. IV a classical random model which will reproduce the expectation values calculated above and will allow us to construct the p.d.f. for the variable $\frac{\delta T}{T}(e_1)\frac{\delta T}{T}(e_2)$. On the other hand, the quantum-mechanical calculations of this Section will shed light on the classical model. As is known, "quantum mechanics helps us understand classical mechanics", see on this subject a paper of Zeldovich signed by the pseudonym Paradoksov [14].

4. Classical Model for the Statistics of the Microwave Background Anisotropies

A distribution of the microwave background temperature over the sky is a real function of the angular coordinates. Assuming that $\delta T/T$ is a sufficiently smooth function on a sphere, one can expand it over the set of orthonormal complex spherical harmonics $Y_{lm}(\theta, \phi)$ [15]:

$$\frac{\delta T}{T}(e) = \sum_{l=0}^{\infty} \sum_{m=-l}^{l} [a_{lm}Y_{lm}(e) + a_{lm}^{*}Y_{lm}^{*}(e)] \quad . \tag{30}$$

We want to formulate a statistical hypothesis about the coefficients a_{lm}, so it is better to write them first in terms of real (r) and imaginary (i) components:

$$
\begin{aligned}
a_{lm} &= a_{lm}^{r} + ia_{lm}^{i}, & a_{lm}^{*} &= a_{lm}^{r} - ia_{lm}^{i} \\
Y_{lm} &= Y_{lm}^{r} + iY_{lm}^{i}, & Y_{lm}^{*} &= Y_{lm}^{r} - iY_{lm}^{i}, \\
\frac{\delta T}{T}(e) &= 2\sum_{l=0}^{\infty}\sum_{m=-l}^{l} [a_{lm}^{r}Y_{lm}^{r}(e) - a_{lm}^{i}Y_{lm}^{i}(e)] \quad .
\end{aligned} \tag{31}
$$

Our statistical hypothesis is as follows: (i) all members of the set of random variables $\{a_{lm}^{r}, a_{lm}^{i}\}$ are statistically independent, (ii) each individual variable is normally distributed and has a zero mean, (iii) all variables with the same index l have the same standard deviation σ_l. All said is expressed by the probability density function (p.d.f.) for individual variables:

$$f(a_{lm}^{r}) = \frac{1}{\sqrt{2\pi}\,\sigma_l}e^{-\frac{(a_{lm}^{r})^2}{2\sigma_l^2}} \quad , \quad f(a_{lm}^{i}) = \frac{1}{\sqrt{2\pi}\,\sigma_l}e^{-\frac{(a_{lm}^{i})^2}{2\sigma_l^2}} \quad , \tag{32}$$

and by the p.d.f. for the entire set of variables, which is simply a product of all p.d.f.'s for all individual variables:

$$f\left(\{a_{lm}^r, a_{lm}^i\}\right) = \sqcap_{l,m} f(a_{lm}^r) f(a_{lm}^i) \quad . \tag{33}$$

Having postulated the p.d.f.'s, we can now compute the expectation values of certain functions of the random variables. Below, the angular brackets will denote the expectation values calculated with the help of the p.d.f. (33), unless other definition is stated.

Obviously, all linear functions have a zero mean:

$$<a_{lm}^r> = 0, \qquad <a_{lm}^i> = 0 \quad . \tag{34}$$

For quadratic combinations we have

$$\begin{aligned}
<a_{l_1 m_1}^r a_{l_2 m_2}^r> &= \sigma_{l_1}^2 \delta_{l_1 l_2} \delta_{m_1 m_2}, & <a_{l_1 m_1}^i a_{l_2 m_2}^i> &= \sigma_{l_1}^2 \delta_{l_1 l_2} \delta_{m_1 m_2}, \\
<a_{l_1 m_1}^r a_{l_2 m_2}^i> &= 0, & <a_{l_1 m_1}^i a_{l_2 m_2}^r> &= 0.
\end{aligned} \tag{35}$$

All triple products have zero means. Among quartic combinations, only those can survive which have four indices (r), or four indices (i), or two indices (r) and two indices (i). Two representative expressions are:

$$\begin{aligned}
<a_{l_1 m_1}^r a_{l_2 m_2}^r a_{l_3 m_3}^r a_{l_4 m_4}^r> &= \sigma_{l_1}^2 \sigma_{l_3}^2 \delta_{l_1 l_2} \delta_{m_1 m_2} \delta_{l_3 l_4} \delta_{m_3 m_4} \\
+ \sigma_{l_1}^2 \sigma_{l_2}^2 \delta_{l_1 l_3} \delta_{m_1 m_3} \delta_{l_2 l_4} \delta_{m_2 m_4} &+ \sigma_{l_1}^2 \sigma_{l_2}^2 \delta_{l_1 l_4} \delta_{m_1 m_4} \delta_{l_2 l_3} \delta_{m_2 m_3} \quad , \tag{36} \\
<a_{l_1 m_1}^r a_{l_2 m_2}^r a_{l_3 m_3}^i a_{l_4 m_4}^i> &= \sigma_{l_1}^2 \sigma_{l_3}^2 \delta_{l_1 l_2} \delta_{m_1 m_2} \delta_{l_3 l_4} \delta_{m_3 m_4} \quad . \tag{37}
\end{aligned}$$

Other quartic combinations can be obtained by the replacement $(r) \leftrightarrow (i)$ in Eqs. (36), (37) (or by permutation of pairs (lm) in case of Eq. (37)). The higher-order correlations can be derived in a similar way, but we will not need them.

In our further calculations related to the random variables $\frac{\delta T}{T}(\mathbf{e})$ and $\frac{\delta T}{T}(\mathbf{e}_1)\frac{\delta T}{T}(\mathbf{e}_2)$ it is easier to deal with the complex coefficients a_{lm}, so we will first translate the above relationships to them. By using the available information one can derive

$$\begin{aligned}
<a_{lm}> &= 0, \quad <a_{l_1 m_1} a_{l_2 m_2}^*> = 2\sigma_l^2 \delta_{l_1 l_2} \delta_{m_1 m_2} \quad , \\
<a_{l_1 m_1} a_{l_2 m_2} a_{l_3 m_3}^* a_{l_4 m_4}^*> &= 4\sigma_{l_1}^2 \sigma_{l_2}^2 (\delta_{l_1 l_3} \delta_{m_1 m_3} \delta_{l_2 l_4} \delta_{m_2 m_4} \\
&\quad + \delta_{l_1 l_4} \delta_{m_1 m_4} \delta_{l_2 l_3} \delta_{m_3 m_3}) \quad . \tag{38}
\end{aligned}$$

The mean values of the complex conjugated quantities are given by the same formulas (38). Other nonvanishing quartic combinations can be obtained from the one in Eq. (38) by the permutation of pairs (lm).

Now, even before deriving the p.d.f.'s for the random variables $\frac{\delta T}{T}(e)$ and $\frac{\delta T}{T}(e_1)\frac{\delta T}{T}(e_2)$, we can find some expectation values. It is clear from the definition (30) and Eq. (38) that

$$\left\langle \frac{\delta T}{T}(e) \right\rangle = 0 \quad . \tag{39}$$

When calculating the angular correlation function one should remember that

$$\sum_{m=-l}^{l} Y_{lm}(e_1)Y_{lm}^*(e_2) = \frac{2l+1}{4\pi}P_l(\cos\delta) \tag{40}$$

(note the origin of the factor $2l+1$ which will accompany us often). By taking the product of two expressions (30) and using Eqs. (38), (40) one can find the angular correlation function

$$\left\langle \frac{\delta T}{T}(e_1)\frac{\delta T}{T}(e_2) \right\rangle = \frac{1}{\pi}\sum_{l=0}^{\infty}\sigma_l^2(2l+1)P_l(\cos\delta) \quad . \tag{41}$$

If the separation angle δ is zero, we obtain

$$\left\langle \frac{\delta T}{T}(e)\frac{\delta T}{T}(e) \right\rangle = \frac{1}{\pi}\sum_{l=0}^{\infty}\sigma_l^2(2l+1) \quad . \tag{42}$$

[One may notice an incidental fact that the mean value of the random variable a_l^2 defined as $a_l^2 = \sum_{m=-l}^{l} a_{lm}a_{lm}^*$ is $<a_l^2> = 2(2l+1)\sigma_l^2$, that is the same expression which enters Eq. (41). This may suggest an interpretation of the quantity $<a_l^4> - <a_l^2>^2$ as the variance of the multipole moments. One should be carefull with this interpretation, see Sec. V.]

We can also find the 4th order expectation values. The product of 4 expressions (30) in conjunction with Eq. (38) gives

$$\left\langle \frac{\delta T}{T}(e_1)\frac{\delta T}{T}(e_2)\frac{\delta T}{T}(e_1)\frac{\delta T}{T}(e_2) \right\rangle - \left\langle \frac{\delta T}{T}(e_1)\frac{\delta T}{T}(e_2) \right\rangle^2$$
$$= \left\langle \frac{\delta T}{T}(e_1)\frac{\delta T}{T}(e_2) \right\rangle^2 + \left\langle \frac{\delta T}{T}(e)\frac{\delta T}{T}(e) \right\rangle^2 . \tag{43}$$

If $\delta = 0$, it follows from Eq. (43) that

$$\left\langle \left[\frac{\delta T}{T}(e)\right]^4 \right\rangle = 3\left\langle \left[\frac{\delta T}{T}(e)\right]^2 \right\rangle^2 \quad . \tag{44}$$

Up to difference in the meaning of the angular brackets, the formulas (39), (43), (44) reproduce the analogous results of the previous Section. Moreover, from comparison of Eqs. (23), (24) with Eqs. (41), (42) we can relate the quantities K_l, derivable from a given cosmological model plus perturbations, with the abstract quantities σ_l.

We can now engage in our major enterprise — the construction of the p.d.f. for the random variable $v \equiv \frac{\delta T}{T}(\mathbf{e}_1)\frac{\delta T}{T}(\mathbf{e}_2)$. We will start from the p.d.f. for the random variable $z \equiv \frac{\delta T}{T}(\mathbf{e})$. When it is necessary to distinguish directions $\mathbf{e}_1$ and $\mathbf{e}_2$, we will use the notations z_1 and z_2.

The variable z is a function of the variables $\{a^r_{lm}, a^i_{lm}\}$ whose p.d.f.'s are known, Eqs. (31), (32). There exist regular methods (see, for example, an excellent book [16]) allowing to derive rigorously the p.d.f. of a function. However, in our case that the function is linear and all p.d.f's are normal, we can partially rely on a guesswork. Combining formulas and guessing we can write

$$f(z) = \frac{1}{\sqrt{2\pi}\,\sigma_z} e^{-\frac{z^2}{2\sigma_z^2}} \quad , \tag{45}$$

where

$$\sigma_z^2 = \frac{1}{\pi}\sum_{l=0}^{\infty} \sigma_l^2(2l+1) \quad . \tag{46}$$

The p.d.f. (45) certainly leads to Eqs. (39), (42), (44). Moreover, it allows us to say that the probability to find z outside of $1\sigma_z$ interval is approximately 32%:

$$P(|z| > \sigma_z) \approx 0.32 \quad .$$

We now introduce two variables, z_1 and z_2, and ask about the p.d.f. in the 2-dimensional space (z_1, z_2). Again, partially relying on a guesswork, we find that

$$f(z_1, z_2) = \frac{1}{2\pi\sigma_z^2\sqrt{1-\rho^2}}\exp\left\{-\frac{1}{2\sigma_z^2(1-\rho^2)}[z_1^2 + z_2^2 - 2\rho z_1 z_2]\right\} \tag{47}$$

where

$$\rho\sigma_z^2 = \frac{1}{\pi}\sum_{l=0}^{\infty}(2l+1)\sigma_l^2 P_l(\cos\delta), \qquad |\rho| \le 1 \quad . \tag{48}$$

(See Eq. (5.11.1) in Ref. [16]). First, we can check that the marginal distributions are correct. For $f(z_1)$, one obtains

$$f(z_1) = \int_{-\infty}^{\infty} f(z_1, z_2)dz_2 = \frac{1}{\sqrt{2\pi}\,\sigma_z}e^{-\frac{z_1^2}{2\sigma_z^2}}$$

and one obtains a similar expression for $f(z_2)$. Second, one can check that

$$\langle z_1^2 \rangle = \sigma_z^2, \qquad \langle z_2^2 \rangle = \sigma_z^2, \qquad \langle z_1 z_2 \rangle = \rho \sigma_z^2$$

where the angular brackets mean the integration with the p.d.f. (47). These equalities are Eqs. (42), (41) which we must have obtained.

Finally, we shall derive the p.d.f. for the variable $v = z_1 z_2$. We will do this in some detail following the prescriptions of [16].

Let us introduce the two new variables (z_1, v) instead of (z_1, z_2) according to the transformation

$$z_1 = z_1, \qquad z_2 = \frac{v}{z_1} \; .$$

The Jacobian of this transformation is $\mathcal{J} = 1/z_1$. The p.d.f. $f(v)$ is the result of the following integration:

$$f(v) = \frac{1}{2\pi\sigma_z^2\sqrt{1-\rho^2}} \int_{-\infty}^{\infty} \frac{1}{|z_1|} \exp\left\{ -\frac{1}{2\sigma_z^2(1-\rho^2)} \left[z_1^2 + \frac{v^2}{z_1^2} - 2\rho v \right] \right\} dz_1 .$$

The integral over z_1 can be taken with the help of 3.471.9 from [17]. The resulting p.d.f. can be written in the form:

$$f(v) = \begin{cases} \frac{1}{\pi\sigma_z^2\sqrt{1-\rho^2}} e^{\frac{\rho v}{\sigma_z^2(1-\rho^2)}} K_0\left(\frac{v}{\sigma_z^2(1-\rho^2)}\right), & \text{for } v > 0 \\[2mm] \frac{1}{\pi\sigma_z^2\sqrt{1-\rho^2}} e^{\frac{\rho v}{\sigma_z^2(1-\rho^2)}} K_0\left(\frac{-v}{\sigma_z^2(1-\rho^2)}\right), & \text{for } v < 0 \end{cases} \qquad (49)$$

where K_0 is the modified Bessel function of its argument [18].

The function $f(v)$ is quite complicated and the distribution is obviously not normal. The function $f(v)$ goes to zero for $v \to \pm\infty$ and diverges logarithmically at the point $v = 0$. Even a verification of the normalization condition

$$\int_{-\infty}^{\infty} f(v) dv = 1 \qquad (50)$$

is not trivial. However, with the help of 6.621.3, 9.131.1, 9.121.7, 1.624.9 and 1.623.2 from [17] one can prove the validity of Eq. (50).

The mean value and the standard deviation of the variable v are known, see Eqs. (41), (43):

$$\langle v \rangle = \rho \sigma_z^2, \qquad \sigma_v = \left[\langle v^2 \rangle - \langle v \rangle^2 \right]^{1/2} = \sigma_z^2 \sqrt{\rho^2 + 1} \; . \qquad (51)$$

We already knew that the standard deviation is big. We now see again that $\sigma_v = \sigma_z^2$ at the separation angles at which the angular correlation function vanishes, $\rho = 0$, and $\sigma_v = \sqrt{2}\sigma_z^2$ at zero separation angle, $\rho = 1$. For other separation angles, σ_v lies between these two numbers.

Now that we know the p.d.f., we can assign probabilities to the different ranges of the variable v. For instance, we can calculate the probability that the measured v will be found, say, outside of the $\lambda\sigma_v$ interval surrounding the mean value of v, where λ is an arbitrary fixed number. The probability of our interest is

$$P(|v - \langle v \rangle| > \lambda\sigma_v) = \int_{-\infty}^{\sigma_z^2(\rho - \lambda\sqrt{\rho^2+1})} f(v)dv + \int_{\sigma_z^2(\rho + \lambda\sqrt{\rho^2+1})}^{\infty} f(v)dv \quad .$$
(52)

To get a qualitative estimate of the associated theoretical uncertainties for the observable v, we will ask a slightly different question. What should the number λ be in order to have the 0.32 chance of finding v outside the $\lambda\sigma_v$ interval and, hence, the 0.68 chance to find it inside the interval?

To evaluate the size of the disaster, we will start from the case $\rho = 0$. In this case, the p.d.f. (49) is symmetric with respect to the origin $v = 0$ (this is why $\langle v \rangle$ is zero in this case) and

$$P(|v| > \lambda\sigma_z^2) = \frac{2}{\pi} \int_{\lambda}^{\infty} K_0(x)dx \quad . \tag{53}$$

We want this number to be approximately equal to 0.32. Judging from the Fig. 9.7 in Ref. [18], a half of the area under the $K_0(x)$ function is accumulated when integrating from approximately $x = 1/2$ and up to infinity. This means that λ should approximately be equal to $1/2$.

If $\rho \neq 0$ the evaluation of P is more complicated. For $\rho \neq 0$, the function (49) is not symmetric with respect to the origin $v = 0$. It has larger values at positive v's if $\rho > 0$ (this is why $\langle v \rangle > 0$ in this case) and it has larger values at negative v's if $\rho < 0$ (this is why $\langle v \rangle < 0$ in this case). The graph of the function $e^x K_0(x)$ plotted on Fig. 9.8 in Ref. [18] is helpful. A qualitative analysis shows again that λ is approximately equal to $1/2$. (More accurate estimates can of course be reached by numerical methods.)

At any rate, the $\frac{1}{2}\sigma_v$ interval gives approximately the same probability estimates as if the distribution (49) were normal.

5. On the "Cosmic Variance"

The set of random variables $\{a_{lm}^r, a_{lm}^i\}$ defined by Eqs. (32), (33) lives its own independent life regardless of whether or not the variables are considered random coefficients in the expansion of some function over spherical harmonics. Being such, it allows introduction of new functions and calculation of their expectation values. One interesting variable is defined by the equation

$$a_l^2 = \sum_{m=-l}^{l} a_{lm} a_{lm}^* = \sum_{m=-l}^{l} |a_{lm}|^2 = \sum_{m=-l}^{l} \left[(a_{lm}^r)^2 + (a_{lm}^i)^2 \right] . \quad (54)$$

By using Eq. (38) one can calculate the expectation value of a_l^2:

$$\langle a_l^2 \rangle = (2l+1) 2 \sigma_l^2 \quad . \quad (55)$$

The factor $2(2l+1)$ reflects the number of independent "degrees of freedom" associated with the index l. One can also introduce the variable a_l^4 and calculate its expectation value:

$$\langle a_l^4 \rangle = (2l+1)(l+1) 8 \sigma_l^4 = \langle a_l^2 \rangle^2 \frac{2(l+1)}{2l+1} \quad . \quad (56)$$

The difference $\langle a_l^4 \rangle - \langle a_l^2 \rangle^2$ is, by definition, the variance of the variable a_l^2. From Eqs. (56), (55) one finds

$$\langle a_l^4 \rangle - \langle a_l^2 \rangle^2 = \frac{1}{2l+1} \langle a_l^2 \rangle^2 \quad . \quad (57)$$

This formula, as it stands, expresses a well-known fact: the variance of the random variable χ^2 defined as the sum of squares of n independent random variables (degrees of freedom) with the same normal density, is $\frac{n}{2}$ times smaller than the square of the mean value of χ^2 [16]. There is nothing "cosmological" or "inflationary" in this fact. Knowing the p.d.f.'s for the set $\{a_{lm}^r, a_{lm}^i\}$ one can calculate the higher-order correlation functions for the variable a_l^2 and its distribution function [23-29]. In the recent literature, formula (57) became known as the "cosmic variance". Formula (57) and a possibility (or lack of) to extract complete information about a stochastic process from its single realization are, in general, different issues. For ergodic processes, the existence of a definitely true relationship (57) prevents in no way the extraction of complete information about the process from a single realization [30].

It is important to realize that it is the mean value of the random variable a_l^2, not a_l^2 itself, that enters the expected angular correlation function in front of the Legendre polynomials and which is often called the multipole moment. The a_l^2 is a random variable and its variance has meaning, the $\langle a_l^2 \rangle$ is a number and its variance has no meaning. Specifically, one can notice that the angular correlation function (41) can be written in the form

$$\left\langle \frac{\delta T}{T}(\mathbf{e}_1) \frac{\delta T}{T}(\mathbf{e}_2) \right\rangle = \frac{1}{2\pi} \sum_{l=0}^{\infty} \langle a_l^2 \rangle P_l(\cos \delta) \quad . \quad (58)$$

On this ground, there may be a temptation to write the random variable $\frac{\delta T}{T}(\mathbf{e}_1)\frac{\delta T}{T}(\mathbf{e}_2)$ in the form

$$\frac{\delta T}{T}(\mathbf{e}_1)\frac{\delta T}{T}(\mathbf{e}_2) = \frac{1}{2\pi}\sum_{l=0}^{\infty} a_l^2 P_l(\cos\delta) \tag{59}$$

and to interpret Eq. (57) as the variance for the multipole moments of the correlation function. One should resist to this temptation.

Let us show that the definition (59) is incorrect despite the fact that it gives correct expectation value (58). It follows from the definition (59) that

$$\frac{\delta T}{T}(\mathbf{e}_1)\frac{\delta T}{T}(\mathbf{e}_2)\frac{\delta T}{T}(\mathbf{e}_1)\frac{\delta T}{T}(\mathbf{e}_2)$$

$$= \frac{1}{4\pi^2}\sum_{l=0}^{\infty} a_l^4 [P_l(\cos\delta)]^2 + \frac{1}{4\pi^2}\sum_{l,l'=0,l\neq l'}^{\infty} a_l^2 P_l(\cos\delta) a_{l'}^2 P_{l'}(\cos\delta) . \tag{60}$$

Using (56), (58) and remembering that a_l^2 and $a_{l'}^2$ are statistically independent for $l \neq l'$, one can find the expectation value of the quantity (60):

$$\left\langle \frac{\delta T}{T}(\mathbf{e}_1)\frac{\delta T}{T}(\mathbf{e}_2)\frac{\delta T}{T}(\mathbf{e}_1)\frac{\delta T}{T}(\mathbf{e}_2) \right\rangle$$

$$= \frac{1}{4\pi^2}\sum_{l=0}^{\infty} \langle a_l^4 \rangle [P_l(\cos\delta)]^2 + \frac{1}{4\pi^2}\sum_{l,l'=0,l\neq l'}^{\infty} \langle a_l^2 \rangle \langle a_{l'}^2 \rangle P_l(\cos\delta) P_{l'}(\cos\delta)$$

$$= \frac{1}{4\pi^2}\sum_{l=0}^{\infty} \langle a_l^4 \rangle [P_l(\cos\delta)]^2 + \frac{1}{4\pi^2}\left[\sum_{l=0}^{\infty} \langle a_l^2 \rangle P_l(\cos\delta)\right]^2$$

$$- \frac{1}{4\pi^2}\sum_{l=0}^{\infty} \langle a_l^2 \rangle^2 [P_l(\cos\delta)]^2$$

$$= \left\langle \frac{\delta T}{T}(\mathbf{e}_1)\frac{\delta T}{T}(\mathbf{e}_2) \right\rangle^2 + \frac{1}{4\pi^2}\sum_{l=0}^{\infty}\frac{1}{2l+1}\langle a_l^2 \rangle^2 [P_l(\cos\delta)]^2 . \tag{61}$$

It follows from (61) that the variance of the variable $\frac{\delta T}{T}(\mathbf{e}_1)\frac{\delta T}{T}(\mathbf{e}_2)$ would read (if (59) were correct):

$$\frac{1}{\pi^2}\sum_{l=0}^{\infty}(2l+1)\sigma_l^4 [P_l(\cos\delta)]^2 . \tag{62}$$

This expression should be compared with the correct variance following from Eq. (43):

$$\frac{1}{\pi^2}\left[\sum_{l=0}^{\infty}(2l+1)\sigma_l^2 P_l(\cos\delta)\right]^2 + \frac{1}{\pi^2}\left[\sum_{l=0}^{\infty}(2l+1)\sigma_l^2\right]^2 . \tag{63}$$

Formulas (62), (63) disagree even for $\delta = 0$, and even in their first, $l = 0$ term. This shows that the *ad hoc* definition (59) is incorrect. The correct definition of the random variable $\frac{\delta T}{T}(\mathbf{e}_1)\frac{\delta T}{T}(\mathbf{e}_2)$ is the one following from the definition (30) and which we have used in this paper.

6. Conclusions

A particular cosmological model plus perturbations gives unambiguous predictions with regard to the expectation values of the measurable quantities. Differing models give different predictions. We want to distinguish them observationally and to learn about physics of the very early Universe. However, the quantum-mechanical origin of the cosmological perturbations is reflected in the theoretical statistical uncertainties surrounding the expectation values. One important measurable quantity is the angular correlation function of the microwave background anisotropies. Its mean value at the zero separation angle was denoted σ_z^2 in this paper. It was shown that the standard deviation for the correlation function is very big. The 68% confidence level corresponds, approximately, to $\frac{1}{2}\sigma_z^2$ at the separation angles where the correlation function vanishes, to $\frac{\sqrt{2}}{2}\sigma_z^2$ at the zero separation angle, and to intermediate numbers for other separation angles.

The angular correlation function has actually been measured. It is presented at Fig. 3 in the paper [19]. The authors surround the measured points by a narrow shaded region which they address as follows: "The shaded region is the 68% confidence region .... including cosmic variance and instrument noise". It is not quite clear what the authors of Ref. [19] (see also Ref. [20]) mean by "cosmic variance", but if they mean the theoretical statistical uncertainties for the correlation function variable v, these uncertainties are significantly larger than what is plotted. According to the calculations presented above, the half-width of the shaded region should be approximately $600\ (\mu K)^2$ near the points where the correlation function vanishes and approximately $840\ (\mu K)^2$ near the point marking the zero separation angle.

The conclusion is a bit disappointing. Apparently, God is telling us something important about the very early Universe by exhibiting the microwave background anisotropies, but the *channel of information* is so noisy that it will be hard to understand the message.

7. Appendix: On the "Standard" Inflationary Formula for Density Perturbations

In the body of this paper we have studied the statistical properties of cosmological perturbations of quantum-mechanical origin and related anisotropies in the microwave background. These properties are essentially universal, they are equally well true for perturbations of any nature - density perturbations, rotational perturbations, or gravitational waves. However, in practice, it is very important to know which sort of perturbations we are actually dealing with, assuming that the observed large-angular-scale anisotropy is indeed caused by perturbations of this origin. In other words, what does theory say about the comparative values of the amplitudes, if we agree on the generating mechanism, equations, and a class of cosmological models of the very early Universe, say, governed by the scale factors (15)? In addition to gravitational waves, density perturbations can also be generated by the same mechanism, if one makes favorable assumptions about the dominant matter in the very early Universe (scalar field) and its coupling to gravity (minimal, the same as for gravitational waves). According to calculations of Ref. [3], density perurbations and gravitational waves will have amplitudes of the same order of magnitude, wheras according to the "standard" inflationary formula the amplitude of density perturbations will be many orders of magnitude larger than the amplitude of gravitational waves, if the expansion rate of the very early Universe was sufficiently close to the archetype inflationary model - the de Sitter expansion.

It is necessarry to say that the quantum-mechanical generating mechanism has become very popular in the context of the inflationary hypothesis. Inflationary literature often speaks about cosmological perturbations being generated "from quantum fluctuations". However, this literature associates the explanation of the phenomenon with such things as ambiguity in the choice of time in the de Sitter universe, horizon temperature, tremendous inflation of scales, and so on. The basic concepts are adjusted accordingly. Instead of *amplification*, with the emphasis on a nonvanishing parametric coupling, increase of amplitude at the expense of energy of the pump field, quantum-mechanical generation of waves (particles) in strictly correlated pairs, etc., inflationary literature speaks about *magnification*, with the emphasis on "stretching the waves" and "crossing the horizons". Apparently for these reasons, inflationists did not get puzzled with their "standard" formula for density perturbations which states that one can produce arbitrarily large amount of density perturbations by practically doing nothing.

The "standard" formula relates the amplitude of density perturbations today with the values of the scalar field during inflation. Let us consider, for definiteness, perturbations of the matter density $\delta\rho/\rho$ with today's wavelengths of the order of today's Hubble radius l_H. The "standard" formula says that

$$\left.\frac{\delta\rho}{\rho}\right|_H \sim \frac{H^2}{\dot{\phi}(t_i)} \tag{64}$$

where the right hand side of this formula is supposed to be evaluated at the time t_i when the wavelengths of our interest were "crossing the horizon" during inflation. Let us agree with the so-called "slow-roll" approximation and assume that the Hubble parameter H was almost constant during that epoch, $|\dot{H}| \ll H^2$. Let us take the numerical value of H during that epoch at the level, say, 20 orders of magnitude smaller than the Planck value of H. For quantum-mechanically generated gravitational waves, this would result in today's amplitude $h \approx 10^{-20}$ and the induced anisotropies of the microwave background $\delta T/T \approx 10^{-20}$ which are much much lower than the level currently discussed in the experiment. However, for density perturbations, according to the "standard" inflationary formula, the situation is totally different. Without changing anything in the curvature of the space-time responsible for the generating process (that is, leaving H almost constant and at the same numerical level 20 orders of magnitude smaller than the Planck value), but simply sending $\dot{\phi}(t_i)$ to zero (which corresponds, due to the Einstein equations, to sending $\dot{H}$ to zero, i.e., making the "slow-roll" approximation better and better, making the expansion law closer and closer to the de Sitter expansion and making the shape of the generated spectrum closer and closer to the scale-invariant form) one produces arbitrarily large $(\delta\rho/\rho)|_H$. Inflationists love to stress that the de Sitter gravitational pump field generates perturbations with the Harrison-Zeldovich (scale-invariant, flat) spectrum. What they do not stress is that, according to the "standard" formula for density perturbations, the amplitudes of the scale-invariant spectrum are infinite, and the amplitudes of the *almost* scale-invariant spectrum are *almost* infinite. Instead of blaming their own formula, inflationists blame the scalar field potentials. This formula is the reason for rejecting certain scalar field potentials on the grounds that they generate "too much" of density perturbations, for claims that the contribution of gravitational waves to $\delta T/T$ is "negligibly small" in the limit of the de Sitter expansion, and even for claims about copious production of black holes during inflation.

Recently, the "standard" formula has been claimed to be confirmed [31] and reconfirmed [32]. In Ref. [31], this formula has been formulated, essentially, as the following "standard result": "... we see that the scalar perturbations can be very strongly amplified" [the increase of numerical value from (almost) zero to (almost) infinity] "in the course of the transition" [the instantaneous change of the cosmological scale factor from one power-law behavior to another power-law behavior]. The authors of the paper [31] assure the trusting reader: "We think that there is nothing strange about this ...".

Here, we will try to understand the origin and mathematical justification for the "standard" inflationary formula.

The early papers which are usually quoted in this connection are the papers [33-35]. We will start from the paper of Hawking [33] which seems to be clearer than others in expressing the basic idea and intentions. The papers [33-35] are similar in many respects.

Hawking considers a scalar field ϕ running slowly down an effective scalar field potential. He discusses the inhomogeneous fluctuations $\phi_1(t, \mathbf{x})$ in the field $\phi = \phi_0(t) + \phi_1(t, \mathbf{x})$ which mean that on a surface of constant time there will be some regions where the ϕ field has run further down the hill than in other regions. He introduces a new time coordinate $\bar{t} = t + \delta t(t, \mathbf{x})$ in such a way that the variations of the field are removed and the surfaces of constant time are surfaces of constant ϕ. Since the scalar field transforms as $\phi_0 + \phi_1 \to \phi_0 + \phi_1 - \dot{\phi}_0 \delta t$, the required condition is achieved by the time coordinate shift $\delta t = \phi_1 / \dot{\phi}_0$. Then Hawking says that the change of time coordinate will introduce inhomogeneous fluctuations in the rate of expansion H. He and other authors take (apparently, on the grounds of dimensionality only) $\delta H \sim H^2 \delta t$. From here they come, implicitly or explicitly, to the dimensionless amplitude of density perturbations

$$\frac{\delta \rho}{\rho} \sim \frac{\delta H}{H} \sim H \delta t \sim \frac{H \phi_1}{\dot{\phi}_0} \quad . \tag{65}$$

Some authors write explicitly $\phi_1 \sim H$ and $\delta \rho / \rho \sim H^2 / \dot{\phi}_0$.

The analysis has been done at the inflationary stage. To obtain the today's amplitude of density perturbations in wavelengths, say, of the order of the today's Hubble radius, it is recommended to calculate the right hand side of Eq. (65) at the moments of time when the scales of our interest were "crossing horizon" during inflationary epoch. In one or another version this formula appears in the most of inflationary literature and because of numerous repetitions it has grown to the "standard" one. According to this formula, the amplitude of density perturbations becomes larger if one takes the $\dot{\phi}_0$ smaller.

The authors of [33-35] work with a specific scalar field potential, so the numerical value of $\dot{\phi}_0$ and the numerical value of $\delta \rho / \rho$ following from Eq. (65) turn out to be dependent on the self-coupling constant in the potential. These authors are concerned about the unacceptably large amplitude of density perturbations that they have produced. But this is not a concern about the fact that the Einstein equations play no role in this argumentation, it is a specific detail in the scalar field potential that the authors of [33-35] do not like.

Now let us show the shortcomings of the argumentation in [33-35]. Let us consider a scalar field ϕ with arbitrary potential. Write the field as $\phi = \phi_0(t) + \phi_1(t)Q$ where Q is the n-th spatial harmonic, $Q^{i}_{,i} + n^2 Q = 0$.

Write the perturbed metric in the form

$$ds^2 = -dt^2 + a^2(t)[(1 + h(t)Q)\delta_{ij} + h_l(t)n^{-2}Q_{,i,j}]dx^i\,dx^j \quad .$$

The de Sitter solution corresponds to $\dot{\phi}_0 = 0$, $a(t) \sim e^{Ht}$, and $H(t) = \dot{a}/a =$ const. It follows from the Einstein equations that the (linear) contribution ϵ_ϕ of the scalar field perturbations to the total energy density $\epsilon = \epsilon_0 + \epsilon_\phi$ can be written as

$$\epsilon_\phi = \dot{\phi}_0 \left\{ \dot{\phi}_1 - \phi_1[\ln(a^3\dot{\phi}_0)]^{\cdot} \right\} Q \quad .$$

The contribution ϵ_ϕ, as well as other components of the perturbed energy-momentum tensor, vanish in the de Sitter limit $\dot{\phi}_0 \to 0$.

Thus, the first conclusion we have to make is that in the de Sitter limit there is no linear density perturbations at all. The scalar field perturbations are uncoupled from gravity, they are not accompanied by linear perturbations of the energy-momentum tensor and they are not accompanied by linear perturbations of the gravitational field. The general solution to the perturbed Einstein equations is a set of purely coordinate solutions which can be produced or totally removed by appropriate coordinate transformations. The scalar field perturbations reduce to a test field whose only role is to identify events in the spacetime. One can still ask about a coordinate system such that the surfaces of constant time τ, $\tau = \phi_1(t, \mathbf{x})$ are surfaces of constant ϕ. But the perturbation of the expansion rate of this new coordinate system will have nothing to do with perturbations in the energy density. Despite the presence of the test scalar field, every space-like hypersurface is a surface of constant energy density.

Now let us assume that $\dot{\phi}_0$ is not zero. Transformation of time $\bar{t} = t + \chi(t)Q$ generates a Lie transformation of the scalar field:

$$\phi_0(t) + \phi_1(t)Q \to \phi_0(t) + [\phi_1(t) - \dot{\phi}_0(t)\chi(t)]Q \quad .$$

If one wants the transformed field to be homogeneous one takes $\chi(t) = \phi_1(t)/\dot{\phi}_0(t)$. The same transformation of time generates Lie transformations of the metric. The transformed g_{oo} component is $\bar{g}_{oo} = -1 + 2\dot{\chi}Q$, the transformed g_{ik} components are described by $\bar{h} = h - 2(\dot{a}/a)\chi$. There appear also the g_{oi} components but they will not participate in our linear analysis. The expansion rate of a given frame of reference is determined by the trace of the deformation tensor [36]:

$$D = \frac{1}{2\sqrt{-g_{oo}}} \frac{\partial(g_{ik} - g_{oi}g_{ok}/g_{oo})}{\partial t} g^{ik} \quad .$$

In the linear approximation and before the transformation,

$$D \approx 3H + \frac{1}{2}(3\dot{h} - \dot{h}_l)Q \quad .$$

After the transformation, $\bar{D} = D - 3\dot{H}\chi Q$.

So, the introduced inhomogeneous fluctuation in the rate of expansion is $\delta H = -\dot{H}\chi Q = \dot{H}\delta t$, not $\delta H = H^2 \delta t$ assumed in Refs. [33-35].

The Einstein equation for energy density $D^2/3 = \kappa\epsilon$ is satisfied before and after the transformation, since the variation of H is balanced by the variation of the energy density. The transformed energy density is

$$\bar{\epsilon} = \epsilon_0 + \epsilon_\phi - \dot{\epsilon}_0\chi Q = \epsilon_0 + \dot{\phi}_0^2 \left(\frac{\phi_1}{\dot{\phi}_0}\right)^{\cdot} Q \quad .$$

Thus, if one makes the $\dot{\phi}_0$ smaller, the energy density perturbation decreases according to the Einstein equations, and it increases according to the conjectures of Refs. [33-35]. The use of the correct expression $\delta H \sim \dot{H}\delta t$ in Eq. (65) makes the $\delta\rho/\rho$ decreasing when the $\dot{\phi}_0$ is decreasing.

The situation becomes even more disturbing if one recalls that the formula (64) has been seemingly confirmed and derived rigorously as a result of more detailed studies. People did really write the perturbed Einstein equations. Moreover, it was done in the framework of the so-called gauge-invariant formalism, the whole purpose of which is to eliminate coordinate solutions and to work exclusively with something "physical". The basic mathematical tool in these studies is the gauge-invariant potentials Φ and Ψ constructed from the components of the perturbed metric.

As we have seen above, density perturbations in the scalar field matter vanish when $\dot{\phi}_0$ goes to zero, and there is no density perturbations at all at the de Sitter stage. This is true irrespective of the wavelength of the perturbation. It can be shorter or much longer than the Hubble radius, that is, it can be "inside" or "outside" the Hubble radius. In particular, this is true of the perturbation whose wavelength is such that it will grow by today to the scale of today's Hubble radius. The gauge-invariant potentials Φ and Ψ are strictly zero at the de Sitter stage. Why does then the today's amplitude of the perturbation go to infinity, according to the "standard" formula, in the limit $\dot{\phi}_0 \to 0$ at the "first horizon crossing"? Because, the inflationary literature explains, the amplitude will be almost infinitely enhanced in "the course of transition" of the background equation of state from the quasi-de Sitter one $p \approx -\epsilon$ to the radiation-dominated $p = \frac{1}{3}\epsilon$ or matter-dominated $p = 0$ one. The favorite concept in this argumentation is the "constancy of ζ". The often quoted papers are Ref. [37], which uses notations and equations of [38], and Ref. [39] which summarizes the previous work and gives a clearer exposition. We will follow equations and notations of the paper [39]. The advantage of this paper is that it contains enough mathematical details to make it possible to follow the spirit and the letter of calculations.

Mukhanov *et al.* [39] work with the gauge-invariant potentials Φ and Ψ. According to one of the perturbed Einstein equations, $\Phi = \Psi$. In terms of the Φ, the basic equation of [39] at the scalar field stage is

$$\Phi'' + 2\frac{(a/\phi_0')'}{(a/\phi_0')}\Phi' - \nabla^2\Phi + 2\phi_0'\left(\frac{\mathcal{H}}{\phi_0'}\right)'\Phi = 0 \qquad (66)$$

where $' = d/d\eta$, $dt = ad\eta$, $\mathcal{H} = a'/a$. Equation (66) is exactly the same equation as the basic equation (2.23) of Ref. [37]. At the perfect fluid stage, the basic equation of [39] for "adiabatic" density perturbations is

$$\Phi'' + 3\mathcal{H}(1 + c_s^2)\Phi' - c_s^2\nabla^2\Phi + [2\mathcal{H}' + (1 + 3c_s^2)\mathcal{H}^2]\Phi = 0 \qquad (67)$$

where $c_s^2 = \delta p/\delta\epsilon = \dot{p}_0/\dot{\epsilon}_0$. Equations (66) and (67) were derived from the original perturbed Einstein equations with the help of manipulations aimed at expressing the equations in terms of the gauge-invariant potentials.

The authors of [39] introduce a new quantity ζ defined as

$$\zeta \equiv \frac{2}{3}\frac{H^{-1}\dot{\Phi} + \Phi}{1 + w} + \Phi \quad, \qquad (68)$$

where $w = p_0/\epsilon_0$. This quantity is simply a new letter. As soon as the function Φ is known, the function ζ can be calculated from the definition (68). Using the definition of ζ, Eq. (66) can be written in the form

$$\frac{3}{2}\dot{\zeta}H(1 + w) = \frac{1}{a^2}\nabla^2\Phi \quad, \qquad (69)$$

and Eq. (67) in the form

$$\frac{3}{2}\dot{\zeta}H(1 + w) = \frac{1}{a^2}c_s^2\nabla^2\Phi \quad. \qquad (70)$$

Mukhanov *et al.* [39] consider perturbations with wavelengths "far outside the Hubble radius for which $\nabla^2\Phi$ can be neglected". Neglecting the right hand sides of Eqs. (69), (70) they arrive, in this approximation, at the equation $\dot{\zeta} \approx 0$ and the "conservation law"

$$\zeta \approx \text{const} \quad. \qquad (71)$$

It is important to note that the derivation of this conservation law did not require any knowledge about the initial data and solutions for Φ. The constant in the right hand side of Eq. (71) emerges as a universal number, irrespective of any particular solution for Φ. Although the "constancy of ζ" is a favorite notion in the inflationary literature, it appears that inflationists have never asked what the origin and numerical value of this constant is. If this constant is a universal number, is it equal to a billion, or one, or zero? We will later show that this particular constant must be equal to zero. However, without specifying the value of this constant, it is often assumeed that it is not zero.

The next step in this argumentation proceeds as follows. Imagine that the background equation of state changes during a short interval of time from the initial (i) $p \approx -\epsilon$ to the final (f) $p = \frac{1}{3}\epsilon$ or $p = 0$, so that $1 + w_i \ll 1$ while $1 + w_f \approx 1$. The authors of [39] consider the evolution of a long-wavelength perturbation from the initial Hubble radius crossing (t_i) to the final Hubble radius crossing (t_f). They make additional assumptions, they assume that $\dot{\Phi}$ vanishes "at very early and very late times". In the definition (68), they drop the term with $\dot{\Phi}$ and simplify the quantity ζ:

$$\zeta(t_i) \approx \frac{5 + 3w(t_i)}{3(1 + w(t_i))} \Phi(t_i)$$

$$\zeta(t_f) \approx \frac{5 + 3w(t_f)}{3(1 + w(t_f))} \Phi(t_f)$$

Then they refer to the constancy of ζ, $\zeta(t_i) = \zeta(t_f)$, and arrive at the formula

$$\Phi(t_f) \approx \frac{1 + w(t_f)}{1 + w(t_i)} \frac{5 + 3w(t_i)}{5 + 3w(t_f)} \Phi(t_i) \approx \frac{1}{1 + w(t_i)} \Phi(t_i) \quad . \tag{72}$$

(which, in fact, is in a conflict with the constatncy of ζ, as we will show later).

According to this formula, and since $1 + w(t_i)$ can be arbitrarily close to zero, the $\Phi(t_f)$ can be made arbitrarily large for any nonvanishing $\Phi(t_i)$. Moreover, since the time of transition from one equation of state to another can be arbitrarily short, the tremendous increase of numerical value of the potential Φ is supposed to happen almost instantaneously. (The author of [32], who reconfirmes the "standard" results, has even plotted a graph for this jump.) The authors of Ref. [39] emphasize that their result (72) (the actually published [39] formula (6.67) contains a misprint: the position of symbols $w(t_i)$ and $w(t_f)$ should be interchanged) is in a full agreement with previous studies and Eq. (64). Formula (72) suggests an arbitrarily large production of density perturbations for no other reason but simply because the $1 + w(t_i)$ was very close to zero. There is something strange with this formula. [I realize well that what I qualify here as strange is certainly considered by others as perfectly allright. Otherwise somebody would raise a voice of protest against the ease with which inflationists generate tremendous amounts of various substances (some of the authors are even claiming that they can "overclose" our Universe). However, judging from the literature, it is not only that there are no voices of protest, but there is rather an element of competition as for who was the first to proclaim the "standard" inflationary results. For instance, the authors of [40] address the inflationary claims about density perturbations as "first quantitatively calculated in [33-35] [and which] have been successfully quantitatively confirmed by the COBE discovery".]

Let us now try to sort out what we are dealing with.

The first point to realize is that the basic equations (66), (67) are, strictly speaking, incorrect. They are incorrect in the following sense: they are constructed from a correct equation and the first time derivative of the correct equation. Transforming the original Einstein equations, the authors of [38,39] have effectively raised the order of differential equations. If the correct equation is satisfied, the constructed equation is satisfied too, but not *vise versa*. To get the feeling of the danger involved, consider a correct equation $\dot{x} = 0$ and a constructed equation $\ddot{x} - A\dot{x} = 0$ where A is arbitrary constant. Solutions to the correct equation do satisfy the constructed equation, but the latter one admits exponentially growing solutions which are not allowed by the original equation.

Let us start our analysis from Eq. (66). Since the potential Φ is an eigenfunction of the Laplace operator, $\nabla^2\Phi = -n^2\Phi$, we will replace $\nabla^2\Phi$ with $-n^2\Phi$. We introduce also a new function of the scale factor $a(\eta)$:

$$\gamma = -\dot{H}/H^2 = 1 + (a/a')'$$

and use the background Einstein equations in order to express the coefficients of Eq. (66) in terms of the scale factor $a(\eta)$ and its derivatives. Now, introduce a new variable μ according to the definition

$$\Phi = \frac{1}{2n^2}\frac{a'}{a}\gamma\left(\frac{\mu}{a\sqrt{\gamma}}\right)' \tag{73}$$

So far, the variable μ is simply a new variable replacing Φ, but the importance of μ is in that the original perturbed Einstein equations require this variable to satisfy the equation

$$\mu'' + \mu\left[n^2 - \frac{(a\sqrt{\gamma})''}{a\sqrt{\gamma}}\right] = 0 \quad . \tag{74}$$

(For those interested in gauge-invariant potentials, I may remark that μ is a genuine gauge-invariant variable in the sense of [38,39].) With the help of Eq. (73), Eq. (66) identically transforms to

$$\left[\mu'' + \mu\left[n^2 - \frac{(a\sqrt{\gamma})''}{a\sqrt{\gamma}}\right]\right]' - \frac{(a\sqrt{\gamma})'}{a\sqrt{\gamma}}\left[\mu'' + \mu\left[n^2 - \frac{(a\sqrt{\gamma})''}{a\sqrt{\gamma}}\right]\right] = 0 . \tag{75}$$

If Eq. (74) is satisfied, Eq. (75) is satisfied too, but not *vice versa*. Equation (75) is totally equivalent to

$$\frac{1}{a^2\gamma}\left[a^2\gamma\left(\frac{\mu}{a\sqrt{\gamma}}\right)'\right]' + n^2\frac{\mu}{a\sqrt{\gamma}} = X \quad , \qquad (76)$$

where X is arbitrary constant. Thus, Eq. (66) involves an arbitrary constant X which, in fact, must be zero according to Eq. (74).

Let us now turn to the perfect fluid equation (67). The situation here is similar to the scalar field case. Introduce a new variable ν according to the definition

$$\Phi = \frac{1}{2n^2}\frac{a'}{a}\frac{\gamma}{c_s^2}\left(\frac{\nu}{a\sqrt{\gamma/c_s^2}}\right)' \quad . \qquad (77)$$

The importance of ν is in that the original perturbed Einstein equations require this variable to satisfy the equation

$$Z \equiv \nu'' + \nu\left[n^2c_s^2 - \frac{(a\sqrt{\gamma/c_s^2})''}{a\sqrt{\gamma/c_s^2}}\right] = 0 \qquad (78)$$

(The function ν is a genuine gauge-invariant variable.) With the help of Eq. (77), Eq. (67) identically transforms to

$$Z' - \frac{(a\sqrt{\gamma c_s^2})'}{a\sqrt{\gamma c_s^2}}Z = 0 \qquad (79)$$

This equation is totally equivalent to

$$\frac{1}{a^2\gamma}\left[\frac{a^2\gamma}{c_s^2}\left(\frac{\nu}{a\sqrt{\gamma/c_s^2}}\right)'\right]' + n^2\frac{\nu}{a\sqrt{\gamma/c_s^2}} = Y \qquad (80)$$

where Y is arbitrary constant. Thus, Eq. (80) does also involve an arbitrary constant Y which, in fact, must be zero according to Eq. (78). The role of these constants X and Y in the constancy of ζ argument we will discuss shortly.

Equations (74), (78) is all we need to solve, in order to find all the perturbed components of the metric tensor and energy-momentum tensor. There is no other way to find the evolution of density perturbations except of doing the hard work of solving these equations, imposing appropriate initial conditions, joing the solutions, etc. (This is what is being done, convincingly or not, in Ref. [3].) Among other things, these solutions will tell you what is the value and time dependence of the functions ζ introduced by the definitions (68), (73), (77). The correct equations do not show anything like enormously large amplitude of today's density perturbations in the limit of the de Sitter expansion. However, we need to return to the concept of "constancy of ζ" which pretends to answer important physical questions without solving any equations at all.

Combine the definitions (68), (73) to show that ζ at the scalar field stage is

$$\zeta = \frac{1}{2n^2} \frac{1}{a^2\gamma} \left[a^2\gamma \left(\frac{\mu}{a\sqrt{\gamma}} \right)' \right]' \, .$$

Combine the definitions (68), (77) to show that ζ at the perfect fluid stage is

$$\zeta = \frac{1}{2n^2} \frac{1}{a^2\gamma} \left[\frac{a^2\gamma}{c_s^2} \left(\frac{\nu}{a\sqrt{\gamma/c_s^2}} \right)' \right]'$$

The term with the Laplacian neglected in Eqs. (66), (67) is the term with n^2 in equations (76) and (80). Let us drop this term as the authors of [39] do. Then, in this approximation, Eq. (76) gives $\zeta \approx X/2n^2 = $ const and Eq. (80) gives $\zeta \approx Y/2n^2 = $ const. These relationships show that ζ can be a universal constant, independent of any particular solution, only if it is supported by the constant X or, correspondigly, by the constant Y. But, as we have shown above, these constants must be equal to zero, if one is willing to work with correct equations. The "constancy of ζ" argument fails at the very beginning, regardless of validity or not of the additional assumptions that have been made on the route to Eq. (72). The conservation law $\zeta(t_i) = \zeta(t_f)$ degenerates to an empty statement $0 = 0$, and nothing can be derived from it.

The idea of a free lunch based on the "constancy of ζ" seems to be so attractive that the following question is often being asked. Suppose that not for all solutions for Φ, but for some of them, suppose that not in the leading order n^{-2}, but in the next order, suppose, nevertheless, that ζ calculated from this specially chosen solution for Φ is approximately a nonzero constant (which is indeed possible). Why cannot we return to Eq. (68) and repeat all the arguments that have seemingly led us from the definition (68) to the result (72)? For instance, the authors of [31] have even considered, along this line, a concrete model consisting of two consecutive power-law scale factors. They speak about an initially small potential Φ being "distributed" after the transition point "into two modes, both very large in amplitude, one which decays, the other yielding" the "standard" result (72). And this is how, they argue, the "scalar perturbations can be very strongly amplified in the course of the transition".

So, we also need to consider this model and explore what is contained in the definition (68) treated as an equation for Φ.

First of all, use the background equation

$$1 + w = -\frac{2}{3}\frac{\dot{H}}{H^2}$$

and write ζ in a more convenient form:

$$\zeta = -\frac{H^2}{a\dot{H}}\left(\frac{a}{H}\Phi\right)^{\cdot} \quad . \tag{81}$$

Integrate this equation to produce the general solution

$$\Phi = \frac{H}{a}\left[C + \int a\zeta\left(\frac{1}{H}\right)^{\cdot}dt\right] \tag{82}$$

where C is arbitrary integration constant. So far, ζ can be an arbitrary function. Now assume that ζ is a constant, $\zeta = \zeta_0 = \text{const}$, and write

$$\Phi = \zeta_0\left[1 - \frac{H}{a}\int_{t_i}^{t} a\,dt\right] + C\frac{H}{a} \quad . \tag{83}$$

[The constants ζ_0 and C can be related to the coefficients in front of two linearly independent solutions to Eq. (74). In the long wavelength approximation,

$$\frac{\mu}{a\sqrt{\gamma}} = C_1\left[1 - n^2\int\frac{1}{a^2\gamma}\left(\int a^2\gamma\,d\eta\right)d\eta\right] + C_2\int\frac{d\eta}{a^2\gamma} + \ldots \quad .$$

Using the definition (73) one can show that $\zeta_0 = -\frac{1}{2}C_1$, $C = \frac{C_2}{2n^2}$.]

As the second step, let us consider, together with Deruelle and Mukhanov [31], an expanding model which decribes a transition from one power-law scale factor to another power-law scale factor. Let us write

$$a_i = a_1 t^{p_1} \quad , \qquad a_f = a_2(t - t_*)^{p_2} \quad . \tag{84}$$

The scale factor and its first time derivative are continuous functions at the transition point $t = t_1$. This requires

$$a_2 = a_1\left(\frac{p_1}{p_2}\right)^{p_2} t_1^{p_1 - p_2} \quad , \qquad t_* = t_1\left(1 - \frac{p_2}{p_1}\right) \quad .$$

To be closer to the inflationary results, one can keep in mind that $p_1 \gg 1$, while p_2 is $1/2$ or $2/3$.

Now we will follow the time evolution of Φ specifically in this model. Let t_i be the time when our perturbation first crosses the Hubble radius: $a(t_i)H(t_i) = n$. Let t_f be the time when the perurbation returns back inside the Hubble radius: $a(t_f)H(t_f) = n$. The functions $a(t)$ and $H(t)$ are continuous all the way from t_i to t_f, and so is the function (83). The initial value of the potential is

$$\Phi(t_i) = \zeta_0 + C\frac{H(t_i)}{a(t_i)} \quad, \tag{85}$$

the final value is

$$\Phi(t_f) = \zeta_0 I + C\frac{H(t_f)}{a(t_f)} \tag{86}$$

where

$$I \equiv 1 - \frac{H(t_f)}{a(t_f)} \int_{t_i}^{t_f} a\,dt \quad.$$

Combining (85) and (86) we find

$$\Phi(t_f) = \Phi(t_i)I + C\left[\frac{H(t_f)}{a(t_f)} - \frac{H(t_i)}{a(t_i)}I\right] \quad. \tag{87}$$

The integral I can be easily calculated for the scale factor (84). Since $a(t_f) \gg a(t_i)$, $p_1 \gg p_2$, and $a(t_1)H(t_1) \gg n$, we have approximately

$$I \approx \frac{1}{p_2 + 1} \quad.$$

Formula (87) reduces to

$$\Phi(t_f) = \frac{1}{p_2 + 1}\Phi(t_i) - C\frac{H(t_i)}{a(t_i)}\frac{1}{p_2 + 1} \quad.$$

The constant C could be set to zero from the very beginning. In any case, the term $C(H(t_i)/a(t_i))$ is of the same order or smaller than $\Phi(t_i)$. By setting $C = 0$, we arrive at

$$\Phi(t_f) \approx \frac{1}{p_2 + 1}\Phi(t_i) \quad.$$

There is nothing like tremendous jumps of Φ at the transition point. The "constancy of ζ" effectively translated into constancy of Φ. [The fact that Φ is approximately constant for wavelengths longer than the Hubble radius follows also from the exact solution to Eq. (74) which can be found in case of power-law scale factors.]

The best what we can say about the "standard" inflationary formula is that it does not follow from correct equations.

It is important to recall that the "standard" inflationary results seem to be an indispensable tool in cosmology of our days (and possibly in the next Millennium too, see for example [41,10]). The expected relative contributions of density perturbations and gravitational waves to the observed microwave background anisotropies are a subject of active study. In the center of discussion are usually the "consistency relations" which state that the ratio of density to gravity-wave contributions goes to infinity when the spectrum of perturbations approaches the most favorite, Harrison-Zeldovich form. In reality, as we have shown above, these "consistency relations" are simply a manifistation of inconsistency of the "standard" inflationary theory from which they are derived.

In conclusion, if the "standard" inflationary results are incorrect and cannot be trusted, what is the amount of density perturbations that can be generated in the early Universe? My part of answer is formulated in Ref. [3,7].

8. Acknowledgments

I appreciate discussions with Yu. V. Sidorov on squeezed states, and collaboration with J. Martin on cosmological perturbations. This work was supported by NASA grants NAGW 2902, 3874 and NSF grant 92-22902.

9. References

1. C. J. Isham, Structural Issues in Quantum Gravity, gr-qc 9510063.
2. G. F. Smoot, *et al.*, Astrophys. J. **396**, L1 (1992).
3. L. P. Grishchuk, Phys. Rev. D **50**, 7154 (1994).
4. L. P. Grishchuk and Y. V. Sidorov, Phys. Rev. D **42**, 3413 (1990); Class. Quantum Grav. **6**, L161 (1989).
5. Special issues: J. Mod. Opt. **34** (6,7) (1987); J. Opt. Soc. Am. **84**, (10) (1987); NASA Conference Publication 3135 (1992).
6. Special issue: Physica Scripta (Eds. W. P. Schleich and S. M. Barnett) **T 48** (1993).
7. L. P. Grishchuk. Cosmological Perturbations of Quantum-Mechanical Origin and Anisotropy of the Microwave Background Radiation. In: *Current Topics in Astrofundamental Physics: The Early Universe* (Eds. N. Sanchez and A. Zichichi) NATO ASI Series C, Vol. 467, p. 205 (Kluwer Academic Publishers, 1995).
8. L. P. Grishchuk. Comment on "Cosmological Perturbations of a Relativistic Condensate". Report WUGRAV-95-12, gr-qc 9506010.
9. J. M. Niedra and A. I. Janis, G.R.G. **15**, 241 (1983).
10. P. J. Steinhardt. Cosmology at the Crossroads. Invited talk at the Snowmass meeting (to be published), astro-ph 9502024.

11. R. K. Sachs and A. M. Wolfe, Astrophys. J. **1**, 73 (1967).

12. L. P. Grishchuk, Phys. Rev. D **48**, 3513 (1993).

13. L. P. Grishchuk, Phys. Rev. D **48**, 5581 (1993).

14. P. Paradoksov, Uspekhi Fiz. Nauk, **89**, 707 (1966) [Sov. Phys. Uspekhi, **9**, 618 (1967)].

15. G. A. Korn and T. M. Korn, *Mathematical Handbook for Scientists and Engineers* (McGraw-Hill Book Co., 1968).

16. M. Fisz, *Probability Theory and Mathematical Statistics* (J. Wiley & Sons, 1967).

17. I. S. Gradshteyn and I. M. Ryzhik, *Table of Integrals, Series, and Products* (Academic Press, 1980).

18. M. Abramowitz and I. A. Stegun, (eds), *Handbook of Mathematical Functions* (Dover Publishing, 1972).

19. C. L. Bennett, *et al.*, Astrophys. J. **436**, 423 (1994).

20. E. L. Wright, *et al.*, Astrophys. J. **436**, 443 (1994).

21. For gravitational waves this formula is in fact known for long time. See Eq. (5b) in L. P. Grishchuk, Ann. N.Y. Acad. Sci. **302**, 439 (1977) where the notational replacement $h(n) = A(n)$ should be made and where $A(n)$ should be interpreted as the *final* (generated) *characteristic* amplitude, $A^2(n) \sim B^2(n)n^{2(1+\beta)}$, while $B(n)$ should be interpreted as the *initial* (zero-point quantum) *characteristic* amplitude, $B(n) \sim n$, in accord with the scale factor, Eq. (15) above, in which the η-time grows from $-\infty$.

22. M. White, D. Scott, and J. Silk, Annu. Rev. Astron. Astrophys. **32**, 319 (1994).

23. L. F. Abbott and M. B. Wise, Astrophys. J. **282**, L47 (1984).

24. J. R. Bond and G. Efstathiou, M.N.R.A.S. **226**, 655 (1987).

25. R. Fabbri, F. Lucchin and S. Matarrese, Astrophys. J. **315**, 1 (1987).

26. R. Scaramella and N. Vittorio, Astrophys. J. **353**, 372 (1990).

27. L. Cayon, E. Martinez-Gonzalez and J. L. Sanz, M.N.R.A.S. **253**, 599 (1991).

28. M. White, L. M. Krauss and J. Silk, Astrophys. J. **418**, 535 (1993).

29. R. Scaramella and N. Vittorio, Astrophys. J. **411**, 1 (1993).

30. A. M. Yaglom, *An Introduction to the Theory of Stationary Random Functions* (Prentice-Hall Inc., 1962).

31 N. Deruelle and V. F. Mukhanov, "On Matching Conditions for Cosmological Perturbation", preprint gr-qc 9503050.

32. R. R. Caldwell, "On the Evolution of Scalar Metric Perturbations in an Inflationary Cosmology", preprint gr-qc 9509027.

33. S. W. Hawking, Phys. Lett. **115B** 295, 1982).

34. A. A. Starobinsky, Phys. Lett. **117B**, 175 (1982).

35. A.H. Guth and S.-Y. Pi, Phys. Rev. Lett. **49**, 1110 (1982).

36. A. L. Zel'manov, Doklady Acad. Nauk USSR **107**, 815 (1956) [Sov. Phys. Doklady **1**, 227 (1956)].
37. J. M. Bardeen, P. J. Steinhardt, and M. S. Turner, Phys. Rev. D **28**, 679 (1983).
38. J. M. Bardeen, Phys. Rev. D **22**, 1882 (1980).
39. V. F. Mukhanov, H. A. Feldman, and R. H. Brandenberger, Phys. Reports **215**, 6 (1992).
40. D. Polarski and A. A. Starobinsky, "Semiclassicality and Decoherence of Cosmological Perturbations", Report LMPM 95-4, gr-qc 9504030.
41. J. A. Frieman, "Cosmological Models at the Millennium", FERMILAB-Conf-95/151-A, 1995.
42. L. P. Grishchuk, "Statistics of the Microwave Background Anisotropics Caused by the Squeezed Cosmological Perturbation", Report WUGRAV-95-6, gr-qc 9504045

COSMOLOGICAL INFLATION

AND

THE NATURE OF TIME

D.S. SALOPEK

Department of Physics, University of Alberta
Edmonton, Canada T6G 2J1

Abstract. Recent advances in observational cosmology are changing the way we view the nature of time. In general relativity, the freedom in choosing a time hypersurface has hampered the implementation of the theory. Fortunately, Hamilton-Jacobi theory enables one to describe all time hypersurfaces on an equal footing. Using an expansion in powers of the spatial curvature, one may solve for the wavefunctional in a semiclassical approximation. In this way, one may readily compare predictions of various inflation models with observations of microwave background anisotropies and galaxy clustering.

1 Introduction

In the standard cosmological formulation of Einstein's theory of gravity, one ordinarily solves the field equations for the 4-metric $g_{\mu\nu}$. In a more elegant although equivalent approach, it has been recommended [1]-[6] that one solve the Hamilton-Jacobi (HJ) equation for general relativity which governs the evolution of the generating functional S. By adopting this method, many advantages have be gained:

(1) Avoiding Gauge Problems. Neither the lapse N nor the shift N_i appear in the HJ equation. The generating functional depends only on the 3-metric γ_{ij} (as well any matter fields that may be present). Hence, the structure of the HJ equation is conceptually simpler than that of the

N. Sánchez and A. Zichichi (eds.), String Gravity and Physics at the Planck Energy Scale, 409–430.
© *1996 Kluwer Academic Publishers.*

Einstein field equations. For instance, one is able to avoid (temporarily) the embarrassing problem of picking a gauge in general relativity. As a corollary, one obtains a deep appreciation for the *Nature of Cosmic Time* [7].

(2) Solution of Constraint Equations. One may solve the constraint equations of general relativity in a systematic manner, even in a nonlinear setting [4], [8]. The momentum constraint is easy to solve using HJ theory: one simply constructs S using integrals of the 3-curvature over the entire 3-geometry. The energy constraint may be solved by expanding in powers of the 3-curvature ('spatial gradient expansion').

(3) Primitive Quantum Theory of Gravity. Solutions of the HJ equation may be interpreted as the lowest order contribution to the wavefunctional for an expansion in powers of $\hbar$ (semiclassical approximation). One may describe some quantum processes such as the initial 'ground state' of the Universe [9]-[10] or tunnelling through a potential barrier [11].

If one accepts that fluctuations for galaxy formation as well as microwave background anisotropies were generated during an inflationary epoch, then it is imperative that one quantize the gravitational field [9]- [12]. However, quantization of the full gravitational field has proven to be elusive. Several possible forms for a viable theory have been advanced, although there is no general consensus. String theory is the most popular candidate, and its status has been reviewed by many of the speakers in this conference [13]-[14]. The goal here will be more modest. Beginning with the HJ equation, I will be content to consider the semiclassical theory of Einstein gravity. In this way, one follows in spirit the historical development of the theory of atomic spectra. Before the development of the quantum theory in 1926, the semiclassical theory of Bohr and Sommerfeld provided a useful although imperfect description of various atoms.

HJ theory has proven to be a particularly powerful tool for the cosmologist. In this article, I will review our current understanding of the theory. In addition, I will discuss the observational status of three models of cosmological inflation:

(1) inflation with an exponential potential ('power-law inflation') [15], [16]

which arises naturally from Induced Gravity [17], [18] or Extended Inflation [19];

(2) inflation with a cosine potential ('natural inflation') [20] where the inflaton is a pseudo-Goldstone boson;

(3) inflation with two scalar fields ('double inflation') [21], [17] where there are two periods of inflation.

2 Hamilton-Jacobi Theory for General Relativity

The Hamilton-Jacobi equation for general relativity is derived using a Hamiltonian formulation of gravity. One first writes the line element using the ADM 3+1 split,

$$ds^2 = \left(-N^2 + \gamma^{ij} N_i N_j\right) dt^2 + 2N_i dt\, dx^i + \gamma_{ij}\ dx^i dx^j\ , \tag{1}$$

where N and N_i are the lapse and shift functions, respectively, and γ_{ij} is the 3-metric. Hilbert's action for gravity interacting with a scalar field becomes

$$\mathcal{I} = \int d^4x \left(\pi^\phi \dot{\phi} + \pi^{ij} \dot{\gamma}_{ij} - N\mathcal{H} - N^i\mathcal{H}_i\right). \tag{2}$$

The lapse and shift functions are Lagrange multipliers that imply the energy constraint $\mathcal{H}(x) = 0$ and the momentum constraint $\mathcal{H}_i(x) = 0$.

The object of chief importance is the generating functional

$$\mathcal{S} \equiv \mathcal{S}[\gamma_{ij}(x), \phi(x)]. \tag{3}$$

For each scalar field configuration $\phi(x)$ on a space-like hypersurface with 3-geometry described by the 3-metric $\gamma_{ij}(x)$, the generating functional associates a complex number. The generating functional is the 'phase' of the wavefunctional in the semiclassical approximation:

$$\Psi \sim e^{i\mathcal{S}}. \tag{4}$$

(The prefactor is neglected here although it has important implications for quantum cosmology [24].) The probability functional, $\mathcal{P} \equiv |\Psi|^2$, is given by the square of the wavefunctional. Replacing the conjugate momenta by functional derivatives of S with respect to the fields,

$$\pi^{ij}(x) = \frac{\delta S}{\delta \gamma_{ij}(x)}, \qquad \pi^{\phi}(x) = \frac{\delta S}{\delta \phi(x)}, \tag{5}$$

and substituting into the energy constraint, one obtains the Hamilton-Jacobi equation [22], [23],

$$\begin{aligned}
\mathcal{H}(x) = \quad & \gamma^{-1/2} \frac{\delta S}{\delta \gamma_{ij}(x)} \frac{\delta S}{\delta \gamma_{kl}(x)} [2\gamma_{il}(x)\gamma_{jk}(x) - \gamma_{ij}(x)\gamma_{kl}(x)] \\
& + \frac{1}{2}\gamma^{-1/2} \left(\frac{\delta S}{\delta \phi(x)} \right)^2 + \gamma^{1/2} V(\phi(x)) \\
& - \frac{1}{2}\gamma^{1/2} R + \frac{1}{2}\gamma^{1/2}\gamma^{ij}\phi_{,i}\phi_{,j} = 0 ,
\end{aligned} \tag{6}$$

which describes how S evolves in superspace. R is the Ricci scalar associated with the 3-metric, and $V(\phi)$ is the scalar field potential. In addition, one must also satisfy the momentum constraint

$$\mathcal{H}_i(x) = -2 \left(\gamma_{ik} \frac{\delta S}{\delta \gamma_{kj}(x)} \right)_{,j} + \frac{\delta S}{\delta \gamma_{lk}(x)} \gamma_{lk,i} + \frac{\delta S}{\delta \phi(x)} \phi_{,i} = 0 , \tag{7}$$

which legislates spatial gauge invariance: S is invariant under reparametrizations of the spatial coordinates. Since neither the lapse function nor the shift function appears in eq.(6) and eq.(7), the temporal and spatial coordinates are *arbitrary*: HJ theory is *covariant*. (Units are chosen so that $c = 8\pi G = \hbar = 1$. The HJ equation for Brans-Dicke gravity has been studied in ref.[25].)

It is quite important that the 3-metric $\gamma_{ij}(x)$ be positive definite otherwise two fatal disasters would occur: (1) the principle of microscopic causality would be violated; (2) points x and y appearing in the generating functional would no longer be space-like, and hence the fields amplitudes $\phi(x)$ and $\phi(y)$ may not be independent —- this would mess up the computation of functional derivatives appearing in eq.(6) and eq.(7).

2.1 Partial Reduction of Einstein's Equations

From a real solution $\mathcal{S}$ of the HJ equation and the momentum constraint, one may construct solutions to Einstein's equations. Provided that the fields evolve according to the definitions of the momenta,

$$\left(\dot{\phi} - N^i \phi_{,i}\right) / N = \gamma^{-1/2} \pi^\phi, \tag{8}$$

$$\left(\dot{\gamma}_{ij} - N_{i|j} - N_{j|i}\right) / N = 2\gamma^{-1/2} \pi^{kl} \left(2\gamma_{jk}\gamma_{il} - \gamma_{ij}\gamma_{kl}\right), \tag{9}$$

one may verify that Einstein's equation are indeed satisfied. (Note that | denotes covariant differentiation with respect to the 3-metric.) As a result, HJ theory allows for a partial reduction of Einstein's equations. The lapse and shift appear only in the definitions of the momenta for gravity and matter, eq.(8) and eq.(9). They are arbitrary and they reflect the observer's freedom in choosing his space-time coordinates. For example, if one chooses time hypersurfaces such that the scalar field is uniform, $\phi = t$, then the lapse is given automatically by eq.(8),

$$N^{-1} = \gamma^{-1/2} \frac{\delta\mathcal{S}}{\delta\phi(x)}. \tag{10}$$

Typically one sets the shift to zero, in which case the spatial coordinates of one time slice are projected orthogonally onto the others. In order to complete the determination of the 4-metric as a function of time ϕ, one need only integrate eq.(9).

2.2 Evolution and Observation of a Gravitational System

The above discussion demonstrates how HJ theory enables one to split the analysis of any gravitational system into two stages:

(1) Gauge-independent evolution of the system which is governed by the HJ equation as well as the momentum constraint equation and

(2) Observation of the system which is gauge-dependent (i.e., it depends on arbitrary choices for N, N_i).

If the generating functional S is complex, then one is describing intrinsically *quantum processes*. The freedom in choosing the lapse and shift then reflect the necessity for gauge-fixing. Otherwise, expectation values for physical observables such as the two-point correlation function [1],

$$< \phi(x)\phi(y) > \equiv \int [d\gamma][d\phi] \; \phi(x)\phi(y) \, |\Psi|^2 \,, \tag{11}$$

would be infinite. In the spirit of Dirac quantization [23], gauge-fixing is performed only after one has determined a solution of the HJ equation (or its quantum analog, the Wheeler-DeWitt equation [26]).

3   Prototype Hamilton-Jacobi Solution

Until recently, it had been thought that the Hamilton-Jacobi equation for general relativity was intractable. As a result, since the pioneering work of DeWitt [26] and Misner [27] in the 1960's and 1970's, much effort was spent studying minisuperspace models where one effectively truncates superspace: one examines only a finite number of variables describing typically a homogeneous universe. For example, in the context of quantum cosmology, these models were studied by Hartle, Hawking, Page, Vilenkin and others in the 1980's. One hoped to understand some qualitative features of semiclassical gravity from these 'toy models'. However, it is generally agreed that such models represent only the initial stages of a much larger program. For example, in order to appreciate more fully the nature of time, one must incorporate the role of inhomogeneities. After all, a time hypersurface represents the arbitrary manner in which one slices a 4-geometry [2]. Many properties of *full superspace* can be understood by employing a series solution method. The prototype solution, the spatial gradient expansion, was suggested by John Stewart and myself [5]. Effectively, one is able to decompose superspace into an infinite but discrete number of minisuperspaces which are tractable.

3.1 Spatial Gradient Expansion

As a first step in solving eq.(6) and eq.(7), one expands the generating functional

$$\mathcal{S} = \mathcal{S}^{(0)} + \mathcal{S}^{(2)} + \mathcal{S}^{(4)} + \dots , \tag{12}$$

in a series of terms according to the number of spatial gradients that they contain. The invariance of the generating functional under spatial coordinate transformations suggests a solution of the form,

$$\mathcal{S}^{(0)}[\gamma_{ij}(x), \phi(x)] = -2 \int d^3x \, \gamma^{1/2} H \left[\phi(x)\right] , \tag{13}$$

for the zeroth order term $\mathcal{S}^{(0)}$. The function $H \equiv H(\phi)$ satisfies the separated HJ equation (SHJE) of order zero [6],

$$H^2 = \frac{2}{3} \left(\frac{\partial H}{\partial \phi}\right)^2 + \frac{1}{3} V(\phi) , \tag{14}$$

which is an ordinary differential equation. (The term 'separated' describes the fact that the metric variables are absent in eq.(14) —- they have been separated out.) Note that $\mathcal{S}^{(0)}$ contains no spatial gradients. The zeroth order term is an excellent approximation for universes where the wavelength of any inhomogeneities is much larger than the Hubble radius H^{-1}. In fact, the notion of *long-wavelength* field is an essential ingredient of any model of cosmic structure formation.

3.2 Higher Order Terms

An important simplification occurs for higher order terms, $\mathcal{S}^{(2n)}$, for $n \geq 1$: they are governed by linear partial differential equations of the inhomogeneous type:

$$-2 \frac{\partial H}{\partial \phi} \frac{\delta \mathcal{S}^{(2n)}}{\delta \phi} + 2H \gamma_{ij} \frac{\delta \mathcal{S}^{(2n)}}{\delta \gamma_{ij}} + \mathcal{R}^{(2n)} = 0 . \tag{15}$$

The remainder term $\mathcal{R}^{(2n)}$ depends on some quadratic combination of the previous order terms [3]. For example, for $n = 1$, it is

$$\mathcal{R}^{(2)} = \frac{1}{2} \gamma^{1/2} \gamma^{ij} \phi_{,i} \phi_{,j} - \frac{1}{2} \gamma^{1/2} R , \tag{16}$$

whereas for $n \geq 2$, the remainder, $\mathcal{R}^{(2n)}(x)$, is given by

$$
\begin{aligned}
\mathcal{R}^{(2n)}(x) = \quad & \gamma^{-1/2} \sum_{p=1}^{n-1} \frac{\delta S^{(2p)}}{\delta \gamma_{ij}(x)} \frac{\delta S^{(2n-2p)}}{\delta \gamma_{kl}(x)} (2\gamma_{jk}\gamma_{li} - \gamma_{ij}\gamma_{kl}) \\
+ \quad & \gamma^{-1/2} \sum_{p=1}^{n-1} \frac{1}{2} \frac{\delta S^{(2p)}}{\delta \phi(x)} \frac{\delta S^{(2n-2p)}}{\delta \phi(x)} .
\end{aligned}
\tag{17}
$$

In order to compute the higher order terms from eq.(15), one introduces a change of variables, $(\gamma_{ij}, \phi) \rightarrow (f_{ij}, u)$:

$$
u = \int \frac{d\phi}{-2\frac{\partial H}{\partial \phi}} , \qquad f_{ij} = \Omega^{-2}(u)\, \gamma_{ij} ,
\tag{18}
$$

where the conformal factor $\Omega \equiv \Omega(u)$ is defined through

$$
\frac{d \ln \Omega}{du} \equiv -2 \frac{\partial H}{\partial \phi} \frac{\partial \ln \Omega}{\partial \phi} = H .
\tag{19}
$$

The equation for $S^{(2n)}$ simplifies considerably:

$$
\left. \frac{\delta S^{(2n)}}{\delta u(x)} \right|_{f_{ij}} + \mathcal{R}^{(2n)}[f_{ij}(x), u(x)] = 0 .
\tag{20}
$$

Eq.(20) has the form of an infinite dimensional gradient. Before integrating it, I will review some elementary results from potential theory.

3.3 Potential Theory

The fundamental problem in potential theory is: given a force field $g^i(u_k)$ which is a function of n variables u_k, what is the potential $\Phi \equiv \Phi(u_k)$ (if it exists) whose gradient returns the force field,

$$
\frac{\partial \Phi}{\partial u_i} = g^i(u_k) \quad ?
\tag{21}
$$

Not all force fields are derivable from a potential. Provided that the force field satisfies the integrability relation,

$$
0 = \frac{\partial g^i}{\partial u_j} - \frac{\partial g^j}{\partial u_i} = \left[\frac{\partial}{\partial u_j}, \frac{\partial}{\partial u_i} \right] \Phi ,
\tag{22}
$$

(i.e., it is curl-free), one may find a solution which is conveniently expressed using a line-integral

$$\Phi(u_k) = \int_C \sum_j dv_j \, g^j(v_l) \, . \tag{23}$$

If the two endpoints are fixed, all contours return the same answer. In practice, one employs the simplest contour that one can imagine: a line connecting the origin to the observation point u_k. Using s, $0 \le s \le 1$, to parameterize the contour, the line-integral may be rewritten as

$$\Phi(u_k) = \sum_{j=1}^n \int_0^1 ds \, u_j \, g^j(su_k) \, . \tag{24}$$

3.4 The Nature of Cosmic Time

In solving eq.(20) for the generating functional $S^{(2n)}$ of order $2n$, one utilizes a line-integral in *superspace*:

$$S^{(2n)} = -\int d^3x \int_0^1 ds \, u(x) \, R^{(2n)}[f_{ij}(x), su(x)] \, . \tag{25}$$

For simplicity, the contour of integration was chosen to be a straight line in superspace. As long as the end points are fixed, the line integral is independent of the contour choice which corresponds to a specific time foliation. This property goes a long way in illuminating the nature of time for semiclassical relativity. (Many questions remain concerning the role of time in a quantum setting [28]; because the above arguments have been quite general, I conjecture that line-integrals in superspace will useful for a full quantum description.)

One may verify the integrability of eq.(20) by explicitly computing the commutator [2],

$$\left[\frac{\delta}{\delta u(x)}, \frac{\delta}{\delta u(y)} \right] S^{(2n)} \equiv \frac{\delta}{\delta u(y)} R^{(2n)}(x) - \frac{\delta}{\delta u(x)} R^{(2n)}(y)$$

$$= [\gamma^{ij}(x) \, \mathcal{H}_j^{(2n-2)}(x) + \gamma^{ij}(y) \, \mathcal{H}_j^{(2n-2)}(y)] \frac{\partial}{\partial x^i} \delta^3(x - y) \, , \tag{26}$$

which assumes by induction that $S^{(2)}, S^{(4)}, \ldots, S^{(2n-2)}$ satisfy eq.(20). The 'integrability condition' of potential theory, eq.(22), demands that the commutator (26) vanish. In the above expression, $\mathcal{H}_j^{(2n-2)}$ is the momentum constraint evaluated using the generating functional of order $(2n-2)$:

$$\mathcal{H}_j^{(2n-2)}(x) \equiv -2\left(\gamma_{jk}\frac{\delta S^{(2n-2)}}{\delta\gamma_{kl}(x)}\right)_{,l} + \frac{\delta S^{(2n-2)}}{\delta\gamma_{kl}(x)}\gamma_{kl,j} + \frac{\delta S^{(2n-2)}}{\delta\phi(x)}\phi_{,i}. \qquad (27)$$

We conclude that $S^{(2n)}$ is indeed integrable provided the term of previous order, $S^{(2n-2)}$, is invariant under reparametrizations of the spatial coordinates: $\mathcal{H}_j^{(2n-2)} = 0$. In general, the integrability condition for the Hamilton-Jacobi equation follows from the Poisson brackets [29] between the energy densities evaluated at the two spatial points x and y:

$$\{\mathcal{H}(x), \mathcal{H}(y)\} = [\gamma^{ij}(x)\mathcal{H}_j(x) + \gamma^{ij}(y)\mathcal{H}_j(y)]\frac{\partial}{\partial x^i}\delta^3(x-y). \qquad (28)$$

Typically, $S^{(2n)}$ is an integral of terms which contain the Ricci tensor and spatial derivatives of the scalar field [3]. For $n = 1$, one determines that

$$S^{(2)}[f_{ij}(x), u(x)] = \int d^3x f^{1/2}\left[j(u)\tilde{R} + k(u)f^{ij}u_{,i}u_{,j}\right]. \qquad (29)$$

$\tilde{R}$ is the Ricci scalar of the conformal 3-metric f_{ij} appearing in eq.(18). The u-dependent coefficients j and k are,

$$j(u) = \int_0^u \frac{\Omega(u')}{2}\,du' + F, \qquad k(u) = H(u)\Omega(u), \qquad (30)$$

where F is an arbitrary constant.

3.5 Characteristics of Cosmic Time

The generalization of the spatial gradient expansion to multiple scalar fields is non-trivial [2]. In this case, one employs the method of characteristics for solving the linear partial differential equation that appears. For a single scalar field ϕ in a HJ description, it is more or less obvious to use some function of ϕ as the integration parameter. In order to facilitate the integration of $S^{(2)}$ for multiple fields, I recommend using the scale factor, $\Omega \equiv \Omega(\phi_a)$, which is a specific function of the scalar fields. A brief summary follows.

One first solves the SHJE of order zero, eq.(14), describing two scalar fields to find a Hubble function

$$H \equiv H(\phi_1, \phi_2; \tilde{\phi}_1, \tilde{\phi}_2) \tag{31}$$

which depends on two homogeneous parameters $\tilde{\phi}_1$ and $\tilde{\phi}_2$. One then makes a change of variables

$$[\phi_1(x), \phi_2(x), \gamma_{ij}(x)] \rightarrow [\Omega(x), e(x), f_{ij}(x)] \tag{32}$$

where the new variables are found by computing partial derivatives of H with respect to the parameters:

$$\Omega(\phi_a) = \left(\frac{\partial H}{\partial \tilde{\phi}_1} \right)^{-1/3} , \tag{33}$$

$$e(\phi_a) = \frac{\partial H}{\partial \tilde{\phi}_1} \bigg/ \frac{\partial H}{\partial \tilde{\phi}_2} , \tag{34}$$

$$f_{ij} = \Omega^{-2}(\phi_a) \, \gamma_{ij} . \tag{35}$$

In terms of the new fields, the generating functional of order two can be computed explicitly:

$$S^{(2)} = \int d^3x \, f^{1/2} \left[j(\Omega, e)\tilde{R} + k_{11}(\Omega, e)\Omega_{;i}\Omega^{;i} + 2k_{12}(\Omega, e)\Omega_{;i}e^{;i} + k_{22}(\Omega, e)e_{;i}e^{;i} \right] . \tag{36}$$

j, k_{11}, k_{12} and k_{22}, are functions of (Ω, e) which are found to be:

$$j(\Omega, e) = \int_0^\Omega d\Omega' \, \frac{1}{2H(\Omega', e)} + j_0(e) , \tag{37}$$

$$k_{11}(\Omega, e) = \frac{1}{H\Omega} , \tag{38}$$

$$k_{22}(\Omega, e) = \int_0^\Omega d\Omega' \, n(\Omega', e) + k_0(e) , \tag{39}$$

$$k_{12}(\Omega, e) = 0 ; \tag{40}$$

$j_0 \equiv j_0(e)$ and $k_0 \equiv k_0(e)$ are arbitrary functions of e; $n(\Omega, e)$ is defined according to

$$n(\Omega, e) = -\frac{1}{2H} \left[\left(\frac{\partial \phi_1}{\partial e} \right)_\Omega^2 + \left(\frac{\partial \phi_2}{\partial e} \right)_\Omega^2 \right] . \tag{41}$$

Details are described in ref.[2].

4 Quadratic Curvature Approximation

In order to describe the fluctuations arising during the inflationary epoch, it is necessary to sum an infinite subset [1], [2], of the terms $\mathcal{S}^{(2n)}$. In this case, one makes an Ansatz which includes all terms which are quadratic in the Ricci tensor $\tilde{R}_{ij}$ of the conformal 3-metric $f_{ij}(x)$:

$$\mathcal{S} = \mathcal{S}^{(0)} + \mathcal{S}^{(2)} + \mathcal{Q}; \tag{42}$$

here the quadratic functional $\mathcal{Q}$ is

$$\mathcal{Q} = \int d^3x f^{1/2} \left[\tilde{R} \ \widehat{S}(u, \tilde{D}^2) \ \tilde{R} + \tilde{R}^{ij} \ \widehat{T}(u, \tilde{D}^2) \ \tilde{R}_{ij} - \frac{3}{8} \tilde{R} \ \widehat{T}(u, \tilde{D}^2) \ \tilde{R} \right], \tag{43}$$

where $\widehat{S}(u, \tilde{D}^2)$ and $\widehat{T}(u, \tilde{D}^2)$ are differential operators which are also functions of u. $\widehat{S}$ and $\widehat{T}$ describe scalar and tensor fluctuations, respectively. $\tilde{D}^2$ is the Laplacian operator with respect to the conformal 3-metric. Terms which are cubic and higher are neglected.

4.1 Multiple Fields: Quadratic Constant Approximation

Once again, the case for two fields [2] is more complicated (after which the extension to any additional fields is straightforward). One replaces the scalar operator $\widehat{S}$ by a matrix operator $\widehat{S}_{ab}$, $a, b = 1, 2$, which is a function of $\Omega(x)$ and $e(x)$, eq.(33) and eq.(34). The scalar operator $\widehat{S}_{ab}$ is then sandwiched between the vector $[\tilde{R}, \tilde{D}^2 e]$ and its transpose in the generalization of eq.(43) to the *quadratic constant approximation*.

5 Comparison with Large-Angle Microwave Background Fluctuations and Galaxy Correlations

Using HJ theory, I will compare the cosmological implications of three inflationary models: Model 1 — power-law inflation; Model 2 — natural inflation, and Model 3 — double inflation. All models will be normalized using large angle microwave background anisotropies determined by COBE [30]: $\sigma_{sky}(10^0) = 30.5 \pm 2.7\mu K$ (68% confidence level).

It is conventional to parametrize the primordial scalar fluctuations arising from inflation by ζ which is proportional to the metric perturbation on a comoving time hypersurface [6]. The power spectra for ζ are shown in Figs.(1a), (2a), (3a), for Models 1, 2, 3, respectively. Both power-law inflation and natural inflation yield power spectra $\mathcal{P}_\zeta(k)$ for ζ that are pure power-laws:

$$\mathcal{P}_\zeta(k) = \mathcal{P}_\zeta(k_0) \left(\frac{k}{k_0}\right)^{n-1} . \tag{44}$$

The spectral index for scalar perturbations is denoted by n, and $n = 1$ describes the flat Zel'dovich spectrum. The simplest models arising from inflation are characterized by $n < 1$. The normalization of the spectra differs for the first two models since gravitational waves may contribute significantly to COBE's signal for the power-law inflation model [16]. The primordial power spectrum for double inflation is not a pure power-law. (In Fig.(3a), the primordial fluctuations for inflation with a single field having a quadratic potential is also shown.)

The power spectra $\mathcal{P}_\delta(k)$ for the linear density perturbation $\delta = \delta\rho/\rho$ at the present epoch are shown in Figs.(1b), (2b), (3b). The data points with error bars are determined from the clustering of galaxies [31]. I have assumed that the evolution of the fluctuations is described by the cold-dark-matter transfer function [32] where the present Hubble parameter is taken to be $H_0 = 50$ km s^{-1}Mpc^{-1}.

For power-law inflation, the best fit is given by $n = 0.9$ (bold curve in Fig.(1b)). However, $n = 0.8$ gives the best fit for natural inflation. From the galaxy data, there is not much difference between the best fits of these two models. One hopes to discriminate these models further when intermediate angle microwave background fluctuations [33] become more precise. Model 3, double inflation model is not particularly attractive since it requires three parameters whereas the previous two models each required one less. For the choice of double inflation parameters advocated by Peter $et\ al$ [34], there is not much advantage over the simpler models of power-law inflation and double inflation [2] (see Fig.(3b)).

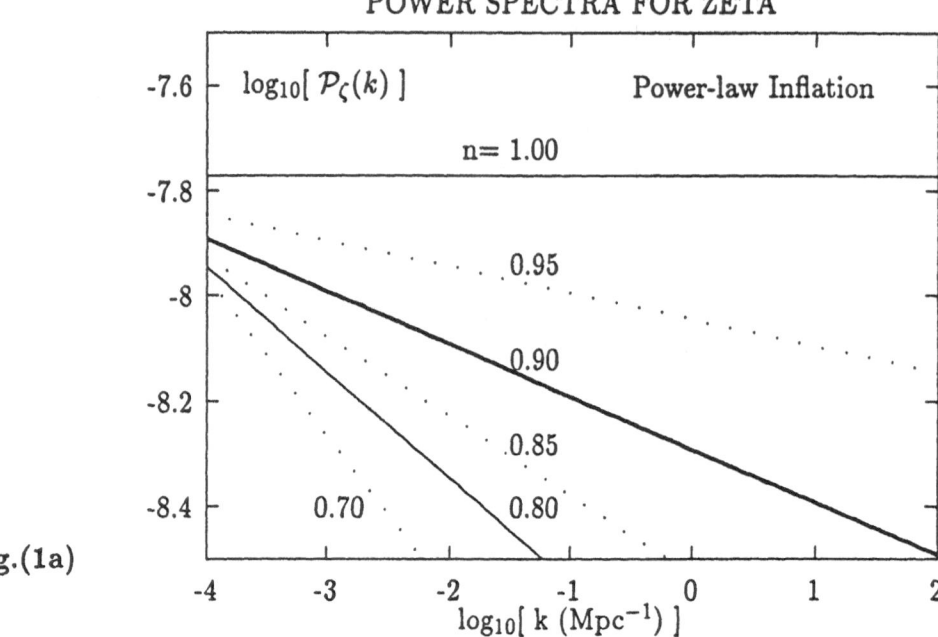

POWER SPECTRA FOR ZETA

Fig.(1a)

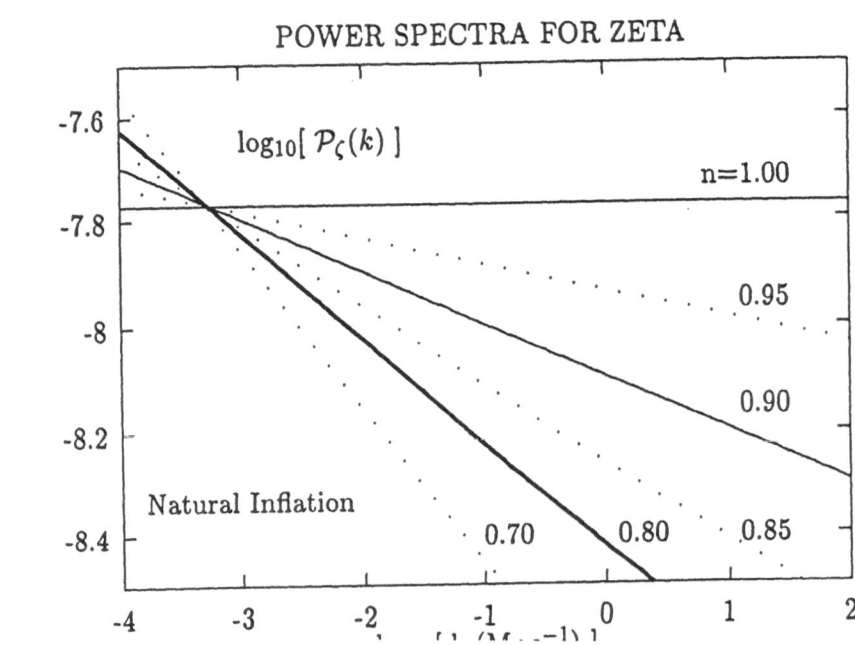

POWER SPECTRA FOR ZETA

Fig.(2a)

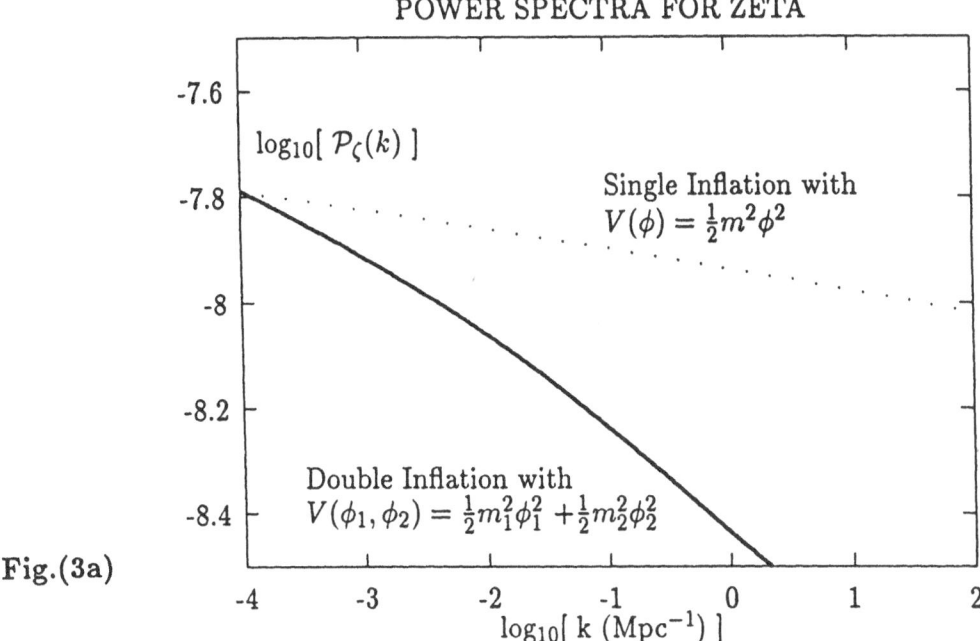

Fig.(3a)

Figs. (1a), (2a), (3a): Shown are the power spectra $\mathcal{P}_\zeta(k)$ for zeta, which describes the primordial scalar perturbations arising from inflation. Three plausible models of inflation are considered: (1) power-law inflation, (2) natural inflation and (3) double inflation. The first two models require two parameters: an arbitrary normalization factor and a spectral index n, where $n - 1$ is the slope of the spectrum in a log-log plot. Double inflation is a three parameter model. For each model, the normalization is fixed by large angle microwave background anisotropies.

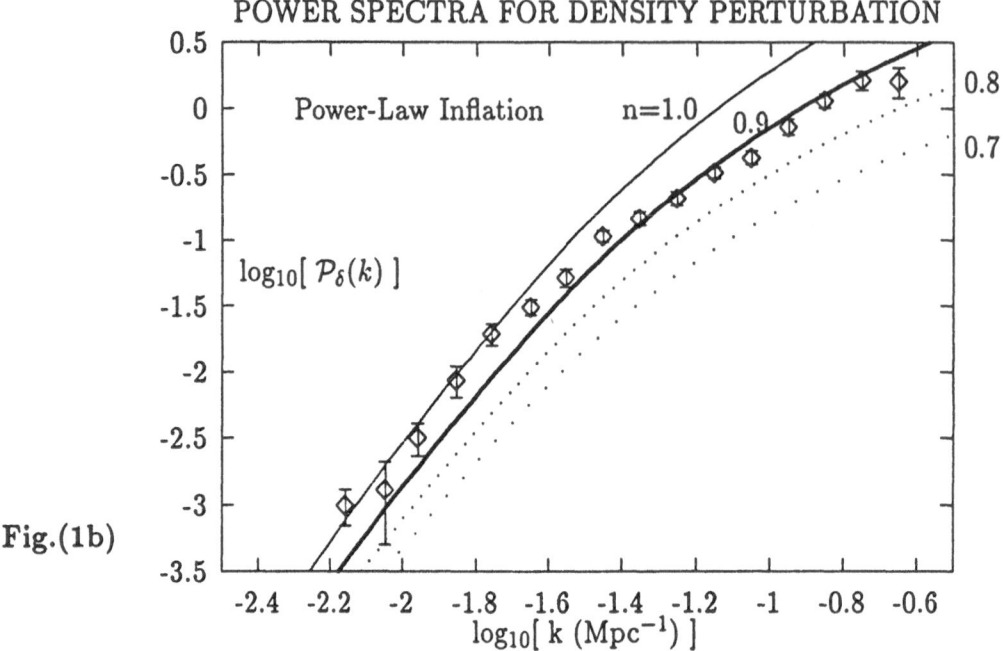

Fig.(1b)

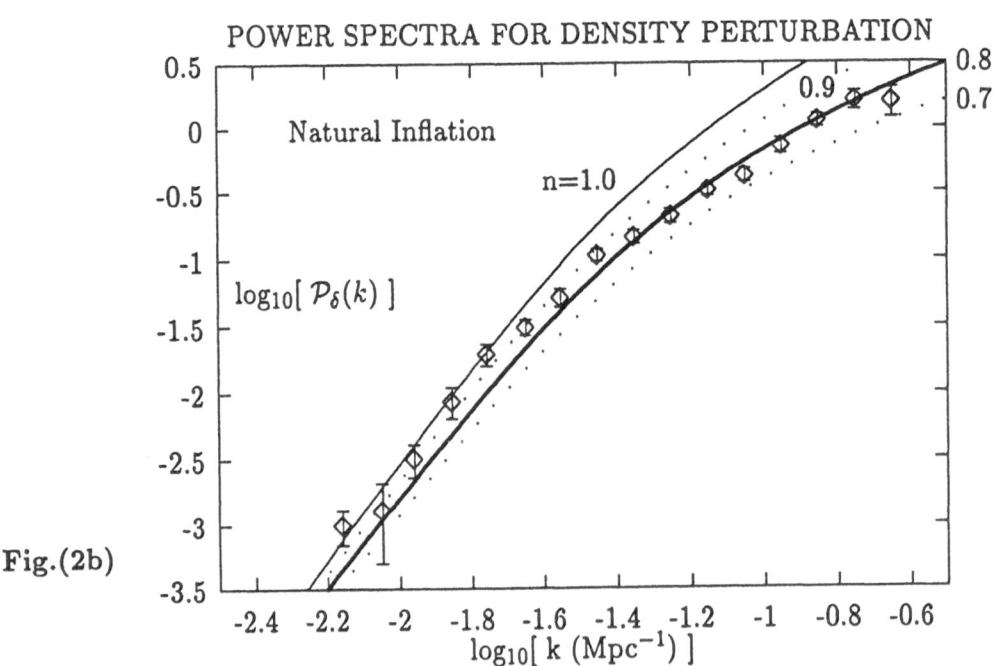

Fig.(2b)

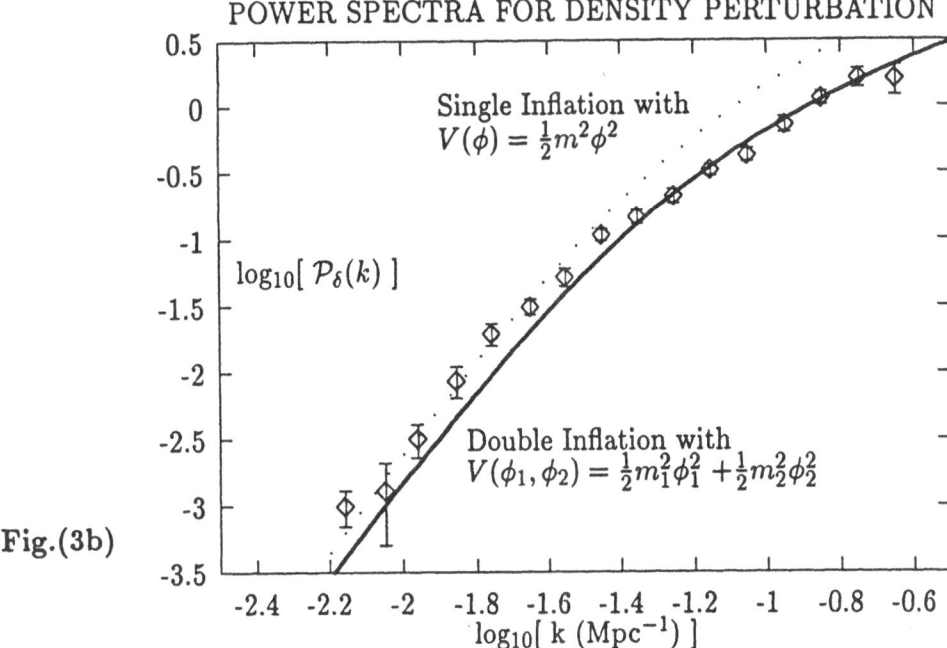

Fig.(3b)

Figs. (1b), (2b), (3b): For the present epoch, the power spectra for the linear density perturbation $\delta\rho/\rho$ are shown for the same models of Fig. (a). The data points with error bars are the observed power spectrum derived from galaxy surveys. For power-law inflation, the best fit (bold curve) is obtained with a spectral index of $n = 0.9$, whereas $n = 0.8$ yields the best fit for natural inflation. (Gravitational waves are important for power-law inflation but not for natural inflation.) Using an additional parameter, double inflation also gives a reasonable fit.

5.1 Disagreement with Grishchuk

In a previous Erice meeting held in the fall of 1994, Grishchuk [35] stated that the tensor fluctuations arising from virtually all inflation models dominate the contribution to the microwave background anisotropy observed by COBE [30]. I disagreed with this claim [36]. More recently, Deruelle and Mukhanov [37] have demonstrated in careful detail that Grishchuk was in error. Here I wish to clarify the issue further.

The main point of disagreement arises during the computation of scalar perturbations and not the tensor perturbations. Once again, the scalar perturbations are parameterized by the variable $\zeta(x)$ which is independent of time when the wavelength of the fluctuation exceeds the Hubble radius. Grishchuk's equation

$$\zeta_{MFB} = \frac{X}{2n^2} \tag{45}$$

on p.225 of ref.[35] is incorrect. It should read

$$\zeta_{MFB} = \left(X - n^2 \frac{\mu}{a\sqrt{\gamma}} \right) \bigg/ (2n^2) \tag{46}$$

Setting $X = 0$, and then taking $n \to 0$, one finds the standard result that

$$\zeta_{MFB} = -\frac{1}{2} \frac{\mu}{a\sqrt{\gamma}} \tag{47}$$

approaches a constant which is not equal to zero as $n \to 0$. (An exact solution of the perturbation equation demonstrates this point quite clearly; see, e.g., ref. [1].) Grishchuk erroneously concluded that $\zeta_{MFB} = 0$ because his method of computation was too crude and he neglected the term described above. The scalar contribution to COBE's signal is indeed proportional to ζ_{MFB}, and for many inflation models, it dominates over the tensor component.

6   Summary

Hamilton-Jacobi theory for general relativity provides some deep insights into the structure of semiclassical superspace which now far exceeds investigations in homogeneous models. Superspace describes an ensemble of evolving universes, and its complexity strains the imagination. However, the gradient expansion allows one to separate superspace into an infinite but discrete number of manageable pieces which are relatively easy to understand.

The question of time choice in general relativity is a difficult one, particularly for the quantum theory [28]. For semiclassical problems of interest to observational cosmology, one may construct a straightforward covariant formalism which treats all time choices on an equal footing. Different contours of integration in superspace correspond to different time foliations and they all yield the same answer for the generating functional provided that spatial gauge invariance is maintained.

Although not quite perfect, reasonable fits of microwave background anisotropies and galaxy clustering may be obtained by power-law inflation with a spectral index of $n = 0.9$, or by natural inflation with a spectral index $n = 0.8$. If one wishes to pay the price for an additional parameter, double inflation is also adequate.

7   Acknowledgments

I thank Prof. Norma Sanchez for organizing yet another beautiful conference in Erice. Many of the Hamilton-Jacobi topics described in this proceedings were developed in collaboration with John Stewart. I thank Alex Vilenkin, Don Page, Werner Israel and Hector DeVega for useful discussions.

This work was supported by NSERC of Canada.

References

[1] D.S. Salopek and J.M. Stewart, *Phys. Rev. D* **51**, 517 (1995).

[2] D.S. Salopek, *Characteristics of Cosmic Time*, accepted for publ. in *Phys. Rev. D* (1995).

[3] J. Parry, D.S. Salopek and J.M. Stewart, *Phys. Rev. D*, **49**, 2872 (1994).

[4] K.M. Croudace, J. Parry, D.S. Salopek and J.M. Stewart, *Astrophys. J.* **423**, 22 (1994); D.S. Salopek, J.M. Stewart and K.M. Croudace, *Mon. Not. Roy. Astr. Soc.*, **271**, 1005 (1994).

[5] D.S. Salopek and J.M. Stewart, *Class. Quantum Grav.* **9**, 1943 (1992).

[6] D.S. Salopek, *Phys. Rev. D* **43**, 3214 (1991); *ibid* **45**, 1139 (1992).

[7] D.S. Salopek, *The Nature of Cosmic Time*, Fields Institute Publication, *Sixth Canadian Conference on General Relativity and Relativistic Astrophysics*, May 25-27, 1995.

[8] Y.B. Zel'dovich, *Astron. & Astrophys.*, **5**, 84 (1970).

[9] J.B. Hartle and S.W. Hawking, *Phys. Rev. D*, **28**, 2960 (1983).

[10] J.J. Halliwell and S.W. Hawking, Phys. Rev. D **31**, 1777 (1985).

[11] A. Vilenkin, *Phys. Lett.* **117B**, 25 (1982); A. Vilenkin, these proceedings.

[12] D.S. Salopek, *Int. J. Mod. Phys. D*, **3**, 257 (1994).

[13] M. Gasperini, N. Sanchez and G. Veneziano, Nucl. Phys. B **364**, 365 (1991); Int. J. Mod. Phys. A, Vol.6, No.21, 3853 (1991).

[14] G. Veneziano; M. Gasperini; T. Banks; A. Tseytlin; H. de Vega; I. Bars; N. Sanchez (these proceedings).

[15] F. Lucchin and S. Matarrese, *Phys. Rev. D* **32**, 1316 (1985).

[16] D.S. Salopek, *Phys. Rev. Lett.*, **69**, 3602 (1992).

[17] D.S. Salopek, J.R. Bond and J.M. Bardeen, *Phys. Rev. D* **40**, 1753 (1989).

[18] A. Zee, *Phys. Rev. Lett.* **42**, 417 (1979).

[19] D. La and P.J. Steinhardt, *Phys. Rev. Lett.* **62**, 376 (1989); E.W. Kolb, D.S. Salopek and M.S. Turner, *Phys. Rev. D* **42**, 3925 (1990); N. Deruelle, C. Gundlach and D. Langlois, *Phys. Rev. D* **45**, 3301 (1992).

[20] F.C. Adams *et al*, *Phys. Rev. D* **47**, 426 (1993).

[21] A.A. Starobinsky, Pis'ma Zh. Eksp. Teor. Fiz. **42**, 124 (1985) (JETP Lett. **42**, 152 (1985)).

[22] A. Peres, *Nuovo Cim.* **26**, 53, (1962).

[23] P.A.M. Dirac, *Lectures on Quantum Mechanics* (Academic, New York, 1965).

[24] A.O. Barvinsky and A.Y. Kamenshchik, Phys. Lett. **B332**, 270 (1994).

[25] J. Soda, Kyoto University preprint, KUCP/U-0081 (1995).

[26] B.S. DeWitt, *Phys. Rev.* **160** 1113 (1967).

[27] C.W. Misner, *Phys. Rev. Lett.* **22**, 1071 (1969).

[28] K.V. Kuchař, in Proc. of 4th Canadian Conference on GR and Relativistic Astrophys., May 16-18 1991, ed. G. Kunstatter *et al*, p.211 (World Scientific, 1992).

[29] V. Moncrief and C. Teitelboim, *Phys. Rev. D* **6**, 966 (1972).

[30] C.L. Bennett *et al*, *Astrophys. J.* **436**, 423 (1994).

[31] J.A. Peacock and S.J. Dodds, *Mon. Not. Roy. Astron. Soc.* **267**, 1020 (1994).

[32] P.J.E. Peebles, *Astrophys. J.* **263**, L1 (1982); J.R. Bond and G. Efstathiou, *Astrophys. J. Lett.* **285**, L45 (1984).

[33] R.L. Davis *et al*, *Phys. Rev. Lett.* **69**, 1856 (1992).

[34] P.Peter, D. Polarski and A.A. Starobinsky, *Phys. Rev. D* **50**, 4827 (1994).

[35] L.P. Grishchuk, in proceedings of *Current Topics in Astrofundamental Physics: The Early Universe*, p.205, Ed. N. Sanchez and A. Zichichi (Kluwer, 1995).

[36] D.S. Salopek, in proceedings of *Current Topics in Astrofundamental Physics: The Early Universe*, p.179, Ed. N. Sánchez and A. Zichichi (Kluwer, 1995).

[37] N. Deruelle and V.F. Mukhanov, Cambridge University preprint (1995).

ASPECTS OF QUANTUM COSMOLOGY

Don N. Page *
CIAR Cosmology Program, Institute for Theoretical Physics
Department of Physics, University of Alberta
Edmonton, Alberta, Canada T6G 2J1

(1995 July 10)

Abstract

Quantum mechanics may be formulated as *Sensible Quantum Mechanics* (SQM) so that it contains nothing probabilistic, except, in a certain frequency sense, conscious perceptions. Sets of these perceptions can be deterministically realized with measures given by expectation values of positive-operator-valued *awareness operators* in a quantum state of the universe which never jumps or collapses. Ratios of the measures for these sets of perceptions can be interpreted as frequency-type probabilities for many actually existing sets rather than as propensities for potentialities to be actualized, so there is nothing indeterministic in SQM. These frequency-type probabilities generally cannot be given by the ordinary quantum "probabilities" for a single set of alternatives. *Probabilism*, or ascribing probabilities to unconscious aspects of the world, may be seen to be an *aesthemamorphic myth*.

No fundamental correlation or equivalence is postulated between different perceptions (each being the entirety of a single conscious experience and thus not in direct contact with any other), so SQM, a variant of Everett's "many-worlds" framework, is a "many-perceptions" framework but not a "many-minds" framework. Different detailed SQM theories may be tested against experienced perceptions by the *typicalities* (defined herein) they predict for these perceptions. One may adopt the *Conditional Aesthemic Principle*: among the set of all conscious perceptions, our perceptions are likely to be typical.

---

*Internet address: don@phys.ualberta.ca

N. Sánchez and A. Zichichi (eds.), String Gravity and Physics at the Planck Energy Scale, 431–450.

1 Basics of Canonical Quantum Cosmology

Quantum cosmology is quantum theory applied to the whole universe. At first sight, this may seem like a strange thing to study, since quantum theory is supposed to apply to the very small, whereas the universe is very large. However, there are at least three motivations for studying quantum cosmology:

(1) Because it's there. Quantum theory certainly seems to apply to parts of the universe (e.g., microscopic systems), and for those parts, the indications are that it is more basic or fundamental than classical theory, which appears to arise as some approximation for certain macroscopic systems. Although we do not know definitely that quantum theory applies to the entire universe, that certainly looks like the simplest possibility consistent with our present knowledge, so it behooves us to investigate the implications if it does.

(2) Our present classical theory of gravity, Einstein's general relativity, contains within itself the seeds for its own destruction in predicting (given certain observations about the present universe and certain reasonable assumptions about the behavior of matter at high densities) that the universe was once so small and highly curved that classical theory should not have been valid. In other words, we believe the universe was once so small that indeed quantum theory should have been needed to understand it.

(3) Present properties of the universe can be described but not explained by classical theory: (a) the flatness of the universe, meaning its large volume and number of particles; (b) the approximate isotropy and homogeneity of the large-scale universe; (c) the particular form of the inhomogeneous structure of the universe on smaller scales; and (d) the thermodynamic arrow of time.

Since the dominant interaction on the largest scales of the universe is gravity, quantum cosmology inevitably involves quantum gravity in a fundamental way. Unfortunately, we do not yet have a complete consistent theory of quantum gravity. Superstring theory seems to be the best current candidate for becoming such a theory, but it is not yet well understood, particularly at the nonperturbative level. Part of the problem of quantum gravity is to get a theory which remains calculable (e.g., can be rendered finite) when one includes arbitrarily high energy fluctuations, and for this problem superstring theory does seem to be remarkably successful, at least at the perturbative level. But another part of the problem are more conceptual issues relating to understanding quantum theory for a closed system, and furthermore a system that is not simply sitting in a fixed background spacetime. These problems have not yet been satisfactorily solved, even by superstring theory, and they tend to be the focus of those of us working in quantum cosmology.

Because in certain limits superstring theory reduces to Einstein's general relativity as an approximation to it, and because many of the more conceptual issues appear to remain, and indeed stand out more clearly, when one makes the approximation to Einsteinian gravity without requiring this gravity to be classical, it is often convenient in quantum cosmology to make the approximation of taking a quantum version of Einsteinian gravity, quantum general relativity. This is not a renormalizable or finite theory, so one would run into trouble using it in multi-loop calculations that allow arbitrarily high energy fluctuations, but fortunately many of the more conceptual issues show up at a cruder approximation, often even at the WKB or 'semiclassical' level with a superposition of fairly classical geometries (though usually not at the complete semiclassical reduction to a single classical geometry that solves the classical Einstein equations with a unique expectation value of the stress-energy tensor as its source).

One standard method of quantizing a spatially closed universe in the approximation of Einsteinian gravity is canonical quantization. Since there exist reviews of this procedure [1, 2], I shall do nothing more here than to give a very brief sketch of the procedure. One starts with the Hamiltonian formulation [3], foliating spacetime into a temporal sequence of closed spatial hypersurfaces with lapse and shift vectors connecting them. Varying the action with respect to these Lagrange multipliers give the momentum and Hamiltonian constraint equations. Then one follows the Dirac method of quantization [4], converting these constraints into operators and requiring that they annihilate the quantum state. When the state is written as a wavefunctional of the three-metric on the spatial hypersurfaces, the momentum constraint (linear in the momenta) for each point of space implies that the wavefunctional is unchanged under any series of infinitesimal diffeomorphisms or coordinate transformations of the three-space [5, 6]. The Hamiltonian constraint (quadratic in the momenta) gives the Wheeler-DeWitt equation [7, 8], actually one equation for each point of space, or one equation for each arbitrary choice of the lapse function.

For each choice of the lapse function, one natural choice for the factoring ordering [7,9-13] of the corresponding Wheeler-DeWitt equation turns it into a Klein-Gordon equation on an indefinite DeWitt metric [7] in the superspace (space of three-metrics), with a potential term that is given by a spatial integral (weighted by the lapse function) of the spatial curvature plus the spatial gradient and potential energy terms that go into the energy density of any matter included in the model. The WKB approximation for this equation gives the Hamilton-Jacobi equation for general relativity, and the trajectories that are orthogonal to the surfaces of constant phase represent classical spacetimes that solve the classical Einstein equations.

In addition to constraint equations (the Wheeler-DeWitt equations in canonically quantized general relativity) that restrict the quantum states, a complete theory of quantum cosmology should specify which solution of these equations describes our universe. There have been various proposals for this in recent years, most notably the Hartle-Hawking 'no-boundary' proposal [14-16] that the wavefunctional evaluated for a compact three-geometry argument is given by a path integral over compact 'Euclidean' four-geometries (with positive-definite metric signatures, as opposed to 'Lorentzian' geometries with an indefinite signature) which reduce to the compact three-geometry argument at its boundary, and the Vilenkin tunneling proposal [17], that the wavefunctional is ingoing into superspace at its regular boundaries and outgoing at its singular boundaries.

Once one has a quantum state or wavefunctional that satisfies the constraints of canonical quantum general relativity, there is the question of how to interpret it to give probabilities. The first stage of this is to find an inner product. DeWitt [7] and many others have proposed using the Klein-Gordon inner product, using the flux of the conserved Klein-Gordon current. However, this is not positive definite and vanishes for real wavefunctionals, such as what one would get from the Hartle-Hawking 'no-boundary' proposal [14-16]. Another approach is to quantize the true physical variables [18], which gives unitary quantum gravity, at least at the one-loop level. However, it is not clear whether this this approach can be carried beyond the perturbative level, even in a minisuperspace approximation. A third approach is third quantization, in which one converts the Wheeler-DeWitt wavefunctional to a field operator on superspace. However, I have not seen any clear way to get testable probabilities out of this.

I myself favor Hawking's approach [16, 13] of using the naïve inner product obtained by taking the integral of the absolute square of a wavefunctional over superspace, with its volume element obtained from the DeWitt metric on it. At least this obviously gives a positive-definite inner product. Unfortunately, it will almost certainly diverge when the square of a physical wavefunctional (i.e., one satisfying the Wheeler-DeWitt equations) is integrated over all of the infinite volume of superspace (which includes directions corresponding to spatial diffeomorphisms and also to time translations). One can try to eliminate the divergences due to the noncompact diffeomorphism group by integrating only over distinct three-geometries (the older meaning of superspace) rather than over all three-metrics that overcount these, though then there is not such an obvious candidate for the volume element. However, one would still have divergences due to the noncompactness of time, which in the canonical quantum gravity of closed universes is encoded in the three-geometry and matter field configuration on the spatial hypersurface. Perhaps one can avoid these divergences by evaluating the inner product only for wavefunctionals that have been acted on by operators, such as projection operators, that do not commute with the constraints and which result in wavefunctionals that are normalizable with the naïve inner product.

2 Basics of Sensible Quantum Cosmology

Before going further with trying to calculate probabilities in quantum cosmology, it may be helpful to ask what probabilities mean in quantum mechanics. If they mean propensities for potentialities to be converted to actualities, then quantum mechanics would seem to be incomplete, since it does not predict *which* possibilities will be actualized, but only the probabilities for each. An interpretation in which quantum mechanics would be more nearly complete would be the Everett or "many-worlds" interpretation [19], in which all possibilities with positive probabilities are actualized, though with measures that depend on the quantum probabilities. However, even this interpretation only seems to work with a single unspecified set of possibilities, whose probabilities add up to one. The arbitrariness of this set seems to indicate that the Everett interpretation is also incomplete, even though it is more nearly complete than the propensity interpretation.

Therefore, I have proposed a version of quantum mechanics, which I call Sensible Quantum Mechanics (SQM) [20-24], in which only a definite set of possibilities have measures, namely, conscious perceptions.

Sensible Quantum Mechanics is given by the following three fundamental postulates [22]:

Quantum World Axiom: The unconscious "quantum world" Q is completely described by an appropriate algebra of operators and by a suitable state σ (a positive linear functional of the operators) giving the expectation value $\langle O \rangle \equiv \sigma[O]$ of each operator O.

Conscious World Axiom: The "conscious world" M, the set of all perceptions p, has a fundamental measure $\mu(S)$ for each subset S of M.

Quantum-Consciousness Connection: The measure $\mu(S)$ for each set S of conscious perceptions is given by the expectation value of a corresponding "awareness operator" $A(S)$, a positive-operator-valued (POV) measure [25], in the state σ of the quantum world:

$$\mu(S) = \langle A(S) \rangle \equiv \sigma[A(S)]. \tag{1}$$

Here a perception p is the entirety of a single conscious experience, all that one is consciously aware of or consciously experiencing at one moment, the total "raw feel" that one has at one time, or [26] a "phenomenal perspective" or "maximal experience."

Since all sets S of perceptions with $\mu(S) > 0$ really occur in SQM, it is completely deterministic if the quantum state and the $A(S)$ are determined: there are no random or truly probabilistic elements. Nevertheless, because SQM has measures for sets of perceptions, one can readily calculate ratios that can be interpreted as conditional probabilities. For example, one can consider the set of perceptions S_1 in which there is a conscious awareness of cosmological data and theory and a conscious belief that the visible universe is fairly accurately described by a Friedman-Robertson-Walker (FRW) model, and the set S_2 in which there is a conscious memory and interpretation of getting reliable data indicating that the Hubble constant is greater than 75 km/sec/Mpc. Then one can interpret

$$P(S_2|S_1) \equiv \mu(S_1 \cap S_2)/\mu(S_1) \tag{2}$$

as the conditional probability that the perception is in the set S_2, given that it is in the set S_1, that is, that a perception included a conscious memory of measuring the Hubble constant to be greater than 75 km/sec/Mpc, given that one is aware of knowledge and belief that an FRW model is accurate.

3   Testing Sensible Quantum Cosmology Theories

The measures for observed (or, perhaps more accurately, experienced) perceptions can be used to test different SQM theories for the universe, thus grounding physics and cosmology in experience. If one had a theory in which only a small subset of the set of all possible perceptions is predicted to occur, one could simply check whether an experienced perception is in that subset. If it is not, that would be clear evidence against that theory. Unfortunately, in almost all SQM theories, almost all sets of perceptions are predicted to have a positive measure, so these theories cannot be excluded so simply. For such many-perceptions theories, the best one can hope for seems to be to find *likelihood* evidence for or against it. Even how to do this is not immediately obvious, since SQM theories merely give measures for sets of perceptions rather than the existence probabilities for any perceptions (unless the existence probabilities are considered to be unity for all existing sets of perceptions, i.e., all those with nonzero measure, but this is of little help, since almost all sets exist in this sense).

In order to test and compare SQM theories, it helps to hypothesize that the set M of all possible conscious perceptions p is a suitable topological space with a prior measure

$$\mu_0(S) = \int_S d\mu_0(p). \tag{3}$$

Then, because of the linearity of positive-valued-operator measures over sets, one can write each awareness operator as

$$A(S) = \int_S E(p) d\mu_0(p), \tag{4}$$

a generalized sum or integral of "experience operators" or "perception operators" $E(p)$ for the individual perceptions p in the set S. Similarly, one can write the measure on a set of perceptions S as

$$\mu(S) = \langle A(S) \rangle = \int_S d\mu(p) = \int_S m(p) d\mu_0(p), \tag{5}$$

in terms of a measure density $m(p)$ that is the quantum expectation value of the experience operator $E(p)$ for the same perception p:

$$m(p) = \langle E(p) \rangle \equiv \sigma[E(p)]. \tag{6}$$

Now one can test the agreement of a particular SQM theory with a conscious observation or perception p by calculating the (ordinary) *typicality* $T(p)$ that the theory assigns to the perception: Let $S_{\leq}(p)$ be the set of perceptions p' with $m(p') \leq m(p)$. Then

$$T(p) \equiv \mu(S_{\leq}(p))/\mu(M). \tag{7}$$

For p fixed and $\tilde{p}$ chosen randomly with the infinitesimal measure $d\mu(\tilde{p})$, the probability that $T(\tilde{p})$ is less than or equal to $T(p)$ is

$$P_T(p) \equiv P(T(\tilde{p}) \leq T(p)) = T(p). \tag{8}$$

In the case in which $m(p)$ varies continuously in such a way that $T(p)$ also varies continuously, this typicality $T(p)$ has a uniform probability distribution between 0 and 1, but if there is a nonzero measure of perceptions with the same value of $m(p)$, then $T(p)$ has discrete jumps. (In the extreme case in which $m(p)$ has one constant value over all perceptions, $T(p)$ is unity for each p.)

Thus the typicality $T_i(p)$ of a perception p is the probability in a particular SQM theory or hypothesis H_i that another random perception will have its measure density and hence typicality less than or equal to that of p itself. One can interpret it as the likelihood of the perception p in the particular theory H_i, not for p to exist, which is usually unity (interpreting all perceptions p with $m(p) > 0$ as existing), but for p to have a typicality no larger than it has.

Once the typicality $T_i(p)$ can be calculated for an experienced perception assuming the theory H_i, one approach is to use it to rule out or falsify the theory if the resulting typicality is too low. Another approach is to assign prior probabilities $P(H_i)$ to different theories (presumably neither propensities nor frequencies but rather purely subjective probabilities, perhaps one's guess for the "propensities" for God to create a universe according to the various theories), say

$$P(H_i) = 2^{-n_i}, \tag{9}$$

where n_i is the rank of H_i in order of increasing complexity (my present favorite choice for a countably infinite set of hypotheses if I could do this ranking, which is another problem I will not further consider here). Then one can use Bayes' rule to calculate the posterior probability of the theory H_i given the perception p as

$$P(H_i|p) = \frac{P(H_i)T_i(p)}{\sum_j P(H_j)T_j(p)}. \tag{10}$$

There is the potential technical problem that one might assign nonzero prior probabilities to hypotheses H_i in which the total measure $\mu(M)$ for all perceptions is *not* finite, so that the right side of Eq. (7) may have both numerator and denominator infinite, which makes the typicality $T_i(p)$ inherently ambiguous. To avoid this problem, one might use, instead of $T_i(p)$ in Eq. (10), rather

$$T_i(p; S) = \mu_i(S_\leq(p) \cap S)/\mu_i(S) \tag{11}$$

for some set of perceptions S containing p that has $\mu_i(S)$ finite for each hypothesis H_i. This is related to a practical limitation anyway, since one could presumably only hope to be able to compare the measure densities $m(p)$ for some small set of perceptions rather similar to one's own, though it is not clear in quantum cosmological theories that allow an infinite amount of inflation how to get a finite measure even for a small set of perceptions. Unfortunately, even if one can get a finite measure by suitably restricting the set S, this makes the resulting $P(H_i|p; S)$ depend on this chosen S as well as on the other postulated quantities such as $P(H_i)$.

Instead of using the particular typicality defined by Eq. (7) above, one could of course instead use any other property of perceptions which places them into an ordered set to define a corresponding "typicality." For example, I might be tempted to order them according to their complexity, if that could be well defined. Thinking about this alternative "typicality" leaves me surprised that my own present perception seems to be highly complicated but apparently not infinitely so. What simple complete theory could make a typical perception have a high but not infinite complexity?

However, the "typicality" defined by Eq. (7) has the merit of being defined purely from the prior and fundamental measures, with no added concepts such as complexity that would need to be defined. The necessity of being able to rank perceptions, say by their measure density, in order to calculate a typicality, is indeed one of my main motivations for postulating a prior measure given by Eq. (3).

Nevertheless, there are alternative typicalities that one can define purely from the prior and fundamental measures. For example, one might define a *reversed typicality* $T_r(p)$ in the following way (again assuming that the total measure $\mu(M)$ for all perceptions is finite): Let $S_\geq(p)$ be the set of perceptions p' with $m(p') \geq m(p)$. Then

$$T_r(p) \equiv \mu(S_\geq(p))/\mu(M). \tag{12}$$

For p fixed and $\tilde{p}$ chosen randomly with the infinitesimal measure $d\mu(\tilde{p})$, the probability that $T_r(\tilde{p})$ is less than $T_r(p)$ is

$$P_{T_r}(p) \equiv P(T_r(\tilde{p}) \leq T_r(p)) = T_r(p), \tag{13}$$

the analogue of Eq. (8) for the ordinary typicality.

In the generic continuum case in which $m(p)$ varies continuously in such a way that there is only an infinitesimally small measure of perceptions whose $m(p)$ are infinitesimally near any fixed value, the reversed typicality $T_r(p)$ is simply one minus the ordinary typicality, i.e., $1 - T(p)$, and also has a uniform probability distribution between 0 and 1. Its use arises from the fact that just as a perception with very low ordinary typicality $T(p) \ll 1$ could be considered unusual, so a perception with an ordinary typicality too near one (and hence a reversed typicality too near zero, $T_r(p) \ll 1$) could also be considered unusual, "too good to be true."

Perhaps one might like to combine the ordinary typicality with the reversed typicality to say that a perception giving either typicality too near zero would be evidence against the theory. For example, one might define the *dual typicality* $T_d(p)$ as the probability that a random perception $\tilde{p}$ has the lesser of its ordinary and its reversed typicalities less than or equal to that of the perception under consideration:

$$T_d(p) \equiv P(\min[T(\tilde{p}), T_r(\tilde{p})] \leq \min[T(p), T_r(p)]) \equiv \mu(S_d(p))/\mu(M), \qquad (14)$$

where $S_d(p)$ is the set of all perceptions $\tilde{p}$ with the minimum of its ordinary and reversed typicalities less than or equal to that of the perception p, i.e., the set with $\min[T(\tilde{p}), T_r(\tilde{p})] \leq \min[T(p), T_r(p)]$. In the case in which $T(p)$, and hence also $T_r(p)$, varies continuously from 0 to 1,

$$T_d(p) = 1 - |1 - 2T(p)|. \qquad (15)$$

Then the dual typicality $T_d(p)$ would be very small if the ordinary typicality $T(p)$ were very near either 0 or 1.

Of course, one could go on with an indefinitely long sequence of typicalities, say making a perception "atypical" if $T(p)$ were very near any number of particular values at or between the endpoint values 0 and 1. But these endpoint values are the only ones that seem especially relevant, and so it would seem rather *ad hoc* to define "typicalities" based on any other values. Since $T_d(p)$ is symmetrically defined in terms of both endpoints (or, more precisely, in terms of both the $\leq$ and the $\geq$ relations for $m(p')$ in comparison with $m(p)$), in some sense it seems the most natural one to use. Obviously, one could use it, or its modification along the lines of Eq. (11), instead of $T(p)$ in the Bayesian Eq. (10).

To illustrate how one may use these typicalities to test different theories, consider the experiment in which one makes a particular measurement of a single particular continuous variable (e.g., the position of a one-dimensional harmonic oscillator in its ground state) which is supposed to have a gaussian quantum distribution. Suppose that in the subset of perceptions S in which this measurement is believed to have been made and the result is known, there is a one-dimensional continuum of perceptions that are linearly related to the measured value of the variable, say labeled by the real number x that is the measured value of the variable, with the prior measure $d\mu_0(p) \propto dx$. (It might be more realistic to suppose that the measuring device can transmit only a discrete set of values to the brain that is doing the perceiving, but if the number of these is large, it is convenient to approximate them by a continuum. As to whether the set of perceptions is discrete or continuous, I know of no strong evidence either way. Even if the possible quantum states of the universe lay in a finite-dimensional Hilbert space, which seems doubtful but is not clearly ruled out by observational evidence, one could easily have a continuum of experience operators $E(p)$ for a continuum set of perceptions p.)

I shall furthermore make the idealized hypothesis that the measure for this subset of perceptions labeled by x is purely given by the quantum distribution of the measured variable and is not further influenced by processes in the brain. One could imagine situations in which certain measured values lead to the release of an anaesthetic that renders the perceiver unconscious and so essentially eliminates the measure for the corresponding perceptions. At first sight such situations that grossly affect the measure for perceptions arising from a quantum measurement seemed contrived, but then I realized that noticeably unusual results of the measuring device (say results several standard deviations from the mean) could very well attract more conscious attention, over a longer time, than results that are not noticeably unusual. It seems highly plausible that the measure for a set of perceptions would increase with their alertness and with the time over which they continually occur, so noticeably unusual events that attract more attention would presumably have a higher measure than one might otherwise naïvely expect. Thus the measure for perceptions arising from the measurement of a variable with a gaussian quantum distribution could well have tails that do not decrease so fast as the original gaussian.

I shall call this effect, whereby the measure for perceptions of unusual events is increased by the attention given them, the Attention Effect. It may well be responsible for the large number of coincidences that one is aware of from anecdotal evidence. (For example, it has occurred to me several times that it was surprising for Canada and the U.S. to have ages in years that were both perfect cubes, 125 and 216 respectively, in 1992.) One can try to combat it, by, e.g., focusing on perceptions in which it is perceived that the quantum measurement was made only a second previous, say, when there would presumably not have been time for a dull result to have receded much from consciousness. So for simplicity in the following discussion I shall make the idealized assumption that the Attention Effect, as well as any other effect that distorts the measure of perceptions from what what one would calculate from the quantum measurement itself, is negligible. (I am grateful for the visit of Jane and Tim Hawking in Edmonton during my writing of the previous paragraph, which gave me the time to realize the importance of the Attention Effect. Was the timing of their arrival another coincidence?)

Assuming no Attention Effect or similar effect, the measure for x would have the gaussian distribution it inherits from the quantum measurement, say

$$m(x) \propto e^{-x^2/(2\sigma^2)}. \tag{16}$$

Within the subset of perceptions S in which this measurement is believed to have been made and the result is known, the ordinary, reversed, and dual typicalities are

$$T(x; S) = \text{erfc}(\sqrt{x^2/(2\sigma^2)}) \equiv 1 - \text{erf}(\sqrt{x^2/(2\sigma^2)}) \equiv 1 - \frac{2}{\sqrt{\pi}} \int_0^{\sqrt{x^2/(2\sigma^2)}} e^{-z^2} dz, \tag{17}$$

$$T_r(x; S) = 1 - T(x; S) = \text{erf}(\sqrt{x^2/(2\sigma^2)}), \tag{18}$$

$$T_d(x; S) = 1 - |1 - 2T(x; S)| = 1 - |1 - 2T_r(x; S)|$$
$$= 1 - |1 - 2\text{erfc}(\sqrt{x^2/(2\sigma^2)})| = 1 - |1 - 2\text{erf}(\sqrt{x^2/(2\sigma^2)})|. \tag{19}$$

Suppose that one does not know the actual standard deviation σ for x but has various hypotheses H_i that it has the various values σ_i. Then one can replace σ with σ_i in Eqs. (17)-(19) above to get the typicalities (restricted to the subset S) of the perception p that gives the perception component x in the corresponding hypothesis. If the typicality one is considering is too low for a certain hypothesis, one might use the perception to exclude (or falsify on a likelihood basis) that hypothesis.

Alternatively, one might assign a prior probability distribution to σ_i and then use the Bayesian Eq. (10), with $T_i(p)$ replaced by $T(x; S)$, $T_r(x; S)$, or $T_d(x; S)$ as given by Eqs. (17)-(19) and with σ replaced by σ_i, to calculate a posterior distribution for σ_i, given x from the perception. For example, one might assign (as a fairly simple concrete choice) the following prior probability distribution for σ_i, which is purely a decaying exponential distribution in its square and thus a slight modification of a gaussian distribution for σ_i itself:

$$P(\sigma_i) = e^{-\sigma_i^2/(2\sigma_0^2)}\sigma_i d\sigma_i/\sigma_0^2. \tag{20}$$

Here σ_0^2 is an arbitrary parameter that one must choose to represent the expectation value of σ_i^2 in the prior distribution. One can then readily calculate [27] that using the ordinary typicality in Eq. (10) gives

$$P(\sigma_i|x; S) = e^{(x/\sigma_0)-\sigma_i^2/(2\sigma_0^2)}\text{erfc}(\sqrt{x^2/(2\sigma^2)})\sigma_i d\sigma_i/\sigma_0^2. \tag{21}$$

This is very strongly damped for small values of σ_i (i.e., for values much less than x), by the complementary error function from Eq. (17) for the typicality, and is damped at large values of σ_i (i.e., for values much greater than σ_0) by the exponentially decaying prior distribution of Eq. (20). However, if one used a different prior distribution that was not significantly damped at large values of σ_i, neither would the posterior distribution be significantly damped at large values of σ_i. Thus using the ordinary typicality T in the Bayesian Eq. (10) is effective in giving a lower limit on the spread of the measure distribution for a number like x assigned to the perceptions (at least if the one used in the calculation is not exactly at the mean value), but it is not effective in giving a better upper limit on the spread than that given by the prior distribution. This is because the ordinary typicality gives a penalty for theories that would predict that an observed result is unlikely or "bad," but it does not give a similar penalty for theories that predict that the observed result is "too good to be likely."

The reversed typicality T_r of Eq. (12) or (18) does indeed penalize theories that predict that the observed result is too good to be likely, but only at the cost of not penalizing at all "bad" results, so it would be worse to use. Better is the dual typicality T_d of Eq. (14) or (19), which penalizes theories that fit the observations either too poorly or too well. Unfortunately, it makes it harder to evaluate the posterior probability distribution Eq. (10) analytically, because of the minimization functions in the definition of the dual typicality, and I have been unable to come up with a simple explicit result for the prior distribution Eq. (20), though one can get an explicit result for a prior distribution that is flat in $n \equiv 1/\sigma_i^2$ [22]. In any case, one can see that with the dual typicality, the posterior distribution for the standard deviation σ_i is more heavily damped at both large and small σ_i than is the prior distribution for σ_i that is assumed.

One can see that if one starts with a smooth continuum prior probability distribution for theories, a Bayesian analysis using one of the typicalities of an experienced perception can give a posterior probability distribution of theories that is more narrow, but it can never lead to a nonzero probability for any single theory out of the continuum of possibilities that are smoothly weighted. If this is the case, one shall never succeed in getting any single final theory that one can say has any significant (e.g., nonzero) probability of being absolutely correct. Of course, one might be able to deduce (after postulating a reasonable continuum prior distribution) fairly narrow ranges for the continuous parameters where most of the posterior probability is concentrated, so the situation could be qualitatively no worse than it is at present for such parameters as the fine structure constant, which is only thought to be likely to be within some very narrow range (and perhaps only within this range for certain components of the quantum state of the universe from which our perceptions get the bulk of their measure).

On the other hand, it is conceivable that theorists will eventually find a discrete set of theories, each with no arbitrary continuous parameters (as in superstring theories), or else with preferred discrete values of such parameters, to each of which they can assign a nonzero prior probability. Then if these are weighted by the likelihoods they predict for the perception of a sufficiently good set of observations, it is conceivable (though at present it might seem somewhat miraculous, but doesn't the order and structure in our world already look miraculous?) that the posterior probabilities will pick out one unique theory with a probability near unity and assign all other theories a total probability much closer to zero. In such a case theorists might well believe that the one unique theory with a probability near unity is indeed *the correct theory* of the universe. Of course, the fact that the other theories would not have a total probability exactly equal to zero (at least for almost any conceivable scenario I can presently imagine) would mean that one could not be sure (at least by the Bayesian analysis outlined above) that one really did have the correct theory for the universe, but if the probability were sufficiently near unity, one could presumably put a great deal of faith in that deduction from theory and observation, just as we presently put a great deal of faith in much smaller pieces of knowledge to which we assign probabilities near unity.

There is also the problem that the prior probabilities seem to be purely subjective, so people could well disagree about whether or not the assignment that led to a posterior probability near unity for one particular theory (if indeed that dream of my present wishful thinking actually does occur) is reasonable. I suspect that such disagreements about prior probabilities are at the heart of many current disagreements (e.g., the existence of God or the truth of superstring theory), though there is also the huge practical problem that unless one has a detailed theory from which one can make the relevant calculations, one cannot even predict what the likelihoods are for the various hypotheses to result in certain observations (e.g., the perception of good and evil or the experience of gravity). However, one could imagine a society which agrees in sufficient broad outline about the prior probabilities, and observational results that sufficiently narrow them down (as often occurs to a fantastic degree in experimental physics, viz. the amazing agreement of experiment with QED), that nearly all members will agree on a unique theory for the universe as having a posterior probability near unity (similar to the fact that most members of the community of physicists believes in QED within its domain of applicability, though QED represents a continuum of theories labeled by the fine structure constant rather than one unique theory).

One can also worry that if Sensible Quantum Mechanics is correct, then presumably our perceptions are completely determined by the detailed theory, and it seems likely that it would only be some sort of an idealization to say that our beliefs are determined by a Bayesian process of modifying prior probabilities to posterior probabilities in the light of the likelihoods of our observations. So it may be purely hypothetical to predict what beliefs we would arrive at if we went though this ideal Bayesian procedure, since it would seem miraculous if we were determined to act in just that way, and there is certainly evidence that most of the time our beliefs are not determined precisely thus. However, idealizations are at the heart of physics, and this Bayesian one does not seem particularly worse than many others. Even if the physics community (or the broader human community) does not actually come to its beliefs by precisely a Bayesian analysis, it is interesting to speculate on what conclusions it might eventually arrive at if it did. Though even such conclusions are not guaranteed to be true (because of the difficulties mentioned above, and various others), it would seem that they would be more likely to be true than whatever conclusions people will actually come up with, especially if they they choose to ignore the Bayesian procedure.

Thus, although Sensible Quantum Mechanics is my current best guess of what is a true framework for a complete description of the universe, my advocacy of a Bayesian analysis to choose between detailed theories within that framework is not a guess of what is truly the way physicists work, but a moral appeal for one way in which I think the search for a detailed theory ought to be conducted.

4 Predictions from Sensible Quantum Cosmology

Although it is certainly an open question whether humans or any other conscious beings within the universe will ever come to some sort of a grasp of a complete theory of the universe (perhaps only as a set of ideas whose logical implications include a complete description of the entire universe, even though it seems extremely unlikely that conscious beings within the universe could ever work out all of these implications in detail), we would like to develop better theories that will enable us to predict more properties of the universe than we can at present.

Vilenkin [28] has recently discussed predictions in quantum cosmology, and I do not have much in detail to add to what he beautifully covered. I agree with him that if the 'constants' of physics that we have measured can actually vary from component to component of the quantum state of the universe, the relevant probability distribution must be obtained by using something like the Principle of Mediocrity that he proposes, that we are a 'typical' civilization within the ensemble of components.

I might personally prefer [22, 24] a slightly different variant of the Weak Anthropic Principle [29-35] which I call the Conditional Aesthemic Principle, that our conscious perceptions are likely to be typical perceptions in the conscious world with its measure. Then the relevant probability distribution for the 'constants' of physics would be weighted by the measures for perceptions. The most basic way to do this would be to use only the perceptions which include a belief in the value of the corresponding 'constant.' However, if the quantum state of the universe is represented by the density matrix ρ, and if the 'constants' are indeed constant over the relative density matrix [22, 24]

$$\rho_p = \frac{E(p)\rho E(p)}{Tr[E(p)\rho E(p)]} \tag{22}$$

for each perception p, then it might be better to weight the probability distribution of the 'constants' in each such relative state by the measure density for the corresponding perception. One might like thus to find out the probability distribution for the cosmological constant, the parameters of the Standard Model of particle physics, and the parameters of an inflaton potential (if any such exists).

One persistent problem that hampers predictions even from simple toy min-isuperspace models in quantum cosmology is the apparent lack of normalizability of most of the quantum states, at least if one takes the positive-definite naïve in-ner product obtained by integrating the absolute square of the wavefunctional over superspace. As discussed above, part of this problem in canonical quantum cosmol-ogy is due to the fact that the wavefunction in the configuration representation is most easily handled as a functional of three-metrics that is invariant under coor-dinate transformations, but this diffeomorphism group is noncompact and leads to divergences when one integrates over it. It seems that this ought to be merely an avoidable technical problem that arises from writing down wavefunctionals of three-metrics rather than of three-geometries, but even if one circumvents it (or avoids it by truncating the configuration space, as in homogeneous minisuperspace models), one next faces the problem of the invariance under the transformations generated by the Wheeler-DeWitt equations, which represent time translations at each point of space. It would seem that then one must face at least the noncompact groups of Lorentz boosts at each point of space, since the hypersurface described by the three-metric argument of the wavefunctional can be tilted rather arbitrarily at each point of space. Even if those divergences are eliminated (as in the minisuperspace models in which the homogeneous three-geometries are prevented from being tilted), one can still get divergences from the infinite range of values allowed by the 'time.'

Perhaps these divergences can be avoided by not seeking to interpret the inte-gral of the absolute square of the physical wavefunctional over the configuration space, but by restricting the interpretation, as in Sensible Quantum Mechanics, to the integrals of wavefunctionals that have been operated on by the experience or perception operators $E(p)$. There seems to be no reason why the resulting wave-functionals should be physical wavefunctionals that obey the constraints such as the Wheeler-DeWitt equation, since a perception could very well be localized to be concentrated near a particular clock time (represented by some property of the three-geometry or matter fields on it) and need not occur over the whole history (or, more accurately, set of histories in the path integral sense) of the universe that is represented by a physical solution of the constraints. Certainly in minisuperspace one can get positive operators, as the experience operators are required to be, that have finite expectation values even for wavefunctionals that are not normalizable, e.g., projection operators to finite ranges of all the configuration space variables.

However, I am still not sure that even this procedure will avoid all divergences.

If one has a model in which an infinite amount of inflation can occur, which would be useful for solving the flatness problem, then in the resulting infinite amount of spatial volume it would seem that the measure for almost any set of perceptions of nonzero prior measure would likely be infinite, since each sufficiently large finite volume would seem likely to give an independent positive contribution to it. If one has infinite measures for almost all nontrivial sets of perceptions, then calculating conditional probabilities and typicalities by taking the ratios of such infinite quantities is meaningless, and I do not yet see how to get testable predictions out from even Sensible Quantum Cosmology.

In conclusion, Sensible Quantum Mechanics seems to give an improved interpretation of quantum theory that is helpful in quantum cosmology for testing different quantum theories of the universe and in making predictions. It seems to ameliorate part of the problem of the lack of normalizability of the wavefunctionals of canonical quantum gravity, but serious problems still seem to remain with this that hamper predictions from quantum cosmology.

People whom I remember to have recently influenced my thoughts on this subject are listed in [22]. I am grateful to Alex Vilenkin for sending me an advance copy of the third paper of [28]. Financial support has been provided by the Natural Sciences and Engineering Research Council of Canada.

References

[1] J. J. Halliwell, in *Quantum Cosmology and Baby Universes*, edited by S. Coleman, J. B. Hartle, T. Piran, and S. Weinberg (World Scientific, Singapore, 1991), p. 159.

[2] D. N. Page, in *Gravitation: A Banff Summer Institute*, edited by R. Mann and P. Wesson (World Scientific, Singapore, 1991), p. 135.

[3] R. Arnowitt, S. Deser, and C. W. Misner, in *Gravitation: An Introduction to Current Research*, edited by L. Witten (Wiley, New York, 1962).

[4] P. A. M. Dirac, Proc. Roy. Soc. (Lond.) **A246**, 326 and 333 (1958); *Lectures on Quantum Mechanics* (Yeshiva University, New York, 1964).

[5] C. W. Misner, Rev. Mod. Phys. **29**, 497 (1957).

[6] P. W. Higgs, Phys. Rev. Lett. **1**, 373 (1959).

[7] B. S. DeWitt, Phys. Rev. **160**, 1113 (1967).

[8] J. A. Wheeler, in *Battelle Rencontres: 1967 Lectures in Mathematics and Physics*, edited by C. DeWitt and J. A. Wheeler (Benjamin, New York, 1968).

[9] C. W. Misner, in *Magic without Magic*, edited by J. R. Klauder (Freeman, San Francisco, 1972).

[10] K. Kuchař, in *Relativity, Astrophysics and Cosmology*, edited by W. Israel (Reidel, Dordrecht, 1973).

[11] M. Henneaux, M. Pilati, and C. Teitelboim, Phys. Lett. **110B**, 123 (1982).

[12] T. Christodoulakis and J. Zanelli, Phys. Rev. **D29**, 2738 (1984); Phys. Lett. **102A**, 227 (1984).

[13] S. W. Hawking and D. N. Page, Nucl. Phys. **B264**, 185 (1986).

[14] S. W. Hawking, in *Astrophysical cosmology: Proceedings of the Study Week on Cosmology and Fundamental Physics* edited by H. A. Brück, G. V. Coyne, and M. S. Longair (Pontificiae Academiae Scientiarum Scripta Varia, Vatican, 1982).

[15] J. B. Hartle and S. W. Hawking, Phys. Rev. **D28**, 2960 (1983).

[16] S. W. Hawking, Nucl. Phys. **B239**, 257 (1984).

[17] A. Vilenkin, Phys. Lett. **117B**, 25 (1982); Phys. Rev. **D30**, 509 (1984); Nucl. Phys. **B252**, 141 (1985); Phys. Rev. **D33**, 3560 (1986); Phys. Rev. **D37**, 888 (1988).

[18] A. O. Barvinsky, in *Proceedings of the Fourth Seminar on Quantum Gravity, May 25-29, 1987, Moscow, USSR*, edited by M. A. Markov, V. A. Berezin, and V. P. Frolov (World Scientific, Singapore, 1988); Nucl. Phys. **B325**, 705 (1989); Phys. Lett. **241B**, 201 (1990); Phys. Rep. **230**, 237 (1993).

[19] H. Everett, III, Rev. Mod. Phys. **29**, 454 (1957); B. S. DeWitt and N. Graham, eds., *The Many-Worlds Interpretation of Quantum Mechanics* (Princeton University Press, Princeton, 1973).

[20] D. N. Page, "Probabilities Don't Matter," to be published in *Proceedings of the 7th Marcel Grossmann Meeting on General Relativity*, edited by M. Keiser and R. T. Jantzen (World Scientific, Singapore 1995) (University of Alberta report Alberta-Thy-28-94, Nov. 25, 1994), gr-qc/9411004.

[21] D. N. Page, "Information Loss in Black Holes and/or Conscious Beings?" to be published in *Heat Kernel Techniques and Quantum Gravity*, edited by S. A. Fulling (Discourses in Mathematics and Its Applications, No. 4, Texas A&M University Department of Mathematics, College Station, Texas, 1995) (University of Alberta report Alberta-Thy-36-94, Nov. 25, 1994), hep-th/9411193.

[22] D. N. Page, "Sensible Quantum Mechanics: Are Only Perceptions Probabilistic?" (University of Alberta report Alberta-Thy-05-95, June 7, 1995), quant-ph/9506010.

450

[23] D. N. Page, "Attaching Theories of Consciousness to Bohmian Quantum Mechanics," to be published in *Bohmian Quantum Mechanics and Quantum Theory: An Appraisal,* edited by J. T. Cushing, A. Fine, and S. Goldstein (Kluwer, Dordrecht, 1996) (University of Alberta report Alberta-Thy-12-95, June 30, 1995), quant-ph/9507006.

[24] D. N. Page, "Sensible Quantum Mechanics: Are Probabilities Only in the Mind?" to be published in *Proceedings of the Sixth Seminar on Quantum Gravity, June 12-19, 1987, Moscow, Russia,* edited by V. A. Berezin and V. A. Rubakov (World Scientific, Singapore, 1996) (University of Alberta report Alberta-Thy-13-95, July 4, 1995), gr-qc/9507024.

[25] E. B. Davies, *Quantum Theory of Open Systems* (Academic Press, London, 1976).

[26] M. Lockwood, *Mind, Brain and the Quantum: The Compound 'I'* (Basil Blackwell, Oxford, 1989).

[27] I. S. Gradshteyn and I. M. Ryzhik, *Table of Integrals, Series, and Products, Corrected and Enlarged Edition,* edited by A. Jeffrey (Academic Press, San Diego, 1980), item 6.284, p. 649.

[28] A. Vilenkin, Phys. Rev. Lett. **74**, 846 (1995); "Making Predictions in Eternally Inflating Universe," gr-qc/9505031; "Predictions from Quantum Cosmology" (Lectures at International School of Astrophysics 'D. Chalonge,' Erice, 1995).

[29] R. H. Dicke, Rev. Mod. Phys. **29**, 355 and 363 (1977); Nature **192** 440 (1961).

[30] B. Carter, in *Confrontation of Cosmological Theories with Observation,* edited by M. S. Longair (Reidel, Dordrecht, 1974), p. 291.

[31] B. J. Carr and M. J. Rees, Nature **278**, 605 (1979).

[32] I. L. Rozental, Sov. Phys. Usp. **23**, 296 (1980).

[33] P. C. W. Davies, *The Accidental Universe* Cambridge University Press, Cambridge, 1982).

[34] J. D. Barrow and F. T. Tipler, *the Anthropic Cosmological Principle* (Clarendon Press, Oxford, 1986).

[35] J. Leslie, Am. Phil. Quart., 141 (April 1982); Mind, 573 (October 1983); in *Current Issues in Teleology,* edited by N. Rescher (University Press of America, Lanham and London, 1983), p. 111; in *Proceedings of the Philosophy of Science Association 1986* (Edwards Bros, Ann Arbor, 1986), vol. 1, p. 87; in *Origin and Early History of the Universe,* edited by J. Demaret (University of Liège, Liège, 1987), p. 439; Mind, 269 (April 1988); *Universes* (Routledge, London and New York, 1989); *Physical Cosmology and Philosophy* (Macmillan, New York, 1990).

NEW ASPECTS OF REHEATING

D. Boyanovsky[(a)], M. D'Attanasio*[(b)(d)], H.J. de Vega[(b)],
R. Holman[(c)] and D.-S. Lee[(e)]

(a) Department of Physics and Astronomy, University of Pittsburgh, Pittsburgh, PA. 15260,
U.S.A.

(b) Laboratoire de Physique Théorique et Hautes Energies[†] Université Pierre et Marie Curie
(Paris VI) , Tour 16, 1er. étage, 4, Place Jussieu 75252 Paris cedex 05, France

(c) Department of Physics, Carnegie Mellon University, Pittsburgh, PA. 15213, U. S. A.

(d) I.N.F.N., Gruppo Collegato di Parma, Italy

(e) Department of Physics and Astronomy, University of North Carolina
Chapel Hill, N.C. 27599, U.S.A.

Abstract

The reheating stage in post-inflationary cosmologies is reanalyzed. New
techniques from non-equilibrium quantum field theory allow a consistent
derivation of the equation of motion including the non-linearity of the dynamics. These offer a rationale for the elementary theory of reheating based
on single particle decay which is seen to be valid only in the linear regime of
coherent oscillations of the scalar field.

A new non-perturbative mechanism of induced amplification of quantum
fluctuations is introduced and studied in detail, both analytically and numerically. This is a non-linear mechanism that is typically a far more efficient
way of transfering energy out of the zero mode and into production of lighter
particles than single particle decay. Thermalization is discussed and we estimate the reheating temperature to be of the order of the inflaton mass, thus
providing a potential solution to the Polonyi and moduli problems.

Typeset using REVTEX

*Della Riccia fellow

N. Sánchez and A. Zichichi (eds.), String Gravity and Physics at the Planck Energy Scale, 451–491.
© 1996 Kluwer Academic Publishers.

I. INTRODUCTION

In typical inflationary scenarios of the early universe, after the many e-folds of inflation necessary to solve the horizon and homogeneity problems, the matter and radiation energy density had been red-shifted to almost zero. However, once inflation has ended, this scenario has to merge with the standard big bang, radiation dominated cosmology. The succesful intertwining of inflation with the standard big bang cosmology thus necessitates some source of energy and entropy to rethermalize, or reheat the universe [1–3].

Models that reheat to temperatures above the GUT scale would have to cope with the monopole and domain wall and inhomogeneity problems associated with the breakdown of their symmetries. However there is a fundamental scale for a reheating temperature: nucleosynthesis, this requires reheating temperatures in excess of 10-100 Mev. If baryogenesis can occur at the electroweak scale then it would be imperative to have reheating temperatures larger than say a few hundred Gev (a baryon over antibaryon excess may be wiped out if this occurs during the fast inflationary era). Therefore, nucleosynthesis and particle physics restrict the reheating temperature but leave a wide window open for the acceptable values of the reheating temperatures: certainly larger than about 100 Mev, and desirably, larger than say about 1 Tev but not close to GUT scale.

The standard picture [2–4] of old and new inflation invokes a scalar field, the inflaton, that produces a phase transition (either second or first order) and that at the end of inflation the expectation value of its zero momentum mode oscillates about the minimum of its (effective) potential. In chaotic inflation [5], the inflaton energy density still drives an inflationary phase, but there need not be any phase transition for this to occur. This inflaton field is constrained [1–3] to have a mass a few orders of magnitude smaller than the Planck mass, but typically much larger than the masses of the particles involved in the particle physics models. In the standard view of inflationary models, the expectation value of the inflaton field oscillates around the minimum of its potential, and its couplings to the lighter particles allow the inflaton field to decay. This decay process is then supposed to induce a damping term in the evolution equation for the inflaton expectation value of the form $\Gamma\dot{\phi}$ with Γ being the total decay rate of the inflaton field [2,6,7]. The standard estimate for the reheating temperature based on single-particle decay [2] is then obtained by comparing the total decay rate of the particle (Γ) to that of the expansion, obtaining $T_r \simeq 0.1\sqrt{\Gamma M_{pl}}$ [2].

Recent investigations of the non-linear quantum dynamics of scalar fields reveal a variety of new and striking phenomena [8–14]. The main relevant implication for the reheating problem being that the particle production induced by the time evolution of the inflaton is significatively **different** from linear estimates.

The non-linear (quantum) effects lead to an extremely effective dissipational dynamics and particle production even in the simplest self-interacting scalar field theory [11,13] in which single particle decay is kinematically forbidden. In the case in which only the *classical* evolution of the expectation value is taken into account in the evolution of the quantum fluctuations, this mechanism corresponds to parametric amplification or parametric resonance, because the classical time evolution is periodic in time [15]. However when back-reaction effects are taken into account, particle production damps out the evolution of the scalar field, and this damped evolution is incorporated in the equations for the quantum fluctuations. In this case the time evolution of the scalar field *including* the dissipative effects is not periodic; this implies that the situation is far different from parametric amplification. We refer to this situation, which includes the back-reaction effects as **induced amplification**, to distinguish it from parametric resonant amplification [15].

Back-reaction effects [11,13] drastically change the picture of particle production. In the case of parametric amplification, particle production never shuts off and the total number of particles created is infinite in Minkowski. It is only when the back-reaction is incorporated that the damping effects on the expectation value feed back to the particle production mechanism, eventually shutting it off. This has been studied in detail analytically and numerically in [13] for a self-interacting scalar field.

Induced amplification is most effective when the amplitude of the expectation value of the scalar field (from the equilibrium value) is large and is, therefore, an intrinsically non-linear process, which we refer to as *non-linear relaxation*.

Particle production via parametric amplification of quantum fluctuations has been studied recently in the semiclassical approximation, but without back-reaction effects, in connection with fermion production by (pseudo) Nambu-Goldstone bosons [16].

Induced amplification and non-linear relaxation will be the primary mechanism for dissipation via particle production in the case of large amplitude of the scalar field. Such is the case in chaotic infation scenarios [5], as well as in the out of equilibrium regime during phase transitions for fields with very flat potentials near the maximum, which is typically required for a long inflationary phase and also occurs for the moduli fields in string theory [17]. In these inflationary models with light moduli fields, with masses typically of the order of a Tev, and couplings of order (Tev/M_{Pl}) the standard reheating theory predicts reheating temperatures of the order of 0.01Mev, well below nucleosynthesis temperatures, this is the Polonyi problem [17–20].

The non-linear mechanism of induced amplification of fluctuations and copious particle production in the large amplitude regime of the scalar field will necessarily modify the standard picture of reheating. The standard reheating scenario was primarily based on the premise that particle production only ocurred when the inflaton oscillates with small amplitude at the bottom of the (effective) potential. Understanding the time scales and non-linear processes of reheating and eventually thermalization acquires further importance with the (speculative) possibility that asymptotic oscillations of the inflaton field around the minima of the (effective) potential may still be present in today's universe in the form of dark matter [11,21].

Thermalization is a process that is fundamentally different from particle production. Typically, as will be seen below, the particles produced by the process of induced amplification will be in a non-thermal distribution [13]. Thermalization is a collisional process, in which the produced particles will exchange energy (and momentum) and eventually achieve a thermal distribution. In principle the time scales for dissipation via particle production may be widely different from the time scales of collisional thermalization. For very weakly coupled theories, if the distribution of produced particles via induced amplification is very far from thermal, many collisions will be necessary to thermalize the system and the time scale for thermalization may very well be large compared to the time scale of dissipation via particle production. This will have implications in cosmology discussed in the conclusions.

The process of thermalization, reheating and relaxation of perturbations, and the ensuing production of entropy is also very relevant in heavy ion collisions [22] and in phase transitions in particle physics, at the electroweak scale within the context of baryogenesis [23], as well as for the quark-gluon plasma, deconfining and chiral phase transitions [22,24].

In the linear relaxation case the important scale is determined by the decay rate Γ, usually referred to as the damping rate. The literature [25] usually identifies $\Gamma \approx \operatorname{Im} \Sigma(\omega, \vec{k})/\omega$ with the "thermalization rate". Here $\operatorname{Im} \Sigma(\omega, \vec{k})$ is the imaginary part of the self-energy. In this article we offer a *real time* critique of this relation by studying the real time evolution and relaxation of linearized perturbations, and point out the following observations:

1. The damping rate is identical to the imaginary part of the self energy *only* when the inflaton (the particle interpolated by the order parameter) is a resonance with Γ being its width. In the complex frequency plane this corresponds to a pole in the second (unphysical) Riemann sheet. Even in this case, this damping rate corresponds to the relaxation of the *expectation value of the scalar field* and is exponential for some time regime. However, eventually relaxation continues with a power law tail. When the imaginary part is zero on-shell and the one particle pole is below the multiparticle thresholds, relaxation is given by a power law and no damping rate can be associated with the imaginary part of the self energy. Moreover, the relation $\Gamma \approx \operatorname{Im} \Sigma(\omega, \vec{k})/\omega$ only holds in the **linear** approximation around equilibrium.

2. Even when relaxation of the expectation value of the scalar field is exponential, this damping rate determines the approach to equilibrium of the *expectation value* of the scalar field but it cannot be immediately inferred that the same time scale describes thermalization, i.e. the approach to a Bose Einstein distribution of an initial off-equilibrium distribution. Thermalization is a rather different process and has to be described with a Boltzmann equation with a collision term. Thus, rather than interpreting this time scale as a thermalization scale, we interpret it as a relaxation scale for the expectation value.

3. We consider the non-linear quantum field evolution (non-linear relaxation) incorporating the order parameter into the effective mass in a self-consistent way. [The self-consistency requirement makes the evolution equations non-linear]. We study the inflaton evolution in real time coupled both to light scalars and fermions, providing exhaustive numerical results. We find that the non-linear relaxation time scales can be **much shorter** than those predicted by linear relaxation. In addition, the particles are produced with a momentum distribution that is very far from thermal and skewed towards low momentum. In the case of fermions we provide numerical evidence for the phenomenon of Pauli blocking that hinders dissipation and production of fermion-antifermion pairs. This phenomenon is very similar to that found by Kluger and collaborators [26] in their studies of fermionic back-reaction in the presence of strong electric fields.

4. We find that the reheating temperature is in between the inflaton mass and the lighter scalar mass. The precise relation is argued to depend on the product of the coupling and the initial amplitude of the inflaton zero mode and also logarithmically on the coupling. The distribution of produced particles is very far from thermal and skewed towards momenta of the order of or smaller than the inflaton mass. We argue that

this implies fairly long thermalization times and a wide separation of the relevant time scales.

Both time scales will have to be understood in detail to provide a *quantitative* and reliable estimate of the reheating temperature.

Although we are ultimately interested in describing reheating and thermalization in Friedmann-Robertson-Walker cosmologies, in this article we will work in Minkowski space *assuming* that the relaxation time scales are much shorter than the expansion time scale. Furthermore, we want to isolate dissipative effects arising from particle production from those resulting from the red-shift in an expanding cosmology.

A more detailed account of the results presented here can be found in ref. [27].

In section II, we introduce the general model that we propose to study and summarize the necessary ingredients of non-equilibrium field theory to provide the reader with the essentials needed to reproduce our calculations, as these do not seem to be part of the standard techniques. In section III, we study linear relaxation in real time and elucidate the rôle of the imaginary parts of self energies, their interpretation in real time and criticize the identification with "thermalization rates". Here we provide a derivation of the reheating temperature based on single particle decay, valid in the linear regime of coherent oscillations of the scalar (inflaton) field. We also discuss the basics of the Polonyi problem.

In section IV we study the process of non-linear relaxation, and particle production both for scalars and fermions, providing exhaustive numerical results. Here we also discuss the issue of thermalization, and the important time scales associated with this collisional process. Under the assumption of thermalization rates larger than the expansion rate we provide an estimate of the reheating temperature based on the total number of produced particles.

Finally we conclude in section V with a discussion of the implications of our results, in particular we emphasize the possibility that the new mechanism for reheating may provide a solution to the Polonyi problem. An Appendix is devoted to a pedagogical exercise in non-equilibrium field theory.

II. THE MODEL AND THE METHODS

We consider the simplest model [2,3] where the inflaton field Φ couples to a scalar σ and to a fermion field ψ. That is,

$$\mathcal{L} = -\frac{1}{2}\Phi\left(\partial^2 + m_\Phi^2 + g\sigma^2\right)\Phi - \frac{\lambda_\Phi}{4!}\Phi^4 - \frac{1}{2}\sigma\left(\partial^2 + m_\sigma^2\right)\sigma - \frac{\lambda_\sigma}{4!}\sigma^4 + \bar{\psi}(i\slashed{\partial} - m_\psi - y\Phi)\psi .$$

$$(2.1)$$

The case $m_\sigma, m_\psi \ll m_\Phi$ will be of particular relevance, since in this case there are open decay channels for the inflaton.

We will investigate the scalar and fermionic couplings independently. Although the situation for reheating corresponds to (almost) zero temperature, we will study the case of linear relaxation at finite temperature. The reason for this is that finite temperature effects reflect the Bose enhancement and Pauli blocking factors that appear whenever there are excitations in the medium. These contributions from the medium will allow us to identify similar physical features in the case of non-linear relaxation.

A. Linear Relaxation: Amplitude and Perturbative expansion

To explore the behavior of the inflaton within the linear regime, we first use the tadpole method to obtain the equation of motion (A3) (see the Appendix for details on how this method is actually applied and reference [28] for an alternative implementation). The next step is to linearize this equation in the inflaton zero mode amplitude and use this to study the relaxational dynamics of the inflaton. We should note that this amplitude expansion is *a priori* different from the standard perturbative expansion in the relevant coupling constant. Later in this work, we will compare the results obtained here with results obtained through a self-consistent, non-perturbative resummation both in the coupling constants *and* the field amplitude.

We arrange the initial conditions to be such that the fields Φ, σ, ψ start to interact at a time that we choose $t = 0$. This can be accomplished by making the coupling "constants" time dependent, i.e. zero for $t < 0$ and different from zero for $t > 0$. To perform the calculations, we will need the non-equilibrium Green's functions and Feynman rules. Since the non-equilibrium generating functionals involve a forward and backward time contour [8,25,29], the number of vertices is doubled. Those in which all the fields are on the forward branch (fields labeled by $(+)$) are the usual interaction vertices, while those in which the fields are on the backward branch (fields labeled by $(-)$) have the opposite sign. The combinatoric factors are the same as in usual field theory. The spatial Fourier transform of the necessary finite (initial) temperature propagators are:

- Bosonic Propagators

$$G_k^{++}(t,t') = G_k^>(t,t')\Theta(t-t') + G_k^<(t,t')\Theta(t'-t) ,$$
$$G_k^{--}(t,t') = G_k^>(t,t')\Theta(t'-t) + G_k^<(t,t')\Theta(t-t') ,$$
$$G_k^{+-}(t,t') = -G_k^<(t,t') ,$$
$$G_k^{-+}(t,t') = -G_k^>(t,t') , \qquad (2.2)$$
$$G_k^>(t,t') = i \int d^3x \, e^{-i\vec{k}\cdot\vec{x}} \, \langle \Phi(\vec{x},t)\Phi(\vec{0},t')\rangle$$
$$= \frac{i}{2\omega_k}\left\{ [1 + n_b(\omega_k)]e^{-i\omega_k(t-t')} + n_b(\omega_k)e^{i\omega_k(t-t')}\right\} ,$$
$$G_k^<(t,t') = i \int d^3x \, e^{-i\vec{k}\cdot\vec{x}} \, \langle \Phi(\vec{0},t')\Phi(\vec{x},t)\rangle$$
$$= \frac{i}{2\omega_k}\left\{ [1 + n_b(\omega_k)]e^{i\omega_k(t-t')} + n_b(\omega_k)e^{-i\omega_k(t-t')}\right\} ,$$
$$\omega_k = \sqrt{\vec{k}^2 + m^2} , \qquad n_b(\omega_k) = \frac{1}{e^{\beta\omega_k} - 1} ,$$

where m is the mass of the boson.

- Fermionic Propagators (Zero chemical potential)

$$S_{\vec{k}}^{++}(t,t') = S_{\vec{k}}^>(t,t')\Theta(t-t') + S_{\vec{k}}^<(t,t')\Theta(t'-t) ,$$
$$S_{\vec{k}}^{--}(t,t') = S_{\vec{k}}^>(t,t')\Theta(t'-t) + S_{\vec{k}}^<(t,t')\Theta(t-t') ,$$
$$S_{\vec{k}}^{+-}(t,t') = -S_{\vec{k}}^<(t,t') ,$$

$$S_{\vec{k}}^{-+}(t,t') = -S_{\vec{k}}^{>}(t,t') \, ,$$

$$S_{\vec{k}}^{>}(t,t') = -i \int d^3x \; e^{-i\vec{k}\cdot\vec{x}} \, \langle \psi(\vec{x},t)\bar{\psi}(\vec{0},t') \rangle \tag{2.3}$$

$$= -\frac{i}{2\omega_k} \left[e^{-i\omega_k(t-t')}(\not{k}+m_\psi)(1-n_f(\omega_k)) + e^{i\omega_k(t-t')}\gamma_0(\not{k}-m_\psi)\gamma_0 n_f(\omega_k) \right] \, ,$$

$$S_{\vec{k}}^{<}(t,t') = i \int d^3x \; e^{-i\vec{k}\cdot\vec{x}} \, \langle \bar{\psi}(\vec{0},t')\psi(\vec{x},t) \rangle$$

$$= \frac{i}{2\omega_k} \left[e^{-i\omega_k(t-t')}(\not{k}+m_\psi)n_f(\omega_k) + e^{-i\omega_k(t-t')}\gamma_0(\not{k}-m_\psi)\gamma_0(1-n_f(\omega_k)) \right] \, ,$$

$$\omega_k = \sqrt{\vec{k}^2 + m_\psi^2} \, , \qquad n_f(\omega_k) = \frac{1}{e^{\beta\omega_k}+1} \, .$$

In the linear amplitude approximation, corresponding to linear relaxation, we find in all cases the following form of the equation of motion for the expectation value (see Appendix)

$$\ddot{\delta}_{\vec{p}}(t) + \Omega_{\vec{p}}^2 \, \delta_{\vec{p}}(t) + \int_0^\infty K_{\vec{p}}(t-t') \, \delta_{\vec{p}}(t') \, dt' = 0 \, , \tag{2.4}$$

$$K_{\vec{p}}(t-t') = \Sigma_{r,\vec{p}}(t-t')\Theta(t-t') \, ,$$

where we have imposed as boundary conditions that the inflaton and the other fields are coupled at time $t = 0$ but uncoupled for previous times, and introduced

$$\delta_{\vec{p}}(t) = \int d^3x \; e^{-i\vec{p}\cdot\vec{x}} \, \phi(\vec{x},t) \, , \quad \Omega_{\vec{p}}^2 = \vec{p}^2 + m_\Phi^2 + \delta m(T) \, , \tag{2.5}$$

where $\delta m(T)$ is the time independent (but temperature dependent) contribution from tadpole diagrams. These contributions renormalize the mass and introduce a temperature dependent effective mass and will be specified later in each particular case. The quantity $\Sigma_{r,\vec{p}}(t-t')\Theta(t-t')$ is the retarded self-energy. It will be computed to dominant order in the couplings for both fermions and bosons. Although the self-energy at finite temperature has been computed before in the literature [30,31], we differ from previous treatments in that we perform our calculations directly in real time, this allows us to study real time relaxation as an initial condition problem.

Eq. (2.4) can be solved by Laplace transform. Define

$$\varphi_{\vec{p}}(s) = \int_0^\infty e^{-st} \, \delta_{\vec{p}}(t) \, dt \, . \tag{2.6}$$

Then, eq. (2.4) becomes

$$s^2\varphi_{\vec{p}}(s) - s\delta_{\vec{p}}(0) - \dot{\delta}_{\vec{p}}(0) + \Omega_{\vec{p}}^2 \, \varphi_{\vec{p}}(s) + \varphi_{\vec{p}}(s) \, \Sigma_{\vec{p}}(s) = 0 \, , \tag{2.7}$$

with $\Sigma_{\vec{p}}(s)$ the Laplace transform of $\Sigma_{r,\vec{p}}(t)$.

For computational purposes, it can be shown that at zero temperature each graph of $\Sigma_{\vec{p}}(s)$ is exactly equal to the corresponding graph of the zero-temperature equilibrium Euclidean quantum field theory, s being the time component of the Euclidean four momentum.

In general $\Sigma_{\vec{p}}(s)$ can be written as a dispersion integral in terms of the spectral density $\rho(p_o,\vec{p},T)$

$$\Sigma_{\vec{p}}(s) = - \int \frac{2 \, p_o \, \rho(p_o, \vec{p}, T)}{s^2 + p_o^2} dp_o \ . \tag{2.8}$$

The imaginary part of the self-energy is found to be

$$\mathrm{Im} \, \Sigma_{\vec{p}}(s = i\omega \pm 0^+) = \pm \Sigma_{I\vec{p}}(\omega) \ ,$$
$$\Sigma_{I\vec{p}}(\omega) = \pi \, \mathrm{sign} \, (\omega) \left[\rho(|\omega|, \vec{p}, T) - \rho(-|\omega|, \vec{p}, T) \right] \ . \tag{2.9}$$

The presence of sign (ω) in the above expression characterizes the *retarded* self-energy.

Let us choose $\delta_{\vec{p}}(0) = \delta_i$, $\dot{\delta}_{\vec{p}}(0) = 0$ for simplicity. We get from eq. (2.7)

$$\varphi_{\vec{p}}(s) = \delta_i \, \frac{s}{s^2 + \Omega_{\vec{p}}^2 + \Sigma_{\vec{p}}(s)} \ . \tag{2.10}$$

The Laplace transform can be inverted through the formula

$$\delta_{\vec{p}}(t) = \int_{-i\infty+\epsilon}^{+i\infty+\epsilon} e^{st} \, \varphi_{\vec{p}}(s) \, \frac{ds}{2\pi i} \ , \tag{2.11}$$

where ϵ is a positive real constant (Bromwich countour). Thus we need to understand the singularities of $[s^2 + \Omega_{\vec{p}}^2 + \Sigma_{\vec{p}}(s)]^{-1}$. We now consider the following cases:

- The inflaton potential admits spontaneous symmetry breaking (SSB), and is only coupled to lighter *scalar* fields.

- The inflaton is coupled to fermions only.

We will also consider the subcases where the initial temperature is taken to be zero, as would be appropriate in the case of evolution in the post inflationary universe, as well as the situation where the initial temperature is non-zero, which would be relevant to the situation of a scalar field starting in an initial (non-equilibrium) but thermal state and evolving out of it.

B. SSB with Coupling to Lighter Scalars Only

If $m_\Phi^2 = -\mu^2 < 0$, then the new minimum is at $\Phi_0 = \sqrt{6\mu^2/\lambda_\Phi}$, and we write $\Phi^\pm(\vec{x}, t) = \Phi_0 + \phi(\vec{x}, t) + \chi^\pm(\vec{x}, t)$. The masses are now shifted to

$$M^2 = 2\mu^2 \ , \qquad M_\sigma^2 = m_\sigma^2 + g^2 \Phi_0^2 \ . \tag{2.12}$$

The contribution from the quartic inflaton self-coupling has been studied previously [13] thus it will not be repeated here.

The tadpole correction to the mass in eq. (2.5) is given by

$$\delta M(T) = -ig \int \frac{d^3k}{(2\pi)^3} G_{k,\sigma}^{++}(t,t) = g \int \frac{d^3k}{(2\pi)^3} \frac{1 + 2 \, n_b(\omega_k)}{2\omega_k} \ ,$$
$$\omega_k = \sqrt{\vec{k}^2 + M_\sigma^2} \ .$$

This mass renormalization does not influence the dynamics. The retarded self energy is found to be at order g^2,

$$K_{\vec{p}}(t - t') = 2ig^2\Phi_0^2 \int \frac{d^3k}{(2\pi)^3} \left[G_{\vec{k},\sigma}^{++}(t - t')G_{\vec{k}+\vec{p},\sigma}^{++}(t - t') - G_{\vec{k},\sigma}^{<}(t - t')G_{\vec{k}+\vec{p},\sigma}^{<}(t - t')\right]. \quad (2.13)$$

With the non-equilibrium Green's functions defined above, we find

$$\Sigma_{r,\vec{p}}(t - t') = -2g^2\Phi_0^2 \int \frac{d^3k}{(2\pi)^3} \frac{1}{2\omega_{\vec{k}}\omega_{\vec{k}+\vec{p}}} \left\{[1 + 2\,n_b(\omega_k)]\sin[(\omega_{\vec{k}} + \omega_{\vec{k}+\vec{p}})(t - t')] \right.$$
$$\left. -2\,n_b(\omega_k)\sin[(\omega_{\vec{k}} - \omega_{\vec{k}+\vec{p}})(t - t')]\right\}. \quad (2.14)$$

The Laplace transform can be written as a dispersion integral in terms of the bosonic spectral density $\rho_b(p_o, \vec{p}, T)$ (see eq. (2.8)). We have to one loop level:

$$\rho_b(p_o, \vec{p}, T) = 2g^2\Phi_0^2 \int \frac{d^3k}{(2\pi)^3 2\omega_k} \int \frac{d^3k'}{(2\pi)^3 2\omega_{k'}} (2\pi)^3\delta^3(\vec{p} - \vec{k} - \vec{k}')$$
$$\times \left[\delta(p_o - \omega_k - \omega_{k'})(1 + 2\,n_b(\omega_k)) - \delta(p_o - \omega_k + \omega_{k'})\,2\,n_b(\omega_k)\right]. $$

The imaginary part of the self-energy is given by eq. (2.9).

We will analyze only the case of a spatially constant order parameter corresponding to $\vec{p} = 0$ because we will later compare with the case of non-linear relaxation which we only study for a homogeneous (translational invariant) expectation value. In this case the spectral density can be written as

$$\rho_b(p_o, \vec{0}, T) = \left[1 + 2\,n_b\left(\frac{p_o}{2}\right)\right]\rho_b(p_o, \vec{0}, 0). \quad (2.15)$$

The spectral density $\rho_b(p_o, \vec{0}, 0)$ is a Lorentz scalar and is proportional to the decay rate of the boson Φ into two σ particles. It is a straightforward exercise to find

$$\rho_b(p_o, \vec{0}, 0) = \frac{g^2\Phi_0^2}{8\pi^2}\left[1 - \frac{4M_\sigma^2}{p_o^2}\right]^{\frac{1}{2}}\Theta\left(p_o^2 - 4M_\sigma^2\right). \quad (2.16)$$

Clearly $\Sigma_{\vec{0}}(s)$ has a logarithmic divergence, independent of s and T. We choose to subtract this divergence at $s = 0$ and absorb the subtraction into a further temperature dependent renormalization of M. In order not to clutter notation, from now on we will refer to M as the fully renormalized mass including the above subtraction, and to $\Sigma(s) \equiv \Sigma_{\vec{0}}(s) - \Sigma_{\vec{0}}(0)$. From equations (2.8, 2.15, 2.16) we find that the self-energy has an imaginary part above the two σ-particles threshold given by eq. (2.9) with

$$\Sigma_I(\omega) = \frac{g^2\Phi_0^2}{8\pi}\left[1 - \frac{4M_\sigma^2}{\omega^2}\right]^{\frac{1}{2}}\left[1 + 2\,n_b\left(\frac{\omega}{2}\right)\right]\Theta\left(\omega^2 - 4M_\sigma^2\right)\,\text{sign}\,(\omega). \quad (2.17)$$

This expression is recognized as the imaginary part of the retarded self-energy (determined by the sign(ω)) and shows the usual Bose enhancement factor [32].

1. Zero Temperature

In this case from 2.14 we find

$$\Sigma(s) = \frac{g^2 \Phi_0^2}{4\pi^2} \left(\sqrt{1 + \frac{4M_\sigma^2}{s^2}} \, \text{ArgTh} \, \frac{1}{\sqrt{1 + \frac{4M_\sigma^2}{s^2}}} - 1 \right) . \tag{2.18}$$

In order to compute the inverse Laplace transform through eq. (2.11) we must first study the analytic structure of $\varphi(s)$ in the s-plane. $\varphi(s)$ has poles at the zeroes of

$$s^2 + M^2 + \Sigma(s) = 0 . \tag{2.19}$$

These correspond to Φ-one-particle states with the mass including one-loop radiative corrections. At zeroth order the poles are purely imaginary

$$s_\pm = \pm iM . \tag{2.20}$$

To find the one-loop correction, we set

$$s_+ = iM + r \tag{2.21}$$

and similarly for s_-. Inserting this in eq. (2.19) yields to order g^2

$$2iMr + \Sigma(iM) = 0 . \tag{2.22}$$

That is,

$$r = i\frac{\Sigma(iM)}{2M} . \tag{2.23}$$

When $M < 2M_\sigma$, $\Sigma(iM)$ is real (see eq. (2.18)) and eq. (2.23) gives a real correction to the Φ mass

$$s_\pm = \pm iM_0 \equiv \pm i \left[M + \frac{g^2 \Phi_0^2}{8\pi^2 M} \left(\sqrt{\frac{4M_\sigma^2}{M^2} - 1} \, \text{ArgTh} \, \frac{1}{\sqrt{\frac{4M_\sigma^2}{M^2} - 1}} - 1 \right) \right] . \tag{2.24}$$

The Laplace transform $\varphi(s)$ also exibits a cut along the imaginary axis starting at $s = i\omega = \pm 2iM_\sigma$. For s in the first Riemann sheet (physical sheet) we obtain

$$\Sigma_{\text{physical}}(i\omega \pm 0^+) = \Sigma_R(\omega) \pm i\Sigma_I(\omega) , \qquad \omega > 2M_\sigma , \tag{2.25}$$

with Σ_R and Σ_I both real and given by

$$\Sigma_R(\omega) = \frac{g^2 \Phi_0^2}{4\pi^2} \left(\sqrt{1 - \frac{4M_\sigma^2}{\omega^2}} \, \text{ArgTh} \, \sqrt{1 - \frac{4M_\sigma^2}{\omega^2}} - 1 \right) , \qquad \Sigma_I(\omega) = \frac{g^2 \Phi_0^2}{8\pi} \sqrt{1 - \frac{4M_\sigma^2}{\omega^2}} > 0 . \tag{2.26}$$

We can now proceed to compute the inverse Laplace transform (2.11) by deforming the contour.

$$\delta(t) = \frac{\delta_i \cos M_0 t}{1 - \frac{\partial \Sigma(iM)}{\partial M^2}} + \frac{2\delta_i}{\pi} \int_{2M_\sigma}^{\infty} \frac{\omega \Sigma_I(\omega) \cos \omega t \, d\omega}{[\omega^2 - M^2 - \Sigma_R(\omega)]^2 + \Sigma_I(\omega)^2} \, . \tag{2.27}$$

For large time $M_\sigma t \gg 1$ the integral over the cut is dominated by the endpoint $\omega = 2M_\sigma$ and goes to zero as

$$\delta_{\text{cut}}(t) \simeq \frac{\delta_i \sqrt{\pi} \, M_\sigma^2 \, g^2 \Phi_0^2}{4\pi^2 (M^2 - 4M_\sigma^2 + \frac{g^2 \Phi_0^2}{4\pi^2})^2} \frac{\cos(2M_\sigma t + \frac{3\pi}{4})}{(M_\sigma t)^{\frac{3}{2}}} \, . \tag{2.28}$$

The $t^{-3/2}$ is completely determined by the behavior of the spectral density at threshold.

The situation changes drastically for $M > 2M_\sigma$. In such case the Φ-particle is unstable and thus $\Sigma(iM)$ becomes complex and its value depends from which side of the cut we approach the imaginary axis. Now the solution for the pole will be complex and r will acquire a *real* part. Due to the discontinuity in Σ_I across the two-particle cut, eq. (2.23) must be written carefully as

$$r = \frac{i}{2M} \Sigma(iM + \text{Re}\,[r]) \, , \tag{2.29}$$

where the (small) real correction inside the argument will determine on which side of the cut the solution resides. Then the real part of r should satisfy

$$\text{Re}\,r = -\frac{1}{2M} \text{Im}\,\Sigma(iM + 0^+ \text{sign}\,(\text{Re}\,[r])) \, . \tag{2.30}$$

From eq. (2.17) we see that $\text{Im}\,\Sigma_{\text{physical}}$ is negative for $\text{sign}\,(\text{Re}\,[r]) < 0$ and positive for $\text{sign}\,(\text{Re}\,[r]) > 0$. Therefore eq. (2.30) *has no solution* in the physical sheet.

The analytic continuation of Σ into the second Riemann sheet is such that [33]

$$\Sigma^{II}(i\omega \pm 0^+) = \Sigma_R(\omega) \mp i\Sigma_I(\omega) \, , \qquad \omega > 2M_\sigma \tag{2.31}$$

and we find the solution

$$\text{Re}\,r = -\frac{1}{2M}\Sigma_I(M) = -\frac{g^2 \Phi_0^2}{16\pi M}\sqrt{1 - \frac{4M_\sigma^2}{M^2}} < 0 \, ,$$

$$\text{Im}\,r = \frac{1}{2M}\Sigma_R(M) = \frac{g^2 \Phi_0^2}{8\pi^2 M}\left(\sqrt{1 - \frac{4M_\sigma^2}{M^2}}\,\text{ArgTh}\sqrt{1 - \frac{4M_\sigma^2}{M^2}} - 1\right) \, .$$

For $M \gg M_\sigma$,

$$\text{Im}\,r = \frac{g^2 \Phi_0^2}{8\pi^2 M}\left(\log\frac{M}{M_\sigma} - 1\right) \, . \tag{2.32}$$

$|\text{Re}\,[r]|$ coincides with the decay rate $\Phi \to 2\sigma$ (as it must be) which is the rate per unit time to produce σ particles. The poles $s_\pm$ move off into the second sheet when M becomes larger than $2M_\sigma$ as expected [33].

Thus when we compute the inverse Laplace transform (2.11) in the unstable case we are left with the integral over the cuts since both poles are in the second Riemann sheet and we find

$$\delta(t) = \frac{2\delta_i}{\pi} \int_{2M_\sigma}^\infty \frac{\omega \Sigma_I(\omega) \cos \omega t \, d\omega}{[\omega^2 - M^2 - \Sigma_R(\omega)]^2 + \Sigma_I(\omega)^2} . \tag{2.33}$$

Since now M is inside the integration region, for weak coupling there is a narrow resonance at $\omega \simeq M$. Thus for weak coupling it takes Breit-Wigner form and we find to a very good approximation,

$$\delta(t) \simeq \delta_i \, A \, e^{-\Gamma t/2} \, \cos(Mt + \alpha) , \qquad \Gamma \ll M , \tag{2.34}$$

where

$$A = 1 + \frac{\partial \Sigma_R(M)}{\partial M^2} , \qquad \Gamma = \frac{g^2 \Phi_0^2}{8\pi M} \sqrt{1 - \frac{4M_\sigma^2}{M^2}} , \qquad \alpha = -\frac{\partial \Sigma_I(M)}{\partial M^2} . \tag{2.35}$$

For $M \gg M_\sigma$,

$$A = 1 + \frac{g^2 \Phi_0^2}{8\pi^2 M^2} + \mathcal{O}\left(\left[\frac{M_\sigma}{M}\right]^2\right) , \qquad \alpha = \frac{g^2 \Phi_0^2 M_\sigma^2}{4\pi M^4} + \mathcal{O}\left(\left[\frac{M_\sigma}{M}\right]^4\right) . \tag{2.36}$$

The Breit-Wigner approximation, however, is valid only for times $\leq \Gamma^{-1} \ln(\Gamma/M_\sigma)$; for longer times the fall off is with a power law $t^{-3/2}$ determined by the spectral density at threshold as before.

2. Non-Zero Temperature

The physical mass gets a finite temperature dependent correction from the tadpole contribution $\delta M(T)$ given by

$$\delta M(T) = g \int_0^\infty \frac{dk}{2\pi^2} \frac{k^2}{\sqrt{k^2 + M_\sigma^2}} \frac{1}{e^{\beta\sqrt{k^2+M_\sigma^2}} - 1} . \tag{2.37}$$

For small β ($T \gg M_\sigma$) this correction takes the form

$$\delta M(T) = g \left\{ \frac{T}{12} - \frac{M_\sigma T}{4\pi} + \frac{M_\sigma^2}{8\pi} \log(\frac{T}{M_\sigma}) + \mathcal{O}(T^0) \right\} . \tag{2.38}$$

We will assume that $gT^2 \ll M^2$ so that we are in the perturbative regime, since otherwise hard thermal loops must be resummed [34,35], a task beyond the scope of this article.

The imaginary part of the self-energy is given in eq. (2.17) and the real part can be obtained from the dispersion integral (2.8) using (2.15, 2.16). The real part of the self-energy is difficult to compute for arbitrary temperature, and we just quote its large T behavior:

$$\Sigma(s,\beta) = g^2 \Phi_0^2 \frac{M_\sigma T}{\pi s^2} \left[1 - \sqrt{1 + \frac{s^2}{4M_\sigma^2}} \right] + \mathcal{O}(T^0) . \tag{2.39}$$

It is interesting to compute $\Sigma(t,\beta)$ in configuration space for large T. We have by inverse Laplace transform

$$\Sigma(t,\beta) = \int_{-i\infty+\epsilon}^{+i\infty+\epsilon} e^{st}\, \Sigma(s,\beta)\, \frac{ds}{2\pi i}\,. \tag{2.40}$$

Upon contour deformation we find

$$\Sigma(t,\beta) = -\frac{g^2\Phi_0^2 T}{\pi^2} \int_1^\infty \frac{dx}{x}\, \sqrt{x^2-1}\,\sin(2M_\sigma xt)\,. \tag{2.41}$$

This function is obviously **not** concentrated at $t = 0$. It can be related to a J_1 Bessel function as

$$\Sigma(t,\beta) = -\frac{g^2\Phi_0^2 T}{2\pi}\left[1 - \frac{4M_\sigma}{\pi}t - 2\,M_\sigma \int_0^t dx\,\left(\frac{t}{x}-1\right)\,J_1(2\,M_\sigma x)\right] \tag{2.42}$$

We find for small and for large t

$$\Sigma(t,\beta) \overset{t\to 0}{=} -\frac{g^2\Phi_0^2 T}{2\pi}\left[1 - \frac{4M_\sigma}{\pi}t + O(t^2)\right]$$

$$\Sigma(t,\beta) \overset{t\to\infty}{=} \frac{g^2\Phi_0^2 T}{2\pi^2}\frac{\sin(2M_\sigma t - \pi/4)}{(M_\sigma t)^{3/2}}\left[1 + O(t^{-1})\right]\,.$$

The behaviour of the kernel $\Sigma(t,\beta)$ shows that it **cannot** be approximated by a phenomenological term $\Gamma\frac{d}{dt}$ even for high T. (As shown in ref. [13] this is not the case either for $T = 0$).

We can now repeat the analysis of the previous section for both cases $M < 2M_\sigma$ and $M > 2M_\sigma$. In the first case, the one particle pole is below the two σ-particles threshold with a (small) finite temperature correction since we are restricted to the perturbative regime in which $gT^2 \ll M$. The long time behavior will be oscillatory with the frequency corresponding to the one particle pole plus long-time power law tails similar to the zero temperature case. The second case is more interesting, since now the scalar "inflaton" is unstable, and the pole moves off into the second Riemann sheet. In the physical sheet there is a resonance with a finite temperature width given by

$$\Gamma(T) = \Gamma(0)\left[1 + 2\,n_b\left(\frac{M}{2}\right)\right]\,, \qquad \Gamma(0) = \frac{g^2\Phi_0^2}{8\pi M}\sqrt{1 - \frac{4M_\sigma^2}{M^2}}\,. \tag{2.43}$$

The Bose enhancement factor increases the rate and therefore enhances dissipation via the production of particles. Although this factor arises from the thermal distribution, we expect in general that whenever there are bosonic excitations present, the relaxation rate will be enhanced as a consequence of bose statistics, independently of whether these excitations are thermally distributed.

C. Unbroken symmetry with coupling to light scalars

In the unbroken symmetry case M_Φ is above all thresholds and hence Φ is a stable particle. The order parameter is again given by a formula like eq. (2.27) except that now the integration starts at $\omega = M + 2M_\sigma$.

The first perturbative contribution to the kernel $\Sigma(t-t')$ in the inflaton equation of motion (2.4) are now the two loop order graphs usually called "setting sun". Since these graphs are quite complicated, we only perform the two loop computation at zero temperature, for which we can exploit the relationship with usual Euclidean field theory. Even with this simplification the computation is hard, and we limit ourselves to the evaluation of the imaginary part of the retarded self-energy near the branch point. We have seen below eq. (2.27) how this actually determines the long time behaviour of the order parameter. In general, if $\Sigma_I(\omega \to \omega_{\text{threshold}})$ vanishes as $(\omega - \omega_{\text{threshold}})^\alpha$, then $\delta_{\text{cut}}(t)$ decays as $t^{-1-\alpha}$ for large times.

We find for the two loops self-energy

$$\Sigma_I(\omega \to M + 2M_\sigma) \simeq \frac{2g^2\pi^2}{(4\pi)^4} \frac{M_\sigma\sqrt{M}}{(M + 2M_\sigma)^{7/2}} [\omega^2 - (M + 2M_\sigma)^2]^2 . \tag{2.44}$$

This yields for $\delta_{\text{cut}}(t)$ a power decay t^{-3} for long times.

For massless σ we find

$$\Sigma_I(\omega \to M) \simeq \frac{g^2\pi}{3M^4(4\pi)^4}(\omega^2 - M^2)^3 , \tag{2.45}$$

yielding a $\delta_{\text{cut}}(t)$ decaying as t^{-4} for long times. In summary, in the unbroken symmetry case $\delta(t)$ is given by the one-particle term oscillating with frequency M_0 plus the power damped cut contribution $\delta_{\text{cut}}(t)$.

D. Inflaton coupled to Fermions

Fermions can be treated similarly in the broken and unbroken symmetry phase. To treat both cases on equal footing we now define M as mass of the inflaton (scalar) field and m as the fermion mass in either case. Using the Feynman rules described in the first section, we obtain to one-loop level the kernel

$$K_{\vec{p}}(t - t') = iy^2 \int \frac{d^3k}{(2\pi)^3} \text{Tr} \left[iS_{\vec{k}}^{++}(t, t')iS_{\vec{k}-\vec{p}}^{++}(t', t) - iS_{\vec{k}}^<(t, t')iS_{\vec{k}-\vec{p}}^>(t', t) \right] ,$$

$$\Sigma_{r,\vec{p}}(t - t') = iy^2 \int \frac{d^3k}{(2\pi)^3} \text{Tr} \left[iS_{\vec{k}}^>(t, t')iS_{\vec{k}-\vec{p}}^<(t', t) - iS_{\vec{k}}^<(t, t')iS_{\vec{k}-\vec{p}}^>(t', t) \right] .$$

We now concentrate on the homogeneous case $\vec{p} = 0$. Using the fermionic Green's functions given in the first section it is straightforward to find the fermionic spectral density to be used in eq.(2.8)

$$\rho_f(p_o, T) = \frac{y^2}{8\pi^2} p_o^2 \left[1 - 2n_f\left(\frac{p_o}{2}\right)\right] \left[1 - \frac{4m^2}{p_o^2}\right]^{\frac{3}{2}} \Theta\left(p_o^2 - 4m^2\right) .$$

From eq.(2.9), the imaginary part of the self energy is then

$$\Sigma_I(\omega) = \frac{y^2}{8\pi}\omega^2 \left[1 - \frac{4m^2}{\omega^2}\right]^{\frac{3}{2}} \left[1 - 2n_f\left(\frac{\omega}{2}\right)\right] \Theta\left(\omega^2 - 4m^2\right) .$$

The finite temperature factor reflects the Pauli blocking term [32]. It is clear that the zero temperature part of $\Sigma(s)$ diverges quadratically and that two subtractions are needed. The first one is independent of s and contributes a (quadratically divergent) mass renormalization. The second one is logarithmic divergent and consists of an s independent term that adds to the mass renormalization and another proportional to s^2 that will be absorbed in wave function renormalization. We choose to subtract at zero temperature and at an arbitrary scale κ. The Laplace transform for the zero momentum component of the equation of motion (2.10) becomes

$$\varphi(s) = \frac{\delta_i \, s}{s^2 \left(1 + \frac{y^2}{4\pi^2} \ln \frac{\Lambda}{\kappa}\right) + M_{1R}^2(T) + y^2 \, \Pi\left(s, T, \kappa\right)} \,, \tag{2.46}$$

where M_{1R}^2 contains the mass renormalization and Π is the twice subtracted kernel, which depends on the renormalization scale. Defining

$$Z_\phi^{-1}(\kappa) = 1 + \frac{y^2}{4\pi^2} \ln \frac{\Lambda}{\kappa} \,,$$

$$y_R(\kappa) = Z_\phi^{\frac{1}{2}} \, y \,,$$

$$M_R(T, \kappa) = Z_\phi^{\frac{1}{2}} \, M_{1R}(T) \,,$$

$$\varphi_R(s, \kappa) = Z_\phi^{-1} \, \varphi(s) \,,$$

we finally obtain the renormalized Laplace transform

$$\varphi_R(s, \kappa) = \frac{\delta_i \, s}{s^2 + M_R^2(T, \kappa) + y_R^2(\kappa) \, \Pi\left(s, T, \kappa\right)} \,. \tag{2.47}$$

The inverse Laplace transform is obtained as in equation (2.11). The result will be the function $\delta_{R,\vec{\sigma}}(t, \kappa)$ which is not a renormalization group invariant. It is clear, however, from the renormalization prescriptions described above that ratios of the amplitude at different times, such as $\delta_{R,\vec{\sigma}}(t, \kappa)/\delta_{R,\vec{\sigma}}(0, \kappa)$ are renormalization group invariant. Now the analysis can proceed as in the previous section. To obtain the inverse Laplace transform we must recognize the singularities in (2.47). If $M_R < 2m$ (m is the fermion mass in the loop), there is a one particle pole (with strength different from one because we decided to renormalize off-shell) and a two-fermion cut at $s^2 = -4m^2$. At long times the amplitude will oscillate with an oscillation period given by the position of the pole, which is perturbatively close to M_R. The contribution of the cut falls-off at long times as $t^{-5/2}$ and is determined by the behavior of the spectral density near the two-fermion threshold.

More interesting is the case in which $M_R > 2m$. As in the bosonic case, the pole moves off the physical sheet into the second sheet. In the physical sheet the spectral density at weak coupling features a sharp peak at $M_R + \mathcal{O}(y^2)$ with width

$$\Gamma(T) = \Gamma(0) \left[1 - 2n_f\left(\frac{M_R}{2}\right)\right] \,, \qquad \Gamma(0) = \frac{y^2}{8\pi} M_R \left[1 - \frac{4m^2}{M_R^2}\right]^{\frac{3}{2}} \,. \tag{2.48}$$

A Breit-Wigner approximation predicts exponential relaxation but eventually at long times (an estimate similar to the bosonic case) a power law relaxation $t^{-5/2}$ ensues, completely determined by the spectral density at threshold.

Pauli blocking makes the resonance narrower and the lifetime of the decaying particle longer. The interpretation of this phenomenon is simple. In the thermal bath, fermionic excited states are filled with the Fermi-Dirac distribution. In order for the scalar field to decay, it must create a fermion-antifermion pair. However, at finite temperature, the available states are already filled with thermal excitations and because of the Pauli exclusion principle, are not available. At infinite temperature (and zero chemical potential), each fermion and antifermion state are populated with occupation 1/2 per spin degree of freedom; in this limit the decay rate goes to zero, the lifetime to infinity and the bosonic particle simply cannot decay because there are no states available to decay into. Even at zero temperature, but in a situation in which excited states are occupied, dissipative processes mediated by the production of fermion-antifermion pairs will be hindered by the Pauli exclusion principle, since states will already be occupied and no longer available in the particle production process. In highly excited states, we expect damping via production of fermion pairs to be strongly suppressed by Pauli blocking. This phenomenon has been seen numerically in the case of fermion pair production in strong electric fields [26] and will be seen numerically in the non-linear relaxation case later.

E. Thermalization or Relaxation?

From the analysis presented above, the following conclusions for the linear regime become very clear:

1. The imaginary part of the self-energy only determines a *relaxation rate* in the case of a **resonance**, that is when there is an imaginary part on-shell for the external particle and the (quasi) particle pole moves off the physical sheet into the second (unphysical) sheet. In this case a Breit-Wigner approximation describes the exponential relaxation for a long time, but eventually the amplitude falls-off with a power law in time, with the power determined by the behavior of the spectral density at threshold. When the on-shell pole is below the two-particle threshold, the imaginary part does not translate to a damping rate. Relaxation is described by a power law with an asymptotic behavior completely determined by the position and residue of the pole.

2. Even in the case of a resonance and exponential damping, the "damping rate" Γ describes exponential relaxation for the *expectation value* of the scalar field. The issue of thermalization is completely different. Thermalization corresponds to the time evolution of the (quasi) particle distribution function towards a thermal distribution which, in principle, has nothing to do with the relaxation of the expectation value of the field.

An alternative way to look at thermalization is as a process of momentum and energy transfer. Thus, the thermalization rate should be identified with the rate of energy and momentum transfer which is not necessarily related to the relaxation rate of the expectation value of the field. In order to understand thermalization, a collisional (quantum) Boltzmann equation must be set up. Although in the Born approximation the collision term includes the scattering cross section that is related to the decay rate, the solution to the Boltzmann equation implies a resummation that is quite different from the resummation of the Dyson series for the propagator. If the initial distribution is very far from thermal and many collisions are necessary for thermalization, the relaxation and thermalization time scales may be widely different and in principle unrelated. Thus we insist that the "damping rate" obtained from the imaginary part of the self-energy *on shell* must be interpreted as the relaxation rate for the expectation value of the scalar field and in principle *not* with the thermalization rate of the particle distribution function. Moreover, this relaxation rate holds in the linear regime (small field amplitude).

F. The Reheating Temperature:

We can use the results obtained above to provide an estimate of the reheating temperature based on single particle decay and valid **only in the linear regime** [2,3]. Furthermore, since the calculations of the relaxation and decay rate were performed in flat Minkowski space these will only be useful for an estimate of the reheating temperature whenever $\Gamma \gg H$ with H being the expansion rate of the universe. Since the calculation of the rate even in flat space relies on a "sharp resonance approximation" (Breit Wigner) this implies that the lifetime of the particle is much longer than its oscillation period. This translates into the constraint $m_\phi \gg \Gamma \gg H$. The inflaton oscillates coherently many times around the minimum of the potential before its oscillation amplitude diminishes appreciably both as a result of decay into lighter particles and because of the red-shift of the energy.

The Friedman equation for the rate of expansion is

$$H^2(t) = \frac{8\pi}{3M_{Pl}^2}\rho(t) \tag{2.49}$$

with $\rho(t)$ the matter energy density.

After the reheating stage is completed it is expected that the light produced particles will be in the form of radiation (ultrarelativistic), thus $\rho(t) \approx T_{reh}^4$. During the period in which the relaxation rate (or rate of particle production) is $\Gamma < H(t)$ the produced particles will be red-shifted in energy and will not contribute to a radiation dominated universe. Only when the expansion rate becomes smaller than the rate of particle production is when the produced particles will remain as a radiation dominated component. Thus the important time scale for which the produced particles will remain as a radiation dominated component in the universe is that for which $H(t_f) = \Gamma$. At this time

$$\rho(t_f) = \frac{3M_{Pl}^2}{8\pi}H^2(t_f) \approx T_{reh}^4 \tag{2.50}$$

leading to the estimate of the reheating temperature within the linear regime:

$$T_{reh} \approx \sqrt{M_{Pl}\Gamma} \tag{2.51}$$

G. The Polonyi problem:

One area of current interest in which this work may have some implications is that of the so-called "post-modern Polonyi problem" concerning flat directions for some of the moduli fields in string theories [17–20]. These are fields with pertubatively flat directions whose degeneracy is lifted by non-perturbative effects. Their masses $m_\phi \approx 10^2 - 10^3$ Gev with vacuum expectation values of order $10^{11} - 10^{12}$ Gev and couplings to normal matter are of order $g \approx m_\phi/M_{Pl} \approx 10^{-17}$. These properties allow the energy density in these fields to dominate that of the radiation in the universe until times well after nucleosynthesis. These extremely weak couplings determine that the decay rate of these fields is of order

$$\Gamma \approx \frac{m_\phi^3}{M_{Pl}^2} \tag{2.52}$$

A calculation of the reheating temperature based on the "old" scenario of one particle decay gives $T_{reh} \approx 0.01 - 0.1$ Mev , well below the nucleosynthesis scale.

III. NON-LINEAR RELAXATION

In this section we study the equation of motion for a homogeneous order parameter beyond the linear approximation. This is achieved as follows [13]. We write the scalar inflaton field as

$$\Phi^\pm(\vec{x}, t) = \phi(t) + \chi^\pm(\vec{x}, t) , \tag{3.1}$$

with $\phi(t)$ the expectation value in the non-equilibrium density matrix and $< \chi^\pm(\vec{x}, t) >= 0$, we consider that $< \sigma^\pm >= 0$ (this is a consistent assumption). The non-equilibrium path integral requires the Lagrangian density

$$\mathcal{L}(\phi; \chi^+; \sigma^+; \chi^-; \sigma^-) = \mathcal{L}(\phi; \chi^+; \sigma^+) - \mathcal{L}(\phi; \chi^-; \sigma^-) ,$$

$$\mathcal{L}(\phi; \chi^+; \sigma^+) = \frac{1}{2} \left[(\partial_\mu \chi^+)^2 - M^2(t) (\chi^+)^2 \right] - \chi^+ \left[\ddot{\phi} + m_\phi^2 \phi + \frac{\lambda}{6} \phi^3 \right] - \frac{\lambda}{6} \phi (\chi^+)^3 - \frac{\lambda}{4!} (\chi^+)^4$$

$$+ \frac{1}{2} \left[(\partial_\mu \sigma^+)^2 - m^2(t) (\sigma^+)^2 \right] - g \phi \chi^+ (\sigma^+)^2 - \frac{g}{2} (\chi^+)^2 (\sigma^+)^2 + \bar{\psi}^+ (i \not{\partial} - m_\psi(t) - y\chi^+)\psi^+ ,$$

$$M^2(t) = m_\phi^2 + \frac{\lambda}{2} \phi^2(t) , \quad m^2(t) = m_\sigma^2 + g \phi^2(t) , \quad m_\psi(t) = m_\psi + y \phi(t) .$$

The difference with the linear relaxation case is that we now incorporate $\phi(t)$ in the definition of the time dependent masses. The equations of motion are obtained as in the previous sections, by treating the *linear*, cubic and quartic terms as perturbations. The necessary Green's functions are constructed from the homogeneous solutions of the quadratic forms. We again treat the cases where the inflaton is coupled to scalars or fermions separately.

A. Inflaton Coupled to Scalars Only

The Green's functions for the scalars are obtained from the mode equations that solve

$$\left[\frac{d^2}{dt^2} + \vec{k}^2 + M^2(t) \right] U_k(t) = 0 , \tag{3.2}$$

$$U_k(0) = 1 , \quad \dot{U}_k(0) = -iW_k = -i\sqrt{\vec{k}^2 + M^2(0)} ,$$

$$\left[\frac{d^2}{dt^2} + \vec{k}^2 + m^2(t)\right] V_k(t) = 0 , \tag{3.3}$$

$$V_k(0) = 1 , \quad \dot{V}_k(0) = -i\,w_k = -i\sqrt{\vec{k}^2 + m^2(0)} ,$$

The initial conditions on the mode functions (3.2, 3.3) correspond to positive frequency solutions at the initial time. In terms of these mode functions, the Green's functions are [13,8]

$$G^>_{\chi,k}(t,t') = \frac{i}{2W_k}\left[(1 + n_b(W_k))\,U_k(t)U_k^*(t') + n_b(W_k)\,U_k^*(t)U_k(t')\right] ,$$
$$G^<_{\chi,k}(t,t') = G^>_{\chi,k}(t',t) ,$$
$$G^>_{\sigma,k}(t,t') = \frac{i}{2w_k}\left[(1 + n_b(w_k))\,V_k(t)V_k^*(t') + n_b(w_k)\,V_k^*(t)V_k(t')\right] ,$$
$$G^<_{\sigma,k}(t,t') = G^>_{\sigma,k}(t',t)$$

and the rest of the Green's functions are given by the relations in equations (2.2). The Green's functions above correspond to the situation in which the initial density matrix is in equilibrium for the positive and negative frequency modes at time $t = 0$. This is determined by the initial conditions on the mode functions above at an initial temperature $T = 1/\beta$. However, for the rest of the analysis we will take $T = 0$. We see that

$$< \chi^2(\vec{x},t) > = \int \frac{d^3k}{(2\pi)^3} \frac{|U_k(t)|^2}{2W_k} \coth\left[\frac{\beta W_k}{2}\right] ,$$
$$< \sigma^2(\vec{x},t) > = \int \frac{d^3k}{(2\pi)^3} \frac{|V_k(t)|^2}{2w_k} \coth\left[\frac{\beta w_k}{2}\right] .$$

Finally the equation of motion to one-loop order for the expectation value is

$$\ddot{\phi}(t) + m_\Phi^2\,\phi(t) + \frac{\lambda}{6}\,\phi^3(t) + \frac{\lambda}{2}\,\phi(t) < \chi^2(\vec{x},t) > + g\,\phi(t) \; < \sigma^2(\vec{x},t) > = 0 \tag{3.4}$$

If one wants to solve the equation of motion (3.4) to order $(\hbar)$ one would expand ϕ in a power series in $\hbar$ and only keep the zeroth order term in the mode equations (3.2, 3.3). As was observed previously [13], such an expansion will result in secular terms and becomes unreliable at long times. A resummation is necessary to capture the long time behavior consistently. We will perform a non-perturbative resummation of the one-loop terms by solving the set of equations (3.2 - 3.4) with the *full* value of ϕ in the mode equations. This then becomes a set of coupled non-linear integro-differential equations that provide a non-perturbative resummation of select one-loop terms as can be seen by looking at a perturbative expansion of the solution to these equations.

The integration of these coupled set of one-loop equations provide the lowest *non-perturbative* self-consistent resummation that is energy conserving (see below for energy conservation). This one loop approximation becomes exact in the large (N,M) limit with N scalar σ fields and M fermionic fields when the σ field self-interaction is neglected.

We want to emphasize this point. By incorporating the full value of ϕ in the evolution equation for the mode functions we are incorporating back-reaction effects self-consistently.

If only the classical evolution of ϕ is used in the mode equations, we would have parametric resonant amplification since the effective frequencies for the mode functions are periodic due to the fact that the classical solution is periodic with constant amplitude. This leads to particle production that never shuts-off. However particle production leads to damping in the evolution of ϕ. Introducing this damped evolution in the mode equations leads to a behavior rather different from parametric resonance: as the evolution of ϕ is damped, the amplitude becomes smaller and particle production should diminish and eventually stop. This was found to be the situation in the case of the scalar field with self-interaction [13].

Clearly the integration of this set of equations will have to be done numerically. Because the self-interacting scalar case was already studied before [13], we will now concentrate on the interaction with the scalar fields in which thresholds are present. But before we do this we must understand the renormalization aspects of this system of equations.

The large-k behavior of the mode functions can be understood from a WKB analysis, the details of which had already been discussed in [8,13]. We find the following divergence structure doing this

$$\int^\Lambda \frac{d^3k}{(2\pi)^3} \frac{|U_k(t)|^2}{2W_k} = \frac{\Lambda}{8\pi^2} - \frac{1}{8\pi^2}\left(m_{\Phi,b}^2 + \frac{\lambda_b}{2}\phi^2(t)\right)\ln\frac{\Lambda}{\kappa} + F_1(t,\kappa) ,$$

$$\int^\Lambda \frac{d^3k}{(2\pi)^3} \frac{|V_k(t)|^2}{2w_k} = \frac{\Lambda}{8\pi^2} - \frac{1}{8\pi^2}\left(m_{\sigma,b}^2 + g_b\phi^2(t)\right)\ln\frac{\Lambda}{\kappa} + F_2(t,\kappa) ,$$

where Λ is an upper momentum cut-off and κ and arbitrary renormalization scale. The subscript b refers to bare quantities and the quantities $F_{1,2}(t,\kappa)$ are finite in the limit $\Lambda \to \infty$. We can now read the mass and coupling constant renormalizations

$$m_{\Phi,R}^2 = m_{\Phi,b}^2 + \frac{\Lambda}{8\pi^2}\left(\frac{\lambda_b}{2} + g_b\right) - \frac{1}{8\pi^2}\left(\frac{\lambda_b}{2}\, m_{\Phi,b}^2 + g_b\, m_{\sigma,b}^2\right)\ln\frac{\Lambda}{\kappa} ,$$

$$\lambda_R = \lambda_b - \frac{3}{4\pi^2}\left(\frac{\lambda_b^2}{4} + g_b^2\right)\ln\frac{\Lambda}{\kappa} .$$

We introduce a further, finite renormalization, by subtracting the functions $F_{1,2}(t=0,\kappa)$, absorbing this subtraction into a (finite) renormalization of the mass. After these renormalizations, we finally arrive at the renormalized set of evolution equations in terms of renormalized quantities. We drop the subscript R for renormalized quantities to avoid cluttering the notation, but with the understanding that all quantities are renormalized at the scale κ:

$$\ddot{\phi}(t) + m_\Phi^2\,\phi(t) + \frac{\lambda}{6}\,\phi^3(t) + \frac{\lambda}{2}\,\phi(t)\int^\Lambda \frac{d^3k}{(2\pi)^3}\frac{|U_k(t)|^2 - 1}{2W_k} + g\,\phi(t)\int^\Lambda \frac{d^3k}{(2\pi)^3}\frac{|V_k(t)|^2 - 1}{2w_k}$$

$$+\frac{1}{8\pi^2}\left(\frac{\lambda^2}{4} + g^2\right)\phi(t)\left[\phi^2(t) - \phi^2(0)\right]\ln\frac{\Lambda}{\kappa} = 0 . \tag{3.5}$$

To this order, we can replace masses and couplings by their renormalized values in the equations for the mode functions (3.2, 3.3). The set of equations (3.2,3.3,3.5) provide a self-consistent, *non-perturbative* set of integro-differential equations with *back-reaction*. This last point is extremely important, a periodic evolution for the scalar field in the mode equations

(3.2, 3.3) would lead to parametric amplification [11] and infinite particle production [37]. However the back-reaction of the quantum fluctuations onto the evolution of the zero mode of the scalar field is determined completely by these mode equations. When the amplitude of the quantum fluctuations grow, this mechanism takes energy of the zero mode, whose evolution now becomes damped in amplitude. This damping , a result of the back reaction, in turn makes the growth of the quantum fluctuations to diminish. Eventually the particle production mechanism must shut-off as a consequence of this back reaction. This is dramatically different from the catastrophic particle production found by Yoshimura [37] precisely because in that analysis back reaction was not taken into account.

We will now chose the renormalization scale to be $\kappa = m_\Phi$ so as to have only one scale in the problem, which makes the numerical evaluation easier. Renormalized non-equilibrium equations had also been obtained by Cooper et. al. [36] within the context of dynamical evolution during the chiral phase transition. The renormalized equation of motion (3.5) may be written without reference to the cutoff Λ which in the end must be taken to infinity, however, numerically the k-integrals must be calculated with an upper momentum cutoff. One must ensure that this numerical cutoff, to be identified with Λ in the evolution equation, be much larger than the masses and amplitudes of the field for the integrals to reach their asymptotics and the cutoff dependence to dissapear. We have been careful to make exhaustive checks that the final results were insensitive (to the working accuracy) to the cutoff which was typically chosen to be $\Lambda \approx 100\,|m_\Phi|$. This implies that the order of magnitude of the error is about $(\frac{m_\Phi}{\Lambda})^2 \simeq 10^{-4}$.

A provision must be made for the initial conditions on the mode functions solution of (3.2, 3.3) in the broken symmetry case. In this case the tree level squared mass is negative $m_\Phi^2 = -\mu^2$ and for initial values of the expectation value $\phi^2(0) < 2\mu^2/\lambda$ the initial configuration is below the classical "spinodal" and imaginary frequencies lead to the spinodal instabilities. We are not interested in this article in studying the time evolution of these instabilities but on the issues of non-linear relaxation. There are two ways to avoid the complex frequencies associated with these instabilities in the initial conditions: (i) one can choose an initial value $\phi^2(0) > 2\mu^2/\lambda$ or (ii) one can impose that the initial frequencies correspond to a positive mass term. This choice corresponds to preparing an initial gaussian density matrix of the modes with this given mass, under time evolution this packet spreads in function space and the time dependence of the width reflects the (quantum and thermal) fluctuations. In our numerical analysis we chose the later possibility with the positive mass squared given by the absolute value of the (negative) mass squared in the Lagrangian.

We introduce now the following dimensionless variables:

$$\eta(\tau) = \sqrt{\frac{\lambda}{6\,|m_\Phi^2|}}\phi(t)\,, \qquad q = \frac{k}{|m_\Phi|}\,, \qquad \tau = |m_\Phi|\,t\,,$$

$$\bar{W}_q = \sqrt{q^2 + \frac{|M^2(0)|}{|m_\Phi^2|}}\,, \qquad \bar{w}_q = \sqrt{q^2 + \frac{m^2(0)}{|m_\Phi^2|}}\,.$$

In terms of which the evolution equation (3.5) becomes

$$\ddot{\eta}(\tau) + \eta(\tau) + \eta^3(\tau) + \frac{\lambda}{8\pi^2}\,\eta(\tau)\int^{\frac{\Lambda}{|m_\Phi|}} q^2 dq\, \frac{|U_q(\tau)|^2 - 1}{\bar{W}_q} + \frac{g}{4\pi^2}\,\eta(\tau)\int^{\frac{\Lambda}{|m_\Phi|}} q^2 dq\, \frac{|V_q(\tau)|^2 - 1}{\bar{w}_q}$$

$$+\frac{\lambda}{8\pi^2}\left(\frac{3}{2}+\frac{6g^2}{\lambda^2}\right)\eta(\tau)\left[\eta^2(\tau)-\eta^2(0)\right]\ln\frac{\Lambda}{|m_\Phi|}=0\,,\tag{3.6}$$

where we chose the renormalization scale $\kappa=|m_\Phi|$.

1. Particle Production

As in any time dependent situation, the concept of particle is ambiguous and has to be specified with respect to some particular state. We choose that state as the equilibrium ensemble at the initial time $t=0$. The initial condition on the mode functions for the fluctuations (3.2, 3.3) naturally determine the set of positive and negative energy states at this initial time. At this time the fluctuation operators may be expanded in this basis. The Fourier components of the fluctuation operators are thus written as

$$\chi_{\vec{k}}(0)=\frac{1}{\sqrt{2W_k}}\left[a_{\vec{k}}(0)-a_{\vec{k}}^\dagger(0)\right]\,,$$

$$\sigma_{\vec{k}}(0)=\frac{1}{\sqrt{2w_k}}\left[b_{\vec{k}}(0)-b_{\vec{k}}^\dagger(0)\right]\,.$$

The number operators at any time t are

$$N_{\chi,k}(t)=\frac{\mathrm{Tr}\,a_{\vec{k}}^\dagger(0)a_{\vec{k}}(0)\rho(t)}{\mathrm{Tr}\,\rho(0)}=\frac{\mathrm{Tr}\,a_{\vec{k}}^\dagger(t)a_{\vec{k}}(t)\rho(0)}{\mathrm{Tr}\,\rho(0)}\,,$$

$$a_{\vec{k}}(t)=U(t)a_{\vec{k}}(0)U^{-1}(t)\,,$$

$$N_{\sigma,k}(t)=\frac{\mathrm{Tr}\,b_{\vec{k}}^\dagger(0)b_{\vec{k}}(0)\rho(t)}{\mathrm{Tr}\,\rho(0)}=\frac{\mathrm{Tr}\,b_{\vec{k}}^\dagger(t)b_{\vec{k}}(t)\rho(0)}{\mathrm{Tr}\,\rho(0)}\,,$$

$$b_{\vec{k}}(t)=U(t)b_{\vec{k}}(0)U^{-1}(t)\,,$$

with $U(t)$ the unitary time evolution operator. Following the arguments presented in reference [13] we find that the creation and annihilation operators at time t are related to those at the initial time $t=0$ by a Bogoliubov transformation. In terms of the dimensionless variables introduced above we find

$$N_{\chi,q}(\tau)=(2\mathcal{F}-1)\,N_{\chi,q}(0)+(\mathcal{F}-1)\,,$$

$$\mathcal{F}=\frac{1}{4}|U_q(\tau)|^2\left[1+\frac{|\dot{U}_q(\tau)|^2}{W_q^2|U_q(\tau)|^2}\right]+\frac{1}{2}\,,$$

$$N_{\sigma,q}(\tau)=(2\mathcal{G}-1)\,N_{\sigma,q}(0)+(\mathcal{G}-1)\,,$$

$$\mathcal{G}=\frac{1}{4}|V_q(\tau)|^2\left[1+\frac{|\dot{V}_q(\tau)|^2}{\bar{w}_q^2|V_q(\tau)|^2}\right]+\frac{1}{2}\,,$$

where $N_{\chi,\sigma}(0)$ are the occupation numbers at the initial time $t=0$ and in the case given by the Bose-Einstein distribution functions, derivatives are with respect to the dimensionless variable τ. Since in this section we will be working at zero temperature, only the last term will contribute to particle production. This term is recognized as the induced contribution.

It can be seen from the initial conditions on the wave functions that the induced contribution vanishes at the initial time.

We will compute numerically the number of particles produced in a correlation volume

$$\mathcal{N}^b(\tau) = \int d^3k N_k(t)/|m_\Phi|^3 . \tag{3.7}$$

2. Energy Conservation:

To one loop order the expectation value of the Hamiltonian (divided by the volume) gives the mean energy density

$$E = \frac{1}{2}\dot{\phi}^2 + \frac{1}{2}m_\phi^2\phi^2 + \frac{\lambda}{4!}\phi^4 + \frac{1}{2}\int\frac{d^3k}{(2\pi)^3}\left\{\frac{1}{2W_k(0)}\left[|\dot{U}_k|^2 + W_k^2(t)|U_k|^2\right] + \right.$$
$$\left. \frac{1}{2w_k(0)}\left[|\dot{V}_k|^2 + w_k^2(t)|V_k|^2\right]\right\} \tag{3.8}$$

$$W_k^2(t) = k^2 + M^2(t) \tag{3.9}$$
$$w_k^2(t) = k^2 + m^2(t) \tag{3.10}$$

Using the one-loop equations for the zero mode and the mode functions it is straightforward to find that the above energy density is conserved in time. Thus this (resummed) one-loop set of self-consistent equations is energy conserving. This is an important consistency check of the set of equations that guarantees that the numerical results are trustworthy.

B. Inflaton Coupled to Fermions Only

The fermionic Green's functions are constructed from the solutions to the homogeneous Dirac equation in the presence of the background field. The main ingredients and treatment is similar to that of fermions in presence of an electric field studied by Kluger et. al. [26]. Writing the independent solutions as

$$\mathcal{U}^{(1,2)}(\vec{x},t) = e^{i\vec{k}\cdot\vec{x}}U_k^{(1,2)}(t) ,$$
$$\mathcal{V}^{(1,2)}(\vec{x},t) = e^{-i\vec{k}\cdot\vec{x}}V_k^{(1,2)}(t) ,$$

the mode functions obey

$$\left[i\gamma_0\frac{d}{dt} - \vec{\gamma}\cdot\vec{k} - m_\psi(t)\right]U_k^{(1,2)}(t) = 0 ,$$

$$\left[i\gamma_0\frac{d}{dt} + \vec{\gamma}\cdot\vec{k} - m_\psi(t)\right]V_k^{(1,2)}(t) = 0 .$$

It is convenient to write the spinors as

$$U_k^{(1,2)}(t) = \left[i\gamma_0 \frac{d}{dt} - \vec{\gamma} \cdot \vec{k} + m_\psi(t) \right] f_k(t) u^{(1,2)} ,$$

$$V_k^{(1,2)}(t) = \left[i\gamma_0 \frac{d}{dt} + \vec{\gamma} \cdot \vec{k} + m_\psi(t) \right] g_k(t) v^{(1,2)} ,$$

with $u^{(1,2)}$, $v^{(1,2)}$ the spinor eigenstates of γ_0 with eigenvalues $+1$, -1 respectively. The functions $f_k(t)$, $g_k(t)$ obey the second order equations

$$\left[\frac{d^2}{dt^2} + \vec{k}^2 + m_\psi^2(t) - i\dot{m}_\psi(t) \right] f_k(t) = 0 ,$$

$$\left[\frac{d^2}{dt^2} + \vec{k}^2 + m_\psi^2(t) + i\dot{m}_\psi(t) \right] g_k(t) = 0 .$$

We now need to append initial conditions. We will consider the situation in which the system was in equilibrium at time $t \leq 0$ with the expectation value of scalar field being $\phi(0)$ and $\dot{\phi}(0) = 0$. Thus the fermion mass is constant and given by $m_\psi(0)$. We can now impose the condition that the modes $f_k(t)$, $g_k(t)$ describe positive and negative frequency solutions for $t \leq 0$ and, normalizing the spinor solutions to the Dirac equation to unity, we impose the following initial conditions

$$f_k(t < 0) = \frac{e^{-ik_0 t}}{\sqrt{2k_0(k_0 + m_\psi(0))}} ,$$

$$g_k(t < 0) = \frac{e^{ik_0 t}}{\sqrt{2k_0(k_0 + m_\psi(0))}} ,$$

$$k_0 = \sqrt{\vec{k}^2 + m_\psi^2(0)} .$$

Equations (3.11) with these boundary conditions imply that

$$g_k(t) = f_k^*(t) . \tag{3.11}$$

The necessary ingredients for the zero temperature fermionic Green's functions are the following

$$S_k^>(t, t') = -i \sum_{\alpha=1,2} U_k^\alpha(t) \bar{U}_k^\alpha(t') ,$$

$$S_k^<(t, t') = i \sum_{\alpha=1,2} V_{-k}^\alpha(t) \bar{V}_{-k}^\alpha(t') .$$

In the standard Dirac representation for the γ matrices, we find

$$S_k^>(t, t') = -i f_k(t) f_k^*(t') \left[\mathcal{W}_k(t)\gamma_0 - \vec{\gamma} \cdot \vec{k} + m_\psi(t) \right] \left(\frac{1 + \gamma_0}{2} \right) \left[\mathcal{W}_k^*(t')\gamma_0 - \vec{\gamma} \cdot \vec{k} + m_\psi(t') \right] ,$$

$$S_k^<(t, t') = -i f_k^*(t) f_k(t') \left[\mathcal{W}_k^*(t)\gamma_0 - \vec{\gamma} \cdot \vec{k} - m_\psi(t) \right] \left(\frac{1 - \gamma_0}{2} \right) \left[\mathcal{W}_k(t')\gamma_0 - \vec{\gamma} \cdot \vec{k} - m_\psi(t') \right] ,$$

$$\mathcal{W}_k(t) = i \frac{\dot{f}_k(t)}{f_k(t)} .$$

With these ingredients the non-equilibrium fermionic Green's functions can be constructed as in section II. It is an important check that the equality $\operatorname{Tr} S_k^{++}(t,t) = \operatorname{Tr} S_k^{--}(t,t) = \operatorname{Tr} S_k^{>}(t,t) = \operatorname{Tr} S_k^{<}(t,t)$ is fulfilled where the trace is over Dirac indices. It is also an important property that

$$\left(-i\dot{f}_k^*(t) + m_\psi(t)f_k^*(t)\right)\left(i\dot{f}_k(t) + m_\psi(t)f_k(t)\right) + k^2 f_k^*(t)f_k(t) = 1 . \tag{3.12}$$

This is a consequence of the conservation of probability and may be checked explicitly.

The evolution equation becomes

$$\ddot{\phi}(t) + m_\Phi^2\,\phi(t) + \frac{\lambda}{6}\,\phi^3(t) + \frac{\lambda}{2}\,\phi(t) < \chi^2(\vec{x},t) > -y\operatorname{Tr} S^{<}(\vec{x},t;\vec{x},t) = 0 ,$$

$$\operatorname{Tr} S^{<}(\vec{x},t;\vec{x},t) = 2\int \frac{d^3k}{(2\pi)^3}\left[1 - 2k^2|f_k(t)|^2\right] . \tag{3.13}$$

The expectation value of the Hamiltonian can again be computed in this (resummed) one-loop approximation. Using the equations of motion for the expectation value of the scalar field and those of the mode functions of the fermions, it can be shown to be conserved similarly to the bosonic case.

The integral in (3.13) is divergent. The divergence structure may be understood in two different ways: carrying out a WKB expansion of the modes, just as was done in the bosonic case (this is a tedious and lengthy exercise), or alternatively calculating the closed loop in the presence of the background field. We carried out both methods using an upper momentum cutoff in the integrals and found

$$2y\int \frac{d^3k}{(2\pi)^3}\left[1 - 2\,k^2|f_k(t)|^2\right] = -\frac{y}{2\pi^2}\,\Lambda^2\,m_\psi(t) + \frac{y}{4\pi^2}[\ddot{m}_\psi(t) + 2\,m_\psi^3(t)]\ln\frac{\Lambda}{\kappa} + \text{finite} ,$$

$$\tag{3.14}$$

where κ is an arbitrary renormalization scale which will be chosen again to be $\kappa = m_\Phi$ for numerical convenience. We will concentrate on the "chiral limit" in which the *bare* mass term for the fermions vanishes and $m_\psi(t) = y\,\phi(t)$. In this limit the discrete symmetry $\phi \to -\phi$ is explicit. The divergent terms in (3.14) are then identified as: mass renormalization (of ϕ) (first term), wave function renormalization (second term) and coupling constant renormalization (third term). From now on, all quantities will be renormalized, and we drop the subscript R to avoid cluttering. For simplicity in the numerical analysis and in order to isolate the fluctuations from the fermionic contribution from those of the scalar sector, we now neglect the contribution from the scalar loop. The scalar fluctuations had already been analyzed in the previous section and in [13]. The final renormalized equation of motion is now

$$\ddot{\phi}(t) + m_\Phi^2\,\phi(t) + \frac{\lambda}{6}\phi^3(t) - 2y\int \frac{d^3k}{(2\pi)^3}\left[1 - 2k^2|f_k(t)|^2\right] - \left\{-\frac{y^2}{2\pi^2}\Lambda^2\,\phi(t)\right.$$

$$\left. + \frac{y^2}{4\pi^2}\left[\ddot{\phi}(t) + 2y^2\,\phi^3(t)\right]\ln\frac{\Lambda}{m_\Phi}\right\} = 0 .$$

With the purpose of a numerical analysis of this equations, it proves convenient to introduce the following dimensionless of variables

$$\eta = y\phi/m_\Phi \,, \quad \tau = m_\Phi t \,, \quad q = k/m_\Phi \,, \quad g' = \lambda_R/6y^2 \,, \quad g = y^2/\pi^2 \,,$$
$$u_q(\tau) = f_k(t)\sqrt{\omega_k^0(\omega_k^0 + y^2\phi^2(0))} \,,$$

in terms of which the equations of motion are

$$\frac{d^2\eta}{d\tau^2} + \eta + g'\eta^3 - g\Sigma(\tau) = 0 \,,$$

$$\left[\frac{d^2}{d\tau^2} + q^2 + \eta^2 - i\dot\eta\right]u_q(\tau) = 0 \,, \qquad u_q(0) = \frac{1}{\sqrt{2}} \,, \qquad \dot u_q(0) = -i\frac{\sqrt{q^2 + \eta^2(0)}}{\sqrt{2}} \,,$$

$$\Sigma(\tau) = \int_0^{\Lambda/m_\Phi} q^2\,dq\left[1 - \frac{q^2|u_q(\tau)|^2}{\sqrt{q^2 + \eta^2(0)}(\sqrt{q^2 + \eta^2(0)} + \eta(0))}\right]$$
$$- \frac{1}{2}\left(\frac{\Lambda}{m_{R,\Phi}}\right)^2\eta + \frac{1}{4}\left(\ddot\eta + 2\,\eta^3\right)\ln\frac{\Lambda}{m_{R,\Phi}} \,. \tag{3.15}$$

Although the equations are finite when the ultraviolet cutoff is taken to infinity and can be written without the introduction of the cutoff by subtracting the integral with a *lower* limit cutoff given by the renormalization scale, numerically these integrals will have to be done by introducing an upper momentum cutoff anyways. Therefore we keep this UV cutoff in the equations.

1. Particle Production

The study of particle production is similar to the scalar case, with the only complication being the spinorial structure of the fermionic fields. At the initial time $t = 0$ the quantized fermion operator is

$$\psi(\vec x, 0) = \int \frac{d^3k}{(2\pi)^3}\sum_{\alpha=1,2}\left[b_k^{(\alpha)}(0)U_k^{(\alpha)}(0) + d_{-k}^{(\alpha)\dagger}(0)V_{-k}^{(\alpha)}(0)\right]e^{i\vec k\cdot\vec x} \,, \tag{3.16}$$

in terms of the mode functions determined above. The number of fermions and antifermions are *defined* as

$$\langle N^f(t)\rangle = \int \frac{d^3k}{(2\pi)^3}\sum_\alpha \frac{\mathrm{Tr}\left[\rho(t)b_k^{(\alpha)\dagger}(0)b_k^{(\alpha)}(0)\right]}{\mathrm{Tr}\,\rho(0)} = \int \frac{d^3k}{(2\pi)^3}\sum_\alpha \frac{\mathrm{Tr}\left[\rho(0)b_k^{(\alpha)\dagger}(t)b_k^{(\alpha)}(t)\right]}{\mathrm{Tr}\,\rho(0)} \,,$$

$$\langle N^{\bar f}(t)\rangle = \int \frac{d^3k}{(2\pi)^3}\sum_\alpha \frac{\mathrm{Tr}\left[\rho(t)d_k^{(\alpha)\dagger}(0)d_k^{(\alpha)}(0)\right]}{\mathrm{Tr}\,\rho(0)} = \int \frac{d^3k}{(2\pi)^3}\sum_\alpha \frac{\mathrm{Tr}\left[\rho(0)d_k^{(\alpha)\dagger}(t)d_k^{(\alpha)}(t)\right]}{\mathrm{Tr}\,\rho(0)} \,.$$

The time dependent coefficients are obtained by projecting the time dependent spinor solutions onto the positive and negative energy solutions at $t = 0$, and are related to the coefficients at $t = 0$ via a Bogoliubov transformation

$$b_k^{(\alpha)}(t) = \sum_\beta\left[\mathcal{F}_{(\beta)k,+}^{(\alpha)}(t)b_k^{(\beta)} + \mathcal{F}_{(\beta)k,-}^{(\alpha)}(t)d_{-k}^{(\beta)\dagger}\right] \,,$$

$$d_{-k}^{(\alpha)\dagger}(t) = \sum_\beta\left[\mathcal{H}_{(\beta)k,+}^{(\alpha)}(t)b_k^{(\beta)} + \mathcal{H}_{(\beta)k,-}^{(\alpha)}(t)d_{-k}^{(\beta)\dagger}\right] \,.$$

The identities

$$\sum_\beta \mid \mathcal{F}^{(\alpha)}_{(\beta)k,+}(t) \mid^2 + \mid \mathcal{F}^{(\alpha)}_{(\beta)k,-}(t) \mid^2 = 1 ,$$

$$\sum_\beta \mid \mathcal{H}^{(\alpha)}_{(\beta)k,+}(t) \mid^2 + \mid \mathcal{H}^{(\alpha)}_{(\beta)k,-}(t) \mid^2 = 1$$

ensure that the transformations preserve the anticommutation relations as they must. It is also a matter of algebra using the equations of motion for the mode functions to prove that the number of fermions minus antifermions is conserved for each $\vec{k}$ mode. In what follows, we only consider the number of fermions produced since fermions and antifermions are created in pairs.

With some algebra, the Bogoliubov coefficients can be expressed as the function of $f_k(t)$ and $f_k^*(t)$. In terms of the dimensionless variables defined above we obtain the number of fermions within a correlation volume $\mathcal{N}^f(\tau) = N^f(\tau)/|m_\Phi|^3$

$$\mathcal{N}^f(\tau) = \frac{1}{2\pi^2} \int q^2 \, dq \mathcal{N}^f_q(\tau)$$

$$= \frac{1}{4\pi^2} \int dq \left[\frac{q^2}{\sqrt{q^2 + \eta^2(0)} \left(\sqrt{q^2 + \eta^2(0)} + \eta(0) \right)} \right]^2$$

$$\times \left[-i\frac{\partial}{\partial \tau} + \eta(\tau) - \sqrt{q^2 + \eta^2(0)} - \eta(0) \right] u_q^*(\tau)$$

$$\times \left[i\frac{\partial}{\partial \tau} + \eta(\tau) - \sqrt{q^2 + \eta^2(0)} - \eta(0) \right] u_q(\tau) . \tag{3.17}$$

IV. NUMERICAL ANALYSIS

The numerical analysis was carried out with a fourth order Runge-Kutta method for the differential equations and a 5-point Bode rule integrator for the k-integrals. The typical step size in time was $1 - 2.10^{-3}$, and the typical step size in k was 10^{-3}. The cutoff $\Lambda/|m_\Phi|$ was varied between 75 and 200; we found no sensitivity to the cutoff in the range of parameters that we tested (see below). The code was very stable within the range of parameters tested.

A. Inflaton Coupled to Scalars: Unbroken Symmetry Case

Since the contribution of the one-loop quantum fluctuations of the inflaton Φ have already been studied previously [13] and we want to study the contribution of the lighter field σ, we will neglect the contribution from the self-interaction in the evolution equation. Thus we study the following equations obtained from equations (3.3, 3.6) in terms of dimensionless variables

$$\ddot{\eta}(\tau) + \eta(\tau) + \eta^3(\tau) + \frac{g}{4\pi^2} \eta(\tau) \int^{\frac{\Lambda}{|m_\Phi|}} q^2 dq \, \frac{|V_q(\tau)|^2 - 1}{\bar{w}_q}$$

$$+ \frac{\lambda}{8\pi^2} \frac{6g^2}{\lambda^2} \eta(\tau) \left[\eta^2(\tau) - \eta^2(0) \right] \ln \frac{\Lambda}{|m_\Phi|} = 0 \, ,$$

$$\left[\frac{d^2}{d\tau^2} + \bar{q}^2 + \frac{m_\sigma^2}{|m_\Phi|^2} + \frac{6g}{\lambda} \eta(\tau) \right] V_q(\tau) = 0 \, ,$$

$$V_k(0) = 1 \, , \quad \dot{V}_q(0) = -i\,\bar{w}_q = -i\sqrt{\bar{q}^2 + \frac{m_\sigma^2}{|m_\Phi|^2} + \frac{6g}{\lambda} \eta^2(0)} \, .$$

Notice that the factor $(3/2)$ multiplying the logarithm in eq. (3.6) and missing from (4.1) arises from the renormalization of the Φ-scalar loop that is not taken into account in (4.1).

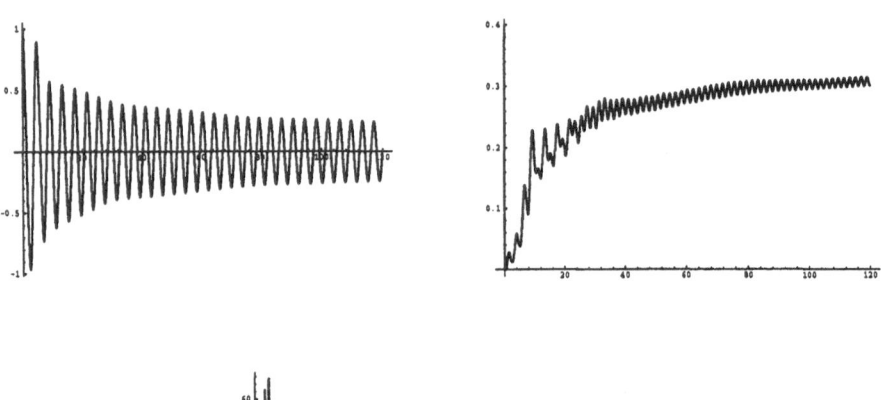

Figs.1 : Scalars: unbroken symmetry for the values of the parameters
$y = 0$; $\lambda/8\pi^2 = 0.2$; $g/\lambda = 1$; $m_\sigma = 0.2\,m_\Phi$; $\eta(0) = 1.0$; $\dot{\eta}(0) = 0$. Upper Left: $\eta(\tau)$ vs τ. Upper Right: $\mathcal{N}_\sigma(\tau)$ vs. τ. Lower: $\mathcal{N}_{q,\sigma}(\tau = 120)$ vs. q

Figures (1.a-c) show the unbroken symmetry case with the quantum fluctuations from the scalar loop of the σ particles for the values $y = 0$; $\lambda/8\pi^2 = 0.2$; $g = \lambda$; $m_\sigma = 0.2\,m_\Phi$; $\eta(0) = 1.0$; $\dot{\eta}(0) = 0$. Figure (1.a) shows $\eta(\tau)$ vs τ, figure (1.b) shows $\mathcal{N}_\sigma(\tau)$ vs τ and figure (1.c) shows $\mathcal{N}_{q,\sigma}$ vs q for $\tau = 120$, similar graphs were obtained with snapshots at different (earlier) times.

Figure (1.a) shows a very rapid, non-exponential damping within few oscillations of the expectation value and a saturation effect when the amplitude of the oscillation is rather small (about 0.1 in this case), the amplitude remains almost constant at the latest times tested. Figure (1.a) and figure (1.b) clearly show that the time scale for dissipation (from figure (1.a) is that for which the particle production mechanism is more efficient (figure

(1.b)). Notice that the total number of particles produced rises on the same time scale as that of damping in figure (1.a) and eventually when the expectation value oscillates with (almost) constant amplitude the average number of particles produced remains constant. These figures clearly show that damping is a consequence of particle production. At times larger than about 40 m_Φ^{-1} (for the initial values and couplings chosen) there is no appreciable damping. The amplitude is rather small and particle production has practically shut off. If we had used the *classical* evolution of the expectation value in the mode equations, particle production would not shut off (parametric resonant amplification), and thus we clearly see the dramatic effects of the inclusion of the back reaction.

In this unbroken symmetry case linear relaxation predicts a slow t^{-3} power law decay to an asymptotic finite amplitude because one particle decay is kinematically forbidden: the self energy contribution is a two-loop effect with one Φ and two σ particle cut and kinematically there is a one-particle pole below the three particle threshold. A slow power law linear relaxation asymptotically cannot be ruled out numerically because we have not continued the integration for longer times but clearly asymptotically the numerical result is compatible with linear relaxation. Figure (1.c) shows the distribution of particles created at the latest time $\tau = 120$ as a function of wave vector. Similar graphs were obtained with snapshots at different earlier times. The distribution is clearly non-thermal and skewed towards small momentum (in units of m_Φ). These figures are qualitatively similar to those obtained in the self-interacting case in [13]. The asymptotic behavior is that of undamped oscillations of small amplitude. This is compatible with the result from the linear relaxation analysis because there is a one-particle pole below the three particle threshold resulting in undamped oscillations at large times. Linear relaxation predicts qualitatively the same results for the self-interacting scalar case and this case in the unbroken phase. The numerical results are consistent with this prediction.

However, for large amplitudes non-linear relaxation via particle production is very effective and dramatically different from linear relaxation.

B. Inflaton Coupled to Scalars: Broken Symmetry Case

In this case, we take $m_\Phi^2 = -|m_\Phi^2|$.

As in the unbroken symmetry case, we only study the effect of the lighter σ fluctuations. In the broken symmetry case, linear relaxation predicts an open decay channel for the inflaton, resulting in exponential relaxation for a long time and eventually relaxation with a power law. Therefore we should not expect a constant amplitude asymptotically but an amplitude that eventually should relax to zero. In this case the renormalized equations for evolution are

$$\ddot{\eta}(\tau) - \eta(\tau) + \eta^3(\tau) + \frac{g}{4\pi^2}\,\eta(\tau) \int^{\frac{\Lambda}{|m_\Phi|}} q^2 dq\, \frac{|V_q(\tau)|^2 - 1}{\bar{w}_q}$$

$$+ \frac{\lambda}{8\pi^2}\frac{6g^2}{\lambda^2}\,\eta(\tau)\left[\eta^2(\tau) - \eta^2(0)\right]\ln\frac{\Lambda}{|m_\Phi|} = 0\,,$$

$$\left[\frac{d^2}{d\tau^2} + \vec{q}^2 + \frac{m_\sigma^2}{|m_\Phi|^2} + \frac{6g}{\lambda}\eta(\tau)\right]V_q(\tau) = 0\,,$$

$$V_k(0) = 1 , \quad \dot{V}_q(0) = -i\,\bar{w}_q = -i\sqrt{\bar{q}^2 + \frac{m_\sigma^2}{|m_\Phi|^2} + \frac{6g}{\lambda}\eta^2(0)} \ .$$

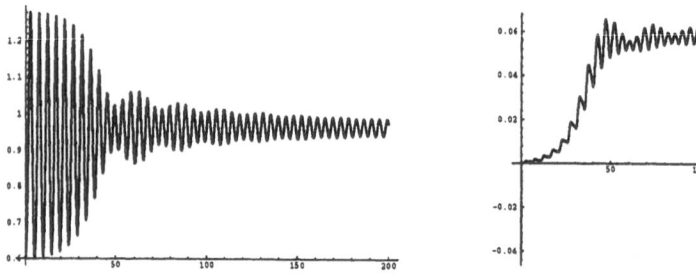

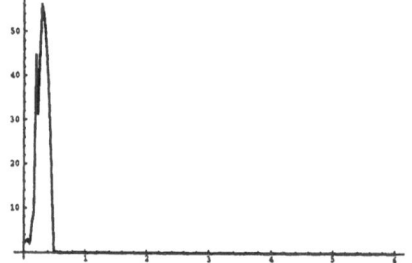

Figs.2 : Scalars: broken symmetry for the values of the parameters ,
$y = 0; \quad \lambda/8\pi^2 = 0.2; \quad g/\lambda = 0.05; \quad m_\sigma = 0.2\,|m_\phi|; \quad \eta(0) = 0.6; \quad \dot{\eta}(0) = 0.$ *Upper Left:* $\eta(\tau)$
vs τ . *Upper Right:* $\mathcal{N}_\sigma(\tau)$ *vs.* τ. *Lower:* $\mathcal{N}_{q,\sigma}(\tau = 120)$ *vs.* q

Figure (2.a-c) show $\eta(\tau)$ vs τ, $\mathcal{N}_\sigma(\tau)$ vs τ and $\mathcal{N}_{q,\sigma}(\tau = 200)$ vs q respectively, for $\lambda/8\pi^2 = 0.2; \quad g/\lambda = 0.05; \quad m_\sigma = 0.2\,|m_\phi|; \quad \eta(0) = 0.6; \quad \dot{\eta}(0) = 0.$ Notice that the mass for the linearized perturbations of the Φ field at the broken symmetry ground state is $\sqrt{2}\,|m_\Phi| > 2m_\sigma$. Therefore, for the values used in the numerical analysis, the two-particle decay channel is open for linear relaxation. For these values of the parameters, linear relaxation predicts exponential decay with a time scale $\tau_{rel} \approx 300$ (in the units used). Figure (2.a) shows very rapid non-exponential damping on time scales about *six times shorter* than that predicted by linear relaxation. The expectation value reaches very rapidly a small amplitude regime, once this happens its amplitude relaxes very slowly. Within our computing time limitations we could not confirm that there is exponential relaxation in the small amplitude regime (for $\tau > 100$) but clearly there is a striking difference with the unbroken symmetry case. The influence of open channels is evident, however in the non-linear regime relaxation is clearly *not* exponential but extremely fast. Although we cannot confirm the exponential (or power law) relaxation numerically in the small amplitude regime, the amplitude at long times seems to relax to the expected value, shifted slightly from the minimum of the tree level potential at $\eta = 1$. This is as expected from the fact that there are quantum fluctuations. Figure

(2.b) shows that particle production occurs during the time scale for which dissipation is most effective, giving direct proof that dissipation is a consequence of particle production. Asymptotically, when the amplitude of the expectation value is small, particle production shuts off. We point out again that this is a consequence of the back-reaction in the evolution equations. Without this back-reaction, as argued above, particle production would continue without indefinitely. Figure (2.c) shows that the distribution of produced particles is very far from thermal and concentrated at low momentum modes $k \leq |m_\Phi|$. This distribution is qualitatively similar to that in the unbroken symmetry case, and points out that the excited state obtained asymptotically is far from thermal.

C. Inflaton Coupled to Fermions Only: Unbroken Symmetry Case

Here we treat the case $y \neq 0$; $g = 0$.

The renormalized evolution and particle production equations in this case are given by eq. (3.15) and eq. (3.17), respectively, in terms of dimensionless quantities. Figures (3.a-c) show $\eta(\tau)$ vs. τ, $\mathcal{N}^f(\tau)$ vs τ and $\mathcal{N}_q^f(\tau = 200)$ respectively for the values of the parameters $m_\psi = 0$; $y^2/\pi^2 = 0.5$; $\lambda/6y^2 = 1.0$; $\eta(0) = 0.6$; $\dot{\eta}(0) = 0$. One observes from these figures that after a rather brief period of initial damping of just a few oscillations of the scalar field, the dissipative mechanism shuts-off. Figure (3.b) shows that during this time scale fermion-antifermion pairs are being produced but then the number of produced particles saturates and oscillates with a small amplitude. Figure (3.c) shows that the distribution of fermions produced is peaked at very low momentum ($k \leq m_\Phi$) and with a maximum value of 2, which is the total number of degrees of freedom per k-wave vector. These numerical results expose the physics of Pauli blocking very clearly; the available low momentum modes are occupied and no more fermion-antifermion pairs can be produced. Pauli blocking shuts off particle production and dissipation very early on. We have obtained snapshots of the particle number as a function of momentum for different times, and they all present the same picture.

This result is markedly different from the prediction of linear relaxation. At zero temperature, eq. (2.48) for linear relaxation predicts exponential damping with a (dimensionless) time scale $\tau_{rel} \approx 10.6$ for the values of the parameters chosen above. The difference between the non-linear evolution and that predicted by linear relaxation is explained by the Pauli blocking phenomenon. Very early in the evolution, the low momentum available fermionic states were filled with produced fermions. Once these states have filled, damping and particle production shuts off. This Pauli blocking effect is explicit in eq. (2.48) but there it appears from finite temperature effects. In the non-linear relaxation case, we began at zero temperature, but an excited state quickly ensues because of particle production.

This analysis reveals that for large amplitudes of the scalar field, fermions will be rather ineffective in dissipation and damping because of Pauli blocking *even at zero temperature*. The time scales for dissipation obtained from the fermion self-energies, which apply to the case of linear relaxation are completely unrelated to the time scales for non-linear dissipative processes, even asymptotically, because the fermionic states are Pauli blocked.

482

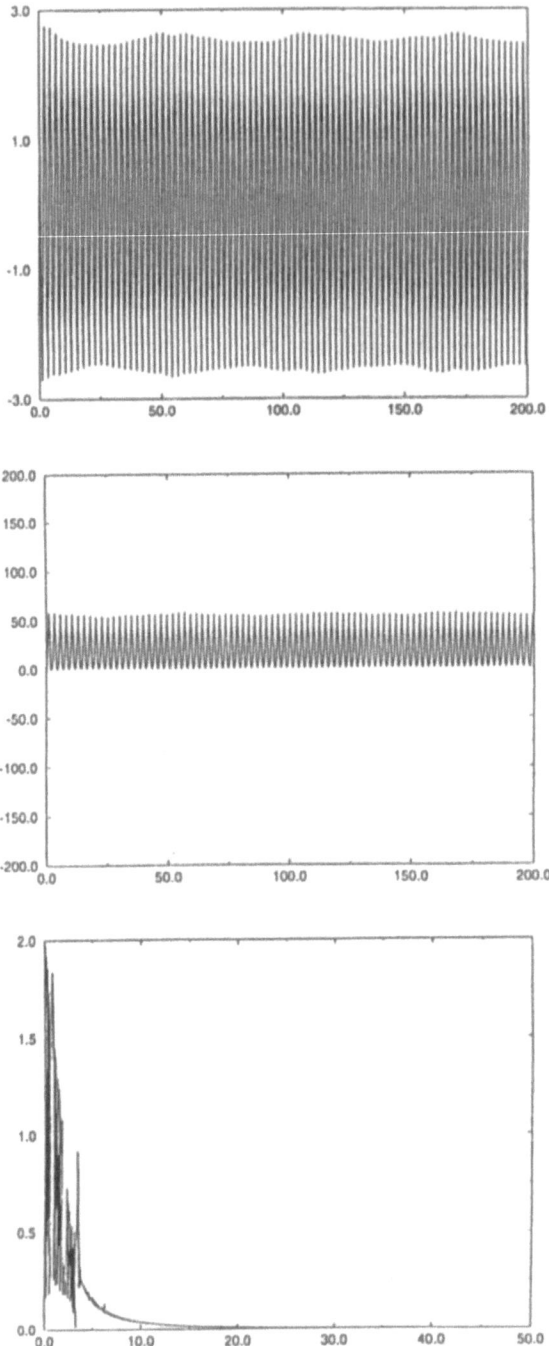

Figs. 3 : Fermions: unbroken symmetry, for the values of the parameters,
$g = 0;\quad y^2/\pi^2 = 0.5;\quad \lambda/6y^2 = 1;\quad m_\psi = 0;\quad \eta(0) = 1.0;\quad \dot{\eta}(0) = 0.$
Upper: $\eta(\tau)$ *vs* τ. *Middle:* $\mathcal{N}^I(\tau)$ *vs.* τ. *Lower:* $\mathcal{N}_q^I(\tau = 200)$ *vs.* q.

D. Thermalization: the argument

The self-consistent resummed one-loop approximation used in our analysis does not incorporate collisional processes. These will appear in the next (two loop order). The distribution of the particles produced is clearly very far from thermal, with typically very many particles in the low momentum modes $k \leq m_\phi$. Although a full study of thermalization will necessarily involve the two loop contribution (and consequently memory effects that will ultimately result in an extensive numerical effort), we can argue on the basis of an effective Boltzmann description. We can take the distribution of particles obtained from the numerical analysis at long times (times longer than the non-linear relaxation scale) as an input in a Boltzmann equation. The integration of this Boltzmann equation will determine the time scales for collisional thermalization. Because the couplings will be typically very small and the initial occupations are rather far from thermal equilibrium these collisional time scales will be very long (as compared to the non-linear relaxation scales). Collisional relaxation does not change the number of particles and is energy conserving (in flat Minkowsky space), just re-distributes the occupation. Therefore a thermal equilibrium state will result with the total number of particles given by the number of produced particles obtained during the fast period of induced amplification and the final reheating temperature will be determined by this number of particles. Including the collisional relaxation via the Boltzmann equation during the period of induced amplification will not modify the time scales substantially. The reason is that initially the occupation numbers are rather small and Boltzmann processes are suppressed. Since the rate of particle production depends on the couplings *and* the initial amplitudes, for large values of the amplitudes the rate of particle production will be much larger than the rate of equilibration via Boltzmann processes (second order in the couplings and independent of the initial amplitude). Thus collisional (Boltzmann) processes will become relevant for thermalization at times much longer (for weak coupling) than the times during which most of the particle production takes place. If the rate of thermalization obtained from this Boltzmann equation is much larger than the expansion rate of the universe, then the reheating temperature can just be estimated from the total number of particles produced during the period of induced amplification. However if this collisional thermalization rate is much smaller than the expansion rate there will be substantial redshifting of the temperature and a full numerical analysis will have to be carried out to obtain a reliable estimate of the reheating temperature.

E. Estimate of the Reheating Temperature

With the arguments on thermalization given above in mind, and *assuming* that the thermalization rate is much larger than the expansion rate (no red-shifting of the temperature), we can use the numerical analysis presented above to obtain an estimate of the reheating temperature.

The important conclusion of the numerical integration is that the density of particles produced during the period of induced amplification is

$$\mathcal{N} = c\, m_a^3 \tag{4.1}$$

where m_a is a mass in between the inflaton and the σ field masses. c a constant of $\mathcal{O}(1-10)$. This constant depends on the *product* of the coupling constants and the *initial amplitudes* as can be seen from the mode equations. It also depends approximately logarithmically on the coupling constants because of the time scales for the shutting-off of the induced amplification mechanism. This can be seen as follows: from the expression for the number of (bosonic) particles produced we find the *rate* of particle production (for example for the σ field)

$$\dot{N}_k(t) = \frac{g\phi^2(0)}{w_k^2(0)} \left[1 - \frac{\phi^2(t)}{\phi^2(0)}\right] \frac{d|V_k(t)|^2}{dt} \tag{4.2}$$

This expression clearly shows that the *rate* of particle production depends on the produc of the coupling and the initial amplitude (squared for the interaction considered), since the mode functions also depend on this combination. The total number of particles is given by this rate integrated up to the time at which the amplification mechanism is shut-off by the back reaction. Un upper bound to this time is estimated as follows: in the early stages, the mode functions increase almost exponentially (because of the resonances) this growth is modified by the damping of the amplitude of the zero mode. Since we find that the back reaction shuts-off the particle production when the one loop term becomes of the order of the tree-level term, this gives a time scale that is logarithmic in the coupling. Thus even for very small couplings but very large initial amplitude of the inflaton zero mode (and this is the situation in cosmological scenarios) the constant c will be of $\mathcal{O}(1-10)$ and can be enhanced by giving the inflaton a larger initial amplitude. Under the assumption of thermalization on time scales shorter than the expansion scale and that the mass of scalars and fermions are much smaller than the inflaton mass an estimate of the reheating temperature is obtained as

$$T_{reh} \approx \mathcal{N}^{\frac{1}{3}} \approx m_a. \tag{4.3}$$

V. CONCLUSIONS AND IMPLICATIONS

The reheating and thermalization processes in inflationary universe models occur very far from equilibrium and must be studied in their full complexity, eventually numerically. The methods developed within real-time non-equilibrium field theory allow a consistent treatment that we used to obtain the equations of evolution for the expectation value of the inflaton field. These equations are non-perturbative and take into account the back reaction effect of both the self couplings of the inflaton, as well as the couplings to lighter scalars and fermions.

We have examined these evolution equations, both in the linear regime, in which the amplitude of the field is small and in the opposite extreme, the non-linear regime, which is intrinsically non-perturbative.

In the linear regime, we have seen that if the inflaton mass is above the two particle threshold, there is some damping behavior for a period of time, and indeed, this damping is governed by the total width of the inflaton. However, at late times the behavior is dominated by *power law* decay rather than exponential damping. Furthermore, we have noted the important distinction between the time scale for the relaxation of the expectation value of the inflaton as opposed to thermalization of the produced particles. It is clear that as treated in this work, relaxation can occur well before thermalization, i.e. interactions that drive the particle distributions towards a thermal one, has a chance to become relevant.

The more interesting case, and the one which would most likely be relevant to the reheating problem in models such as chaotic inflation [5], is that of when the inflaton is in the non-linear regime. Unlike the elementary approach to reheating, in which the "decay" of the coherent inflaton oscillations occured a time $t_{\text{decay}} \sim \Gamma^{-1}$, where Γ is the inflaton decay width, after the end of the slow-roll regime, particle production in the non-linear regime begins at very early times and then shuts off. Indeed, by the time the inflaton expectation value reaches its asymptotic state of oscillations around its minimum, the period of particle production is essentially over. This will be the stage of thermalization via collisions, however we find that the spectrum of produced particles is extremely non-thermal, and collisional relaxation towards an equilibrium thermal state may take a long time.

An inflationary model that cannot reheat the universe to at least nucleosynthesis temperatures is wrong. Thus it is clearly important to accurately compute the reheating temperature in inflationary models. What our analysis here shows is that this computation is of necessity much more complicated than has previously been thought but perhaps more importantly that there are different time scales. Although so far we have studied the situation in flat Minkowski space, recent investigations by Kaiser [38] reveal that these non-linear processes will still be very effective in inflationary cosmologies.

The first stage which occurs rather fast is that of particle production and damping of oscillations as a result of induced amplification. The second stage, that begins typically when the amplitude of the inflaton is rather small and it oscillates with almost constant amplitude at the bottom of the potential is that of one-particle decay and of thermalization via collisional relaxation.

The thermalization time will have to be studied setting up a (quantum) Boltzmann equation using the distributions of produced particles at the end of the induced amplification stage as input for the Boltzmann evolution. Since these initial distributions are very far from equilibrium, it may take many collisions to relax to a thermal equilibrium state. If the couplings are very small (and this will clearly depend on the models) the thermalization time may be very large.

Following the particle numbers to the point at which they become thermal would then allow us to pick off the final reheating temperature. In the weak coupling regime, as mentioned above, the thermalization rate may be smaller than the expansion rate, allowing for significant redshifting of the total energy density, leading to a low reheat temperature, perhaps of the order of the inflaton mass or less.

The particular case of large thermalization rates compared to the expansion rate, allows for a quick estimate of the reheating temperature. In this case we can assume that thermalization occurs without red-shifting of the energy which is then conserved during the thermalization time. By taking the energy density at the beginning of this stage to be αT^4 with α the Stephan Boltzmann accounting for the particle statistics one can then obtain an estimate of the reheating temperature, but the justification for this should ultimately arise from a deeper understanding of the time scales involved.

Our main point in all of this, though, is that self-consistent calculations taking into account the non-linearity of the dynamics are necessary in order to extract the reheating temperature in a reliable way. Our work clarifies what the relevant particle production mechanisms are in these theories and provide a consistent and implementable framework of calculation. We are now extending these studies to FRW cosmologies.

We emphasized in this study that thermalization and particle production are fundamentally different processes and in the non-linear regime likely to occur on widely different time scales.

A. A possible solution to the Polonyi problem:

As mentioned before a problem of much current interest in which this work may have some implications is the "post-modern Polonyi problem" concerning flat directions for some of the moduli fields in string theories [17]. In these scenarios the Polonyi fields are very light (of order of a Tev) with couplings to the particle physics sector that is extremely small (order of 10^{-17}). A calculation of the reheating temperature based on the "old" scenario of one particle decay give $T_{reh} \approx 0.01$ Mev , well below the nucleosynthesis scale.

This may not necessarily be true given our analysis. Since particle production occurs not just during oscillations around the minimum now but throughout the evolution of the the moduli field. Even for couplings $\mathcal{O}(m_\phi/M_{pl})$ since the initial amplitudes of these moduli field are $\mathcal{O}(M_{pl})$ then the mechanism of particle production via induced amplification will be very efficient and our analysis leads to an estimate of the reheating temperature $T_{reh} \approx m_\phi \approx 1$ Tev that is larger than the electroweak scale, alleviating the Polonyi problem. Thus this mechanism may very well provide a solution to the Polonyi problem reconciling moduli field cosmology with Standard Big Bang cosmology.

ACKNOWLEDGMENTS

D. B. would like to thank F. Cooper, and R. Pisarski for illuminating conversations. D.B. and D.-S. Lee thank the N.S.F. for support under grant awards: PHY-9302534 and INT-9216755. D.B., R.H. and D.-S. Lee would like to thank E. Mottola and A. Linde for interesting discussions. H. J. de V. would like to thank I. Krichever , A. Linde and N. Sánchez for useful discussions. D.S. thanks H.L. Yu and B.L. Hu for enlightening conversations. H. J. de V. has been supported in part by a CNRS-NSF binational grant. R. Holman was supported in part by DOE contract DE-FG02-91-ER40682. M. D'A. would like to thank LPTHE (Paris VI-VII) for kind hospitality.

APPENDIX A: A PEDAGOGICAL EXERCISE

In this appendix we show explicitly how to implement the tadpole method for the case of the inflaton coupled to a lighter scalar field via a simple trilinear coupling. We will not worry about renormalization issues here, though they were, of course, dealt with in the text. The formalism for non-equilibrium quantum field theory has already been described several times in the literature [29]. A path integral representation involves an integration along a path in the complex time plane with forward, backward branches and if the initial density matrix was that of an equilibrium system at an (initial) temperature T also an imaginary time branch. For real time correlation functions the imaginary time branch does not contribute and only determines the boundary conditions on the Green's functions.

We use the tadpole method [8,13,28] to study the time evolution of the expectation value of the inflaton.

$$\phi(\vec{x},t) \equiv \frac{\text{Tr}\left[\Phi^+(\vec{x})\rho(t)\right]}{\text{Tr}\,\rho(0)} = \frac{\text{Tr}\left[\Phi^-(\vec{x})\rho(t)\right]}{\text{Tr}\,\rho(0)}, \tag{A1}$$

where $\Phi^\pm$ are the fields defined on the forward and backward branches respectively. We set

$$\Phi^\pm(\vec{x},t) = \phi(\vec{x},t) + \chi^\pm(\vec{x},t), \tag{A2}$$

where $\chi^\pm$ are field fluctuation operators defined along the respective branches.

The effective evolution equation for the background field $\phi(\vec{x},t)$ follows from the condition.

$$< \chi^\pm(\vec{x},t) > = 0. \tag{A3}$$

Treating the *linear* and non-linear terms in $\chi^\pm$ as interactions and imposing the condition (A3) consistently in a perturbative or loop expansion, one obtains expressions of the form (here we quote the equation obtained from $< \chi^+(\vec{x},t) > = 0$)

$$\int d\vec{x}'dt' \left[< \chi^+(\vec{x},t)\chi^+(\vec{x}',t') > \mathcal{O}^{++}(\vec{x}',t') + < \chi^+(\vec{x},t)\chi^-(\vec{x}',t') > \mathcal{O}^{+-}(\vec{x}',t') \right] = 0 \tag{A4}$$

and similarly for $< \chi^-(\vec{x},t) > = 0$. The $\mathcal{O}^{\pm\pm}$ are in general integro-differential operators acting on the background field.

Because the Green's functions $< \chi^+(\vec{x},t)\chi^+(\vec{x}',t') >$, etc. are all independent one obtains the equations of motion in the form

$$\mathcal{O}^{++}(\vec{x}',t') = 0, \quad \mathcal{O}^{+-}(\vec{x}',t') = 0, \quad \mathcal{O}^{-+}(\vec{x}',t') = 0, \quad \mathcal{O}^{--}(\vec{x}',t') = 0. \tag{A5}$$

It is a consequence of the properties of the non-equilibrium Green's function (see below) and ultimately a consequence of unitarity that all the integro-differential operators $\mathcal{O}$ are the same.

Consider the Lagrangian density

$$\mathcal{L}(\Phi,\sigma) = \mathcal{L}_0(\Phi) + \mathcal{L}_0(\sigma) + g(t)\,\Phi\,\sigma^2, \tag{A6}$$

with the $\mathcal{L}_0$ being the free field Lagrangian density (with respective mass terms) and have allowed the coupling to depend on time. The non-equilibrium path integral requires $\mathcal{L}(\Phi^+,\sigma^+) - \mathcal{L}(\Phi^-,\sigma^-)$. In the tadpole method we write $\Phi^\pm(\vec{x},t) = \chi^\pm(\vec{x},t) + \phi(t)$ and identify $\phi(t)$ as the (non-equilibrium) expectation value of the field Φ . This identification then requires that $< \chi(\vec{x},t) > = 0$ where the expectation value is in the non-equilibrium density matrix with the path integral representation along the contour in complex time. After this shift, the action reads:

$$L = \int d^3x dt \left\{ \mathcal{L}_0(\chi^+) + \mathcal{L}_0(\sigma^+) + \chi^+ \left[-\ddot{\phi} - m_\Phi^2\phi\right] + g(t)\phi(t)(\sigma^+)^2 + g(t)\chi^+(\sigma^+)^2 \right.$$
$$\left. - \left(\chi^+,\sigma^+ \to \chi^-,\sigma^-\right)\right\}.$$

The linear term in $\chi^{\pm}$ is included as a perturbation. The first contribution to the equation of motion is obtained from this linear term, from the condition $< \chi^{+}(\vec{x},t) >= 0$ one obtains to this order

$$\int dt' \left\{ < \chi^{+}(t)\chi^{+}(t') > (i) \left[-\ddot{\phi} - m_{\Phi}^2 \phi \right] - < \chi^{+}(t)\chi^{-}(t') > (i) \left[-\ddot{\phi} - m_{\Phi}^2 \phi \right] \right\} = 0 , \quad (A7)$$

where we have suppressed the spatial arguments. Because the correlation functions $< \chi^{+}(t)\chi^{+}(t') >$, $< \chi^{+}(t)\chi^{-}(t') >$ are independent, one obtains the tree level equations of motion. It is straightforward to see that the same is obtained by imposing $< \chi^{-}(\vec{x},t) >= 0$. In an amplitude expansion (an expansion in powers of $\phi(t)$) the one loop correction to the equation of motion is obtained by expanding (the exponential of) $g\phi(t)(\sigma^{+})^2 + g\chi^{+}(\sigma^{+})^2 - (+ \rightarrow -)$ to first and second order. The first order gives a tadpole contribution, the second order needs one vertex with χ, the other with ϕ. One obtains

$$\int dt' < \chi^{+}(t)\chi^{+}(t') > \left\{ (i)\left(-\ddot{\phi} - m_{\Phi}^2 \phi\right) + (ig(t')) < (\sigma^{+}(t'))^2 > + \right.$$
$$\int dt''(ig(t''))^2 \left[< (\sigma^{+}(t'))^2(\sigma^{+}(t''))^2 > - < (\sigma^{+}(t'))^2(\sigma^{-}(t''))^2 > \right] \phi(t'') \right\} -$$
$$\int dt' < \chi^{+}(t)\chi^{-}(t') > \left\{ (i)\left(-\ddot{\phi} - m_{\Phi}^2 \phi\right) + (ig(t')) < (\sigma^{-})^2(t') > - \right.$$
$$\int dt''(-ig(t''))^2 \left[< (\sigma^{-}(t'))^2(\sigma^{-}(t''))^2 > \phi(t'') - < (\sigma^{-}(t'))^2(\sigma^{+}(t''))^2 > \right] \phi(t'') \right\} = 0 .$$

The expectation values are computed using Wick's theorem and using the free-field Green's functions of section II. The tadpole (time independent) is absorbed in a shift of the expectation value. The coefficient of $< \chi^{+}\chi^{+} >$, $< \chi^{+}\chi^{-} >$ must vanish independently because these Green's functions are independent and must vanish at all times. From the Green's functions (and more generally from the formal time contour integral) it is seen that the equations obtained are identical. If the expectation value is translational invariance, the spatial integrals set the momentum transfer to zero. For example using the zero temperature Green's functions of section I, one finds that the term that has the non-local (in time) correlation functions (last term) becomes

$$\int dt''(ig(t''))^2 \left[< (\sigma^{+}(t'))^2(\sigma^{+}(t''))^2 > - < (\sigma^{+}(t'))^2(\sigma^{-}(t''))^2 > \right] \phi(t'') =$$
$$2 \int dt''(ig(t''))^2 \phi(t'')\Theta(t' - t'') \int \frac{d^3k}{(2\pi)^3}(-2i)\frac{\sin\left[2\omega_k(t' - t'')\right]}{4\omega_k^2} .$$

For the resummed one-loop approximation the term $g\phi(t)(\sigma^{\pm})^2$ is absorbed in the mass term for the σ field and only the $\chi\sigma^2$ terms are considered as perturbation. In this case the 1-loop contribution is obtained from the tadpole term $< (\sigma^{\pm}(t))^2 >$, which is now time dependent and obtained from the non-equilibrium Green's functions constructed with the mode functions for the time-dependent mass as in section IV. One now finds the evolution equations

$$\ddot{\phi}(t) + m_{\Phi}^2 \phi(t) + g(t) < \sigma^2(\vec{x},t) >= 0 ,$$
$$< \sigma^2(\vec{x},t) >= -i \int \frac{d^3k}{(2\pi)^3} G_{k,\sigma}^{++}(t,t) = \int \frac{d^3k}{(2\pi)^3} \frac{|V_k(t)|^2}{2w_k} \coth\left[\frac{\beta w_k}{2}\right] ,$$

$$\left[\frac{d^2}{dt^2} + \vec{k}^2 + m^2(t)\right] V_k(t) = 0 \ ,$$

$$V_k(0) = 1 \ , \quad \dot{V}_k(0) = -iw_k = -i\sqrt{\vec{k}^2 + m^2(0)} \ ,$$

$$m^2(t) = m_\sigma + g(t)\phi(t) \ .$$

REFERENCES

† Laboratoire Associé au CNRS UA280.

[1] For early reviews on the inflationary scenario see for example: R. H. Brandenberger, Rev. of Mod. Phys. 57, 1 (1985); Int. J. Mod. Phys. A2, 77 (1987); L. Abbott and S. Y. Pi, Inflationary Cosmology, World Scientific, 1986; A. D. Linde, Rep. Prog. Phys. 47, 925 (1984).

[2] A. D. Linde, Particle Physics and Inflationary Cosmology, Harwood, Chur, Switzerland, 1990; Lectures on Inflationary Cosmology, in Current Topics in Astrofundamental Physics, 'The Early Universe', Proceedings of the Chalonge Erice School, N. Sánchez and A. Zichichi Editors, Nato ASI series C, vol. 467, 1995, Kluwer Acad. Publ. "Quantum Cosmology and the Structure of Inflationary Universe" gr-qc/9508019 (1995).

[3] E. Kolb and M. S. Turner, The Early Universe, Addison-Wesly, Redwood city, 1990.

[4] A. D. Dolgov and A. D. Linde, Phys. Lett. 116B, 329 (1982).

[5] A. D. Linde, Phys. Lett. 129B, 177 (1983).

[6] L. F. Abbott, E. Farhi and M. B. Wise, Phys. Lett. 117B, 29 (1982).

[7] A. D. Dolgov and D. P. Kirilova, Sov. J. Nucl. Phys. 51, 273 (1990).

[8] D. Boyanovsky and H. J. de Vega, Phys. Rev. D47, 2343 (1993); D. Boyanovsky, D.-S. Lee and A. Singh, Phys. Rev. D48, 800 (1993).

[9] D. Boyanovsky, H. J. de Vega and R. Holman, Phys. Rev. D49, 2769 (1994).

[10] See for a review, D. Boyanovsky, H. J. de Vega and R. Holman, in the Proceedings of the Second Paris Cosmology Colloquium, Observatoire de Paris, June 1994, p. 127-215, H. J. de Vega and N. Sánchez Editors, World Scientific, 1995.

[11] L. Kofman, A. D. Linde, and A. A. Starobinsky, Phys. Rev. Lett. 73, 3195 (1994) and " Non-Thermal Phase Transitions After Inflation", hep-th-9510119 (1995).

[12] Y. Shtanov, J. Traschen and R. Brandenberger, Phys. Rev. D51, 5438 (1995).

[13] D. Boyanovsky, H. J. de Vega, R. Holman, D.-S. Lee and A. Singh, Phys. Rev. D51, 4419 (1995).

[14] D. Boyanovsky, H. J. de Vega and R. Holman, in Advances in Astrofundamental Physics, Erice Chalonge Course, N. Sánchez and A. Zichichi Editors, World Scientific, 1995.

[15] L. D. Landau and E. M. Lifshits, Mechanics, Pergamon Press, London, 1958.

[16] A. Dolgov and K. Freese, Phys. Rev. D51, 2693 (1995).

[17] T. Banks, D. Kaplan, A. Nelson, Phys. Rev. D49, 779 (1994); L. Randall and S. Thomas, "Solving the Cosmological Moduli Problem with Weak Scale Inflation", MIT preprint MIT-CCP-2331, hep-ph/9407248; T. Banks, M. Berkooz and P.J. Steinhardt "The Cosmological Moduli Problem, Supersymmetry Breaking and Stability in Postinflationary Cosmology", Rutgers preprint RU-94-92, hep-ph/9501053.

[18] See the contribution of Tom Banks to these proceedings

[19] D. Lyth and E. Stewart, "Thermal inflation and the moduli problem" hep-ph- 9510204 (1995).

[20] L. Randall, "Flat directions and baryogenesis in supersymmetric theories" hep-ph-9507266 (1995).

[21] C. J. Hogan, Phys. Rev. Lett. 74, 3105 (1995).

[22] B. Mueller in "Particle Production in Highly Excited Matter", p. 11, H. H. Gutbrod and J. Rafelski Editors, Plenum Press, N.Y., 1993; B. Mueller, Rep. of Prog. in Phys. 58, 611 (1995).

[23] For a recent review see: A. G. Cohen, D. B. Kaplan and A. E. Nelson, Ann. Rev. Nucl. Part. Sci. 43 (1993).

[24] D. Boyanovsky, H. J. de Vega and R. Holman, Phys. Rev. D51, 734 (1995).

[25] See for example: N. P. Landsman and Ch. G. van Weert, Phys. Rep. 145, 141 (1987).

[26] Y. Kluger, J. M. Eisenberg, B. Svetitsky, F. Cooper and E. Mottola, Phys. Rev. D45, 4659 (1992); Phys. Rev. Lett. 67, 2427 (1991).

[27] D. Boyanovsky, M. D'Attanasio, H. J. de Vega, R. Holman, and D.-S. Lee, "Linear vs. Non-Linear Relaxation: Consequences for Reheating and Thermalization" (to appear in Phys. Rev. D 1995) hep-ph-9507414.

[28] P. Elmfors, K. Enqvist and I. Vilja, Nucl. Phys. B422, 521 (1994).

[29] see for example: E. Calzetta and B. L. Hu, Phys. Rev. D35, 495 (1988); *ibid* D37, 2838 (1988); J. P. Paz, Phys. Rev. D41, 1054 (1990); *ibid* D42, 529 (1990); B. L. Hu, in the Proceedings of the Second Paris Cosmology Colloquium, Observatoire de Paris, June 1994, p.111, H. J. de Vega and N. Sánchez Editors, World Scientific, 1995; G. Zhou, Z. Su, B. Hao and L. Yu, Phys. Rep. 118, 1 (1985); A. Niemi and G. Semenoff, Ann. of Phys. (N.Y.) 152, 105 (1984); Nucl. Phys. B230, 181 (1984).

[30] H. A. Weldon, Phys. Rev. D28, 2007 (1983).

[31] W. Keil, Phys. Rev. D40, 1176 (1989).

[32] H. A. Weldon, Ann. of Phys. (N.Y.) 228, 43 (1993).

[33] R. J. Eden, P. V. Landshoff, D. I. Olive and P. C. Polkinghorne, The Analytic S-Matrix, Cambridge Univ. Press, 1966.

[34] R. D. Pisarski, Phys. Rev. Lett. 63, 1129 (1989).

[35] E. Braaten and R. D. Pisarski, Nucl. Phys. B337, 569 (1990); *ibid* B339, 310 (1990).

[36] F. Cooper, S. Habib, Y. Kluger, E. Mottola and J. P. Paz, Phys. Rev. D50, 2848, (1994); F. Cooper, Y. Kluger, E. Mottola and J. P. Paz. Phys. Rev. D51, 2377 (1995).

[37] M. Yoshimura, "Catastrophic particle production under periodic perturbation", hep-th-9506176 and "Particle production in dissipative cosmic field", hep-ph-9508378 (1995).

[38] David I. Kaiser, "Post-Inflation Reheating in an Expanding Universe" astro-ph-9507108 (1995).

THE ELECTROWEAK PHASE TRANSITION:

A STATUS REPORT

LAURENCE G. YAFFE*
University of Washington
Department of Physics
Seattle, Washington 98195-1560 USA

The possibility that the observed cosmological baryon density might be a consequence of non-perturbative dynamics at the electroweak phase transition is one of the most exciting topics at the interface of particle physics and cosmology. However, determining the viability of electroweak baryogenesis in (minimal extensions of) the standard model requires knowledge of the equilibrium behavior of the electroweak phase transition, non-equilibrium dynamics around expanding bubble walls, and non-perturbative baryon violating processes in both the high and low temperature phases [1].

This talk will examine the current understanding of the electroweak phase transition. Determining the order of the phase transition, and the magnitudes of the latent heat, correlation lengths, and the baryon violation rate at the transition are some of the key questions. Viable baryogenesis scenarios require a first order transition with rapid suppression of baryon violating transitions in the low temperature phase [1, 2, 3].

Performing reliable quantitative calculations of electroweak transition properties is challenging, in part because there is important dynamics, both perturbative and non-perturbative, on many differing length scales. Approximation methods which have been applied to this problem include weak coupling perturbation theory (or mean field theory) [2, 4, 5], ϵ-expansions [6, 7, 8], and numerical simulations [9, 10, 11, 12]. Each of these approaches have significant (but differing) limitations; for example, ordinary perturbation theory is valid if the Higgs mass is much lighter than the W-boson mass, but does not appear to be trustworthy for the experimentally allowed range of possible Higgs masses. Nevertheless, all of these approaches

*Research supported in part by Department of Energy grant DE-FG06-91ER40614.

493

N. Sánchez and A. Zichichi (eds.), String Gravity and Physics at the Planck Energy Scale, 493–506.
© *1996 Kluwer Academic Publishers.*

have helped to elucidate the rich interplay of the physics at the electroweak phase transition.

I will focus on the behavior of the minimal standard model, despite the fact that electroweak baryogenesis in the strictly minimal model cannot explain the observed cosmological baryon density. There are two basic problems with baryogenesis in the minimal model. For experimentally allowed values of the Higgs mass, most of the baryon excess which is produced at the phase transition is destroyed shortly thereafter [2]. And even if that weren't the case, the baryon excess produced in the minimal standard model would still be far too small to agree with present day observations [1]. The first problem will be discussed further below. The second problem is a consequence of the nature of CP violation in the minimal model. Without fundamental CP violation, no baryon excess can be dynamically produced. But in the minimal standard model, CP violation arises only through a phase in the CKM matrix elements, and can only appear in processes involving all three generations of fermions. Because of the freedom to redefine the relative phases of fermion fields, one may show that any CP violating transition must involve a reparameterization invariant combination of Yukawa couplings and mixing angles which is tiny [3],

$$\delta_{\rm CP} \lesssim 10^{-20}. \tag{1}$$

Given this suppression, simple estimates show that it is essentially impossible to produce a baryon-to-photon ratio of the required 10^{-10} magnitude [1].

Both of these difficulties can be avoided in extensions of the minimal standard model which incorporate, for example, a single additional scalar field. This allows explicit CP violation in the scalar sector, and can increase the sphaleron mass and thereby slow down the rate of baryon-violating reactions just after the transition. However, the challenges involved in analyzing the electroweak phase transition are, for the most part, generic to any electroweak theory. Consequently, the minimal standard model provides a useful toy model (with the fewest adjustable parameters) which may be used as a testing ground for quantitative calculations of electroweak phase transition properties.

1. Perturbation Theory

Since the standard model is weakly coupled (at scales of several hundred GeV), perturbation theory is the most obvious approach for studying the electroweak phase transition. A one-loop calculation of the effective potential for the Higgs field is elementary and gives (schematically)

$$V(\phi) = V_{\rm tree}(\phi) + T \int d^3k \, \ln\left(1 - e^{-E(k)/T}\right) \tag{2}$$

$$= \left(-\mu^2 + a\, g^2 T^2\right) \phi^2 - b\, g^3 T\, \phi^3 + \lambda \phi^4 + (\phi\text{-independent}) \quad (3)$$

For simplicity, only the thermal W-boson contribution is indicated in (2). The W-boson energy $E(k) \equiv \sqrt{k^2 + M_W(\phi)^2}$ with $M_W(\phi) \equiv \frac{1}{2} g\phi$, and a and b are dimensionless constants which I will henceforth suppress.

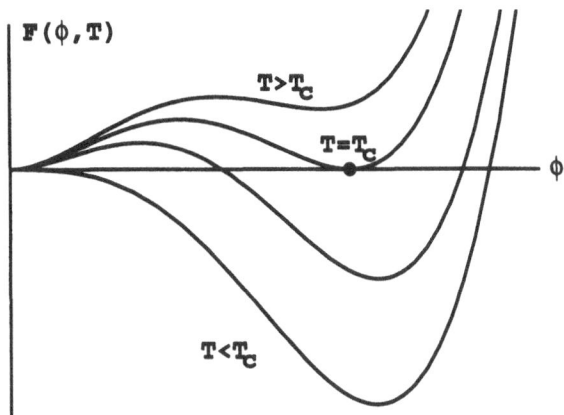

Figure 1. The form of the free energy, as a function of ϕ, for different temperatures.

The $O(T^2)$ thermal correction to the quadratic term in the potential is responsible for driving the transition from the Higgs phase, with $\langle \phi \rangle \neq 0$, to an unbroken symmetry phase with $\langle \phi \rangle = 0$, as the temperature is increased. If corrections to this one-loop result are negligible, then the presence of the $O(g^3 T)$ cubic term will cause the transition to be first-order, with a barrier separating co-existing broken and unbroken symmetry phases at the transition temperature T_c. (See Fig. 1.) In the Higgs phase near the transition, the quadratic, cubic, and quartic terms in the potential (3) are all comparable in size. Consequently, $\phi_c \sim g^3 T/\lambda$, or

$$\left. \frac{M_W}{g^2 T} \right|_{T_c} \sim \frac{g^2}{\lambda} \sim \left. \frac{M_W^2}{M_H^2} \right|_{T=0}. \quad (4)$$

These ratios are important for several reasons. First, the rate of non-perturbative baryon violating reactions in the low temperature Higgs phase is controlled by a thermal activation energy given by the (temperature dependent) sphaleron mass,

$$\Gamma_{\Delta B} \sim T^4\, e^{-M_{\rm sph}(T)/T}, \quad (5)$$

and the sphaleron mass $M_{\rm sph}(T)$ equals $16\pi M_W(T)/g^2$ (to within a factor of 2). Hence, the relation (4) implies that the baryon violation rate, just after the transition, is exponentially sensitive to the zero temperature ratio of Higgs and W-boson masses,

$$\ln \Gamma_{\Delta B} \sim - \left. \frac{M_W^2}{M_H^2} \right|_{T=0}. \tag{6}$$

In order for electroweak baryogenesis to be viable, baryon violating reactions after the transition must turn off sufficiently rapidly so that the baryon asymmetry produced during the transition can survive to the present day. The simple result (6) implies that this will only be possible if the Higgs to W mass ratio is sufficiently small. More detailed analysis, based on the one-loop effective potential in the minimal standard model, produces an estimate of 35 Gev for the *upper* limit on the Higgs mass if baryon violating rates after the transition are to be acceptably small [2]. This, of course, is inconsistent with the current experimental *lower* bound on the Higgs mass of about 65 GeV.

As noted above, one can circumvent this "no-go" result if minor extensions (such as adding a second scalar field) are made to the minimal standard model. However, one should first ask whether the 35 GeV bound in the minimal model, or any other conclusion based on the one-loop analysis (including the predicted first order nature of the transition!) is reliable. To answer this, one must understand whether higher order corrections to the one loop results are significant.

Life would be simple if the temperature T were the only relevant scale for physics at the phase transition. If this were the case, then higher order corrections would automatically be suppressed by powers of g^2 (or α_W) at the scale T, which is, in fact, small.[1] However, life is not so simple. Consider, for example, any effective potential diagram containing only gauge field lines. The contribution of each loop may be estimated as[2]

$$g^2 T \int d^3 q \left(\mathbf{q}^2 + M_W(T)^2 \right)^{-2} \sim g^2 T / M_W(T). \tag{7}$$

Therefore the real loop expansion parameter is not g^2, but rather equals $g^2 T / M_W(T)$. Consequently, the reliability of perturbation theory, in the Higgs phase near the transition, is controlled by the the same ratio of

$$\frac{g^2 T_c}{M_W(T_c)} \sim \left. \frac{M_H^2}{M_W^2} \right|_{T=0}$$

[1] This assumes that the Higgs is light enough to be weakly coupled.

[2] Recall that Euclidean space frequencies are discrete at finite temperature, $q^0 = \omega_n \equiv (2\pi n T)$. Convergent diagrams are dominated by the static $n=0$ frequency component.

which governs the baryon violation rate. Perturbation theory is not reliable unless the physical Higgs is sufficiently light.

Of course, the above estimate does not determine whether perturbation theory really breaks down at $M_H = M_W/2$, or at $M_H = 2M_W$. For a quantitative answer, one must actually compute higher order corrections to various quantities of interest. Fortunately, a number of two-loop calculations of phase transition properties have now been performed. Fig. 2 illustrates the result for the effective potential at a Higgs mass of 35 GeV [4, 5]. The magnitude of $\langle\phi\rangle$ in the Higgs phase at the transition shifts by only about 20%, but the height of the barrier changes by almost a factor of three!

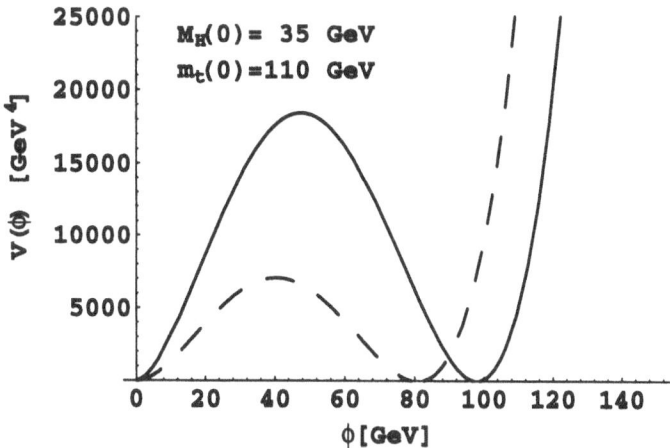

Figure 2. The effective potential at the critical temperature for $M_H(0) = 35$ GeV and $m_t(0) = 110$ GeV. The dashed and solid lines are the one-loop and two-loop results respectively. [Why a 110 GeV top mass? Because this is an old graph. But the results aren't particularly sensitive to m_t.]

Since both the barrier height, and the Higgs expectation value, are gauge dependent, one might legitimately wonder whether truly physical quantities will be better behaved. However, a two-loop evaluation of the dimensionless latent heat, $\Delta Q/T_c^4$, and surface tension, σ/T_c^3, at a Higgs mass of 50 GeV finds 50–100% changes over the one-loop results [12]. Consequently, there appears to be good reason to expect a Higgs mass somewhere between 35 and 50 GeV to be the upper limit for the reliability of perturbation theory for most quantities.[3]

---

[3] Note, however, that the authors of ref. [11] claim that using the 3d renormalization group to improve some logarithmic corrections can delay the breakdown of perturbation theory.

Finally, even when the Higgs to W mass ratio is small, so that perturbation theory in the Higgs phase is well behaved, the reliability of perturbative calculations of phase transition properties is limited by the presence of non-perturbative physics in the unbroken symmetry phase. The long distance physics of the unbroken phase is described by a three dimensional non-Abelian theory with a confinement scale of $g^2 T$, and has a free energy density of order $(g^2 T)^3$ which is incalculable in perturbation theory. Consequently, for small values of the Higgs fields, where $M_W(\phi) \lesssim g^2 T$, any perturbative approximation to the effective potential will have a systematic uncertainty of order $g^6 T^4$. This will generate an uncertainty in the value of the transition temperature, or any other phase transition property. However, when λ/g^2 is small, this uncertainty is of the same order as the unknown four-loop contributions to the free energy in the Higgs phase [4].

2. Numerical Simulations

The electroweak phase transition can be also be studied numerically by performing stochastic simulations in a lattice version of the theory. Simulating the full theory (with chiral fermions and $SU(3) \times SU(2) \times U(1)$ gauge fields) is not practical. However, since one is interested in high temperature physics, all non-zero frequency Fourier components of fields have very short correlation lengths (of order T^{-1}) and may be integrated out perturbatively. Fermion fields, being antiperiodic, have no static components and may be completely eliminated. Once the fermions are gone, the $SU(3)$ gauge field is decoupled and may be dropped, as may the $U(1)$ gauge field if the small weak mixing angle is treated as perturbation. This leaves an $SU(2)$-Higgs theory, which is straightforward to define on a lattice.

During the past two years, several groups have performed large scale simulations of the finite temperature phase transition in $SU(2)$-Higgs theory. The authors of refs. [9, 12] have chosen to simulate a four-dimensional $SU(2)$-Higgs theory with a periodic "time" dimension consisting of $N_t = 2, 3, \ldots$ lattice spacings. In contrast, the authors in refs. [10, 11] begin with the effective three dimensional theory which results from integrating out all non-static Fourier components. Spatial lattice sizes in these simulations are substantial (typically 16^3 to 32^3).

I will not attempt to summarize all the results of these lattice simulations (which include data on the transition location and character, latent heat, surface tension, and correlation lengths). Interested readers should refer to the latest papers. However, if one asks what lessons can be extracted from the results of these simulations, the the following points appear to be fairly well established:

a. Reasonably good calculations can be performed, at least for light Higgs masses. Finite lattice size and non-zero lattice spacing errors are (at least beginning to be) under reasonable control.

b. For Higgs masses below M_W, a first order phase transition is clearly seen. The discontinuity at the transition becomes steadily weaker with increasing Higgs mass.

c. When $M_H < 20$ GeV, the lattice results agree quite well with perturbative predictions. When $M_H \gtrsim 50$ GeV, the strength of the transition appears to be larger than two-loop perturbative predictions.

d. Simulations (of sufficient accuracy to study a weakly discontinuous transition) become progressively more difficult as the Higgs mass grows. So far, no convincing conclusions can be drawn about the nature of the transition when $M_H > M_W$.

3. ϵ-expansions

The ϵ-expansion provides an alternative systematic approach for computing the effects of (near)-critical fluctuations. It is based on the idea that instead of trying to solve a theory directly in three spatial dimensions, it can be useful to generalize the theory from 3 to $4-\epsilon$ spatial dimensions, solve the theory near four dimensions (when $\epsilon \ll 1$), and then extrapolate to the physical case of 3 spatial dimensions. Specifically, one expands physical quantities in powers of ϵ and then evaluates the resulting (truncated) series at $\epsilon = 1$ [13]. This can provide a useful approximation when the relevant long distance fluctuations are weakly coupled near 4 dimensions, but become sufficiently strongly coupled that the loop expansion parameter is no longer small in three dimensions.

Scalar ϕ^4 theory (or the Ising model) is a classic example. In four dimensions, the long distance structure of a quartic scalar field theory is trivial; this is reflected in the fact that the renormalization group equation $\mu(d\lambda/d\mu) = c\,\lambda^2 + O(\lambda^3)$, has a single fixed point at $\lambda = 0$. In $4-\epsilon$ dimensions, the canonical dimension of the field changes and the renormalization group equation acquires a linear term,

$$\mu\frac{d\lambda}{d\mu} = -\epsilon\,\lambda + c\,\lambda^2 + O(\lambda^3)\,.$$

This has a non-trivial fixed point (to which the theory flows as μ decreases) at $\lambda^* = \epsilon/c + O(\epsilon^2)$. The fixed point coupling is $O(\epsilon)$ and thus small near four dimensions, but grows with decreasing dimension and becomes order one when $\epsilon = 1$. Near four dimensions, a perturbative calculation in powers of λ is reliable and directly generates an expansion in powers of ϵ.

The existence of an infrared-stable fixed point indicates the presence of a continuous phase transition as the bare parameters of the theory are varied. Performing conventional (dimensionally regularized) perturbative calculations and evaluating the resulting series at the fixed point, one finds, for example, that the susceptibility exponent (equivalent to the anomalous dimension of ϕ^2) has the expansion [13, 14]

$$\gamma = 1 + 0.167\,\epsilon + 0.077\,\epsilon^2 - 0.049\,\epsilon^3 + O(\epsilon^4)\,. \tag{8}$$

Adding the first three non-trivial terms in this series, and evaluating at $\epsilon = 1$, yields a prediction which agrees with the best available result to within a few percent [15, 16].

Inevitably, perturbative expansions in powers of λ are only asymptotic; coefficients grow like $n!$, so that succeeding terms in the series begin growing in magnitude when $n \gtrsim O(1/\lambda)$. Expansions in ϵ are therefore also asymptotic, with terms growing in magnitude beyond some order $n \gtrsim O(1/\epsilon)$. If one is lucky, as is the case in the pure scalar theory, $O(1/\epsilon)$ really means something like three or four when $\epsilon = 1$ and the first few terms of the series will be useful. If one is unlucky, no terms in the expansion will be useful. Whether or not one will be lucky cannot be determined in advance of an actual calculation.

To apply the ϵ-expansion to electroweak theory, one begins with the full 3+1 dimensional finite temperature Euclidean quantum field theory (in which one dimension is periodic with period $\beta = 1/T$) and integrates out all non-static Fourier components of the fields. The integration over modes with momenta of order T or larger may be reliably performed using standard perturbation theory in the weakly-coupled electroweak theory. This reduces the theory to an effective 3-dimensional SU(2)-Higgs theory.[4] The effective theory depends on three relevant renormalized parameters:

$$g_1(T)^2 \quad - \quad \text{the SU(2) gauge coupling,}$$

$$\lambda_1(T) \quad - \quad \text{the quartic Higgs coupling,}$$

$$m_1(T)^2 \quad - \quad \text{the Higgs mass (squared).}$$

Next, one replaces the 3-dimensional theory by the corresponding $4-\epsilon$ dimensional theory (and scales the couplings so that g_1^2/ϵ and λ_1/ϵ are held fixed). This is the starting point for the ϵ-expansion. When ϵ is small, one may reliably compute the renormalization group flow of the effective

---

[4]Fermions, having no static Fourier components, are completely eliminated in the effective theory. For simplicity, the effects of a non-zero weak mixing angle and the resulting perturbations due to the U(1) gauge field are ignored. Finally, one may also integrate out the static part of the time component of the gauge field, since this field acquires an $O(gT)$ Debye-screening mass.

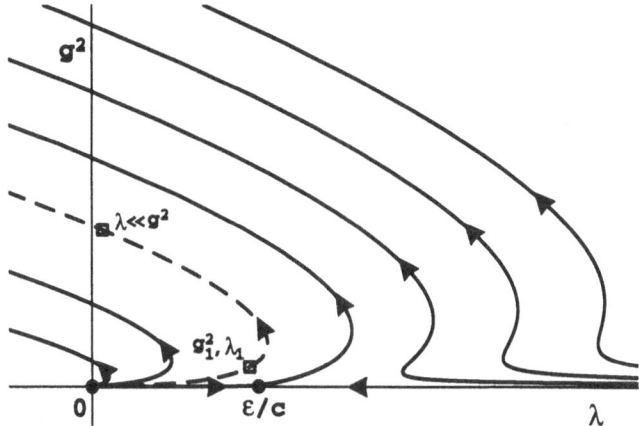

Figure 3. The renormalization group flow for an SU(2)-Higgs theory. Arrows indicate the direction of decreasing renormalization point. The dashed line is the trajectory which flows from an initial set of couplings (g_1^2, λ_1) into the region where $\lambda \ll g^2$.

couplings. The renormalization group equations have the form

$$\mu \frac{d\lambda}{d\mu} = -\epsilon \lambda + (a\, g^4 + b\, g^2 \lambda + c\, \lambda^2) + \cdots , \qquad (9)$$

$$\mu \frac{dg^2}{d\mu} = -\epsilon\, g^2 + \beta_0\, g^4 + \cdots . \qquad (10)$$

The precise values of the coefficients (and the next order terms) may be found in reference [6]. These equations may be integrated analytically, and produce the flow illustrated in figure 3.

Note that a non-zero gauge coupling renders the Ising fixed point at $\lambda = \epsilon/c$ unstable, and that no other (weakly coupled) stable renormalization group fixed point exists. Trajectories with $g^2 > 0$ eventually cross the $\lambda = 0$ axis and flow into the region where the theory (classically) would appear to be unstable. Such behavior is typically indicative of a first-order phase transition [17]. To determine whether this is really the case, one must be able to perform a reliable calculation of the effective potential (or other physical observables). As discussed earlier, the loop expansion parameter for long distance physics is $\lambda(\mu)/g^2(\mu)$. Consequently, the best strategy is to use the renormalization group to flow from the original theory at $\mu = T$, which may have $\lambda(T)/g^2(T)$ large, to an equivalent theory with $\mu \ll T$ for which $\lambda(\mu)/g^2(\mu)$ is small. This is equivalent to the condition that one decrease the renormalization point until it is comparable to the relevant scale for long distance physics, specifically, the gauge boson mass, M. By doing so, one eliminates large factors of $[(M/\mu)^\epsilon - 1]/\epsilon$ which would otherwise spoil the reliability of the loop expansion. (This, of course, is nothing other than the transcription to $4-\epsilon$ dimensions of the usual story in 4 dimensions, where appropriate use of the renormalization group allows one to sum up large logarithms which would otherwise spoil the perturbation expansion.)

For small ϵ, the change of scale required to flow from an initial theory where $\lambda_1/g_1^2 = O(1)$ to an equivalent theory with $\lambda(\mu)/g^2(\mu) \ll 1$ is exponentially large; the ratio of scales is

$$ s \equiv \frac{T}{\mu} \sim e^{\lambda_1/g_1^4} \sim e^{O(1/\epsilon)} . $$

Given the parameters $g^2(\mu)$, $\lambda(\mu)$ and $m^2(\mu)$ of the resulting effective theory, one may use the usual loop expansion to compute interesting physical quantities. Because the change in scale is exponentially sensitive to $1/\epsilon$, the result for a typical physical quantity will have the schematic form

$$ \mathcal{O} = f[g^2(\mu), \lambda(\mu), m^2(\mu)] \left(\frac{\mu}{T}\right)^{\#} \tag{11} $$

$$ \sim \epsilon^{\#} \left(1 + O(\epsilon) + \cdots\right) \exp\left[\frac{\#}{\epsilon} \left(1 + O(\epsilon) + \cdots\right)\right] . \tag{12} $$

In general, a calculation accurate to $O(\epsilon^n)$ requires an n-loop calculation in the final effective theory, together with $n+1$ loop renormalization group evolution.

To obtain predictions for the original theory in three spatial dimensions, one finally truncates the expansions at a given order and then extrapolates from $\epsilon \ll 1$ to $\epsilon = 1$. Just as for the simple ϕ^4 theory, the reliability of the resulting predictions at $\epsilon = 1$ can only be tested *a-posteriori*.

This procedure has been carried out for a variety of observables characterizing the electroweak phase transition at both leading and next-to-leading order in the ϵ-expansion [6]. Leading order results are available for the scalar correlation length in both symmetric and asymmetric phases, the free energy difference $\Delta F(T)$ between the symmetric and asymmetric phases, the latent heat $\Delta Q = -T(d\Delta F/dT)|_{T_c}$, the surface tension σ between symmetric and asymmetric phases at T_c, the bubble nucleation rate $\Gamma_N(T)$ below T_c, and the baryon violation (or sphaleron) rate $\Gamma_{\Delta B}(T_c)$.

The lowest order ϵ-expansion predictions differ from the results of standard one-loop perturbation theory (performed directly in three space dimensions) in several interesting ways. First, the ϵ-expansion predicts a *stronger* first order transition than does one-loop perturbation theory (as long as $M_H < 130$ GeV). The correlation length at the transition is smaller, and the latent heat larger, than the perturbation theory results. The size of the difference depends on the value of the Higgs mass (see reference [6] for quantitative results). Naively, one would expect that a stronger first order transition would imply a smaller baryon violation rate, since a larger effective potential barrier between the co-existing phases should decrease the likelyhood of thermally-activated transitions across the barrier. This expectation is wrong (in essence, because it unjustifiably assumes that the shape of the barrier remains unchanged). Along with predicting a strengthing of the transition, the ϵ-expansion predicts a *larger* baryon violation rate. This occurs because the baryon violation rate is exponentially sensitive to the sphaleron action (or mass),

$$\Gamma_{\Delta B} \propto \exp -S_{\text{sphaleron}},$$

and the sphaleron action depends inversely on ϵ,

$$S_{\text{sphaleron}} = \frac{\#}{g^2(\mu)} = O(1/\epsilon).$$

Hence, unlike other observables, the exponential sensitivity to $1/\epsilon$ in the baryon violation rate does not arise solely from an overall power of the scale factor μ/T.

Note that an increase in the baryon violation rate (compared to standard perturbation theory) makes the constraints for viable electroweak baryogenesis more stringent; specifically, the (lowest order) ϵ-expansion suggests that the minimal standard model bound $M_H \lesssim 35$–40 GeV derived using one-loop perturbation theory in ref. [2] should be even lower, further ruling out electroweak baryogenesis in the minimal model.

As emphasized earlier, in general there is no way to know, in advance of an actual calculation, how many terms (if any) in an ϵ-expansion will be useful when results are extrapolated to $\epsilon = 1$. Therefore, to assess the reliability of an ϵ-expansion one must be able to test predictions for actual physical quantities. For the electroweak theory, two types of tests are possible:

A. $\lambda \ll g^2$. In the limit of a light (zero temperature) Higgs mass, or equivalently small λ_1/g_1^2, the loop expansion in three dimensions is reliable. Hence, although this is not a realistic domain, one may easily test the reliability of the ϵ-expansion in this regime by comparing with direct three-dimensional perturbative calculations. Table 1 summarizes the fractional error for various physical quantities produced by truncating the ϵ-expansion at leading, or next-to-leading, order before evaluating at $\epsilon = 1$, in the light Higgs limit. Although the lowest-order results often error by a factor of two or more, all but one of the next-to-leading order results are correct to better than 10%. (The free energy difference at the limit of metastability, $\Delta F(T_0)$, has the most poorly behaved ϵ-expansion. However, if one instead computes the logarithm of this quantity, then the next-to-leading order result is correct to within 17%. The baryon violation rate is not shown because, due to the way its ϵ-expansion was constructed, the result is trivially the same as the three-dimensional answer when $\lambda_1 \ll g_1^2$. See ref. [6] for details.)

observable ratio		LO	NLO
asymmetric correlation length	ξ_{asym}	0.14	-0.06
symmetric correlation length	ξ_{sym}	0.62	-0.08
latent heat	ΔQ	-0.23	0.04
surface tension	σ	-0.40	-0.02
free energy difference	$\Delta F(T_0)$	-0.76	-0.44

TABLE 1. The fractional error in the ϵ-expansion results, when computing prefactors through leading order (LO) and next-to-leading order (NLO) in ϵ, when $\lambda_1 \ll g_1^2$.

B. $\lambda \gtrsim g^2$. When λ/g^2 is $O(1)$, the three-dimensional loop expansion is no longer trustworthy. However, one may still test the stability of ϵ-expansion predictions by comparing $O(\epsilon^n)$ and $O(\epsilon^{n+1})$ predictions — provided, of course, one can evaluate at least two non-trivial orders in the ϵ-expansion. For most physical quantities this is not (yet) possible; determining the lowest-order behavior of the prefactor in expansion (12) requires a one-loop calculation in the final effective theory together with a two-loop evaluation of the (solution to the) renormalization group equations. A consistent next-to-leading order calculation requires a two-loop calculation in the final theory together with three-loop renormalization group evolution. Althouth two-loop results for the effective potential and beta functions are known, three loop renormalization group coefficients in the scalar sector are not currently available. Nevertheless, by taking suitable combinations of physical quantities one can cancel the leading dependence on the scale ratio μ/T and thereby eliminate the dependence (at next-to-leading order) on the three loop beta functions. For example, the latent heat depends on the scale as $\Delta Q \sim (\mu/T)^{2+\epsilon}$ while the scalar correlation length $\xi \sim (\mu/T)^{-1}$. Therefore, the combination $\xi^2 \Delta Q$ cancels the leading $\mu/T \sim e^{O(1/\epsilon)}$ dependence and thus requires only two-loop information for its next-to-leading order evaluation. The result of the (rather tedious) calculation may be put in the form

$$\xi^2_{\text{asym}} \Delta Q = T^{1-\epsilon} f(f_1^2, \lambda_1) \left[1 + \delta + O(\epsilon^2) \right] , \tag{13}$$

where δ, the relative size of the next-to-leading order correction, is plotted in figure 4. The correction varies between roughly ±30% for (zero temperature) Higgs masses up to 150 GeV. This suggests that the ϵ expansion is tolerably well behaved for these masses. For larger masses the correction does not grow indefinitely, but is bounded by 80%, suggesting that the ϵ expansion may remain qualitatively useful even when it does not work as well quantitatively.

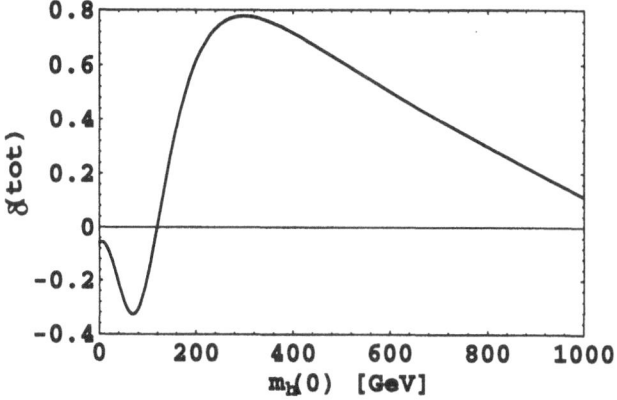

Figure 4. The relative size of the next-to-leading order correction to $\xi^2_{\text{asym}} \Delta Q$ in the ϵ-expansion. The values are given as a function of the (tree-level) zero-temperature Higgs mass in minimal SU(2) theory ($N = 2$) with $g = 0.63$.

4. Conclusions

The most useful technique for studying the electroweak phase transition clearly depends on the range of value of the Higgs mass. For sufficiently light Higgs ($M_H \lesssim 40$ GeV) perturbation theory is adequate. At intermediate masses (40 GeV $\lesssim M_H \lesssim 80$ GeV) numerical simulations have been quite effective. For heavier Higgs masses, the ϵ-expansion appears promising.

Although considerable progress has been made toward a quantitative understanding of the electroweak phase transition, much remains to be done. With continuing efforts, numerical results will undoubtedly improve, particularly for Higgs masses above M_W. Calculations of additional physical quantities at next-to-leading order in the ϵ-expansion should definitely be performed to further confirm the reliability of the method.

Nevertheless, it should be noted that quantitative understanding of the phase transition is already reasonably good in the range of Higgs masses

506

for which the transition is sufficiently discontinuous to be compatible with baryogenesis. Hence, the major source of uncertainty about the viability of electroweak baryogenesis appears to be the lack of knowledge about which theory is correct (extra Higgs, supersymmetry, *etc.*) not the limitations in our ability to compute the thermodynamics of these theories.

References

1. See, for example, A. Cohen, D. Kaplan and A. Nelson, *Annu. Rev. Nucl. Part. Sci.* **43**, 27 (1988), and references therein.

2. M. Dine, R. Leigh, P. Huet, A. Linde and D. Linde, *Phys. Lett.* **B238**, 319 (1992); *Phys. Rev.* **D46**, 550 (1992).

3. M. Shaposhnikov, *JETP Lett.* **44**, 465 (1986); *Nucl. Phys.* **B287**, 757 (1987); *Nucl. Phys.* **B299**, 797 (1988).

4. P. Arnold and O. Espinosa, *Phys. Rev.* **D47**, 3546 (1993).

5. J. Bagnasco and M. Dine, *Phys. Lett.* **B303**, 308 (1993).

6. P. Arnold and L. Yaffe, Phys. Rev. **D49**, 3003–3032 (1994).

7. M. Alford and J. March-Russell, *Nucl. Phys.* **B417**, 527 (1993);

8. M. Gleisser and E. Kolb, *Phys. Rev.* **D48**, 1560 (1993).

9. B. Bunk, E.-M. Ilgenfritz, J. Kripfganz and A. Schiller, *Phys. Lett.* **B284**, 371 (1992).

10. K. Kajantie, K. Rummukainen, and M. Shaposhnikov, *Nucl. Phys.* **B407**, 356 (1993).

11. K. Farakos, K. Kajantie, K. Rummukainen, and M. Shaposhnikov, *Phys. Lett.* **B336**, 494 (1994); *Nucl. Phys.* **B425**, 67 (1994); *Nucl. Phys.* **B442**, 317 (1995).

12. Z. Fodor, J. Hein, K. Jansen, A. Jaster and I. Montvay, *Nucl. Phys.* **B439**, 147 (1995); W. Buchmüller, Z. Fodor and A. Hebecker, DESY preprint DESY 95-028 (hep-ph/9502321), Feb. 1995; F. Ciskor, Z. Fodor, J. Hein, J. Heitger, CERN preprint CERN-TH-95-170 (hep-lat/9506029), June 1995;

13. K. Wilson and M. Fischer, *Phys. Rev. Lett.* **28**, 40 (1972); K. Wilson and J. Kogut, *Phys. Reports* **12**, 75–200 (1974), and references therein.

14. S. Gorishny, S. Larin, F. Tkachov, *Phys. Lett.* **101A**, 120 (1984).

15. J. Le Guillou, J. Zinn-Justin, *Phys. Rev. Lett.* **39**, 95 (1977); *ibid.*, *J. Physique Lett.* **46**, L137 (1985); *ibid.*, *J. Physique* **48**, 19 (1987); *ibid.*, *J. Phys. France* **50**, 1365 (1989); B. Nickel, *Physica* **A177**, 189 (1991).

16. C. Baillie, R. Gupta, K. Hawick and G. Pawley, *Phys. Rev.* **B45**, 10438 (1992); and references therein.

17. B. Halperin, T. Lubensky, and S. Ma, *Phys. Rev. Lett.* **32**, 292 (1974); J. Rudnick, *Phys. Rev.* **B11**, 3397 (1975); J. Chen, T. Lubensky, and D. Nelson, *Phys. Rev.* **B17**, 4274 (1978); P. Ginsparg, *Nucl. Phys.* **B170** [FS1], 388 (1980).

ASPECTS OF COSMIC-RAY ASTROPHYSICS IN THE GALAXY AND BEYOND

Maurice M. Shapiro
University of Maryland,
College Park, Maryland

ABSTRACT: We review some salient aspects of cosmic-ray astrophysics. Recent advances are discussed, and several outstanding problems are sketched.

1. Introduction

The XXIV International Conference on Cosmic Radiation convened in Rome from August 28 to September 8, 1985. It provided an overview of recent experimental findings and of new insights in this discipline. Perennially, the problem of cosmic-ray (CR) origin was discussed, notably such questions as injection, accelerating processes, and sites of CR production.

In solving these puzzles, perhaps the most useful set of clues is the composition of the CR at the source(s). This set encompasses both the elemental and isotopic abundances. In the latter category, experimental advances have been noteworthy, with progress in measuring the abundances of ever heavier nuclides, up into the iron group. Propagation theory is required to extrapolate the *observed* composition back to the sources, and the semi-empirical cross-sections of Silberberg and Tsao continue to be an important input into the calculations.

The role of shock acceleration in various environments has continued to be a flourishing theoretical discipline. Reacceleration has now been shown to work for cosmic-ray electrons, as well as for nuclei.

High-energy neutrino (Hev) astrophysics will contribute vitally to our knowledge of the dominant processes in the CR sources. Construction of several undersea observatories (and one under Antarctic ice) designed to initiate this new field have been started, but progress has been slow and difficult.

The lowest-energy cosmic rays have been the subject of

507

N. Sánchez and A. Zichichi (eds.), String Gravity and Physics at the Planck Energy Scale, 507–519.
© 1996 *Kluwer Academic Publishers.*

intense study, using satellites, space probes and balloons. With the Ulysses spacecraft, our knowledge of cosmic rays in the Heliosphere has expanded into the solar-polar domain.

New results were reported on the highest-energy cosmic rays, and the extensive air showers they generate. A proposal developed in a workshop at the University of Chicago was described. Its aim is to build two arrays of detectors, some 3,000 km^2 in area.

This lecture will be devoted mainly to the following topics: Sec. 2 will discuss the galactic dynamics of cosmic rays; Sec. 3 will deal with the composition and propagation of cosmic rays; Sec. 4 with injection and acceleration; Sec. 5 will describe the expected impact of Heν astrophysics on our understanding of cosmic rays; Sec. 6 will outline some unresolved aspects of cosmic ray astrophysics.

2. Galactic Dynamics of Cosmic Rays

The cosmic rays may be treated as a relativistic gas pervading the Galaxy. The lines of force of the magnetic fields in the Galactic disk are aligned approximately along the spiral arms. From observations of Faraday rotation, a magnetic field strength B of 3 to 6 microgauss (μG) has been deduced. The field is frozen into the interstellar (thermal) gas, but the CR gas is also tied to the field.

To appreciate how well most cosmic rays are confined by the magnetic field, we note that, for B = 3 μG, a CR proton of 10 GeV has a Larmor radius r of 10^{13} cm, comparable to one astronomical unit. However, the scale height of the Galactic disk is ~ 10^7-10^8 times greater. So particles of characteristic cosmic-ray energies (~10 GeV) are effectively trapped by the magnetic field.

The field is confined to the disk by the weight of the thermal interstellar gas. In order to maintain quasi-equilibrium perpendicular to the disk, the pressures exerted by the field, the thermal gas, and the cosmic-ray "gas" must be balanced by the gravity of the thermal gas.

$$\frac{d}{dz} \left(p_{gas} + \frac{B^2}{8\pi} + p_{CR} \right) = -\rho g$$

where the distance z is measured perpendicular to the galactic plane, the three pressures refer, respectively, to the thermal gas, the magnetic field and the cosmic-ray gas, ρ is the mean density of the thermal gas, and g is the gravitational acceleration normal to the disk; see Fig. 1.

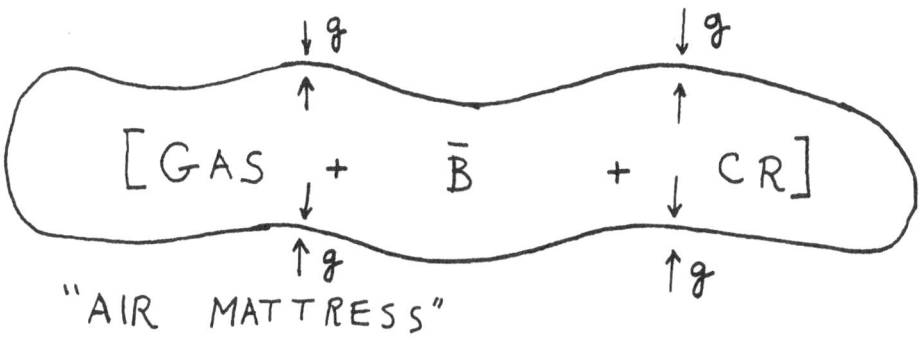

Fig. 1. Parker has aptly compared the disk to an air mattress: the gas, the field, and the cosmic-ray pressures tend to expand the disk, while gravity exerts the countervailing pressure.

It is known that the mean free path of most cosmic-ray nuclei for escape is shorter than the mean free path for nuclear collision. But how do the cosmic rays manage to escape from the Galactic disk? According to Parker's picture, this occurs thanks to instabilities in the interstellar medium (ISM), i.e., to perturbations in the gas density near a "surface" of the disk. The thermal gas is constrained to move along lines of force. (See Fig. 2).

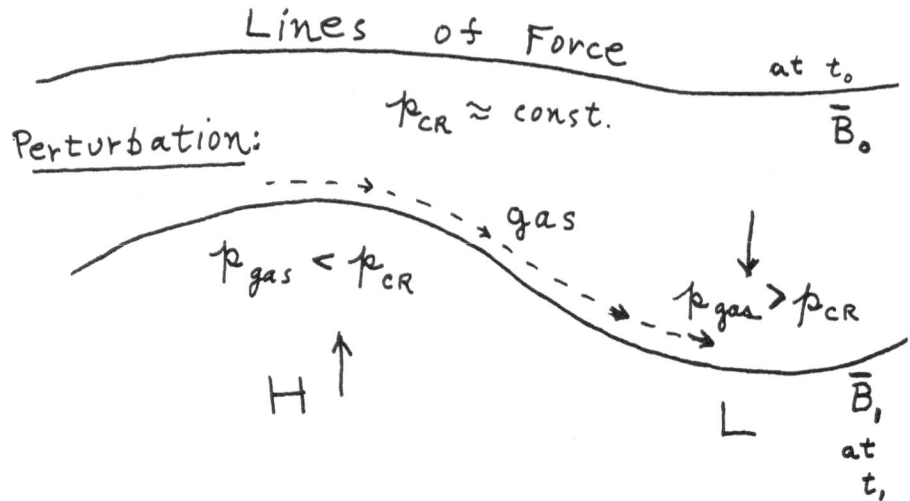

Fig. 2. If the lines of force are perturbed, e.g., by raising them at H and lowering them at L, then the gas tends to slide from high (H) to low (L) by gliding along the lines of force, burdening the low places and unloading the high places, where the field is then free to expand upward. (After E. N. Parker).

The formation of a cosmic-ray magnetic bubble near the surface of the Galactic disk, and the escape of cosmic rays are illustrated in Fig. 3. Thus, while the Galactic disk is a magnetic trap, it is a trap with escape hatches.

3. Composition and Propagation of Cosmic Rays

The mutability of cherished physical theories is exemplified by the early history of the cosmic-ray discipline. Widely held views as to the identity of the penetrating "primaries" changed from decade to decade. First they were gamma rays, then, successively, electrons (and positrons), protons, heavier nuclei up to iron, culminating in the periodic table of the elements. Paradoxically, the progression came full circle with the advent of gamma-ray astronomy, albeit the relevant fluxes are a thousand times lower than those of the nuclei. High-energy *cosmic* neutrinos are yet to be observed, but their detection in the near future seems likely. Antiprotons have been detected. The flux of these particles is consistent with their production as secondaries of CR collisions in space.

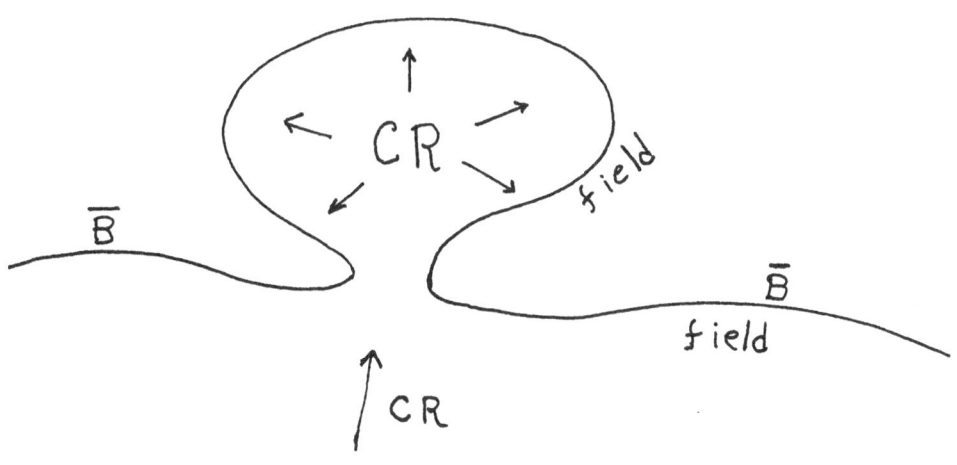

Fig. 3. The light fluids--the field and the cosmic rays--
bubble up through the heavy thermal gas. The CR inflate
the field, and push their way out, producing bubbles. When
these burst, the CR are released. (After E. N. Parker).

The multifarious transformations of cosmic-ray nuclei
in the interstellar medium provide a veritable Rosetta
Stone. They have unveiled the "*secrets*" of propagation,
such as path-length distribution, storage time in the
Galaxy, source composition, and selective injection. They
have also imposed constraints on theories of early and late
acceleration. Comparisons of relative abundances in the
cosmic rays with those in the sun (especially in solar
flares), and in local Galactic regions have provided useful
astrophysical insights.

3.1 Isotopic Composition

A triumph of propagation theory, supplemented by
measurements of cross sections, has been the prediction of

the "arriving" isotopic composition. This differs completely from the "universal" abundances.

In considering the origin of cosmic rays, a central question is this: what are the relative cosmic-ray abundances when the particles are just leaving the sources? This source composition provides a starting point for calculating the production and propagation of secondary nuclides. To do so requires the solution of diffusion equations in which the path length distribution, the partial cross sections for breakup, and the particle energies are the main parameters.

The composition at the cosmic-ray sources is calculated by adjusting the values of the source-abundances in the diffusion equation until the *calculated* arriving composition agrees with the *observed* composition of cosmic-ray nuclei at the top of the earth's atmosphere. The calculations require detailed knowledge of the many modes of cosmic-ray fragmentation and their cross sections as a function of energy. Until about 1980 only a few of the relevant cross sections had been measured and others were crudely estimated. The production of unstable secondaries such as ^7Be and ^{11}C, had been measured by radiochemists working with high-energy proton beams. Partial cross sections for production of *stable* nuclides were scarce.

In recent years, progress has been made both in measurements and in calculations of breakup cross sections. The group at the Bernas Laboratory in Orsay applied mass spectrometry to this problem, while the group at the Naval Research Laboratory (NRL) devised semiempirical relationships based on nuclear systematics and the available measured values. Meanwhile, relativistic beams of heavy ions were developed at Berkeley, Dubna, and other heavy-ion accelerators, with energies exceeding those hitherto available by two orders of magnitude. Thus, it became possible to measure the production rates of many more stable secondaries and the predictive value of the semiempirical formulations could be tested and confirmed. Moreover, the formulae themselves could be further refined as additional measurements became available.

These diverse developments enhanced our confidence in applying the propagation model to such problems as deducing the relative abundances of the cosmic-ray elements at the sources. Once a source composition has been derived (by a best fit to the arriving *elemental* source composition), the procedure can be turned around to estimate the arriving *isotopic* composition. Starting with their calculated elemental source composition, the NRL group computed the expected *isotopic* distribution of cosmic rays at the top of the earth's atmosphere. To do so, it was also necessary to adopt a model for the isotopic abundances of the various elements at the sources. The latter were taken to be the same as those in the solar system, according to Cameron. Some typical results are depicted in Figure 4 that compares some of the predicted cosmic-ray abundances with the corresponding nuclidic abundances in the general (thermal) composition found in the solar system.

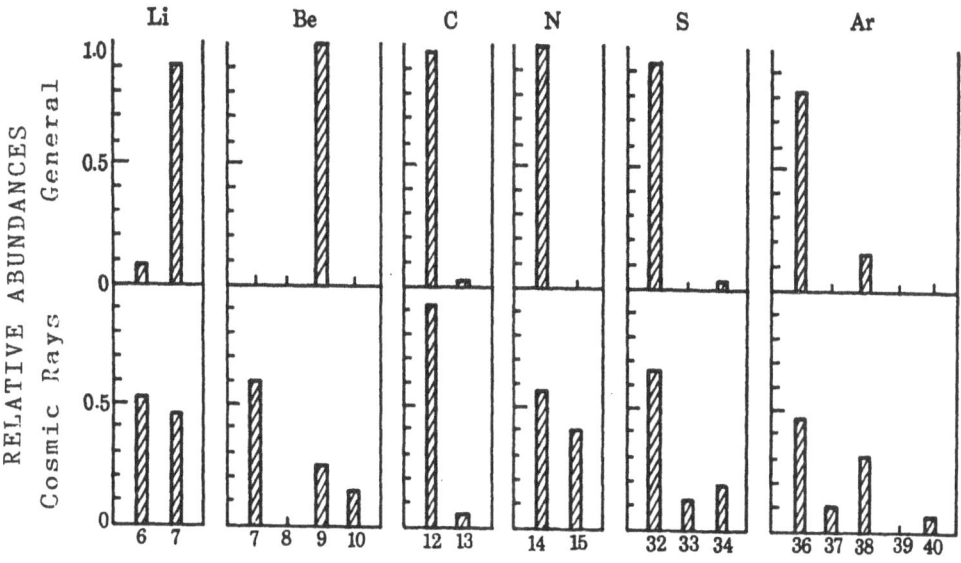

MASS NUMBER

Fig. 4. Comparison of the *calculated* isotopic abundances for six elements in the arriving cosmic rays with observed abundances in solar-system material. The ordinates are adjusted so that the total for each element is normalized to unity.

For most of the elements the two distributions are distinctly different. For example, in the general composition, 93 percent of lithium consists of ^7Li; in the cosmic rays, about half of it is ^6Li. Normal beryllium is almost entirely ^9Be, while in the arriving cosmic rays, only about 30 percent of this element should be ^9Be, about

60 percent ^{7}Be, and < 10 percent ^{10}Be. (Nuclei like ^{7}Be, that decay only by electron capture, are expected to survive in cosmic rays.) ^{15}N comprises only 4×10^{-3} of ordinary nitrogen, but it constitutes nearly one half of the element in cosmic rays.

Any significant isotopic *differences* that may be discovered between predicted values of the local cosmic rays and future observations, would show that the isotopic composition at the sources must differ from that elsewhere in nature. Such discrepancies would provide guidelines for modifying the initially adopted source model; they would tell us about special conditions at the sources, and the processes of nucleosynthesis--and perhaps of selective acceleration--operating there. They might also lead to conjectures about special sources, e.g., certain active stars, which could give rise to anomalies in composition.

4. Injection and Acceleration

According to current theories of acceleration, cosmic rays are energized by shocks in the interstellar medium (ISM), notably by those in the expanding shells of supernova remnants (SNR). That SNR are the most likely agents of CR origin has long been recognized. The early assumption that the CR are the direct, immediate products of supernova nucleosynthesis is passé; it generates an unsatisfactory GCR composition. To be sure, expanding SN shock fronts are vital as the engines for acceleration, providing the required energy. However, it is thought today that the *particles* being promoted to CR energies reside in the ISM, where hey are picked up and accelerated by the expanding shock fronts. Theories of shock-wave acceleration have generally assumed that the thermal ions in the ISM are *directly* accelerated, i.e., that no prior injection of supra-thermal ions is required. This, however, yields a flawed composition.

It is my view that supra-thermal seed particles *are* needed prior to the main acceleration, owing to the nature of the CR source composition. The latter is rather well established for energies up to ~10^3 GeV/amu. (Numerically, < 10^{-4} of the GCR nuclei have higher energies; we are here concerned with the preponderant bulk of GCR at E < 10^3 GeV/amu.) A viable theory of GCR origin must not conflict with the known relative abundances. These are rather well ordered according to the first-ionization potential (FPI) of the elements.

In a scheme of origin that I have discussed, the seed particles destined to become CR nuclei are injected mainly from flares on red-dwarf stars, principally dMe and dKe stars. These flares are very powerful and frequent compared with solar flares. Estimates of the numbers and energies of the seed particles, and their probability of encountering a shock, seem to fit salient observational requirements, e.g., the composition and the energy budget of cosmic rays in the Galaxy.

Among the considerations that support this hypothesis are the following: (a) Some 90 per cent of stars are red dwarfs. (b) dM_e and dK_e stars flare powerfully (e.g., 10^4 x solar). (c) Flares occur often--daily, and even hourly. (d) Solar energetic flare particles provide a useful model. (e) Radio, X-ray and optical emissions suggest copious particle production. (f) "FIP" selection works in red-dwarf chromospheres--so the observed composition does not conflict. (g) Energy budget is modest (cf. that for acceleration), and available. (h) The "seed particles" are very likely to be hit by a strong shock.

5. High-Energy Neutrino Astrophysics and Cosmic Rays

From cosmic regions where high-energy processes are dominant, decisive information can be elicited through the nascent science of high-energy neutrino astronomy and astrophysics. Neutrinos--"invented" by Pauli and later observed by Reines and Cowan--have played a vital role in nuclear and elementary-particle physics. At high energies (HE), progress was facilitated by the intense neutrino beams available at large accelerators. At modest (MeV) energies, astrophysical processes involving neutrinos have been explored theoretically; thus, we could compare the predicted fluxes of solar neutrinos and of neutrinos from supernova explosions (SN 1987a) with those observed. The deficit of solar neutrinos found by Davis has been confirmed, and efforts to explain it continue unabated.

By contrast, extraterrestrial neutrinos at high energies (Hev) say, $> 10^{11}$ MeV, have eluded observation (although Hev_μ produced in the Earth's atmosphere by cosmic rays have been observed, mainly via upward-moving muon secondaries). The detection of cosmic Hev_μ requires a large array of sensors coupled to a very massive target since the interaction cross sections of neutrinos are exceedingly low and backgrounds are troublesome. Despite these difficulties, several major projects are under way to develop HEν astronomy. The motivation is to open a new window upon celestial objects and regions--especially those where HE processes are prominent.

The Heν resulting from the production and decay of charged pions and kaons seem to be the most unmistakable tracers of cosmic rays. They should ultimately provide more definitive answers than we now have to questions such as these: By what mechanism(s) and how efficiently do cosmic rays get their prodigious energies (are shock waves the only accelerators)? In what Celestial objects or regions does the acceleration occur--especially at energies $> 10^{14}$ eV? What energy spectra are produced? And in analyzing the energetic processes giving rise to the observed neutrinos, what can we learn about the nature and distribution of matter and fields in the source regions?

6. Some of the Unresolved Problems
[This section is adapted from the author's concluding remarks in the Victor F. Hess Memorial Lecture presented earlier this month at the XXIV International Cosmic-Ray Conference in Rome.]

6.1. The Highest-Energy Cosmic Rays

When Pierre Auger discovered extensive air showers (EAS) in the late thirties, it was realized that these cascades are produced by primary cosmic-ray particles of energies $\geq 10^{14}$ eV. Since that time the observed shower energies have climbed by six orders of magnitude. Today we are especially puzzled by the highest-energy cosmic rays, notably those with energies exceeding 10^{19} eV. What is their composition--mainly protons, or heavier nuclei? Where do they originate, and how do they get their fantastic particle energies? Can we reconcile tentative evidence for the expected Zatsepin-Kuzmin-Greisen cutoff with persistent indications of spectral flattening at the very highest energies?

These questions have eluded reliable answers despite many years of observation with large EAS arrays and Fly's Eye detectors. The main reason, of course, is the paucity of these super-energetic particles. Yet there are compelling motives for trying to solve the puzzles. For example, the information could yield special insights into mysteries of the most powerful AGN (active galactic nuclei) recently detected in space observatories.

It seems reasonable to suppose that the highest-energy cosmic rays may originate in AGN. Hence it is important to seek definitive evidence for the supermassive black holes-- the putative power-houses of these prodigious galaxies.

Recent reports from two EAS arrays have revealed showers having energies of 2×10^{20} eV (Akeno) and 3×10^{20} eV (Fly's Eye) respectively. These have provided further impetus for the proposed Auger Project, led by James Cronin and Alan Watson. It aims to set up two gigantic air shower arrays, one each in the Northern and Southern hemispheres, each covering an area of 3,000 sq. km. The arrays would exploit a variety of detectors, including large water-Cerenkov tanks and Fly's Eyes. Communication between the instrument stations would be accomplished by radio links. A recent workshop in Chicago produced a detailed design, and a search for suitable locations has yielded some promising candidate sites.

6.2 Cosmic Rays "Unseen" in the Solar System

Very low-energy CR, 10-1000 MeV/amu, are excluded from the heliosphere or severely slowed down by the solar wind. We cannot directly observe their ISM fluxes or energy spectra even with far-ranging spacecraft. It may, however, be possible to get one or more handles on this problem.

Radioastronomy has opened up the rich field of space chemistry which involves complex networks of chemical reactions occurring in interstellar clouds. These reactions are believed to be catalyzed by highly ionizing cosmic rays, mainly of low energy. Already there have been efforts to deduce cosmic-ray fluxes in the ISM from the observation of molecules in space. If the hypothesis of cosmic-ray injection by dMe flare stars proves viable, then studies of the seed particles may provide clues to the spectra of the low-energy cosmic rays pervading the Galaxy.

A promising approach toward elucidation of this problem has recently come from the observation of gamma-ray lines originating in the Orion complex. These reveal the collisions of low-energy light nuclei that apparently produce the line emission.

6.3 Gamma-Ray Bursts

The BATSE (Burst and Transient Space Experiment) aboard the Compton Gamma-Ray Observatory has increased by more than an order of magnitude the rate of detection of gamma-ray bursts. These display a fantastic variety of features, deepening the mystery of what they are and where they originate. An origin at cosmological distances seems likely, but alternative hypotheses have not yet been ruled out. Current schemes for the rapid discovery of counterpart objects may help to clear up the puzzle.

6.4 Deficit of Hydrogen and Helium

Although they are the most prominent elements in the CR, H and He are severely underabundant relative to the local galactic composition. They also have slightly flatter spectra than do the heavier nuclides. I therefore suggested some years ago that these elements may have an origin different from that of the heavier CR. A model designed to account for this deficit invoked supernovae of Type 1 which are characterized by a lack of H and He. The scenario involved supernovae of Type 2 as well.

It was assumed that the CR nuclei comprise two components: (A) the ^{1}H and ^{4}He, together with "normal" fluxes of the heavy nuclei, probably injected into the (ISM) from a variety of stars, and (B) the bulk of the heavier nuclei, assumed to be injected from supernova explosions of Type 1. The main acceleration, following injection, is assumed to occur in the ISM, mostly at shock fronts. Elaborating on this hypothesis, Yanagita et al. added a third source component--the interstellar medium-- to determine whether a suitable mix of nucleosynthetic processes could yield a good match to the CR source composition.

None of the ad hoc assumptions succeeded in providing a satisfactory fit to the source composition. So the H-He underabundance is still in search of a good explanation.

6.5 Wave-Particle Interactions

In the vicinity of an expanding supernova renmant and its associated shock front, a streaming motion of cosmic rays can generate Alfvén waves. These, in turn, scatter the cosmic rays. The wave-particle interactions give rise to non-linearities that complicate calculations of acceleration and diffusion. Is it possible that mathematical techniques of chaos theory could be applied to this problem?

7. Acknowledgment

I am grateful to Professor Norma Sanchez and Professor Antonino Zichichi for inviting me to participate in this stimulating workshop.

QUANTUM FIELD THEORY FOR DYNAMICAL SYSTEMS WITH CURVED PHASE SPACE

E. S. FRADKIN

P.N. Lebedev Physical Institute
Russian Academy of Sciences
Leninsky Prospect 53
117 924 Moscow
RUSSIA

A gauge-invariant approach to geometric quantization is developed. It yields a complete quantum description for dynamical systems with non-trivial geometry and topology of the phase space.
The method is a global version of the gauge-invariant approach to quantization of second-class constraints developed by Batalin, Fradkin and Fradkina (BFF). Physical quantum states and quantum observables are respectively described by covariantly constant sections of the Fock bundle and the bundle of hermitian operators over the phase space with a flat connection defined by the nilpotent BFV-BRST operator. Perturbative calculation of the first non-trivial quantum correction to the Poisson brackets leads to the Chevalley cocycle known in deformation quantization. Consistent conditions lead to a topological quantization condition with metaplectic anomaly.

References: E.S. Fradkin and V. Ya. Linetsky, Nucl. Phys. B431 (1994) 569, and preprint (1995).

N. Sánchez and A. Zichichi (eds.), String Gravity and Physics at the Planck Energy Scale, 521.
© 1996 Kluwer Academic Publishers.

Photographs of the Institute

Photos LAZZARI, C. MENDIS, L. POMARA

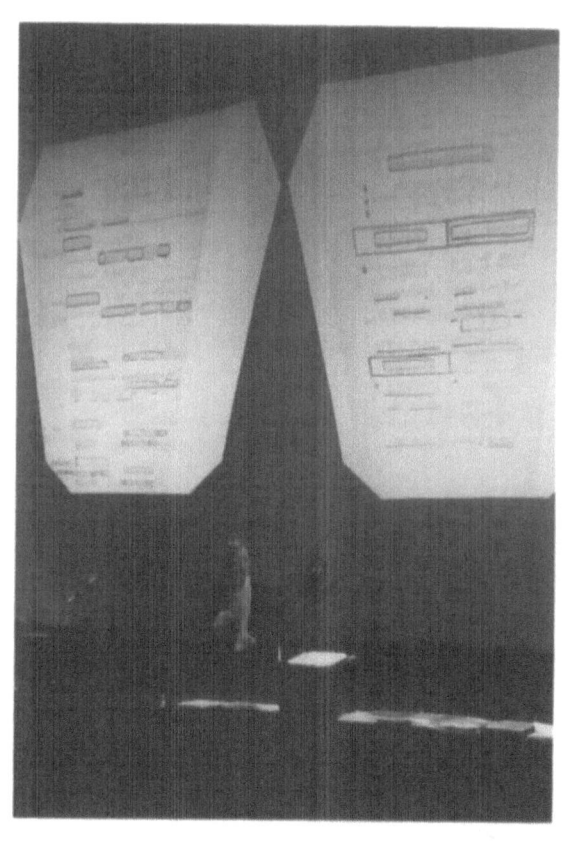

AUTHOR INDEX

ASHTEKAR, A.: 255

BANKS, T. : 277

BARS, I. : 151

BOYANOVSKY, D. : 451

CHAMBLIN, A. : 233

DE VEGA, H. J. : 11 and 451

DI VECCHIA, P. : 105

FRADKIN, E. S. : 521

FROLOV, V. P. : 187

GASPERINI, M. : 305

GIBBONS, G.W. : 233

GRISHCHUK, L. P. : 369

HU, B. L. : 219

ISRAEL, W. : 171

LARSEN, A. L. : 65

NOVIKOV, S. P. : 1

PAGE, D. N. : 431

SALOPEK, D. S. : 409

SANCHEZ, N. : 11 and 65

SHAPIRO, M. M. : 507

STROMINGER, A. : 209

TSEYTLIN, A. A. : 121

VENEZIANO, G. : 285

VILENKIN, A. : 345

YAFFE, L. G. : 493